Remote Sensing of Watershed

Remote Sensing of Watershed

Editors

Jingzhe Wang
Zhongwen Hu
Yangyi Wu
Jie Zhang

MDPI • Basel • Beijing • Wuhan • Barcelona • Belgrade • Manchester • Tokyo • Cluj • Tianjin

Editors

Jingzhe Wang
Shenzhen Polytechnic
Shenzhen, China

Zhongwen Hu
Shenzhen University
Shenzhen, China

Yangyi Wu
Wuhan University
Wuhan, China

Jie Zhang
China Agricultural University
Beijing, China

Editorial Office
MDPI
St. Alban-Anlage 66
4052 Basel, Switzerland

This is a reprint of articles from the Special Issue published online in the open access journal *Remote Sensing* (ISSN 2072-4292) (available at: https://www.mdpi.com/journal/remotesensing/special_issues/rswatershed).

For citation purposes, cite each article independently as indicated on the article page online and as indicated below:

LastName, A.A.; LastName, B.B.; LastName, C.C. Article Title. *Journal Name* **Year**, *Volume Number*, Page Range.

ISBN 978-3-0365-8130-9 (Hbk)
ISBN 978-3-0365-8131-6 (PDF)

Cover image courtesy of Jingzhe Wang

Contents

About the Editors

Jingzhe Wang

Jingzhe Wang is currently a lecturer with the School of Artificial Intelligence, Shenzhen Polytechnic, Shenzhen, China. He received his B.Sc. degree from Anqing Normal University, Anqing, China in 2010 and his Ph.D. degree from Xinjiang University, Urumqi, China in 2019. His research interests focus on earth observation and remote sensing, spectral modeling, quantitative estimation of soil properties, digital soil mapping, GIS, spatial analysis, and environmental sustainability.

Zhongwen Hu

Zhongwen Hu is currently an associate professor with the School of Architecture and Urban Planning, Shenzhen Univeristy, Shenzhen, Guangdong, China. He received his B.Sc. degree in remote sensing and his Ph.D. degree in photogrammetry and remote sensing from Wuhan University, Wuhan, China, in 2008 and 2013, respectively. His research interests include high spatial resolution remote sensing image analysis, urban remote sensing and coastal remote sensing.

Yangyi Wu

Yangyi Wu is currently an associate researcher (Assistant Professor) at the School of Urban Design, Wuhan University, Wuhan, Hubei, China. He received his B.E. degree from Wuhan University, Wuhan, China, in 2013; his M.S. degree from the University of Texas at Dallas, Dallas, U.S.A.; and his Ph.D. degree from the University of Utah, Salt Lake City, U.S.A. His research interests are concentrated on analyzing uneven urban development relying on advanced spatial analysis and statistics in the United States and China.

Jie Zhang

Jie Zhang is currently an associate professor with the College of Information and Electrical Engineering, Agricultural University, Beijing, China. She received her B.S. degree from Wuhan University, Wuhan, China in 2008 and her Ph.D. degree from Tsinghua University, Beijing, China in 2018. Her research interests include environmental monitoring and modelling using remote sensing and multiple models (crop models, hydrology models, and environmental models), big data and intelligent analysis for soil, water quality, and CNP cycles, and agriculture and regional sustainable development.

Preface to "Remote Sensing of Watershed"

The watershed, a fundamental unit in the global terrestrial water cycle, plays a crucial role not only in the Earth's natural systems but also in human societies. Its implications in water accessibility, agriculture, and human settlements have led to a surge in research interest. Moreover, in the context of the increasingly pressing issues of climate change and water security, understanding and managing watersheds effectively have become more important than ever.

Advancements in remote sensing technologies have provided a major boost to this research area. As a robust tool for geographical analysis, remote sensing offers the capacity to monitor Earth surface processes, track human activities, and assess ecosystem services at both the macro and micro scales. In turn, watersheds have become key research objects in hydrology and remote sensing studies, providing insights into the intricate balance between natural processes and human-induced impacts.

This book, "Remote Sensing of Watershed", brings together 15 articles that represent the latest developments in the field. Contributed by leading experts in their respective areas, these papers present an array of novel methods and findings across various aspects of watershed research. The topics explored range from natural resource monitoring and mapping to watershed management, environmental modeling, and analyses of spatiotemporal variations within the watershed.

The intent of this compilation is not just to present a series of independent research works but to provide a cohesive narrative that can offer comprehensive insights into the current state and future direction of watershed studies. Each article in this collection contributes a unique perspective and adds a new layer to our collective understanding of watersheds.

We believe that this book will serve as a valuable resource for researchers, practitioners, and policymakers in their pursuits to further understand, manage, and protect watersheds. Through the exploration of these articles, we hope to foster deeper insights and stimulate further research in this critical area. We invite you to delve into this book and join us in the exciting journey of understanding and conserving our watersheds through the lens of remote sensing.

Jingzhe Wang, Zhongwen Hu, Yangyi Wu, and Jie Zhang
Editors

remote sensing

MDPI

Editorial

Remote Sensing of Watershed: Towards a New Research Paradigm

Jingzhe Wang [1,2], Yangyi Wu [3,4,*], Zhongwen Hu [5] and Jie Zhang [6,7]

1 School of Artificial Intelligence, Shenzhen Polytechnic, Shenzhen 518055, China; jzwang@szpt.edu.cn
2 State Key Laboratory of Resources and Environmental Information System, Institute of Geographic Sciences and Natural Resources Research, Chinese Academy of Sciences, Beijing 100101, China
3 School of Urban Design, Wuhan University, Wuhan 430072, China
4 Hubei Habitat Environment Research Centre of Engineering and Technology, Wuhan 430072, China
5 MNR Key Laboratory for Geo-Environmental Monitoring of Great Bay Area & Guangdong Key Laboratory of Urban Informatics & Guangdong–Hong Kong-Macau Joint Laboratory for Smart Cities & Shenzhen Key Laboratory of Spatial Smart Sensing and Services, Shenzhen University, Shenzhen 518060, China; zwhoo@szu.edu.cn
6 College of Information and Electrical Engineering, China Agricultural University, Beijing 100085, China; jiezhang@cau.edu.cn
7 National Innovation Center for Digital Agricultural Products Circulation, China Agricultural University, Beijing 100085, China
* Correspondence: yangyi.wu@whu.edu.cn

1. Introduction

Watersheds are critical natural systems that serve as the foundation for sustaining life on Earth [1]. They play a vital role in the hydrological cycle, supporting water supply, agriculture, ecosystems, and biodiversity [2–6]. However, Anthropocene has presented challenges to watersheds, including deforestation, land-use changes, pollution, and climate change [7,8]. To maximize the resilience of watersheds, it is essential to implement sustainable land-use practices and effective watershed management strategies [9]. Consequently, understanding watersheds' complex dynamics and their response to natural and anthropogenic stressors is essential for sustainable development and the well-being of human societies.

The advent of remote sensing has revolutionized watershed research, providing unprecedented insights into watershed dynamics and spatiotemporal patterns [10,11]. These techniques offer breakthrough advantages over traditional field-based methods, including covering large areas with low cost and high efficiency, monitoring remote and inaccessible regions, and obtaining data at different spatial, spectral, and temporal resolutions [12–14]. More specifically, remote sensing of watersheds encompasses a wide range of applications that involve acquiring and utilizing hydrological, ecological, and environmental information in watersheds [15–17], including watershed mapping [18], monitoring of underlying surface elements in the watershed [19], inversion of water cycle parameters [20], water resource assessment [21], acquisition of watershed ecological environment parameters [22], monitoring natural disasters [23], analyzing upstream and downstream [24], supporting water governance priorities [25], water resource development [26], and irrigation water management [27]. Recent research trends show that the application of remote sensing techniques has been transiting from mapping, monitoring, and assessment to governance and management [28], calling for deeper and wider explorations on related research focus.

Based on these research contexts, several shortages in terms of current studies appear. First, the concept of the watershed is usually multiscale [29,30]. Studies vary at the global scale, concentrating on large watersheds (for example, oceans) [31], global trends (for example, global warming) [32], and international cooperation [33], to the local scale, concentrating on a single watershed (for example, lake or wetland) [34], local specific context (for example, pollution or urbanization) [35], and local governance [36]. There is still a lack of multiscalar understanding of watersheds. Second, the watershed is a unique

Citation: Wang, J.; Wu, Y.; Hu, Z.; Zhang, J. Remote Sensing of Watershed: Towards a New Research Paradigm. *Remote Sens.* **2023**, *15*, 2569. https://doi.org/10.3390/rs15102569

Received: 3 May 2023
Accepted: 5 May 2023
Published: 14 May 2023

nature–human system that contains multiple interacting elements [37]. However, current studies usually concentrate on a single element or process of watersheds. Hence, a comprehensive view of the "total" element of watersheds is needed. Third, while remote sensing techniques are currently widely applied in environmental studies, the scope is largely limited to natural sciences, and interdisciplinary research, especially from the scope of social sciences, is urgently needed [38]. Fourth, concentrating on remote sensing techniques, the methodology is most targeted on a single platform and imagery and may not meet the multiscalar, multidimensional, and interdisciplinary research demands of the current and prospective watershed studies.

Drawing upon these research contexts, this Special Issue is dedicated to scientific reports on the remote sensing of watersheds. This issue focuses on applying remote sensing techniques in watersheds in terms of hydrology, ecology, environment, and human activities. It aims to contribute to the current understanding of integrating research scopes and developing advanced methods.

2. Key Findings of the Special Issue

This Special Issue includes 14 articles that focus on various aspects of watershed remote sensing and aim to address the gaps and challenges in the field. The articles cover various topics, including hydrology, land use/land cover, vegetation, soil, and topography. These articles present innovative research methods and techniques for utilizing remote sensing data to enhance our understanding of watershed processes and support resource management decision making.

In terms of research objective and scope, the majority of papers on this Special Issue are focused on natural phenomena, hydrological processes [39], air [40,41], and land use/vegetation [42–44]. Further related it to human activities, such as carbon emission [45], agriculture [46,47], ecosystem service [48], and even culture diffusion [49]. Furthermore, the research encompasses diverse geographical areas, including rivers [39], lakes [40,41], drylands [50], countries [45], and regions [43,46].

In terms of methods and data, a wide range of methods and data sources were employed in the studies featured in this Special Issue. The methods vary from the comparison and validation of different methods combining the research object and local context [40,48], spatial simulation models based on the research object integrating remote sensing data [47,48], and developing advanced methods or indices to understand research phenomena [39] precisely. Some studies also discussed underlying data uncertainty issues [48].

In terms of data sources, data from platforms, including satellite-based and drone-based, are applied [41,43]. Many studies applied multisource data. For example, Duan et al. (2022) aim to solve the missing data issues by integrating multiple satellite sources [40], and others rely on multisource data to gain a comprehensive understanding [51].

In general, this Special Issue has revealed the multiscalar, multidimensional, and interdisciplinary nature of the application of remote sensing in watershed studies. The studies propose a series of advanced data, models, and strategies to serve watershed research better. However, there are still limitations and areas for improvement. Future studies are encouraged to consider and address these shortcomings, further advancing the field of remote sensing in watershed studies.

3. Future Perspectives: Towards a New Paradigm

3.1. Integration of Multisource Data

The integration of multisource data is a promising direction in remote sensing for watersheds [52]. Combining data from various sensors, platforms, and resolutions can provide a more comprehensive understanding of watershed dynamics. For example, integrating optical, thermal, and radar imagery can offer complementary information on land cover, water resources, and vegetation. Similarly, merging remote sensing data with in situ measurements, socio-economic data, and other geospatial information can enhance the accuracy and reliability of watershed analyses. Developing robust data fusion methodologies and interoperable platforms is crucial for maximizing the potential of multisource data in watershed management.

3.2. Multiscale Modeling and Analysis

Watershed processes occur at various spatial and temporal scales, necessitating multiscale modeling and analysis approaches in remote sensing. By considering the interactions between processes at different scales, researchers can develop more accurate and representative models of watershed dynamics. For instance, combining high-resolution imagery for detailed local assessments with coarse-resolution imagery for regional analyses can offer valuable insights into land use and hydrological processes. Developing scalable remote sensing methods that can be applied across various spatial and temporal scales is essential for addressing the complex challenges associated with watershed management.

3.3. Analysis of the "Total Environment"

A comprehensive understanding of watershed dynamics requires the consideration of all elements of the "total environment," including physical, biological, and socio-economic components. Remote sensing techniques can contribute to this understanding by providing information on land use, water resources, ecosystems, and human activities. Future research should focus on developing integrated frameworks and methodologies that can analyze the interactions between these components and assess their combined impacts on watershed health. Such holistic approaches are crucial for developing sustainable watershed management strategies that balance the competing demands of various stakeholders.

3.4. Data Barriers and Data Sharing

Addressing data barriers and promoting data sharing is a critical perspective in remote sensing for watershed management. Ensuring that remote sensing data, derived products, and analytical tools are accessible to researchers, decision makers, and stakeholders can foster collaboration and facilitate informed decision making. This requires the development of open data platforms, standardized data formats, and data-sharing policies that encourage the exchange of information and knowledge. Overcoming data barriers can also help bridge the gap between scientific research and practical applications in watershed management.

3.5. Targeting Industrial Demands and Serving Decision Making

Remote sensing for watershed management should be oriented toward addressing the specific needs of the industry and decision makers. This includes developing tailored products, tools, and methodologies that can support decision-making processes in various sectors, such as water resource management, agriculture, urban planning, and environmental conservation. By focusing on practical applications and providing actionable insights, remote sensing can contribute to the development of evidence-based policies and strategies for sustainable watershed management.

In conclusion, the future of remote sensing for watershed management lies in the integration of multisource data, multiscale modeling and analysis, comprehensive assessments of the total environment, overcoming data barriers and sharing, and targeting industrial demands to serve decision-making processes. By embracing these perspectives, remote sensing can continue to play a pivotal role in advancing our understanding of watershed dynamics and informing sustainable management practices.

Author Contributions: Conceptualization, J.W. and Y.W.; writing—original draft preparation, J.W., Y.W., Z.H. and J.Z. All authors have read and agreed to the published version of the manuscript.

Funding: This work was jointly supported by the Natural Science Foundation of Guangdong Province (2023A1515011273 and 2020A1515111142), Basic Research Program of Shenzhen (20220811173316001), Shenzhen Polytechnic Research Fund (6023310031K), Post-doctoral Later-stage Foundation Project of Shenzhen Polytechnic (6023271008K) and a grant from State Key Laboratory of Resources and Environmental Information System.

Data Availability Statement: Not applicable.

Acknowledgments: As the Guest Editors, we would like to thank all of the authors who submitted their research results and ideas in this Special Issue. Special thanks to all anonymous reviewers involved in the SI and helped the authors to improve their manuscripts. Thanks also to the editorial staff of Remote Sensing for supporting the idea of this SI.

Conflicts of Interest: The authors declare no conflict of interest. The funders had no role in the design of the study; in the collection, analyses, or interpretation of data; in the writing of the manuscript; or in the decision to publish the results.

References

1. Chung, M.G.; Frank, K.A.; Pokhrel, Y.; Dietz, T.; Liu, J. Natural infrastructure in sustaining global urban freshwater ecosystem services. *Nat. Sustain.* **2021**, *4*, 1068–1075. [CrossRef]
2. Branca, G.; Lipper, L.; Neves, B.; Lopa, D.; Mwanyoka, I. Payments for Watershed Services Supporting Sustainable Agricultural Development in Tanzania. *J. Environ. Dev.* **2011**, *20*, 278–302. [CrossRef]
3. Luck, G.W.; Chan, K.M.A.; Fay, J.P. Protecting ecosystem services and biodiversity in the world's watersheds. *Conserv. Lett.* **2009**, *2*, 179–188. [CrossRef]
4. Rockström, J.; Folke, C.; Gordon, L.; Hatibu, N.; Jewitt, G.; Penning de Vries, F.; Rwehumbiza, F.; Sally, H.; Savenije, H.; Schulze, R. A watershed approach to upgrade rainfed agriculture in water scarce regions through Water System Innovations: An integrated research initiative on water for food and rural livelihoods in balance with ecosystem functions. *Phys. Chem. Earth Parts ABC* **2004**, *29*, 1109–1118. [CrossRef]
5. Li, X.; Cheng, G.; Ge, Y.; Li, H.; Han, F.; Hu, X.; Tian, W.; Tian, Y.; Pan, X.; Nian, Y.; et al. Hydrological Cycle in the Heihe River Basin and Its Implication for Water Resource Management in Endorheic Basins. *J. Geophys. Res. Atmos.* **2018**, *123*, 890–914. [CrossRef]
6. Wu, Y.; Wei, Y.D. Regional Drinking Water Supply in Pakistan: Regional Disparity, Inequality, and Development Pattern. *Prof. Geogr.* **2023**, 1–17. [CrossRef]
7. Frei, R.J.; Abbott, B.W.; Dupas, R.; Gu, S.; Gruau, G.; Thomas, Z.; Kolbe, T.; Aquilina, L.; Labasque, T.; Laverman, A.; et al. Predicting Nutrient Incontinence in the Anthropocene at Watershed Scales. *Front. Environ. Sci.* **2020**, *7*, 200. [CrossRef]
8. Waldman, J.R.; Quinn, T.P. North American diadromous fishes: Drivers of decline and potential for recovery in the Anthropocene. *Sci. Adv.* **2022**, *8*, eabl5486. [CrossRef]
9. Davenport, M.A.; Seekamp, E. A Multilevel Community Capacity Model for Sustainable Watershed Management. *Soc. Nat. Resour.* **2013**, *26*, 1101–1111. [CrossRef]
10. Wang, D.; Gong, J.; Chen, L.; Zhang, L.; Song, Y.; Yue, Y. Spatio-temporal pattern analysis of land use/cover change trajectories in Xihe watershed. *Int. J. Appl. Earth Obs. Geoinf.* **2012**, *14*, 12–21. [CrossRef]
11. Ma, C.; Xie, Y.; Duan, S.-B.; Qin, W.; Guo, Z.; Xi, G.; Zhang, X.; Bie, Q.; Duan, H.; He, L. Characterization of spatio-temporal patterns of grassland utilization intensity in the Selinco watershed of the Qinghai-Tibetan Plateau from 2001 to 2019 based on multisource remote sensing and artificial intelligence algorithms. *GIScience Remote Sens.* **2022**, *59*, 2217–2246. [CrossRef]
12. Cucho-Padin, G.; Loayza, H.; Palacios, S.; Balcazar, M.; Carbajal, M.; Quiroz, R. Development of low-cost remote sensing tools and methods for supporting smallholder agriculture. *Appl. Geomat.* **2020**, *12*, 247–263. [CrossRef]
13. Song, C.; Huang, B.; Ke, L.; Richards, K.S. Remote sensing of alpine lake water environment changes on the Tibetan Plateau and surroundings: A review. *ISPRS J. Photogramm. Remote Sens.* **2014**, *92*, 26–37. [CrossRef]
14. Navalgund, R.R.; Jayaraman, V.; Roy, P. Remote sensing applications: An overview. *Curr. Sci.* **2007**, *93*, 1747–1766.
15. Shao, Z.; Fu, H.; Li, D.; Altan, O.; Cheng, T. Remote sensing monitoring of multi-scale watersheds impermeability for urban hydrological evaluation. *Remote Sens. Environ.* **2019**, *232*, 111338. [CrossRef]
16. Zhai, L.; Cheng, S.; Sang, H.; Xie, W.; Gan, L.; Wang, T. Remote sensing evaluation of ecological restoration engineering effect: A case study of the Yongding River Watershed, China. *Ecol. Eng.* **2022**, *182*, 106724. [CrossRef]
17. Xian, G.; Crane, M. Assessments of urban growth in the Tampa Bay watershed using remote sensing data. *Remote Sens. Environ.* **2005**, *97*, 203–215. [CrossRef]
18. Rather, A.F.; Ahmed, R.; Wani, G.F.; Ahmad, S.T.; Dar, T.; Javaid, S.; Ahmed, P. Mapping of groundwater potential zones in Pohru Watershed of Jhelum Basin-Western Himalaya, India using integrated approach of remote sensing, GIS and AHP. *Earth Sci. Inform.* **2022**, *15*, 2091–2107. [CrossRef]
19. Xu, C.; Zhang, W.; Wang, S.; Chen, H.; Azzam, A.; Zhang, B.; Xu, Y.; Nie, N. Spatiotemporal green water dynamics and their responses to variations of climatic and underlying surface factors: A case study in the Sanjiang Plain, China. *J. Hydrol. Reg. Stud.* **2023**, *45*, 101303. [CrossRef]
20. Jiang, D.; Wang, K. The Role of Satellite-Based Remote Sensing in Improving Simulated Streamflow: A Review. *Water* **2019**, *11*, 1615. [CrossRef]
21. Papa, F.; Crétaux, J.-F.; Grippa, M.; Robert, E.; Trigg, M.; Tshimanga, R.M.; Kitambo, B.; Paris, A.; Carr, A.; Fleischmann, A.S.; et al. Water Resources in Africa under Global Change: Monitoring Surface Waters from Space. *Surv. Geophys.* **2023**, *44*, 43–93. [CrossRef] [PubMed]

22. Song, W.; Song, W.; Gu, H.; Li, F. Progress in the Remote Sensing Monitoring of the Ecological Environment in Mining Areas. *Int. J. Environ. Res. Public Health* **2020**, *17*, 1846. [CrossRef] [PubMed]
23. Pangali Sharma, T.P.; Zhang, J.; Koju, U.A.; Zhang, S.; Bai, Y.; Suwal, M.K. Review of flood disaster studies in Nepal: A remote sensing perspective. *Int. J. Disaster Risk Reduct.* **2019**, *34*, 18–27. [CrossRef]
24. Wu, T.; Sang, S.; Wang, S.; Yang, Y.; Li, M. Remote sensing assessment and spatiotemporal variations analysis of ecological carrying capacity in the Aral Sea Basin. *Sci. Total Environ.* **2020**, *735*, 139562. [CrossRef] [PubMed]
25. Vörösmarty, C.; Hoekstra, A.; Bunn, S.; Conway, D.; Gupta, J. What scale for water governance. *Science* **2015**, *349*, 478–479. [CrossRef]
26. Zhou, C.; Gong, H.; Chen, B.; Gao, M.; Cao, Q.; Cao, J.; Duan, L.; Zuo, J.; Shi, M. Land Subsidence Response to Different Land Use Types and Water Resource Utilization in Beijing-Tianjin-Hebei, China. *Remote Sens.* **2020**, *12*, 457. [CrossRef]
27. Calera, A.; Campos, I.; Osann, A.; D'Urso, G.; Menenti, M. Remote Sensing for Crop Water Management: From ET Modelling to Services for the End Users. *Sensors* **2017**, *17*, 1104. [CrossRef]
28. Thakkar, A.K.; Desai, V.R.; Patel, A.; Potdar, M.B. Impact assessment of watershed management programmes on land use/land cover dynamics using remote sensing and GIS. *Remote Sens. Appl. Soc. Environ.* **2017**, *5*, 1–15. [CrossRef]
29. Zhai, J.; Ren, J.; Xi, M.; Tang, X.; Zhang, Y. Multiscale watershed landscape infrastructure: Integrated system design for sponge city development. *Urban For. Urban Green.* **2021**, *60*, 127060. [CrossRef]
30. Chen, D.; Shen, H.; Hu, M.; Wang, J.; Zhang, Y.; Dahlgren, R.A. Chapter Five—Legacy Nutrient Dynamics at the Watershed Scale: Principles, Modeling, and Implications. In *Advances in Agronomy*; Sparks, D.L., Ed.; Academic Press: Cambridge, MA, USA, 2018; Volume 149, pp. 237–313.
31. Breitburg, D.; Levin, L.A.; Oschlies, A.; Grégoire, M.; Chavez, F.P.; Conley, D.J.; Garçon, V.; Gilbert, D.; Gutiérrez, D.; Isensee, K.; et al. Declining oxygen in the global ocean and coastal waters. *Science* **2018**, *359*, eaam7240. [CrossRef]
32. Koutroulis, A.G. Dryland changes under different levels of global warming. *Sci. Total Environ.* **2019**, *655*, 482–511. [CrossRef] [PubMed]
33. Olmstead, S.M.; Sigman, H. Damming the Commons: An Empirical Analysis of International Cooperation and Conflict in Dam Location. *J. Assoc. Environ. Resour. Econ.* **2015**, *2*, 497–526. [CrossRef]
34. Liu, H.; Benoit, G.; Liu, T.; Liu, Y.; Guo, H. An integrated system dynamics model developed for managing lake water quality at the watershed scale. *J. Environ. Manag.* **2015**, *155*, 11–23. [CrossRef] [PubMed]
35. He, Y.; Wang, X.; Chen, H.; Yuan, X.; Wu, N.; Zhang, Y.; Yue, J.; Zhang, Q.; Diao, Y.; Zhou, L. Effect of watershed urbanization on N2O emissions from the Chongqing metropolitan river network, China. *Atmos. Environ.* **2017**, *171*, 70–81. [CrossRef]
36. Simms, R.; Harris, L.; Joe, N.; Bakker, K. Navigating the tensions in collaborative watershed governance: Water governance and Indigenous communities in British Columbia, Canada. *Geoforum* **2016**, *73*, 6–16. [CrossRef]
37. Herzog, C.P. A multifunctional green infrastructure design to protect and improve native biodiversity in Rio de Janeiro. *Landsc. Ecol. Eng.* **2016**, *12*, 141–150. [CrossRef]
38. Rochon, G.; Szlag, D.; Daniel, F.; Chifos, C. Remote sensing applications for sustainable watershed management and food security. In *Observing Our Environment from Space*; CRC Press: Boca Raton, FL, USA, 2021; pp. 285–291.
39. Chen, C.; Fu, R.; Ai, X.; Huang, C.; Cong, L.; Li, X.; Jiang, J.; Pei, Q. An Integrated Method for River Water Level Recognition from Surveillance Images Using Convolution Neural Networks. *Remote Sens.* **2022**, *14*, 6023. [CrossRef]
40. Duan, P.; Zhang, F.; Jim, C.-Y.; Tan, M.L.; Cai, Y.; Shi, J.; Liu, C.; Wang, W.; Wang, Z. Reconstruction of Sentinel Images for Suspended Particulate Matter Monitoring in Arid Regions. *Remote Sens.* **2023**, *15*, 872. [CrossRef]
41. Amantai, N.; Ding, J. Analysis on the Spatio-Temporal Changes of LST and Its Influencing Factors Based on VIC Model in the Arid Region from 1960 to 2017: An Example of the Ebinur Lake Watershed, Xinjiang, China. *Remote Sens.* **2021**, *13*, 4867. [CrossRef]
42. Wu, X.; Zhang, R.; Bento, V.A.; Leng, S.; Qi, J.; Zeng, J.; Wang, Q. The Effect of Drought on Vegetation Gross Primary Productivity under Different Vegetation Types across China from 2001 to 2020. *Remote Sens.* **2022**, *14*, 4658. [CrossRef]
43. Leng, S.; Huete, A.; Cleverly, J.; Yu, Q.; Zhang, R.; Wang, Q. Spatiotemporal Variations of Dryland Vegetation Phenology Revealed by Satellite-Observed Fluorescence and Greenness across the North Australian Tropical Transect. *Remote Sens.* **2022**, *14*, 2985. [CrossRef]
44. Wang, C.; Wang, J.; Naudiyal, N.; Wu, N.; Cui, X.; Wei, Y.; Chen, Q. Multiple Effects of Topographic Factors on Spatio-Temporal Variations of Vegetation Patterns in the Three Parallel Rivers Region, Southeast Qinghai-Tibet Plateau. *Remote Sens.* **2022**, *14*, 151. [CrossRef]
45. Li, Y.; Mi, W.; Zhang, Y.; Ji, L.; He, Q.; Wang, Y.; Bi, Y. Change of Human Footprint in China and Its Implications for Carbon Dioxide (CO2) Emissions. *Remote Sens.* **2023**, *15*, 426. [CrossRef]
46. Wei, G.; Cao, J.; Xie, H.; Xie, H.; Yang, Y.; Wu, C.; Cui, Y.; Luo, Y. Spatial-Temporal Variation in Paddy Evapotranspiration in Subtropical Climate Regions Based on the SEBAL Model: A Case Study of the Ganfu Plain Irrigation System, Southern China. *Remote Sens.* **2022**, *14*, 1201. [CrossRef]
47. Shen, Q.; Niu, J.; Sivakumar, B.; Lu, N. Effects of Mulching on Maize Yield and Evapotranspiration in the Heihe River Basin, Northwest China. *Remote Sens.* **2022**, *14*, 700. [CrossRef]
48. Sun, M.; Hu, J.; Chen, X.; Lü, Y.; Yang, L. Comparison of Five Models for Estimating the Water Retention Service of a Typical Alpine Wetland Region in the Qinghai–Tibetan Plateau. *Remote Sens.* **2022**, *14*, 6306.

49. Wang, Y.; Wang, N.; Zhao, X.; Liang, X.; Liu, J.; Yang, P.; Wang, Y.; Wang, Y. Field Model-Based Cultural Diffusion Patterns and GIS Spatial Analysis Study on the Spatial Diffusion Patterns of Qijia Culture in China. *Remote Sens.* **2022**, *14*, 1422. [CrossRef]
50. Li, M.; Yao, J.; Guan, J.; Zheng, J. Vegetation Browning Trends in Spring and Autumn over Xinjiang, China, during the Warming Hiatus. *Remote Sens.* **2022**, *14*, 1298. [CrossRef]
51. Xu, Y.; Hu, Z.; Zhang, Y.; Wang, J.; Yin, Y.; Wu, G. Mapping Aquaculture Areas with Multi-Source Spectral and Texture Features: A Case Study in the Pearl River Basin (Guangdong), China. *Remote Sens.* **2021**, *13*, 4320. [CrossRef]
52. Parsian, S.; Amani, M.; Moghimi, A.; Ghorbanian, A.; Mahdavi, S. Flood Hazard Mapping Using Fuzzy Logic, Analytical Hierarchy Process, and Multi-Source Geospatial Datasets. *Remote Sens.* **2021**, *13*, 4761. [CrossRef]

remote sensing

MDPI

Article

Reconstruction of Sentinel Images for Suspended Particulate Matter Monitoring in Arid Regions

Pan Duan [1,2], Fei Zhang [3,4,*], Chi-Yung Jim [5], Mou Leong Tan [6], Yunfei Cai [1], Jingchao Shi [7], Changjiang Liu [8], Weiwei Wang [1] and Zheng Wang [9]

1 College of Geography and Remote Sensing Sciences, Xinjiang University, Urumqi 830046, China
2 College of Hydrology and Water Resources, Hohai University, Nanjing 210024, China
3 College of Geography and Environmental Sciences, Zhejiang Normal University, Jinhua 321004, China
4 Key Laboratory of Oasis Ecology, Xinjiang University, Urumqi 830046, China
5 Department of Social Sciences, Education University of Hong Kong, Lo Ping Road, Tai Po, Hong Kong SAR, China
6 GeoInformatic Unit, Geography Section, School of Humanities, Universiti Sains Malaysia, Penang 11800, Malaysia
7 Departments of Earth Sciences, The University of Memphis, Memphis, TN 38152, USA
8 Xinjiang Institute of Technology, Aksu 843000, China
9 College of Geography Science and State Key Laboratory Cultivation Base of Geographical Environment Evolution (Jiangsu Province), Nanjing Normal University, Nanjing 210023, China
* Correspondence: zhangfei3s@xju.edu.cn

Abstract: Missing data is a common issue in remote sensing. Data reconstruction through multiple satellite data sources has become one of the most powerful ways to solve this issue. Continuous monitoring of suspended particulate matter (SPM) in arid lakes is vital for water quality solutions. Therefore, this research aimed to develop and evaluate the performance of two image reconstruction strategies, spatio-temporal fusion reflectance image inversion SPM and SPM spatio-temporal fusion, based on the measured SPM concentration data with Sentinel-2 and Sentinel-3. The results show that (1) ESTARFM (Enhanced Spatio-temporal Adaptive Reflection Fusion Model) performed better than FSDAF (Flexible Spatio-temporal Data Fusion) in the fusion image generation, particularly the red band, followed by the blue, green, and NIR (near-infrared) bands. (2) A single-band linear and non-linear regression model was constructed based on Sentinel-2 and Sentinel-3. Analysis of the accuracy and stability of the model led us to the conclusion that the red band model performs well, is fast to model, and has a wide range of applications (Sentinel-2, Sentinel-3, and fused high-accuracy images). (3) By comparing the two data reconstruction strategies of spatio-temporal fused image inversion SPM and spatio-temporal fused SPM concentration map, we found that the fused SPM concentration map is more effective and more stable when applied to multiple fused images. The findings can provide an important scientific reference value for further expanding the inversion research of other water quality parameters in the future and provide a theoretical basis as well as technical support for the scientific management of Ebinur Lake's ecology and environment.

Keywords: Ebinur Lake; spatio-temporal fusion model; Sentinel-2; images reconstruction; suspended particulate matter (SPM)

Citation: Duan, P.; Zhang, F.; Jim, C.-Y.; Tan, M.L.; Cai, Y.; Shi, J.; Liu, C.; Wang, W.; Wang, Z. Reconstruction of Sentinel Images for Suspended Particulate Matter Monitoring in Arid Regions. *Remote Sens.* **2023**, *15*, 872. https://doi.org/10.3390/rs15040872

Academic Editor: Richard Gloaguen

Received: 14 November 2022
Revised: 17 January 2023
Accepted: 29 January 2023
Published: 4 February 2023

1. Introduction

Ebinur Lake in Xinjiang Uygur Autonomous Region (hereinafter referred to as Xinjiang), Northwest China, presents a critical ecological problem. The continual increase in socio-economic development, cultivated land, and industrial and agricultural wastewater have exacerbated its water quality [1]. Furthermore, Ebinur Lake has become one of the primary sources of sand and salt dust in Western China, with an extension to Middle-East China, severely threatening the ecological quality of the arid lands [2]. Suspended particulate matter (SPM) is an essential indicator in lake water quality monitoring, with a direct influence on the transparency and turbidity of water bodies [3,4]. Therefore, the continuous dynamic monitoring of SPM concentration is crucial for environmental management.

Recent advances in remote sensing technology have fostered increasingly accurate water quality monitoring. Despite improved acquisition conditions of remote sensing data, high temporal-cum-spatial resolution images still cannot be obtained jointly. To upgrade image quality and make up for data deficiency, many remote sensing spatio-temporal fusion models have been developed to permit image reconstruction [5]. They are divided into three categories: the transformation model, the pixel reconstruction model, and the dictionary learning model. The transformation model mainly uses principal component analysis and wavelet transformation methods. Shevyrnogov et al. extracted the brightness component of multi-spectral satellite (MSS) data based on principal component analysis and fused it with NOAA (National Oceanic and Atmospheric Administration) NDVI (Normalized Difference Vegetation Index) to generate data with high spatio-temporal resolution [6]. Malenovsky et al. were the first to use wavelet transformation to fuse MODIS (Moderate-resolution Imaging Spectroradiometer) and TM (Thematic Mapper) [7].

The pixel reconstruction model mainly includes filtering and unmixing methods. The spatio-temporal fusion model based on filtering predicts high-resolution images by introducing neighbor information [8], including Spatio-temporal Adaptive Reflectance Fusion Model (STARFM) [9], Enhanced Spatio-temporal Adaptive Reflection Fusion Model (ESTARFM) [10], and Spatio-temporal Non-local Filter-Based Fusion Model (STNLFFM) [11]. Spatio-temporal fusion models based on disaggregation include the Spatio-temporal Data Fusion Approach (STDFA) [12], Unmixing-Based Spatio-temporal Adaptive Reflectance Fusion Model (USTARFM) [13], and Flexible Spatio-temporal Data Fusion (FSDAF) [14].

The spatio-temporal fusion model based on dictionary learning constructs the corresponding relationship between high and low resolution to predict the high-resolution images on the prediction date [15]. Huang et al. proposed a sparse representation based on a Spatio-temporal Reflectance Fusion Model (SPSTFM) [16]. With the rise of deep learning, the method has been applied to spatio-temporal fusion [17]. Song et al. established Spatio-temporal Fusion by a Deep Convolutional Neural Network (STFDCNN) [18].

Remote sensing has been combined with modeling technology to form an inversion model of SPM, which include empirical or semi-empirical and analytical or semi-analytical models [19]. In building empirical models, the statistical relationship between measured SPM and image data is first established, and then the value of SPM is extrapolated. This method is widely used in multi-spectral satellite image water quality monitoring. It selects a single band or a band combination to build a regression model [20,21].

The semi-empirical model uses the spectral characteristics of SPM for statistical analysis and selects the best band to estimate the parameter contents. It relies heavily on hyperspectral remote sensing techniques [22,23]. The semi-analytical model is based on the radiative transfer equation to build the functional relationship between reflectance and the inherent optical characteristics of water [24]. There are three main semi-analytical methods: (1) the Nechad model [25], (2) the quasi-analytical algorithm (QAA) [26,27], and (3) the semi-empirical radiative transfer (SERT) [28].

Theoretically, the analytical model has high inversion accuracy and versatility and does not need a large amount of measured SPM. It is based on the known spectral characteristics of pure water and its components [29,30]. As the spectral characteristics of each component need to be measured, involving complex procedures and equipment, this method is rarely applied [31].

Spatio-temporal fusion algorithms have been widely adopted [10,14]. The application requirements differ notably concerning research objectives. Different from the global large-scale SPM monitoring research [32,33], this research mainly used multi-source, high spatio-temporal resolution, and time-continuous SPM monitoring on a regional scale. The specific aims were (1) to determine a better spatio-temporal fusion algorithm, (2) to establish a stable and widely applicable SPM inversion model, and (3) to develop a reliable SPM image reconstruction strategy to provide scientific reference for further water quality data reconstruction research.

2. Overview of the Research Area

Ebinur Lake is located in Xinjiang, Northwest China (44°54′~45°08′N, 82°35′~83°10′E), and is a broken subsidence basin formed by the Himalayan orogeny (Figure 1) [34]. The lake basin is the lowest depression, with an elevation of about 190 m. Surrounded by mountains on the west, south, and north, it is located in the heart of the Eurasian continent, with little precipitation, intense evaporation, and abundant sunlight and heat. The climate is typical temperate continental, with an annual average temperature of 6.6~7.8 °C and annual precipitation of 116.0~169.2 mm. Northwest of Ebinur Lake is the famous gale mouth of Alashankou, noted as having a maximum wind speed of over 55.0 m/s for 164 days/year [35]. The lake has an average depth of 1.4 m, with a lake surface water density of about 1.079 g/cm^3, pH 8.49, and a mineralization degree of 112.4 g/L [36].

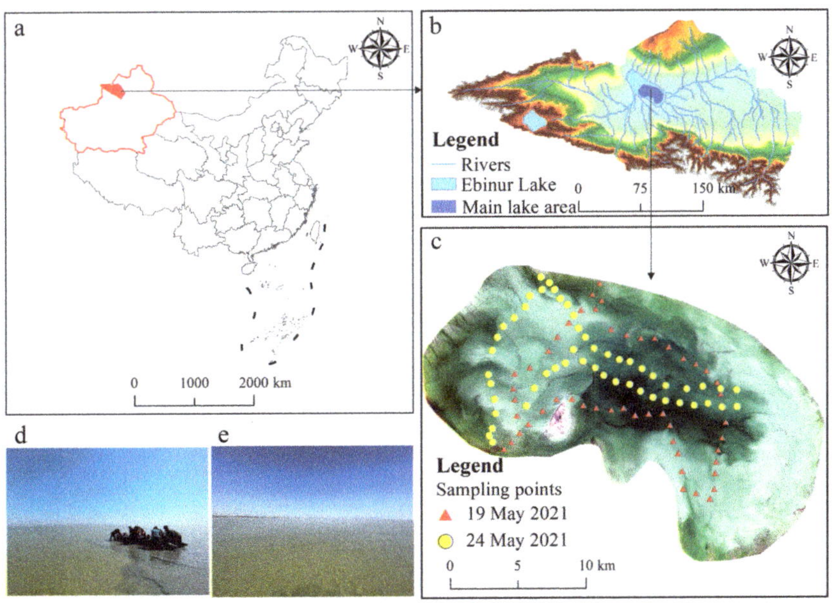

Figure 1. The research area. (**a**) The Ebinur Lake Basin is located in Xinjiang, Northwest China; the administrative division is derived from the National Geographic Information Resource Catalogue Service System https://www.webmap.cn/ (accessed on 5 March 2022). (**b**) It is located at the center of the basin. (**c**) Distribution of sampling points in Ebinur Lake performed on 19 and 24 May 2021. (**d**) The inflatable kayak used for sampling from the lake. (**e**) Landscape of the central portion of Ebinur Lake.

3. Data Source and Processing

3.1. Water Sample Collection and Laboratory Analysis

On-site acquisition was the principal way of obtaining the basic raw data in this research, acquired on 19 and 24 May 2021, at 103 sampling points. Sampling was conducted at 11:00~16:00, Beijing time. Sentinel-2 images were taken on 19 and 24 May 2021 at 12:26 and Sentinel-3 images were taken on 19 May 2021 from 12:00 to 12:03 and 24 May from 12:10 to 12:13, with the on-site data collected at the same time as the satellite transit (±4 h), strictly following the principle of interstellar synchronization [37].

Figure 1c indicates the sampling point design at about 1.5~2 km apart. The points cover different parts of the water body. Field sampling data included GPS location, water depth, salinity, temperature, DO, and pH. Inflatable kayaks were used to move around in the lake, and 2 L water samples were collected at 0.1 to 0.3 m depth using a fine mouth polyethylene bottle. The collected water samples were kept in cold storage (<4 °C) before conducting laboratory experiments to reduce the changes in physicochemical attributes in water [38].

3.2. Images and Preprocessing

Sentinel-2, consisting of two stars, A and B, is an environmental monitoring satellite launched by the European Space Agency (ESA), capable of providing ground-based observations with high spatial and temporal resolution. It has an orbital height of 768 km and a width of 290 km, with a single-star revisit period of 10 d and a binary revisit period of 5 d. The satellite is equipped with a push-scan multispectral imager (MSI) to obtain 13 band images with spatial resolutions of 10, 20, and 60 m. Detailed band information is shown in Table 1. Obtained from the ESA platform https://scihub.copernicus.eu/dhus/#/home (accessed on 10 December 2021), the images were an atmospheric apparent reflectivity product with orthotropic and geometric precision corrections.

Table 1. Sentinel-2 image band information.

Band	Description	S2A Center Wavelength (nm)	S2B Center Wavelength (nm)	Band Width (nm)	Spatial Resolution (m)
B1	Coastal aerosol	442.7	442.2	20	60
B2	Blue	492.4	492.1	65	10
B3	Green	559.8	559.0	35	10
B4	Red	664.6	664.9	30	10
B5	Red-edge1	704.1	703.8	15	20
B6	Red-edge2	740.5	739.1	15	20
B7	Red-edge3	782.8	779.7	20	20
B8	NIR	832.8	832.9	115	10
B8a	Narrow NIR	864.7	864.0	20	20
B9	Water vapor	945.1	943.2	20	60
B10	Cirrus	1373.5	1376.9	30	60
B11	SWIR1	1613.7	1610.4	90	20
B12	SWIR2	2202.4	2185.7	180	20

In this research, the Sentinel-2 images were pretreated by the Dark Spectrum Fitting (DSF) atmospheric correction algorithm, which is especially suitable for turbid water [39–41]. The Acolite provided by the DSF method was used to conduct atmospheric correction preprocessing https://github.com/acolite/acolite (accessed on 25 December 2021). Sentinel-2A/B data were corrected for atmosphere, processed in a batch program, exported to TIFF standard format, and then banded for synthesis. Since the Sentinel images lost three 60 m resolution bands for water vapor and SWIR-Cirrus after processing as 10 m resolution images, 11 bands were retained. They included B1-Coastal aerosol, B2-Blue, B3-Green, B4-Red, B8-NIR, B8a-Narrow NIR, B11-SWIR1, B12-SWIR2, and B5, B6, and B7 for vegetation red-edge bands. The output images were trimmed to cover the whole research area.

The Sentinel-3 satellite monitors the global ocean and land in real time. Among them, sea temperature, sea color, and sea level height data can be used to monitor climate change, ocean pollution, biological productivity [42–44], terrestrial forest fires, terrestrial vegetation health, and water levels of lakes and rivers [45–47]. The satellite has an orbital altitude of 800–830 km and a revisit period of less than 2 days. It carries a Sea-Land Surface Temperature Radiometer (SLSTR) and a Sea-Land Chromaticity Instrument (OLCI). This research mainly used the OLCI sensors, with 21 bands and a spatial resolution of 300 m, with detailed band information listed in Table 2. The data were obtained from the ESA data platform, which is an atmospheric apparent reflectivity product, using DSF atmospheric correction for Sentinel-3 OLCI [48,49]. These data were reprojected and resampled to 10 m to coincide with the image numbers of the research area in Sentinel-2 to form spatio-temporal fused data pairs to facilitate the research.

Table 2. Sentinel-3 OLCI image band information.

Band	Center Wavelength (nm)	Wave Width (nm)	Noise-Signal Ratio
Oa1	400	15	2188
Oa2	412.5	10	2061
Oa3	442.5	10	1811
Oa4(Blue)	490	10	1541
Oa5	510	10	1488
Oa6(Green)	560	10	1280
Oa7	620	10	997
Oa8(Red)	665	10	883
Oa9	673.5	7.5	707
Oa10	681.25	7.5	745
Oa11	708.75	10	785
Oa12	753.75	7.5	605
Oa13	761.25	7.5	232
Oa14	764.38	3.75	305
Oa15	767.5	2.5	330
Oa16	778.75	15	812
Oa17(NIR)	865	20	666
Oa18	885	10	395
Oa19	900	10	308
Oa20	940	20	203
Oa21	1020	40	152

4. Methods

4.1. Spatio-Temporal Fusion Algorithm

ESTARFM, an enhanced spatio-temporal fusion algorithm proposed by Zhu et al., was used to generate fusion images of Ebinur Lake's surface [10]. This algorithm is suitable for the lake's constantly changing SPM, considering that the ground reflectivity may change over time. The image processing mainly required the two-stage Sentinel-2 and -3 image pairs before and after the reconstruction date and one Sentinel-3 image on the reconstruction date. The Sentinel-2 images of the day were fused through the ESTARFM model.

ESTARFM thoroughly considered the spatial heterogeneity of the Sentinel-3 (coarse spatial and high temporal resolution) image and introduced the conversion coefficient to improve the fusion simulation results. The simulated image as the central image was used to build a relatively large moving window. The average image element with similar spectral features and the central image were calculated and selected to assign the value weight. Finally, the central value was calculated. The central image value was computed by Equation (1):

$$L_b(x_{w'/2}, y_{w'/2}, T) = L_b(x_{w'/2}, y_{w'/2}, T') + \sum_{i=1}^{n} W_i \times v_i \times (M_b(x_i, y_i, T) - M_b(x_i, y_i, T')) \quad (1)$$

where L_b and M_b represent band b of Sentinel-2 (fine space, low time resolution image) and Sentinel-3 images, respectively; w' is the moving window size; $(x_{w'/2}, y_{w'/2})$ represents analog image position; T and T' represent time; W_i is the first image similar to analog image weight; v_i represents the first and analog spectral image conversion coefficient; n is an analog which is similar to analog image number; and (x_i, y_i) represents the image position.

This study selected t_1, t_3 of Sentinel-2/3 images as model data input, and t_2 of Sentinel-3 simulated t_2 of Sentinel-2: $L_{b,t}(x_{w'/2}, y_{w'/2}, t_2)$ $(t = t_1, t_2)$. The t_2 time images simulated at t_1, t_3 were weighted to obtain more accurate t_2 time simulation images, and the weight ε_t calculation is shown in Equation (2). Equation (3) was used to calculate the simulated central image value and obtain the simulated remote sensing images with high spatio-temporal resolution.

$$\varepsilon_t = \frac{1/\left|\sum_{j=1}^{w'}\sum_{i=1}^{w'} M_b(X_i, Y_j, t) - \sum_{j=1}^{w'}\sum_{i=1}^{w'} M_b(X_i, Y_j, t_2)\right|}{\sum_t \left(1/\left|\sum_{j=1}^{w'}\sum_{i=1}^{w'} M_b(x_i, y_j, t) - \sum_{j=1}^{w'}\sum_{i=1}^{w'} M_b(x_i, y_j, t_2)\right|\right)}, t = t_1, t_3 \quad (2)$$

$$L_{b,t}(x_{w'/2}, y_{w'/2}, t_2) = \sum_t \varepsilon_t \times L_{b,t}(x_{w'/2}, y_{w'/2}, t), t = t_1, t_3 \quad (3)$$

FSDAF can predict regions of the ground type well [14]. It first classifies high-score, low-frequency images in known periods based on unsupervised classification methods. We adopted the K-means unsupervised classification method and set the classification number to four categories, a, b, c, and d, and then calculated the richness of each category in each high-frequency, low-scoring pixel using Equation (4).

$$f(X, c) = N(X, c)/m \quad (4)$$

where $f(X,c)$ is the richness of category c in high-frequency and low-frequency pixels X in known periods, $N(X,c)$ is the number of high-score and low-frequency pixels in pixel X and category c, and m is the number of high-score and low-frequency images in high-frequency and low-frequency pixel X. After we selected the high-frequency, low-score pixel with the highest richness in various categories, we found the difference between these high-frequency, low-score pixels in the known period and the prediction period and then fit the change value of the high-score, low-frequency pixel in each category by the least squares method.

4.2. Spatio-Temporal Fusion Strategy

To reconstruct the Sentinel-2 reflectance image, we needed to perform a spatio-temporal fusion operation based on the Sentinel-3 reflectance image. Among them, blue, green, red, and NIR are important visible light bands to monitor lake SPM [49]. As shown in Figure 2, they are also bands within the common spectral range of Sentinel-2 and Sentinel-3, with similar central wavelengths, and are an important remote sensing data basis for establishing spatio-temporal fusion models [14].

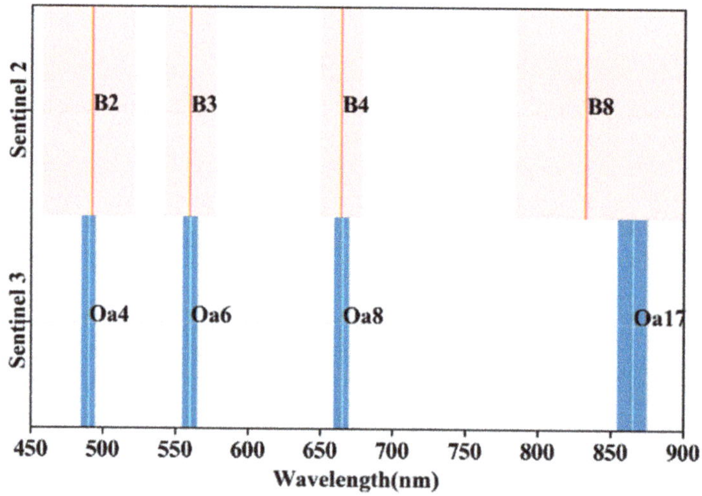

Figure 2. Sentinel-2/3 blue, green, red, and NIR band wavelengths and central wavelengths.

In this research, the model was tested by using two image data pairs to further reduce information loss in the fusion results with different input image pairs that were used. There are a total of six fusion results for the (a) and (b) fusion strategies in Figure 3.

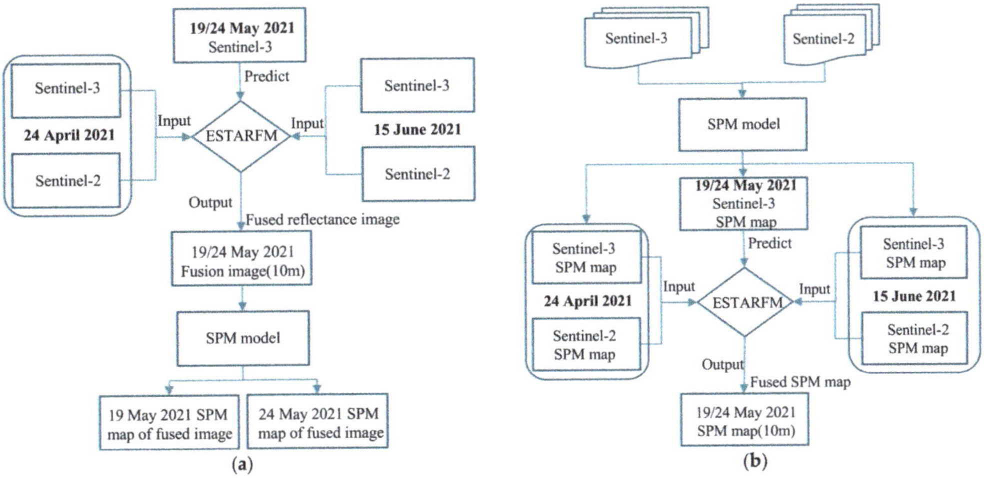

Figure 3. Flow chart of (**a**) SPM concentration map of fused reflectance image and (**b**) fused SPM concentration map.

- Fusion Strategies: The images (pixel 10 m) of Ebinur Lake on 19 and 24 May 2021 were used as reconstruction targets, and data pairs from different time points were used as inputs for ESTARFM and FSDAF models to reconstruct the optimal reflectance images. Firstly, Sentinel-2 and -3 images presented on 15 June and 24 April 2021 served as input image pairs for the ESTARFM model. According to the Sentinel-3 image on 19 and 24 May 2021, the ESTARFM fusion remote sensing image with a spatial resolution of 10 m was predicted. Secondly, the Sentinel-2 and -3 image pairs from 15 June and 24 April 2021 were used as input to the FSDAF model. The Sentinel-3 images on 19 and 24 May 2021 were used to predict FSDAF fusion images on the same date. Thirdly, the fused ESTARFM, FSDAF0424, and FSDAF0516 images were analyzed, validated, and compared with the original Sentinel-2 reference images on both sampling days (Figure 4), The small color differences in the fused images are mainly caused by errors in the fused bands. Finally, the SPM concentration inversion was performed on the fused images.

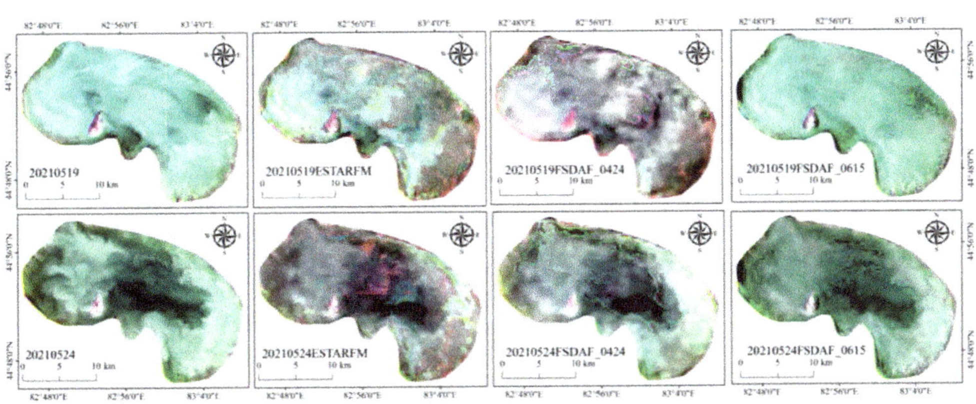

Figure 4. The fusion images of the ESTARFM and FSDAF models on 19 and 24 May 2021.

- Fusion Strategies: The Ebinur Lake SPM concentration map on 19 and 24 May 2021 was used as the reconstruction target. The Sentinel-2 and -3 SPM concentration inversion maps on 15 June and 24 April 2021 were used as the input image pairs for the ESTARFM model. ESTARFM fused SPM maps with a spatial resolution of 10 m that were predicted based on the Sentinel-3 images of 19 and 24 May 2021.

4.3. Spatio-Temporal Fusion Images Evaluation Indicators

To quantitatively research the quality of spatial and reference image quality, the Pearson correlation coefficient (R) was selected [50], along with normalized root mean square error (RMSE), peak signal-to-noise ratio (PSNR), and structural similarity (SSIM) to evaluate image quality [51]. The reference image I size was $m \times n$ and the spatio-temporal fusion image was F; $I_{(i,j)}$ was the reference image value and $F_{(i,j)}$ was the spatio-temporal fusion image value.

R evaluated the degree of consistency between spatio-temporal fusion images and reference images using Equation (5):

$$\mathrm{R} = \frac{\sum_{i=1}^{m}\sum_{j=1}^{n}(I_{(i,j)} - \overline{I})(F_{(i,j)} - \overline{F})}{\sqrt{\sum_{i=1}^{m}\sum_{j=1}^{n}(I_{(i,j)} - \overline{I})^2}\sqrt{\sum_{i=1}^{m}\sum_{j=1}^{n}(F_{(i,j)} - \overline{F}^2)}} \tag{5}$$

RMSE refers to the square root of the deviation between the spatio-temporal fusion image and the reference image, calculated by Equation (6):

$$\mathrm{RMSE} = \sqrt{\frac{\sum_{i=1}^{m}\sum_{j=1}^{n}(I_{(i,j)} - F_{(i,j)})^2}{mn}} \tag{6}$$

PSNR was used to evaluate the amount of fusion image information. A large value represents less image information loss. The mean square error MSE was computed by Equation (7) and PSNR by Equation (8):

$$MSE = \frac{1}{mn}\sum_{i=0}^{m-1}\sum_{j=0}^{n-1}[I(i,j) - F(i,j)]^2 \tag{7}$$

$$\mathrm{PSNR} = 10 \cdot \log_{10}\left(\frac{MAX_{(i,j)}^2}{MSE}\right) \tag{8}$$

where $MAX_{(i,j)}$ is the maximum image value of the image.

SSIM evaluated the structural similarity between spatio-temporal fusion images and reference images, calculated by Equation (9):

$$\mathrm{SSIM} = \frac{(2u_I u_F + C_1)(2\sigma_{I\cdot F} + C_2)}{(u_I{}^2 + u_F{}^2 + C_1)(\sigma_I{}^2 + \sigma_F{}^2 + C_2)}) \tag{9}$$

where u_I and u_F represent the mean, σ_I and σ_F represent the variance between the reference image and the fused image, and $\sigma_{I\cdot F}$ represents the covariance between the two images. C_1 and C_2 are two constants close to 0 used to stabilize the results. An SSIM value close to 1 indicates high structural similarity between the two images.

4.4. SPM Evaluation Indicators

The coefficient of determination (R^2), root mean square error (RMSE), mean absolute percentage error (MAPE), and deviation (bias) test whether the predicted and measured values are consistent. R^2 describes the degree to which the independent variable (remote sensing reflectance) explains the dependent variable (SPM concentration) [52,53]. It is usually used for auxiliary evaluation of model performance. A higher R^2 value should not be excessively pursued in building this high-precision model because it is prone to overfitting the modeling data. Moreover, accuracy may drop if the independent validation data are used to verify the overfitted model. Therefore, a larger R^2 does not mean that the model is invariably applicable. In balance, multiple indicators are needed to evaluate the model's reliability.

R^2 evaluated the degree of consistency between predictions and true values using Equation (10):

$$R^2 = 1 - \frac{\sum_{i=1}^{n} (y_i - \hat{y})^2}{\sum_{i=1}^{n} (y_i - \overline{y})^2} \tag{10}$$

RMSE is the square root of the ratio of the forecast and the image matrix, computed by Equation (11):

$$\text{RMSE} = \sqrt{\frac{\sum_{i=1}^{n} (y_i - \hat{y})^2}{n}} \tag{11}$$

MAPE represents the percentage of the absolute value of the predicted value, calculated by Equation (12):

$$MAPE = \frac{1}{n}\sum_{i=1}^{i-n}\left|\frac{y_i - \hat{y}_i}{y_i}\right| \times 100\% \tag{12}$$

Bias represents the deviation from the predicted value and the true value, computed by Equation (13):

$$Bias = \frac{1}{n}\sum_{i=1}^{n}|y_i - \hat{y}_i| \tag{13}$$

5. Results and Analysis

5.1. Spatio-Temporal Fusion Reflectance Image Reconstruction and Evaluation

Figure 4 compares the ESTARFM and FSDAF fusion images with the reference images. By using blue, green, and red true color band channels to display the lake images, the visual discrimination of the spatio-temporal fusion model can better realize the generation of predicted date images.

Four evaluation indicators were applied to assess the fusion images to further quantitatively evaluate the image quality and retention of spectral information (Figure 5). The overall image quality of the ESTARFM fusion image on 19 May 2021 was better than that of the FSDAF. The evaluation indicators R, RMSE, PSNR, and SSIM in the blue band verified that the ESTARFM fusion image had the best quality; the FSDAF0615 ranked second, and the FSDAF0424 was the last. Using the green band indicators, the ESTARFM fusion image remained the best, and the FSDAF0615 accuracy was slightly higher than FSDAF0424. For the red band indicators, the ESTARFM fusion image also had the best quality (R was 0.72, RMSE was 0.0140, PSNR was 37.09, and SSIM was 0.93). Finally, in the NIR band, the accuracy of the FSDAF0424 fusion image was relatively poor, whereas the accuracy difference between FSDAF0615 and ESTARFM was small.

The concentration of the plotted points in a small core area along the 1:1 line and limited dispersion away from the core to the periphery signify a good match between the fusion and reference images (Figure 5). In the blue band, the ESTARFM model with the smallest point spread and the most prominent concentration indicated the best distribution and effects of the three. The FSDAF0424 model showed relatively more point scattering in the core and peripheral areas. The FSDAF0615 model demonstrated a concentration in the core area, with large prediction errors in the low-value area. In the green band, the FSDAF0424 model had quite bundled points, but the FSDAF0615 model displayed more errors in the low-value area. The red band had the best effects compared with other bands, with more points clustering along and adjacent to the 1:1 line. The ESTARFM model registered the best performance, and the FSDAF0424 had a relatively more scattered distribution in the core and peripheral areas. Although the FSDAF0615 model demonstrated a concentration in the core area, it had more fusion errors in the low-value area. For the NIR band, the three graphs showed heavy crowding in the lowest-value part adjoining the origin. However, some points were dispersed to the medium- and high-value areas. Such patterns indicated relatively large discrepancies between the fusion and reference images.

Using the 24 May 2021 images, we further investigated the accuracy of fusion images with the same evaluation indicators (Figure 6). The overall image quality of ESTARFM was better than FSDAF. Among these indicators, the results of R evaluation conflict with those jointly evaluated by RMSE, PSNR, and SSIM. In other words, the accuracy of RMSE, PSNR, and SSIM of ESTARFM model is higher than that of FSDAF0615, while R is on the contrary, reflecting that R cannot evaluate image quality to a certain extent. Therefore, we applied three indicators to evaluate the blue band; the ESTARFM fusion image quality was the best, followed by FSDAF0615 and then FSDAF0424. However, in the green band, the ESTARFM fusion image had the best quality effect, whereas FSDAF0615 and FSDAF0424 were of poor quality. In the red band, the ESTARFM fusion image had the best quality effect. In the NIR band, FSDAF0424 and FSDAF0615 fusion images had relatively poor accuracy, and ESTARFM had relatively good accuracy.

From the scatter distribution analysis in Figure 6, we found that in the blue band, the ESTARFM model denoted pronounced point concentration and suitable accuracy and that the FSDAF0424 model had scattered to the high-value area versus the FSDAF0615 model's scattering to the low-value area. In the green band, the prediction errors of FSDAF0424 also appeared in the high-value area, and the FSDAF0615 had prediction errors mainly in the middle-value area. In the red band, the three fusion images demonstrated the best distribution, with a concentration near the 1:1 line, compared with other bands. The ESTARFM model showed the best effect. For the NIR band, the results were similar to 19 May 2021. Most points were bundled in the lowest-value area around the graph origin. A notable number of points dispersed to the middle and high-value areas, signifying considerable errors in image prediction.

5.2. Construction of the SPM Inversion Models for Sentinel-2 and Sentinel-3

The light absorption and scattering properties of various substances in the lake water determine the spectral reflection characteristics of the water body [54]. Changes in the composition and concentration of SPM in Ebinur Lake trigger corresponding changes in spectral reflection characteristics. From Figure 7, the SPM reflectance information of Sentinel-2 and Sentinel-3 is higher in the red, green, and blue bands and lower in the NIR band, and the reflectance information of SPM sampling points is best separated in the red band. These results provided a basis to establish the Sentinel-2 and Sentinel-3 regression models.

Section 5.1 shows that the fusion red band had the highest accuracy. Therefore, the red band was chosen for the inversion of SPM in this research. In the SPSS software, the red band was used as the independent input variable in the modeling, and the measured SPM concentration was regression-analyzed as the dependent variable. The 73 (70%) random samples of the 103 matched sample pairs were used to build the regression model. The remaining 30 (30%) sample pairs were used to test model accuracy. Mathematical models were built for the red bands, with the regression coefficients solved and R^2 determined (Table 3).

The sample distribution in the scatter plot of the red band (Figure 8) showed that the SPM concentration fit the reflectance well, but some samples were dispersed in the high-value region. Table 3 indicates that the fitting effects of the models were suitable. The R^2 of the Sentinel-2 exponential model was the highest at 0.63, and the linear model had the lowest R^2 at 0.47. The R^2 of the Sentinel-3 exponential model was the highest at 0.73, and the linear model had the lowest R^2 at 0.65.

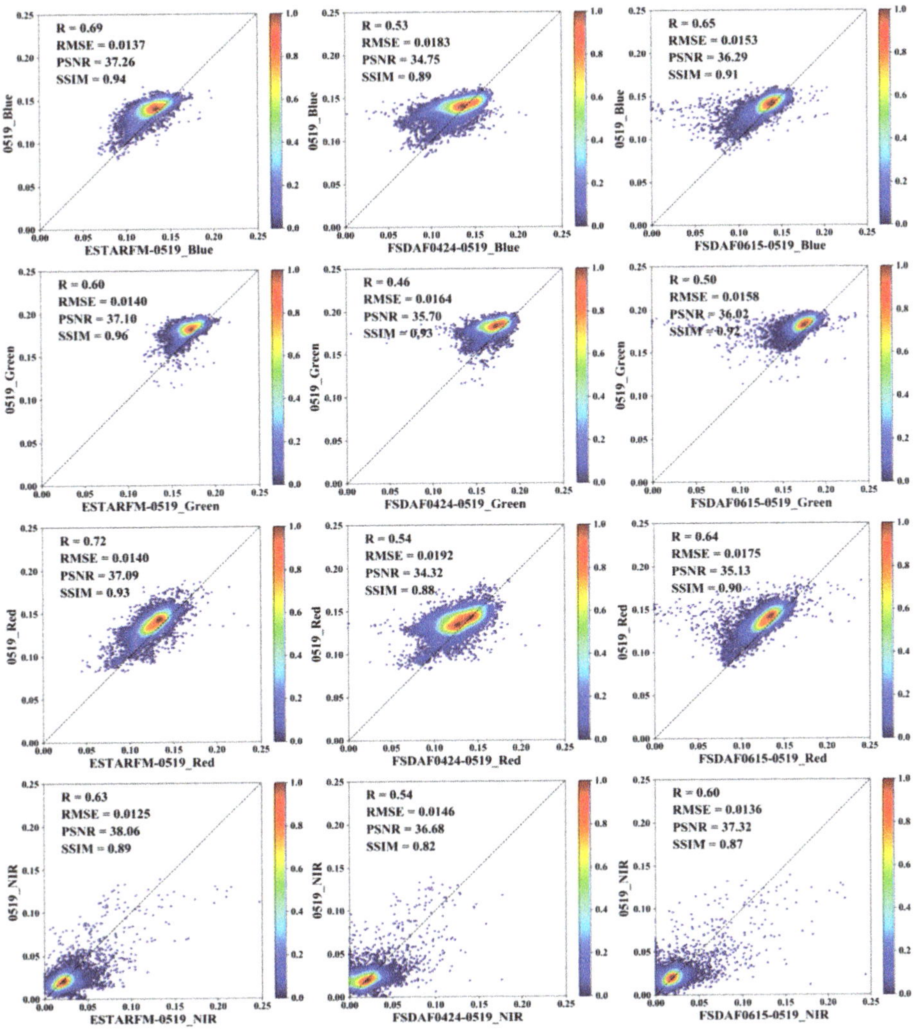

Figure 5. The scatter density plots of the fusion and reference images of the ESTARFM and FSDAF models were on 19 May 2021.

To quantify the accuracy of these models, the common evaluation indicators RMSE, MAPE, and bias were applied for comparison. Figure 9 indicates that the minimum RMSE of the red band model based on Sentinel-2 was 35.47 mg/L, and the minimum MAPE was 15.30%. The minimum bias based on the polynomial model was −1.42 mg/L.

It can be seen from Figure 10 that the minimum RMSE and the minimum MAPE of the red band polynomial model based on Sentinel-3 were 43.59 mg/L and 16.05%, respectively. The minimum bias based on the red band linear model was −19.33 mg/L.

5.3. SPM Images Reconstruction Strategy

5.3.1. Estimation of SPM Using the Spatio-Temporal Fusion Reflectance Image

The results were compared with the red band SPM concentration estimates of the reference image (Figure 11). The three fusion images recorded on both sampling days could reflect the general trend of the SPM concentration distribution.

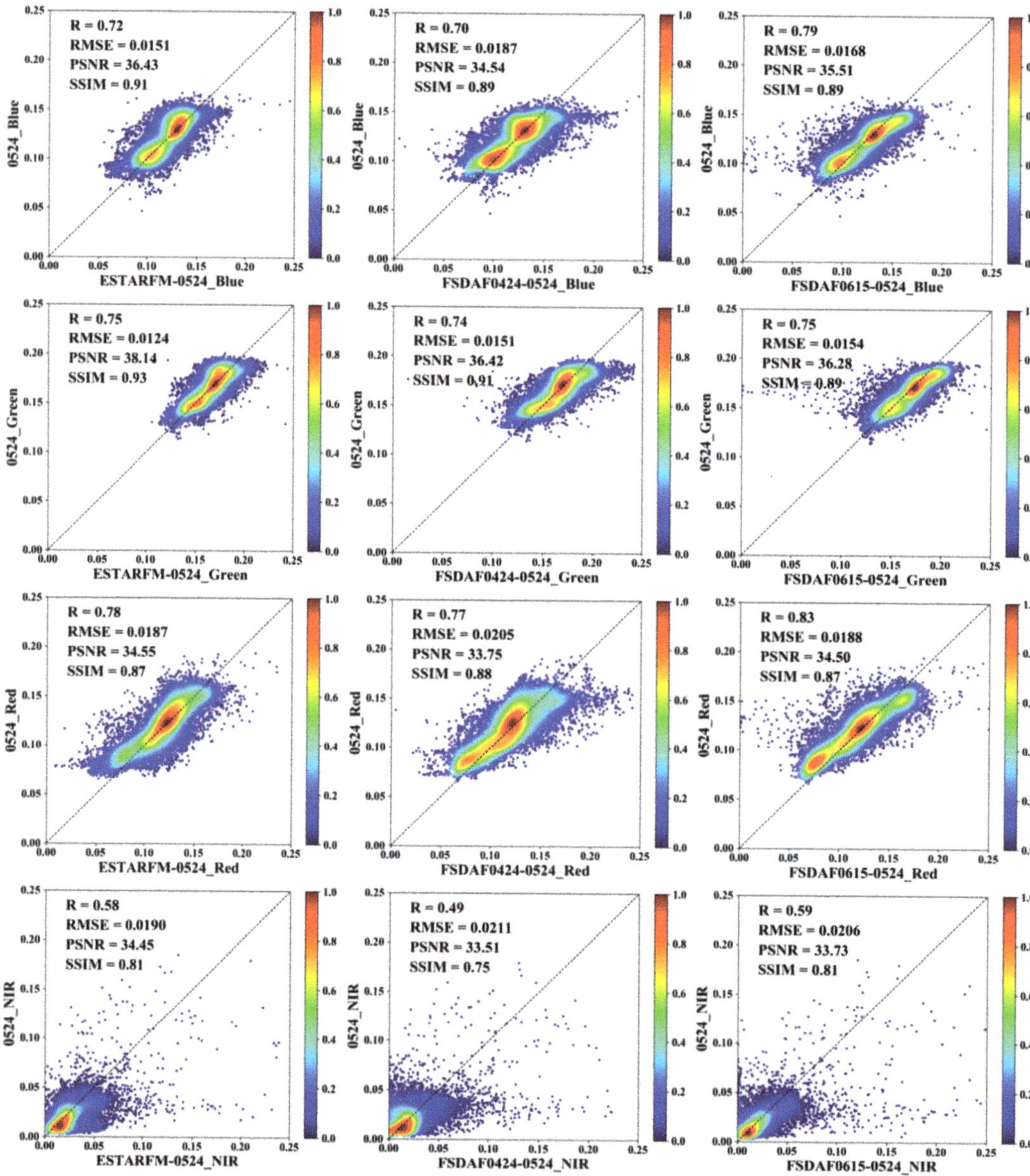

Figure 6. The scatter density plots of the fusion and the reference images of the ESTARFM and FSDAF models on 24 May 2021.

The accuracy estimation results of the SPM concentrations of different fusion models on both sampling days were compared (Figure 12). The overall evaluation indicators showed that the fusion image estimate on 24 May 2021 was better than that on 19 May 2021. Among the fusion models, the SPM concentration accuracy of the ESTARFM image inversion was the best on 19 May 2021. The evaluation indicators on 24 May 2021 showed that the ESTARFM image inversion had lower accuracy than FSDAF0615, but the RMSE difference was small.

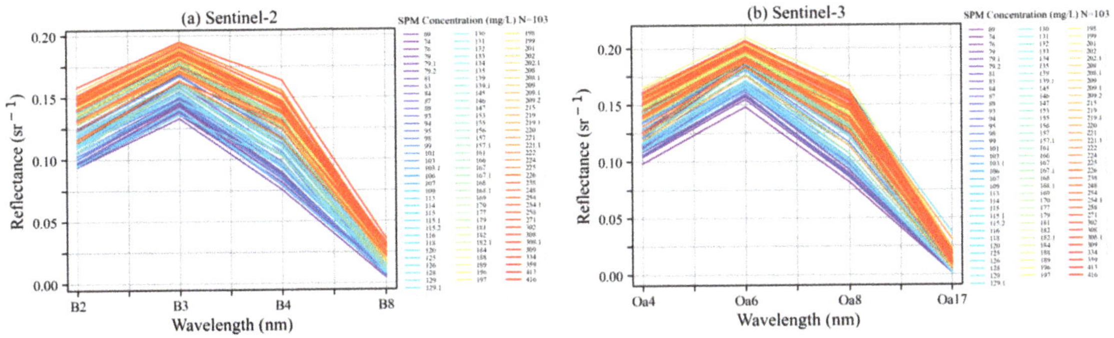

Figure 7. Remote sensing reflectance of SPM sample sites for Sentinel-2 and Sentinel-3 (reflectance conversion method based on Catherine et al. [37]).

Table 3. The Sentinel-2 and Sentinel-3 red band regression model and validation.

	Model	Regression Equation	R^2	p
Sentinel-2	Linear	$y = 2482.17x - 135.52$	0.47	<0.001
	Polynomial	$y = 8728.28x^2 + 385x - 13.26$	0.47	<0.001
	Power	$y = 7386.98x^{1.83}$	0.62	<0.001
	Exponential	$y = 23.05 \times e^{15.56x}$	0.63	<0.001
Sentinel-3	Linear	$y = 1727.50x - 73.45$	0.65	<0.001
	Polynomial	$y = 9493.85x^2 - 629.63x + 66.99$	0.66	<0.001
	Power	$y = 3095.03x^{1.50}$	0.72	<0.001
	Exponential	$y = 28.40e^{12.39x}$	0.73	<0.001

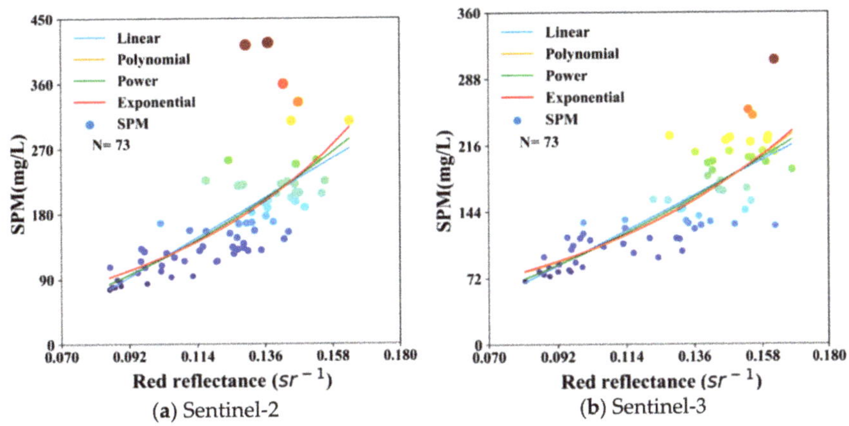

Figure 8. The scatter plots of the regression fitting between the measured SPM and the reflectance in the Sentinel-2 and Sentinel-2 Red band (The Power in Figure an overlap mostly with the Polynomial and can be distinguished at the lowest part of the curve. The darker the colored dot represents a larger SPM value).

The ESTARFM fusion image estimate yielded the best results among the three scatter density graphs of the SPM concentration for 19 May 2021. The FSDAF0424 and FSDAF0615 models showed different distribution patterns. For the 24 May 2021 graph, the ESTARFM fusion image estimates were the best, with a high concentration in the core area and less dispersion compared with the FDSAF0615 and FSDAF0424 models.

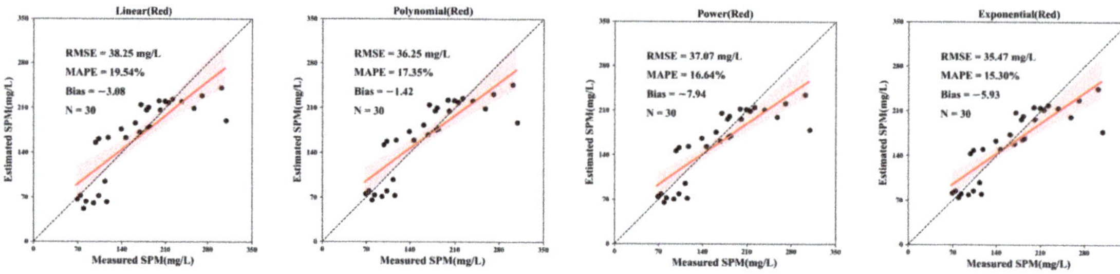

Figure 9. The scatter plots of the Sentinel-2 Red band regression model validation.

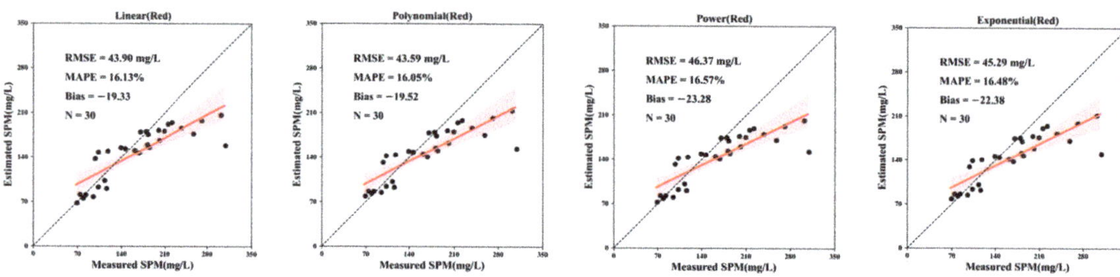

Figure 10. The scatter plots of the Sentinel-3 red band regression model validation.

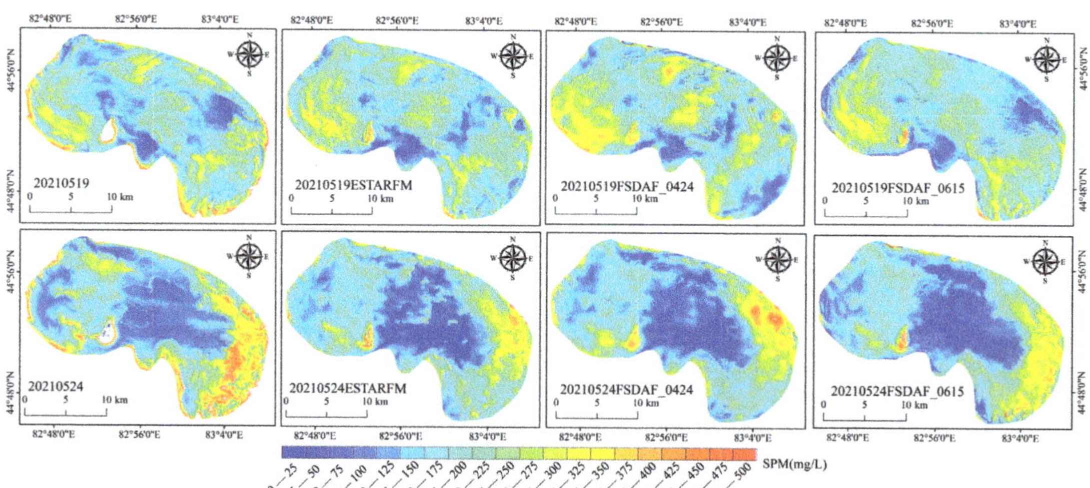

Figure 11. The SPM concentration inversion maps of the red band linear models of the reference and fusion images.

In sum, the fusion image quality of the ESTARFM model satisfied various evaluation indicators and had adequate stability. The FSDAF model showed some uncertainty, with accuracy sometimes depending on the input images. Therefore, a reliable ESTARFM model was adopted in the subsequent analysis.

5.3.2. Spatio-Temporal Fusion SPM

The raw images from Sentinel-2 and -3 were used to invert SPM concentrations (Sections 5.1 and 5.2). The SPM concentration inversion maps on 24 April and 15 June 2021 were used as the data pairs. The SPM concentration inversion maps on 19 May and 24 May

2021 were used as data sources. The estimated results were compared with the reference image results (Figure 13). The fusion images on both sampling days could better reflect the general SPM concentration trend. Analysis of different fusion strategies indicated that the SPM spatio-temporal fusion accuracy was higher (Figure 14). The image performance based on the evaluation indicators alone was sufficient. Analysis of different data sources signified that the fusion SPM estimate on 24 May 2021 was better than that on 19 May 2021.

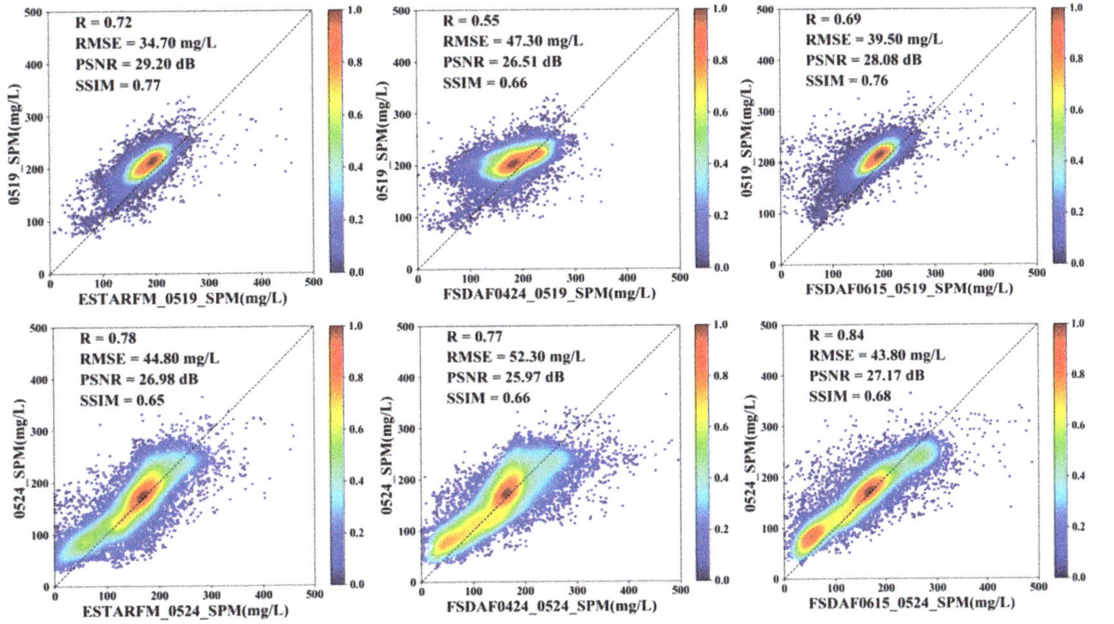

Figure 12. The scatter density plots of SPM concentrations in the red band linear model of the reference and fusion images.

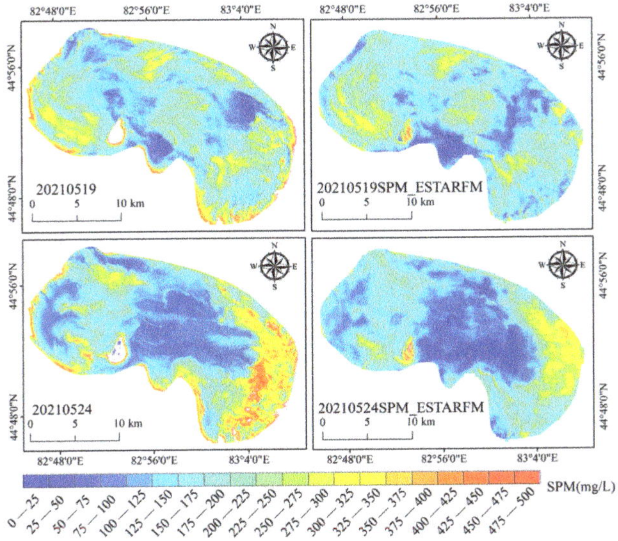

Figure 13. The original and ESTARFM fusion SPM concentration inversion maps.

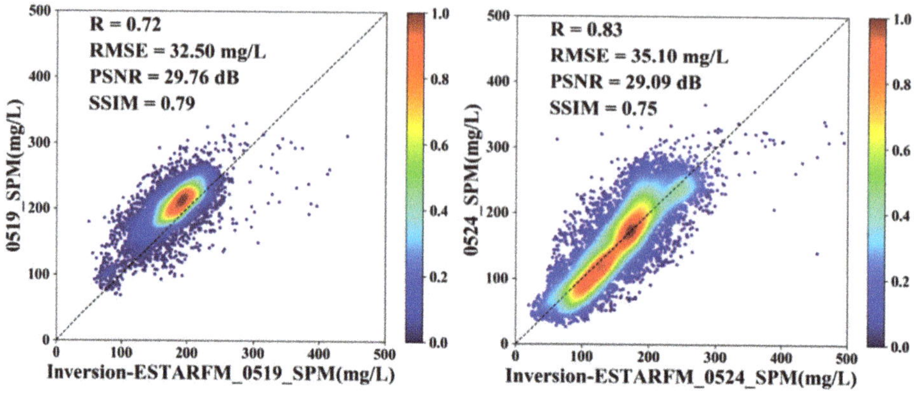

Figure 14. The scatter density plots of the raw and ESTARFM fusion SPM concentrations.

Figure 14 further shows that the SPM inversion accuracy was higher before fusion. The ESTARFM model and the after-fusion inversion strategy were adopted to generate spatio-temporal fusion images to complement the missing high-resolution SPM distribution maps.

6. Discussion

6.1. Spatio-Temporal Fusion Algorithm

Spatio-temporal fusion models have been widely used to research land use cover, vegetation, soils, water bodies, and other spectral reflectance imagery [8]. Han et al. briefly compared the accuracy of four spatio-temporal fusion algorithms (STARFM, ESTARFM, FSDAF, and FITFC) with R ranging from 0.621 to 0.907 and RMSE ranging from 0.019 to 0.08 using MODIS (500 m) Landsat-8 and Sentinel-2 as input data pairs, and concluded that their fused reflectance image accuracies all met their research requirements [55]. In this research, a comparison of two time points was used as a predictor to test the stability of the model. The ESTARFM model had R ranging from 0.58 to 0.78 and RMSE from 0.019 to 0.0124, while the FSDAF model had R ranging from 0.46 to 0.83 and RMSE from 0.0211 to 0.0136. As a result, it can be seen that the ESTARFM model in this research had a smaller and more stable margin of error. A comparative analysis of ESTARFM and FSDAF models was carried out to apply the spatio-temporal fusion model to lake SPM research, further demonstrating the feasibility of applying the spatio-temporal fusion model to the inversion of SPM in water bodies.

6.2. Accuracy of SPM Models

In this research, a more convenient and fast regression model was used for SPM inversion modeling, providing that sufficient sample points were collected from the lake [56–58]. Firstly, the inverse model has suitable interpretability. Secondly, the focus of this research was to conduct a spatio-temporal fusion strategy study, and the use of a convenient and fast modeling approach helped to reduce the transmission of errors and the mitigation of uncertainty in modeling. When modeling with the four regression models, it was found through Figure 6 that the SPM spectral information in the red band was better separated and therefore modeled with red. However, after the four models were built, their validation revealed that when the models were applied to Sentinel-2 (Figure 8), the power and exponential models, although better modeled with R^2 values of 0.62 and 0.63, respectively, had larger bias for validation, while the accuracy of the polynomial model was slightly better than the linear model, and both could be inverted for SPM concentration. When the model was applied to Sentinel-3 (Figure 9), it was found that the power and exponential models were better modeled, with R^2 values of 0.72 and 0.73, respectively, again with larger bias and RMSE at validation, while the linear model had better bias metrics than the polynomial model when the other metrics were not too different from the polynomial model. In summary, to avoid systematic error transfer due to inconsistencies in the modeling model formulations, given the small differences in model accuracy, Sentinel-2 and Sentinel-3 both use a linear model with a simpler model structure and the highest accuracy requirements for the later spatio-temporal fusion strategy.

6.3. Spatio-Temporal Fusion Strategy

The spatio-temporal fusion algorithm is primarily based on the fusion of remote sensing images and then the inversion of the target SPM. In this way, during the fusion of the target SPM, the target SPM information is more or less lost due to differences in the quality and timing of the input image pairs and the calculation methods of different spatio-temporal fusion models.

The accuracy validation evaluation of Figures 12 and 14 reveals that both strategies a and b from Figure 3 can have suitable fusion accuracy and that the closer the time is to the fusion target, the better the accuracy of the fused image, with R ranging from 0.55 to 0.84, RMSE from 52.30 to 34.70 mg/L, PSNR from 25.97 to 29.20 dB, and SSIM from 0.65 to 0.77 for strategy a, while strategy b has R ranging from 0.72 to 0.83, RMSE ranging from 35.10 to 32.50 mg/L, PSNR from 29.09 to 29.76 dB, and SSIM from 0.75 to 0.79. It can be seen that the fusion results for strategy a with high and low accuracy are generated by the FSDAF model, indicating that there is high instability in the FSDAF model, which is detrimental to our future extension applications. At the same time, the accuracy advantage of the ESTARFM model fusion results for strategy b over a is also all-encompassing, which better informs future similar studies and further clarifies the specific impact of the different spatio-temporal fusion strategies on the results. Another important reason for the greater stability of ESTARFM may also be that the input data pairs are two pairs, whereas FSDAF has only one pair of input data pairs, which also provides an idea for future research. Thus, we will explore the effect of models with two input pairs or one input pair on the results in the next step. However, for this research, the use of different fusion strategies plays a decisive role in the reconstruction of SPM images, and fusing target water quality images has an all-around better accuracy performance than fusing reflection images.

7. Conclusions

Based on the measured SPM concentration data and Sentinel-2 and -3 images, the optimal SPM inversion model and data reconstruction strategy with adequate capabilities and suitability were identified. From the results, we drew the following conclusions:

1. The ESTARFM fusion of blue, green, red, and NIR bands was the best, among which the red band had the highest accuracy.
2. The red band was determined to be the best choice for regression modeling based on an accurate assessment of the measurements and model stability analysis.
3. The fused SPM concentration map proved to be better and more stable.

In future research, we could use more accurate physical models or semi-physical models to carry out the research. In the meantime, we are prepared to adopt the incorporation or improvement of more spatio-temporal fusion algorithms for comparative studies to further enhance the applicability and scalability of our research.

Author Contributions: Conceptualization, P.D.; methodology, P.D.; software, P.D.; validation, P.D.; formal analysis, F.Z.; investigation, Y.C.; resources, F.Z. and Y.C.; data curation, P.D., Y.C., C.L., W.W. and Z.W.; writing—original draft preparation, P.D.; writing—review and editing, P.D., C.-Y.J., M.L.T. and J.S.; visualization, P.D., M.L.T. and J.S.; supervision, F.Z.; project administration, F.Z.; funding acquisition, F.Z. All authors have read and agreed to the published version of the manuscript.

Funding: This research was funded by the National Natural Science Foundation of China (Grant Number 42261006), the State Key Laboratory of Lake Science and Environment (Grant Number 2022SKL007, and the Tianshan Talent Project (Phase III) of the Xinjiang Uygur Autonomous Region.

Data Availability Statement: The data presented in this study are available on request from the corresponding author. The data are not publicly available due to confidentiality.

Acknowledgments: We appreciate the helpful comments offered by the anonymous reviewers and editors to improve our manuscript.

Conflicts of Interest: The authors declare no conflict of interest.

References

1. Liu, C.; Zhang, F.; Wang, X.; Chan, N.W.; Rahman, H.A.; Yang, S.; Tan, M.L. Assessing the factors influencing water quality using environment water quality index and partial least squares structural equation model in the Ebinur Lake Watershed, Xinjiang, China. *Environ. Sci. Pollut. Res.* **2022**, *29*, 29033–29048. [CrossRef]
2. Liu, D.; Abuduwaili, J.; Lei, J.; Wu, G.; Gui, D. Wind erosion of saline playa sediments and its ecological effects in Ebinur Lake, Xinjiang, China. *Environ. Earth Sci.* **2011**, *63*, 241–250. [CrossRef]
3. Sagan, V.; Peterson, K.T.; Maimaitijiang, M.; Sidike, P.; Sloan, J.; Greeling, B.A.; Maalouf, S.; Adams, C. Monitoring inland water quality using remote sensing: Potential and limitations of spectral indices, bio-optical simulations, machine learning, and cloud computing. *Earth-Sci. Rev.* **2020**, *205*, 103187. [CrossRef]
4. Liu, C.; Duan, P.; Zhang, F.; Jim, C.Y.; Tan, M.L.; Chan, N.W. Feasibility of the spatiotemporal fusion model in monitoring Ebinur Lake's suspended particulate matter under the missing-data scenario. *Remote Sens.* **2021**, *13*, 3952. [CrossRef]
5. Li, J.; Li, Y.; He, L.; Chen, J.; Plaza, A. A new sensor bias-driven spatio-temporal fusion model based on convolutional neural networks. *Sci. China Inf. Sci.* **2020**, *63*, 1–17. [CrossRef]
6. Shevyrnogov, A.; Trefois, P.; Vysotskaya, G. Multi-satellite data merge to combine NOAA AVHRR efficiency with Landsat-6 MSS spatial resolution to study vegetation dynamics. *Adv. Space Res.* **2000**, *26*, 1131–1133. [CrossRef]
7. Malenovsky, Z.; Bartholomeus, H.M.; Acerbi-Junior, F.W.; Schopfer, J.T.; Painter, T.H.; Epema, G.F.; Bregt, A.K. Scaling dimensions in spectroscopy of soil and vegetation. *Int. J. Appl. Earth Obs. Geoinf.* **2007**, *9*, 137–164. [CrossRef]
8. Sun, L.; Gao, F.; Xie, D.H.; Anderson, M.; Chen, R.; Yang, Y.; Yang, Y.; Chen, Z. Reconstructing daily 30m NDVI over complex agricultural landscapes using a crop reference curve approach. *Remote Sens. Environ.* **2020**, *253*, 112156. [CrossRef]
9. Gao, F.; Masek, J.; Schwaller, M.; Hall, F. On the blending of the Landsat and MODIS surface reflectance: Predicting daily Landsat surface reflectance. *IEEE Trans. Geosci. Remote Sens.* **2006**, *44*, 2207–2218.
10. Zhu, X.; Chen, J.; Gao, F.; Chen, X.; Masek, J.G. An enhanced spatial and temporal adaptive reflectance fusion model for complex heterogeneous regions. *Remote Sens. Environ.* **2010**, *114*, 2610–2623. [CrossRef]
11. Cheng, Q.; Liu, H.; Shen, H.; Wu, P.; Zhang, L. A spatial and temporal nonlocal filter-based data fusion method. *IEEE Trans. Geosci. Remote Sens.* **2017**, *55*, 4476–4488. [CrossRef]
12. Wu, M.; Huang, W.; Niu, Z.; Wang, C. Generating daily synthetic Landsat imagery by combining Landsat and MODIS data. *Sensors* **2015**, *15*, 24002–24025. [CrossRef] [PubMed]
13. Xie, D.; Zhang, J.; Zhu, X.; Pan, Y.; Liu, H.; Yuan, Z.; Yun, Y. An improved STARFM with help of an unmixing-based method to generate high spatial and temporal resolution remote sensing data in complex heterogeneous regions. *Sensors* **2016**, *16*, 207. [CrossRef] [PubMed]
14. Zhu, X.; Helmer, E.H.; Gao, F.; Liu, D.; Chen, J.; Lefsky, M.A. A flexible spatiotemporal method for fusing satellite images with different resolutions. *Remote Sens. Environ.* **2016**, *172*, 165–177. [CrossRef]
15. Wang, L.; Wang, X.; Wang, Q.; Atkinson, P.M. Investigating the influence of registration errors on the patch-based spatio-temporal fusion method. *IEEE J. Sel. Top. Appl. Earth Obs. Remote Sens.* **2020**, *13*, 6291–6307. [CrossRef]
16. Huang, B.; Song, H. Spatiotemporal reflectance fusion via sparse representation. *IEEE Trans. Geosci. Remote Sens.* **2012**, *50*, 3707–3716. [CrossRef]
17. Li, W.; Zhang, X.; Peng, Y.; Dong, M. Spatiotemporal fusion of remote sensing images using a convolutional neural network with attention and multiscale mechanisms. *Int. J. Remote Sens.* **2021**, *42*, 1973–1993. [CrossRef]
18. Song, H.; Liu, Q.; Wang, G.; Hang, R.; Huang, B. Spatiotemporal satellite image fusion using deep convolutional neural networks. *IEEE J. Sel. Top. Appl. Earth Obs. Remote Sens.* **2018**, *11*, 821–829. [CrossRef]
19. Tan, Z.; Cao, Z.; Shen, M.; Chen, J.; Song, Q.; Duan, H. Remote estimation of water clarity and suspended particulate matter in qinghai lake from 2001 to 2020 using MODIS images. *Remote Sens.* **2022**, *14*, 3094. [CrossRef]
20. Du, Y.; Song, K.; Liu, G.; Wen, Z.; Fang, C.; Shang, Y.; Zhao, F.; Wang, Q.; Du, J.; Zhang, B. Quantifying total suspended matter (TSM) in waters using Landsat images during 1984–2018 across the Songnen Plain, Northeast China. *J. Environ. Manag.* **2020**, *262*, 110334. [CrossRef]
21. Ford, R.T.; Vodacek, A. Determining improvements in Landsat spectral sampling for inland water quality monitoring. *Sci. Remote Sens.* **2020**, *1*, 100005. [CrossRef]
22. Liang, Z.; Zou, R.; Chen, X.; Ren, T.; Su, H.; Liu, Y. Simulate the forecast capacity of a complicated water quality model using the long short-term memory approach. *J. Hydrol.* **2020**, *581*, 124432. [CrossRef]
23. Flink, P.; Lindell, L.T.; Östlund, C. Statistical analysis of hyperspectral data from two Swedish lakes. *Sci. Total Environ.* **2001**, *268*, 155–169. [CrossRef]
24. Rotta, L.; Alcântara, E.; Park, E.; Bernardo, N.; Watanabe, F. A single semi-analytical algorithm to retrieve chlorophyll-a concentration in oligo-to-hypereutrophic waters of a tropical reservoir cascade. *Ecol. Indic.* **2021**, *120*, 106913. [CrossRef]

25. Nechad, B.; Ruddick, K.G.; Park, Y. Calibration and validation of a generic multisensor algorithm for mapping of total suspended matter in turbid waters. *Remote Sens. Environ.* **2010**, *114*, 854–866. [CrossRef]
26. Alcântara, E.; Curtarelli, M.; Ogashawara, I.; Rosan, T.; Kampel, M.; Stech, J. Developing QAA-based retrieval model of total suspended matter concentration in Itumbiara reservoir. In Proceedings of the Brazil//2015 IEEE International Geoscience and Remote Sensing Symposium (IGARSS), Milan, Italy, 26–31 July 2015; pp. 711–714.
27. Sun, D.; Qiu, Z.; Hu, C.; Wang, S.; Wang, L.; Zheng, L.; Peng, T.; He, Y. A hybrid method to estimate suspended particle sizes from satellite measurements over Bohai Sea and Yellow Sea. *J. Geophys. Res. Ocean.* **2016**, *121*, 6742–6761. [CrossRef]
28. Lei, S.; Xu, J.; Li, Y.; Li, L.; Lyu, H.; Liu, G.; Chen, Y.; Lu, C.; Tian, C.; Jiao, W. A semi-analytical algorithm for deriving the particle size distribution slope of turbid inland water based on OLCI data: A case study in Lake Hongze. *Environ. Pollut.* **2021**, *270*, 116288. [CrossRef]
29. Salama, M.S.; Verhoef, W. Two-stream remote sensing model for water quality mapping: 2SeaColor. *Remote Sens. Environ.* **2015**, *157*, 111–122. [CrossRef]
30. Liu, D.; Duan, H.; Yu, S.; Shen, M.; Xue, K. Human-induced eutrophication dominates the bio-optical compositions of suspended particles in shallow lakes: Implications for remote sensing. *Sci. Total Environ.* **2019**, *667*, 112–123. [CrossRef]
31. Kishino, M.; Tanaka, A.; Ishizaka, J. Retrieval of chlorophyll a, suspended solids, and colored dissolved organic matter in Tokyo Bay using ASTER data. *Remote Sens. Environ.* **2005**, *99*, 66–74. [CrossRef]
32. Wei, J.W.; Wang, M.H.; Jiang, L.D.; Yu, X.; Mikelsons, K.; Shen, F. Global Estimation of Suspended Particulate Matter From Satellite Ocean Color Imagery. *J. Geophys. Res. Ocean.* **2021**, *126*, e2021JC017303. [CrossRef] [PubMed]
33. Liu, X.; Wang, M. Global daily gap-free ocean color products from multi-satellite measurements. *Int. J. Appl. Earth Obs. Geoinf.* **2022**, *108*, 102714. [CrossRef]
34. Zhu, S.D.; Zhang, F.; Zhang, Z.Y.; Kung, H.; Yushanjiang, A. Hydrogen and oxygen isotope composition and water quality evaluation for different water bodies in the Ebinur Lake Watershed, Northwestern China. *Water* **2019**, *11*, 2067. [CrossRef]
35. Wang, L.; Li, Z.; Wang, F.; Li, H.; Wang, P. Glacier changes from 1964 to 2004 in the Jinghe River basin, Tien Shan. *Cold Reg. Sci. Technol.* **2014**, *102*, 78–83. [CrossRef]
36. Liu, C.J.; Zhang, F.; Johnson, V.C.; Duan, P.; Kung, H.T. Spatio-temporal variation of oasis landscape pattern in arid area: Human or natural driving? *Ecol. Indic.* **2021**, *125*, 107495–107509. [CrossRef]
37. Catherine, K.; Aline, D.M.V.; Nick, W.; Luke, L.; Henrique, O.S.; Milton, K.; Jeffrey, R.; Philipp, S.; John, C.; Rob, S.; et al. Performance of Landsat-8 and Sentinel-2 surface reflectance products for river remote sensing retrievals of chlorophyll-a and turbidity. *Remote Sens. Environ.* **2019**, *224*, 104–118.
38. Wen, Z.; Wang, Q.; Liu, G.; Jacinthe, P.A.; Wang, X.; Lyu, L.; Tao, H.; Ma, Y.; Duan, H.; Shang, Y.; et al. Remote sensing of total suspended matter concentration in lakes across China using Landsat images and Google Earth Engine. *ISPRS J. Photogramm. Remote Sens.* **2022**, *187*, 61–78. [CrossRef]
39. Vanhellemont, Q.; Ruddick, K. Atmospheric correction of Sentinel-3 OLCI data for mapping of suspended particulate matter and chlorophyll-a concentration in Belgian turbid coastal waters. *Remote Sens. Environ.* **2021**, *256*, 112284. [CrossRef]
40. Tavares, M.H.; Lins, R.C.; Harmel, T.; Fragoso, C.R., Jr.; Martínez, J.M.; Motta-Marques, D. Atmospheric and sunglint correction for retrieving chlorophyll-a in a productive tropical estuarine-lagoon system using Sentinel-2 MSI imagery. *ISPRS J. Photogramm. Remote Sens.* **2021**, *174*, 215–236. [CrossRef]
41. Vanhellemont, Q. Adaptation of the dark spectrum fitting atmospheric correction for aquatic applications of the Landsat and Sentinel-2 archives. *Remote Sens. Environ.* **2019**, *225*, 175–192. [CrossRef]
42. Rieu, P.; Moreau, T.; Cadier, E.; Raynal, M.; Clerc, S.; Donlon, C.; Borde, F.; Boy, F.; Maraldi, C. Exploiting the Sentinel-3 tandem phase dataset and azimuth oversampling to better characterize the sensitivity of SAR altimeter sea surface height to long ocean waves. *Adv. Space Res.* **2021**, *67*, 253–265. [CrossRef]
43. Xu, W.; Wooster, M.J.; Polehampton, E.; Yemelyanova, R.; Zhang, T. Sentinel-3 active fire detection and FRP product performance-Impact of scan angle and SLSTR middle infrared channel selection. *Remote Sens. Environ.* **2021**, *261*, 112460. [CrossRef]
44. Xu, J.; Zhao, Y.; Lyu, H.; Liu, H.; Dong, X.; Li, Y.; Cao, K.; Xu, J.; Li, Y.; Wang, H.; et al. A semianalytical algorithm for estimating particulate composition in inland waters based on Sentinel-3 OLCI images. *J. Hydrol.* **2022**, *608*, 127617. [CrossRef]
45. Zarei, A.; Shah-Hosseini, R.; Ranjbar, S.; Hasanlou, M. Validation of non-linear split window algorithm for land surface temperature estimation using Sentinel-3 satellite imagery: Case study; Tehran Province, Iran. *Adv. Space Res.* **2021**, *67*, 3979–3993. [CrossRef]
46. Gou, J.; Tourian, M.J. Riwi. SAR-SWH: A data-driven method for estimating significant wave height using Sentinel-3 SAR altimetry. *Adv. Space Res.* **2022**, *69*, 2061–2080. [CrossRef]
47. Odebiri, O.; Mutanga, O.; Odindi, J. Deep learning-based national scale soil organic carbon mapping with Sentinel-3 data. *Geoderma* **2022**, *411*, 115695. [CrossRef]
48. Pahlevan, N.; Smith, B.; Schalles, J.; Binding, C.; Cao, Z.; Ma, R.; Alikas, K.; Kangro, K.; Gurlin, D.; Hà, N.; et al. Seamless retrievals of chlorophyll-a from Sentinel-2 (MSI) and Sentinel-3 (OLCI) in inland and coastal waters: A machine-learning approach. *Remote Sens. Environ.* **2020**, *240*, 111604. [CrossRef]
49. Pahlevan, N.; Smith, B.; Alikas, K.; Anstee, J.; Barbosa, C.; Binding, C.; Bresciani, M.; Cremella, B.; Giardino, C.; Gurlin, D.; et al. Simultaneous retrieval of selected optical water quality indicators from Landsat-8, Sentinel-2, and Sentinel-3. *Remote Sens. Environ.* **2022**, *270*, 112860. [CrossRef]

50. Wang, Z.; Bovik, A.C.; Sheikh, H.R.; Simoncelli, E.P. Image quality assessment: From error visibility to structural similarity. *IEEE Trans. Image Process.* **2004**, *13*, 600–612. [CrossRef]
51. Hore, A.; Ziou, D. Image quality metrics: PSNR vs. SSIM. In Proceedings of the 2010 20th International Conference on Pattern Recognition, Istanbul, Turkey, 23–26 August 2010; pp. 2366–2369.
52. Thomas, G.W. The relationship between organic matter content and exchangeable aluminum in acid soil. *Soil Sci. Soc. Am. J.* **1975**, *39*, 591. [CrossRef]
53. Klein, G.A. A recognition-primed decision (RPD) model of rapid decision making. *Decis. Mak. Action Model. Methods* **1993**, *5*, 138–147.
54. Cao, Q.; Yu, G.; Qiao, Z. Application and recent progress of inland water monitoring using remote sensing techniques. *Environ. Monit. Assess.* **2023**, *195*, 1–16. [CrossRef] [PubMed]
55. Han, L.; Ding, J.; Ge, X.; He, B.; Wang, J.; Xie, B.; Zhang, Z. Using spatiotemporal fusion algorithms to fill in potentially absent satellite images for calculating soil salinity: A feasibility study. *Int. J. Appl. Earth Obs. Geoinf.* **2022**, *111*, 102839. [CrossRef]
56. Li, P.; Ke, Y.; Wang, D.; Ji, H.; Chen, S.; Chen, M.; Lyu, M.; Zhou, D. Human impact on suspended particulate matter in the Yellow River Estuary, China: Evidence from remote sensing data fusion using an improved spatiotemporal fusion method. *Sci. Total Environ.* **2021**, *750*, 141612. [CrossRef] [PubMed]
57. Song, K.; Ma, J.; Wen, Z.; Fang, C.; Shang, Y.; Zhao, Y.; Wang, M.; Du, J. Remote estimation of Kd (PAR) using MODIS and Landsat imagery for turbid inland waters in Northeast China. *ISPRS J. Photogramm. Remote Sens.* **2017**, *123*, 159–172. [CrossRef]
58. Yu, X.; Lee, Z.; Shen, F.; Wang, M.; Wei, J.; Jiang, L.; Shang, Z. An empirical algorithm to seamlessly retrieve the concentration of suspended particulate matter from water color across ocean to turbid river mouths. *Remote Sens. Environ.* **2019**, *235*, 111491. [CrossRef]

remote sensing

MDPI

Article

Change of Human Footprint in China and Its Implications for Carbon Dioxide (CO_2) Emissions

Yuan Li [1,2], Wujuan Mi [2], Yuheng Zhang [2], Li Ji [1], Qiusheng He [3], Yuanzhu Wang [4] and Yonghong Bi [2,*]

1 School of Environment and Resources, Taiyuan University of Science and Technology, Taiyuan 030024, China
2 State Key Laboratory of Freshwater Ecology and Biotechnology, Institute of Hydrobiology, Chinese Academy of Sciences, Wuhan 430072, China
3 Institute of Intelligent Low Carbon and Control Technology, Taiyuan University of Science and Technology, Taiyuan 030024, China
4 Central Southern Safety & Environmental Technology Institute Company Limited, Wuhan 430072, China
* Correspondence: biyh@ihb.ac.cn; Tel.: +86-27-68780016

Abstract: Humans have altered the earth in unprecedented ways, and these changes have profound implications for global climate change. However, the impacts of human pressures on carbon dioxide (CO_2) emissions over long time scales have not yet been clarified. Here, we used the human footprint index (HF), which estimates the ecological footprint of humans in a given location, to explore the impacts of human pressures on CO_2 emissions in China from 2000 to 2017. Human pressures (+13.6%) and CO_2 emissions (+198.3%) in China are still on the rise during 2000–2017 and are unevenly distributed spatially. There was a significant positive correlation between human pressures and CO_2 emissions in China, and northern China is the main driver of this correlation. The increase of CO_2 emissions in China slowed down after 2011. Although human pressures on the environment are increasing, high-quality development measures have already had noticeable effects on CO_2 emission reductions.

Keywords: carbon dioxide emissions; global climate change; human footprint index; human pressures; macro control

Citation: Li, Y.; Mi, W.; Zhang, Y.; Ji, L.; He, Q.; Wang, Y.; Bi, Y. Change of Human Footprint in China and Its Implications for Carbon Dioxide (CO_2) Emissions. *Remote Sens.* **2023**, *15*, 426. https://doi.org/10.3390/rs15020426

Academic Editor: Ashraf Dewan

Received: 21 November 2022
Revised: 3 January 2023
Accepted: 4 January 2023
Published: 10 January 2023

1. Introduction

Humans have been reshaping the world for millennia. The current geological epoch has been renamed the "Anthropocene" [1] because of the major environmental impacts associated with human activities, especially the conversion of large-scale natural habitats to cropland and the construction of land to feed and house our burgeoning population [2]. Human activities affect the stability of ecosystems and have a multitude of deleterious effects on the environment, including nutrient pollution [3]; modifications of land surface hydrology [4]; loss of the productivity, composition, and diversity of terrestrial ecosystems [5]; introduction of alien species; and alterations of the biogeochemical cycles of carbon [6]. Scientists have expressed much concern over the extent of human-induced environmental destruction, which is rapidly approaching catastrophic levels.

As a consequence of the cumulative impacts of anthropogenic activities on the planet in recent decades, the concentrations of greenhouse gases including carbon dioxide (CO_2), methane (CH_4), and nitrous oxide (N_2O), have been continuously increasing [7]. The accumulation of greenhouse gases prevents the loss of heat from the earth and increases the temperature of the earth's surface, which results in an increase in the frequency and intensity of extreme weather events, including drought, flood, heat waves, and freezing stress [8,9]. The 2015 Paris Agreement ambitiously aims to limit warming to between 1.5 °C and 2 °C by mid-century, which will require achieving a balance between anthropogenic emissions and the removal of longer-lived greenhouse gases such as CO_2 [10]. Many countries, such as China and the United States, have made net-zero pledges and have started to reduce emissions [11].

The extent and intensity of anthropogenic changes are spatially heterogeneous. Many anthropogenic pressures, such as population growth, fossil fuel combustion, industrial emissions, agricultural production, livestock farming, and land-use change, can interact in diverse ways [12,13]. Currently used mapping approaches often fail to capture many lower-intensity forms of human pressures, such as our extensive networks of roads, grazing lands, and low-density human settlements, the effects of which are more insidious than outright habitat conversion [14]. For example, land-use cover has major effects on the global extent and distribution of terrestrial carbon emissions [15]. Changes in land-use type from high-vegetation to low-vegetation biomass usually result in carbon emissions; land-use management, such as measures to control wildfires, pests, and diseases, can also affect carbon storage. Land-use change and management have together been estimated to contribute approximately one-third of all anthropogenic carbon emissions since the industrial revolution [16,17]. Non-settlement areas can also potentially result in population displacement and enrichment [2]; however, the role of low stress factors on the environment is often ignored in such areas. A comprehensive human stress indicator of multiple factors for studying the spatiotemporal patterns of CO_2 emissions is often more effective for characterizing regions with high emissions [18]. Furthermore, key variables, including technology, infrastructure, environment and finance, have important practical significance in achieving carbon neutrality [19–21]. For example, Labzovskii et al., (2017) [22] have projected that a beneficial policy would result in 24%, 80%, 166% less CO_2 emissions in East Asia by 2020, 2025 and 2030, respectively.

Improvements in remote sensing geographic data and geographic information systems permit the construction of detailed maps displaying human activities, such as the human footprint index (HF) [23]. HF captures the total ecological footprint of the human population in a given location [24]. Venter et al., constructed a global human footprint map from 1993 to 2009 and found that intense human pressures have had significant effects on native biodiversity [25]. This study also revealed that environmental pressures were lower in the wealthiest countries. A series of studies have also shown that human activities have had substantial effects on marine and lake ecosystems through the construction of human footprint maps [26,27]. However, the macro-scale effects of human pressures on CO_2 emissions have not yet been thoroughly studied.

China has experienced rapid economic development since its reform and opening in the late 1970s, which has resulted in the emission of large amounts of CO_2. A better understanding of spatiotemporal patterns in human pressures and their effects on CO_2 emissions is needed to respond to calls for rapid action to limit CO_2 emissions. Here, we used the latest multi-source remote sensing data to quantify the effects of human pressures on China's regional and sectoral CO_2 emissions between 2000 and 2017. Remote sensing data on (1) land-use cover, (2) roads, (3) railways, (4) human population density, (5) grazing density, and (6) night-time lights were used. We explored the spatiotemporal dynamics in human pressures and CO_2 fluxes and analyzed the correlations between human pressures and CO_2 fluxes. Our results reveal that CO_2 emissions in China have been on the rise from 2000 to 2017 and that there is a strong correlation between CO_2 emissions and human pressures. Nevertheless, increases in CO_2 emissions have slowed down after 2011, indicating that the effects of high-quality development and national measures to reduce greenhouse gas emissions have started to have impacts on CO_2 emission reductions.

Our results enhance the understanding of the impacts of human pressures on the environment and have implications for the formulation of effective and environmentally friendly strategies. The results have a positive guiding significance for carbon emission reduction policies, including urban land planning, population size control, industrial structure optimization, industrial technology upgrading, etc. In this paper, Section 2 provides a brief overview of the study methods, data sources, and model construction. Then, the spatio-temporal characteristics of HF and CO_2, and their correlation analysis during 2000–2017 in China are provided in Section 3. The cause of spatial distribution of HF and CO_2, and the response and driving factors of CO_2 emissions to human pressure changes are discussed in Section 4. Finally, Section 5 outlines the potential extension, outlook, and the summary of this study.

2. Methods and Data

2.1. Overview

The human footprint (HF) is a global map of human influence on the land surface [28]. The human footprint can reflect human disturbance to the natural environment and is directly or indirectly related to human fossil fuel combustion, fertilizer use, and industrial activity, thus establishing a certain relationship with CO2 emissions [29]. We used the HF for data from 2000 to 2020 to facilitate comparison across human pressures. The human pressures considered were (1) land-use cover, (2) roads, (3) railways, (4) human population density, (5) night-time lights, and (6) grazing density. We performed buffer analysis and assigned scores ranging from 0 to 10 to various spatial data layers according to the intensity of each human pressure; the assigned layers were then overlaid and normalized by partitioning to obtain human footprint data that indicate the extent of terrestrial human activity. We used ArcGIS 10.2 to integrate spatial data on human pressures at the 1 km^2 resolution for China's land areas. For any grid cell, the human footprint ranged between 0 and 60. We classified HF into three categories to reflect the degree of human pressures: low (HF < 20), moderate (HF 20–30), and high (HF 30–60).

2.2. Land-Use Cover

Change in land-use cover reflects the long-term impact of human activities on the environment [30]. Anthropogenic land-use cover change has a pronounced effect on regional climate change [31]. For example, land-use change (e.g., deforestation, increases in cropland, livestock rearing, and urban and industrial land expansion) and land management (wildfires, pests, and diseases) can directly or indirectly increase CO_2 emissions, especially in developing and poor countries [32–35]. Overall land-use change and land management have contributed approximately 1.45 Pg of the total carbon released from 1990 to 2010 [36]. We downloaded the database files of land-use cover (accurate to 1 km) from the Resource and Environment Science and Data Center (http://www.resdc.cn/, accessed on 1 January 2022). We assigned a pressure score of 10 to construction land, 7 to agricultural land, 4 to grasslands, and 0 to all other types of land-use cover. We used the land-use cover data of 2000, 2005, 2010, 2015, and 2020 to approximate the data for 2000–2003, 2004–2007, 2008–2011, 2012–2015, and 2016–2017, respectively.

2.3. Roads and Railways

Transport infrastructure is a fundamental physical foundation of societies; it thus plays a key role in supporting socio-economic activities and affects both the local and global environment [37,38]. Road transport contributes more than 60% of the CO_2 emissions of all transport activities [39], and a unidirectional causality has been noted between railway infrastructure and energy consumption [40]. An increasing number of transport infrastructure projects have been implemented in newly urbanizing regions.

We acquired data on the distribution of roads and railways for the years 2000, 2005, 2010, and 2012–2017 from OpenStreetMap. The OpenStreetMap database represents the most comprehensive publicly available database on roads and railways. We excluded all trails and private roads and used provincial major highways to denote roads. We used the road network data of 2000, 2005, and 2010 to approximate the data for 2000–2003, 2004–2007, and 2008–2011, respectively. We evaluated the direct and indirect pressures of roads by designating a pressure score of 10, 8, and 4 for 0.5, 0.5–1.5, and 1.5–2.5 km of distance on either side of the roads. The direct pressure of railways was assigned a score of 8 for a distance of 0.5 km on either side of them.

2.4. Population Density

Population density is an important indicator of the intensity of the interaction between human activities and ecosystems, and there is a strong correlation between population density and CO_2 emissions. Population density has significant positive effects on increases in CO_2 emissions. A previous study has shown that population density increases of 1% lead to total net CO_2 emissions increases of 4% [41,42].

Population density was mapped using the gridded population data published by WorldPop. The data set provides a 1 km × 1 km gridded summary of population census data from 1990 to 2017. Random forest-based asymmetric redistribution was used for mapping. The impact of population density on ecosystems is logarithmic. The population density data were assigned a pressure score ranging from 0 to 10. The maximum value of population density was 366,587 persons/km^2, and the population density pressure score was logarithmically scaled using the following formula:

$$\text{Population score} = 2.21398 \times \log\left(\text{Population density} + 1\right) \tag{1}$$

2.5. Grazing Density

Overgrazing (i.e., heavy grazing) associated with the rapid sharp growth of the human population and food demand in recent years is a major contributor to increases in greenhouse gas fluxes [43].

We tallied the number of large livestock animals, including cattle, horses, donkeys, mules, and camels, kept in each provincial administrative region of China using information published by the National Bureau of Statistics of China from 2000 to 2017. The impact of grazing density on ecosystems is logarithmic. The grazing density data were assigned a pressure score of 0–10. The maximum value of grazing density was 93.327 heads/km^2 (calculated by provincial administrative unit), and the grazing density pressure score was logarithmically scaled using the following formula:

$$\text{Livestock score} = 2.51531 \times \log\left(\text{grazing density} + 1\right) \tag{2}$$

2.6. Night-Time Lights

Night-time light satellite imagery is correlated with socioeconomic parameters such as urbanization, economic activity, and population density [44–46]. Night-time light imagery of 2000–2013 was obtained from the Defense Meteorological Satellite Program–Operational Linescan System (DMSP–OLS). The night-time light imagery of 2014–2020 was obtained from the launch of the NASA/NOAA's Suomi—Visible Infrared Imaging Radiometer Suite (VIIRS) sensor.

We used the annual mean synthesis algorithm without the effect of moonlight and cloud cover, and the spatial resolution of the images was 1 km × 1 km. Because of the lack of in-orbit radiometric calibration and correction facilities, the nocturnal radiometric signals on all light images were discretized into digitized radiometric brightness values (hereafter referred to as DN values, with scores ranging from 0 to 63). The synthetic product of this series of stable night-light signals eliminates the effect of short-time radiation sources so that the parts covered by the high brightness DN values usually correspond to high-density human settlements and activities. DN values of 8 or less were excluded from consideration before inter-calibration of data, as the shape of the quadratic function leads to severe distortion of very low DN values. The inter-calibrated DN data were then rescaled using an equal quintile approach into a 0–10 scale.

2.7. CO_2 Fluxes

The emission inventory of CO_2 fluxes was obtained from the latest energy data revision (2015) of the China National Bureau of Statistics from Scientific Data (https://doi.org/10.6084/m9.figshare.c.5136302.v2, accessed on 1 January 2022) [47,48]. The carbon emission unit is accurate to the county administrative region. We collected data on the CO_2 emissions of 2689 county-level administrative regions in China from 2000 to 2017 (Figure S1). The CO_2 emission fluxes were classified into three categories: low (0–10 Mt), moderate (10–30 Mt), and high (30–60 Mt).

2.8. Statistical Analyses

We performed all simple linear regression analysis, including the determination of 95% confidence intervals, with SigmaPlot 14.0 (Systat Software, San Jose, CA, USA). r^2 indicates the fit of a one-dimensional linear regression, and $p < 0.0001$ was the threshold for statistical significance. Standard deviations of CO_2 emissions and HF by different regions were calculated using SigmaPlot 14.0 software. We evaluated bivariate comparisons of continuous data measurements using analysis of variance (ANOVA) tests. CO_2 emissions and HF by different regions were expressed as average and standard error (SE). We tested the errors of HF calculation using a Monte Carlo simulation with 1000 iterations. We used histograms, conducted linear regressions, and built line plots. We used Origin Pro 2021 (Microcal Software, Seattle, WA, USA) to make violin plots and stacked area plots. All the spatial patterns of CO_2 emissions and HF in China were analyzed by ArcGIS 10.2 (ESRI, Redlands, CA, USA) [49].

3. Results

3.1. Spatio-Temporal Pattern of Human Footprint in China

HF in China was mainly concentrated in the 0–20 range, spanning 482 $\times$ 10^4 km^2 (Figure 1a). The area of high (30–60), moderate (20–30), and low HF (0–20) in China was 5.53 $\times$ 10^4 km^2, 100.11 $\times$ 10^4 km^2, and 684.09 $\times$ 10^4 km^2, respectively (Table S1). There was high spatial variation in HF (Figure 2a; Figure S2). The largest and smallest areas of high HF were observed in eastern (1.57 $\times$ 10^4 km^2) and northwestern China (0.22 $\times$ 10^4 km^2) during 2000–2017. The mean HF in China was 20.89 and gradually increased from 19.84 in 2000 to 22.54 in 2017 (Table 1; Figure 1b). HF in China increased 13.6% during 2000–2017. The rate of change of HF was the highest (16.87%) and the lowest (9.63%) in southern and northern China, respectively (Table 2). The area of low HF was 699.50 $\times$ 10^4 in 2000 and 645.05 $\times$ 10^4 km^2 in 2017 (a decrease of 7.78%), and the area of high HF was 2.66 $\times$ 10^4 in 2000 and 12.84 $\times$ 10^4 km^2 in 2017 (an increase of 382.71%) (Table S1). The area of high HF was 0.07 $\times$ 10^4 in 2000 and 0.99 $\times$ 10^4 km^2 in 2017 (an increase of 1314.29%,) and the area of high HF was 0.67 $\times$ 10^4 in 2000 and 1.97 $\times$ 10^4 km^2 in 2017 (an increase of 194.03%). Some of the low HF areas were converted into high HF areas (Table S1; Figure 3).

Table 1. Average human footprint index in different regions of China during 2000–2017.

Region/Human Footprint	Nationwide	North	South	Northeast	Northwest	Southwest	Central	East
2000	19.84	22.22	17.75	20.04	15.60	17.51	21.10	22.96
2001	20.30	22.62	18.25	20.50	15.86	17.97	21.60	23.56
2002	20.39	22.71	18.32	20.60	15.89	18.07	21.72	23.62
2003	20.30	22.57	18.23	20.52	15.86	18.01	21.57	23.58
2004	20.66	22.99	18.72	20.89	16.07	18.16	21.82	24.22
2005	20.51	22.91	18.48	20.79	16.08	18.12	21.74	23.76
2006	20.59	22.94	18.56	20.73	16.14	18.14	21.78	24.09
2007	20.62	22.89	18.61	20.84	16.14	18.20	21.82	24.07
2008	21.00	23.18	19.53	21.20	14.68	17.26	22.32	26.65
2009	20.36	22.45	18.15	21.52	15.94	18.03	21.40	23.58
2010	20.80	22.83	18.58	21.82	16.20	18.28	21.93	24.41
2011	20.62	22.58	18.56	21.53	16.14	18.13	21.77	24.13
2012	20.66	22.63	18.62	21.55	16.15	18.11	21.77	24.28
2013	21.01	23.01	19.08	21.82	16.42	18.26	22.13	24.80
2014	21.76	23.56	19.89	22.44	17.17	19.18	22.92	25.58
2015	21.94	23.75	20.05	22.45	17.39	19.47	23.10	25.73
2016	22.06	23.86	20.23	22.40	17.47	19.60	23.28	25.84

Table 1. *Cont.*

Region/Human Footprint	Nationwide	North	South	Northeast	Northwest	Southwest	Central	East
2017	22.54	24.36	20.75	22.61	17.95	20.17	23.78	26.38
Mean	20.89	23.00	18.91	21.35	16.29	18.37	22.08	24.51
Increment	2.70	2.14	3.00	2.56	2.36	2.67	2.68	3.43
Chang rate (%)	13.60	9.63	16.87	12.79	15.10	15.22	12.70	14.93

Table 2. Average CO_2 emissions in different regions of China during 2000–2017.

Region/CO_2 (Mt)	Nationwide	North	South	Northeast	Northwest	Southwest	Central	East
2000	3194.81	667.21	356.84	400.53	229.78	330.06	476.33	734.06
2001	3217.34	669.42	350.19	386.20	226.75	319.82	461.73	803.24
2002	3479.13	719.42	381.91	421.16	243.48	352.55	507.16	853.46
2003	4096.50	837.24	450.23	485.05	281.49	411.74	590.76	1039.99
2004	4563.96	938.00	495.90	528.05	312.21	451.57	649.31	1188.92
2005	5436.70	1146.47	566.70	599.41	376.44	519.93	755.62	1472.13
2006	6108.38	1288.76	633.53	666.01	428.81	583.65	852.48	1655.14
2007	6531.10	1392.01	676.34	696.40	463.74	609.99	906.38	1786.25
2008	6998.28	1517.54	718.40	728.45	515.83	644.30	969.04	1904.71
2009	7544.90	1624.49	775.05	798.30	562.53	707.13	1050.15	2027.24
2010	8255.42	1797.20	842.24	857.27	645.66	775.00	1153.96	2184.08
2011	9203.94	2034.41	956.37	880.08	836.84	855.11	1326.43	2314.70
2012	9392.04	2070.30	976.45	905.84	846.57	877.53	1354.86	2360.50
2013	9435.14	2040.80	992.97	942.87	932.23	884.33	1379.38	2262.56
2014	9627.14	2060.61	1019.12	958.79	955.57	905.33	1408.05	2319.68
2015	9094.56	1955.65	962.94	903.13	884.74	837.85	1316.03	2234.21
2016	9370.32	1955.91	995.40	934.62	907.05	863.46	1355.86	2318.02
2017	9531.09	1998.36	1022.76	904.80	999.15	909.80	1395.81	2300.41
Accumulated value	125,081	26,714	13,173	12,997	10,649	11,839	17,909	31,759
Increment	6336	1331	666	504	769	580	919	1566
Chang rate (%)	198.33	199.51	186.61	125.90	334.83	175.65	193.03	213.38

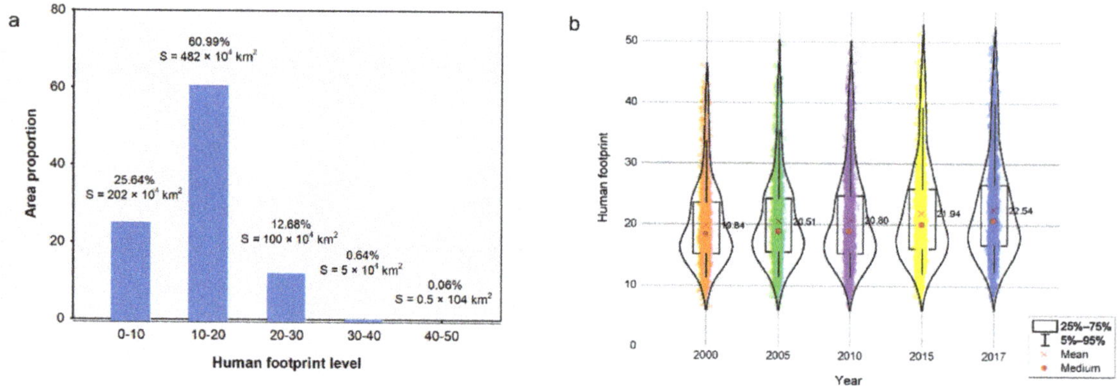

Figure 1. *Cont.*

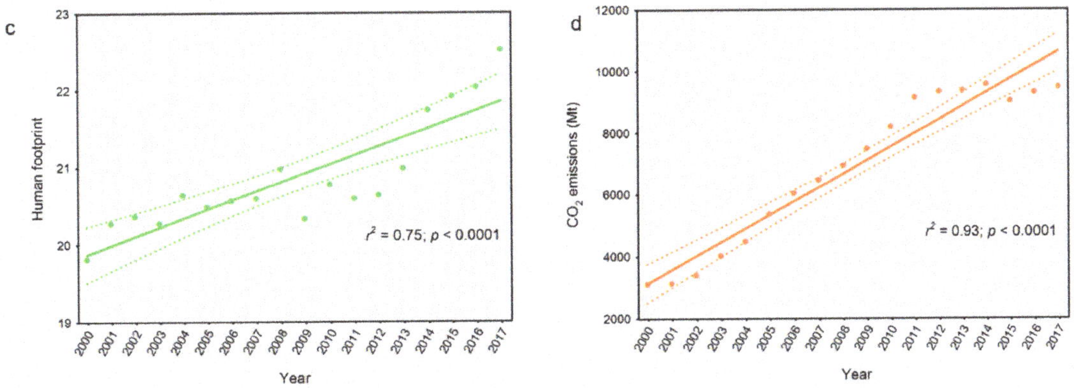

Figure 1. Features of human footprint and CO_2 emissions in China including (**a**) area of different human footprint classes, (**b**) variation in the average human footprint in different years, (**c**) regression analysis of human footprint and time, and (**d**) regression analysis of CO_2 emissions and time.

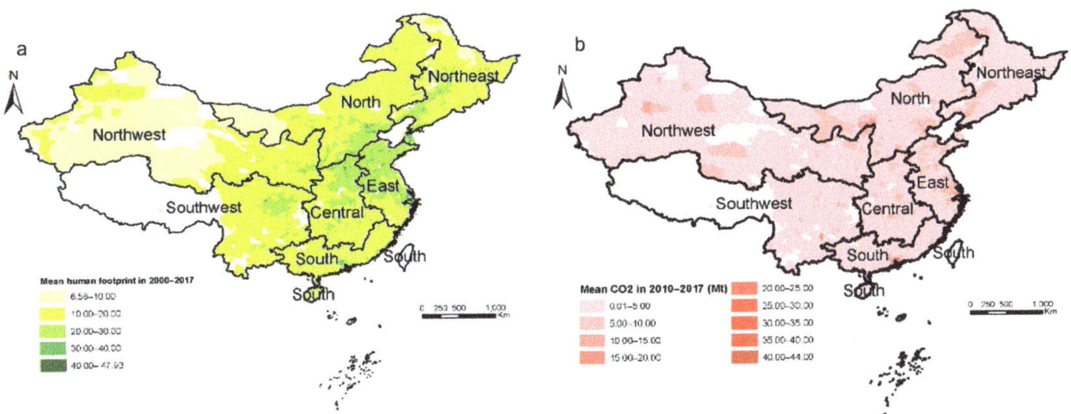

Figure 2. Spatial pattern of the average (**a**) human footprint and (**b**) CO_2 emissions in different regions in China during 2000–2017.

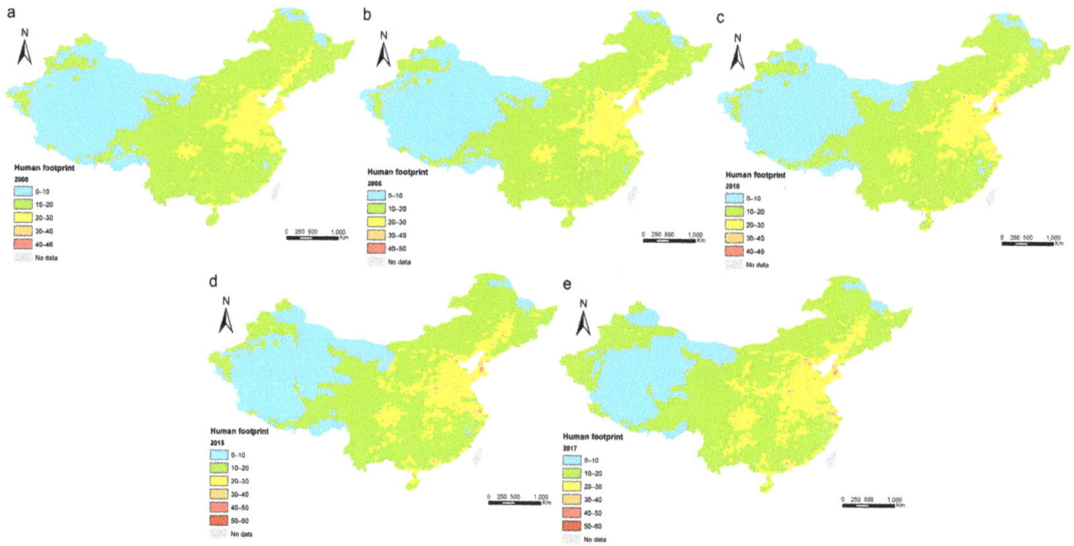

Figure 3. Spatio-temporal pattern of human footprint in China in (**a**) 2000, (**b**) 2005, (**c**) 2010, (**d**) 2015, and (**e**) 2017.

3.2. Spatio-Temporal Dynamics of CO_2 Emission Fluxes in China

China emitted a total of 125,081 Mt CO_2 during 2000–2017 (Table 2). CO_2 emissions in China gradually increased from 3194.81 in 2000 to 9531.09 Mt in 2017 (an increase of 198.33%). There was pronounced spatial variation in CO_2 emission fluxes in different regions (Figure 2b). CO_2 emissions contributed by eastern and northwestern China were the highest (31,759 Mt) and lowest (11,839 Mt), respectively (Table 2; Figure 4). CO_2 emission fluxes in China increased by 6336 Mt; the largest increase was observed in eastern China (1556 Mt), and the lowest increase was observed in northeastern China (504 Mt). CO_2 emissions in all regions of China increased by more than 100%. The rate of change in CO_2 emissions for northwestern (334.83%), eastern (213.38%), and northern China (199.51%) exceeded the national average (Table 2). The high (30–60), moderate (10–30), and low (0–10) CO_2 emissions in China increased 122.05, 1625.53, and 4588.70 Mt, respectively (Table S2; Figure S3). The high CO_2 emissions in southern (44.03 Mt) and eastern China (43.77 Mt) increased the most. Some of the low CO_2 emission areas were converted into high CO_2 emission areas (Table S2; Figure 5). CO_2 emissions were shown to have peaked in 2011 (+948.52 Mt) and were lowest in 2015 (−532.58 Mt) (Figure 6a). There was a gradual decrease in the magnitude of the increases of average CO_2 emissions in China during 2000–2017.

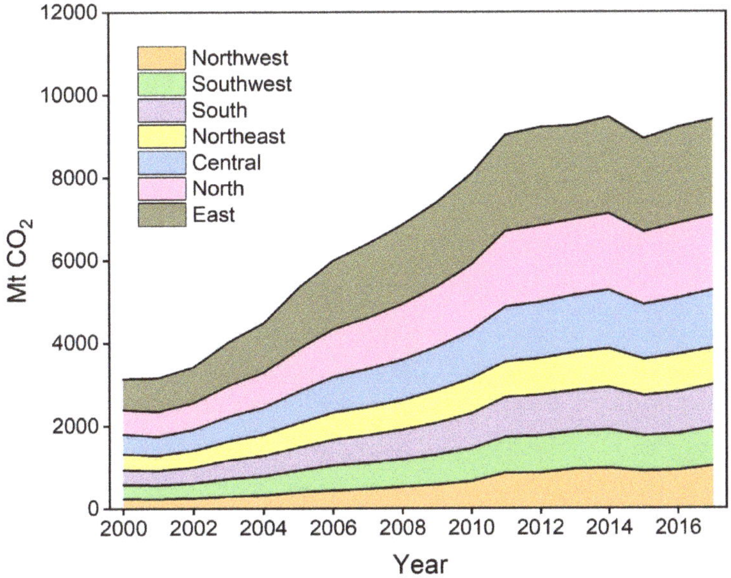

Figure 4. Stacked area map of annual CO_2 emissions in different regions in China during 2000–2017.

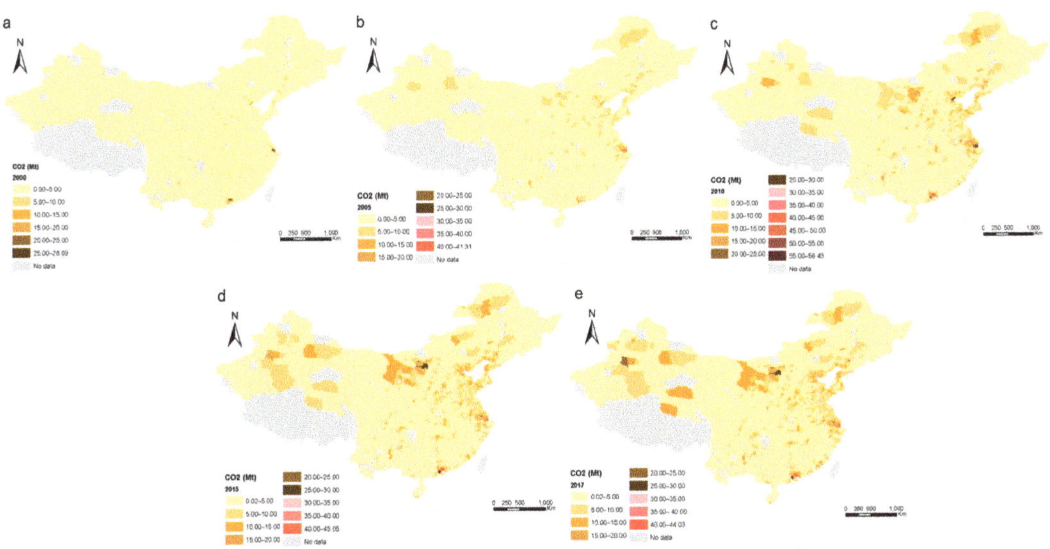

Figure 5. Spatio-temporal pattern of CO_2 emissions in China in (**a**) 2000, (**b**) 2005, (**c**) 2010, (**d**) 2015, and (**e**) 2017.

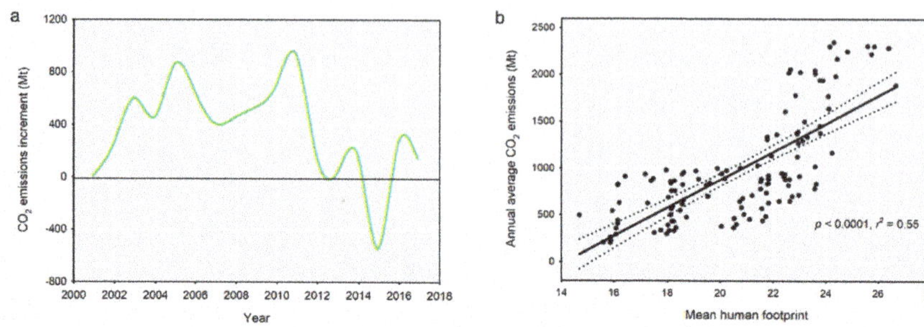

Figure 6. Trends in the (**a**) annual increase in CO_2 emissions and the (**b**) relationship between the average human footprint and average annual CO_2 emissions.

CO_2 emissions were higher in the provinces of Shandong (577.08 Mt y^{-1}), Jiangsu (497.00 Mt y^{-1}), and Hebei (492.35 Mt y^{-1}), which accounted for 23% of total emission fluxes in China (Figure S4). Global Moran's *I* was calculated to measure the spatial autocorrelation in CO_2 emissions. Its value was 0.14 ($p < 0.01$), which indicates a significant positive spatial autocorrelation in CO_2 emissions. The local Moran's *I* was used to further study the spatial characteristics of CO_2 emissions in 2000 (Figure 7a) and 2017 (Figure 7b) in China. There was a significant expansion of high-high cluster areas from 2000 to 2017, especially into eastern and northern China. The low–low cluster areas were mainly located in southwest China.

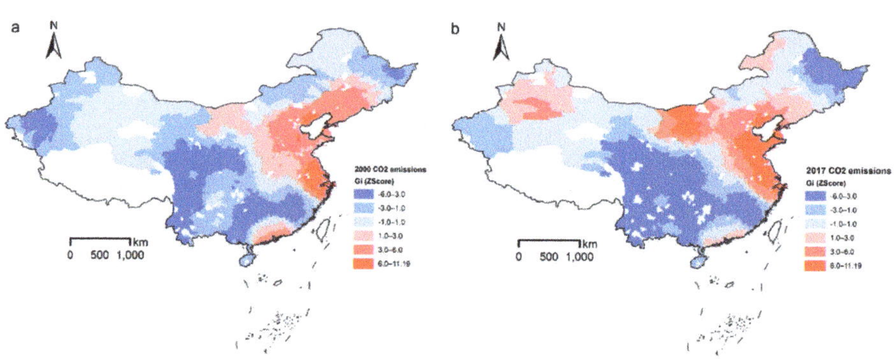

Figure 7. Spatial autocorrelation analysis (local Moran's *I* clustering) of CO_2 emissions in China in (**a**) 2000 and (**b**) 2017. Red and blue areas indicate positive and negative spatial correlation, respectively.

3.3. *Correlations between CO_2 Emissions and Human Pressures*

HF and CO_2 emissions in different regions of China were significantly positively correlated based on a Spearman's correlation coefficient for ranked data (correlation coefficient = 0.49, $p < 0.0001$) (Figure 6b). HF ($n = 18$; $r^2 = 0.75$; $p < 0.0001$) and CO_2 emissions ($n = 18$; $r^2 = 0.93$; $p < 0.0001$) in China increased significantly over time during the period 2000–2017 (Figure 1c; Figure 1d). The HF was highest (24.51) and lowest (16.29) in eastern and northwestern China, respectively (Table 2). CO_2 emissions were highest (1484 Mt) and lowest (592 Mt) in eastern and northwestern China, respectively (Table 2). During this 18-year period, the annual CO_2 emissions in each region were significantly positively correlated with HF ($n = 126$; $r^2 = 0.55$; $p < 0.0001$).

4. Discussion

4.1. Characteristics of Human Pressures in China

The human footprint provides a spatially explicit and temporally consistent quantitative measure of the magnitude of human pressures on the environment. Human pressures alter natural environments and harm natural systems [50]. These pressures include land-use cover (e.g., agricultural land and construction land), transportation (e.g., roads, railways, and navigable waterways), agricultural land and construction land, and trade (e.g., imports, exports, and international exports of energy-intensive industries) [51]. Human pressures (indicated by HF) could also be frequent, small in magnitude, and thus overlooked, such as the effects of extensive road systems and pasture lands. Spatial patterns of HF and human pressures were closely related. China has been the world's second largest economy since 2010. Spatial heterogeneity in regional human activities leads to spatial heterogeneity in human pressures in China. Eastern China is located between the mainland and the sea; it is characterized by gentle terrain and features agriculturally productive soils. In addition, this region is rich in aquatic products, oil, iron ore, salt, and other resources, the labor force is large, and the industrial and agricultural base is strong. By the end of 2017, eastern China accounted for only 5.25% of the country's area, but 23.28% of its population, 32.70% of its GDP, 32.99% of its road area, and 33.01% of its construction land. The average urbanization rate of eastern China was 64.56%, which was higher than the national rate of 58.52%; Shanghai had the highest urbanization rate in the country (87.70%). The high density of the population in this region has increased the demand for farmland, housing, and energy [52]. The extraction of environmental resources and the discharge of pollutants has had deleterious effects on the environment and has resulted in the loss of biodiversity and other ecosystem services, soil degradation, and the disruption of hydrological cycles [53]. Humans are constantly modifying the environment to accommodate their needs, yet this land-use change is also an important driver of human stress [54]. Over the past decades, China has experienced varying extents of urbanization. Intense urbanization improves social development in various ways but often creates substantial human pressures that could have negative effects on human welfare [55].

Human pressures in China have expanded over time. The main contributors to the HF model were construction land, areas of high population density, and road surroundings. From 2000–2017, China's population (1.27–1.40 billion, +0.6%/year), GDP (12.7–80.3 trillion, +31.0%/year), urbanization rate (36.0%–58.5%, +3.5%/year), farmland area (128.3–134.9 square kilometer, +0.3%/year), road area (2000–8000 square kilometer, +16.7%/year), livestock output (0.7–2.9 trillion, +16.5%/year) increased rapidly, which may be the reason that human pressure has increased by 13.6%. Human pressures from eastern, central, northeastern, and southwestern China have been increasing significantly over time, and the potential environmental implications caused by expanding human pressures associated with economic development require increased attention [56]. The control of these high-pressure areas is needed to relieve anthropogenically induced environmental damage.

4.2. CO_2 Emissions in China

Given China's large size and the heterogeneity in its economic development, lifestyles, resources, and economies vary among the provinces [57]. Eastern and northern China are the main contributors to CO_2 emissions and account for 45.39% of total emissions in China based on the 18-year average. The provinces of Shandong (2017 GDP: 1.12 trillion USD) and Jiangsu (2017 GDP: 1.38 trillion USD) in eastern China are two of the most developed regions in China and account for approximately 20% of the national total GDP. They also account for 15.71% of total CO_2 emissions given their large industrial areas and advanced technology. In contrast, the provinces of Inner Mongolia, Shanxi, and Hebei in northern China are less developed regions that account for only 7.86% of the country's GDP but 16.94% of its CO_2 emissions. Similar patterns were observed for northeastern and northwestern China.

CO_2 emissions continuously increased from 2000 to 2014 and declined in 2015–2017 by −5.52%, −3.10%, and −1.85% from the previous year, respectively. The economy of China has grown rapidly since its reform and opening in the 1970s. This economic expansion has led to the emission of a large amount of CO_2. From 2000 to 2017, China's GDP increased by 724.83% from 1.55 trillion US dollars (USD) to 12.80 trillion USD. Infrastructure construction and energy consumption have been the major drivers of the rapid growth of China's economy and emissions since 2002 [56]. The economy relies on carbon-intensive industries such as thermal power generation, steel, cement, and vehicle production. Due to increased greenhouse gas emissions, a tight energy supply, and severe air pollution, the government of China has begun to implement a series of strategies to conserve energy and mitigate emissions. The 11th and 12th Five-Year Plan (2006–2015) strategies have reduced emissions of CO_2 and coal by 3 billion tons and 1.4 billion tons, respectively [58]. From 2015–2017, the effect of slowing population growth (1.38–1.40 billion, +0.5%/year) and the optimization of industrial structure (2014, +4% CO_2) may result in a decrease in carbon emissions reduction and air pollution improvement. According to the report of the Netherlands Environmental Assessment Agency (PBL), China's carbon emissions increased by only 0.9% in 2014, despite its economic growth of 7% [59]. The slowdown in economic growth, coupled with the shift to cleaner energy and the reduction of energy intensive manufacturing has reduced the energy intensity of the country's economy. Control of the growth of CO_2 emissions in economically underdeveloped regions such as northwestern China will require the restructuring and upgrading of industry and industrial technology, respectively.

4.3. Impacts of Human Pressures on CO_2 Emissions

The CO_2 emissions per unit of economic output and per capita in the northern provinces are mostly higher than the national average, which indicates that economic development in northern China disproportionately contributes to CO_2 emissions compared with other regions of the country (Figure 8). For example, Shanghai led the country in CO_2 emissions per capita in 2000 (7.91 t per capita^{-1}), while Ningxia was first in 2017 (25.16 t per capita^{-1}) and has seen its CO_2 emissions increase more than three-fold since 2000 (5.97 t per capita^{-1}). The population and GDP are higher, and industrial development is more advanced in Shanghai compared with Ningxia. Several factors have caused the CO_2 emissions per capita to stabilize (2017, Shanghai, 7.95 t per capita^{-1}). The mismatch in the increase in GDP and technology levels in western China is an important factor affecting CO_2 emissions per capita. The high degree of economic development and the large human population in eastern China are the main reasons for the expansion of the high–high clustering area of CO_2 emissions. However, increasing energy consumption in northern China is the main factor underlying the emergence of the high–high clustering region [60]. Three northern provinces were among the top five emitters in 2000 (Figure 8a); in 2017 however, the top five emitters in the country were all northern provinces (Figure 8b). The carbon intensity (CO_2 per unit of GDP) of China decreased from 2000 to 2017 due to changes in economic development. The less economically developed provinces were mainly located in northern and western China. The carbon intensity of these regions mostly exceeded the national average in 2000 (49.86 t per USD^{-1}) (Figure 8c) and 2017 (17.00 t per USD^{-1}) (Figure 8d). Heavy industries are central to the economies of the northern provinces, and these products are mainly exported to other provinces [61]. Such provinces do not have adequate human resources to upgrade their technologies and equipment, and the economies of these provinces are mainly based on energy-intensive industries, such as mining, metals, electricity, and chemical products; consequently, the resource use efficiency of these provinces is low, and their greenhouse gas emissions are high [56]. China's economy is largely dependent on primary energy resources, and these resources are mainly located in less developed regions. Technology transfer and optimization of the industrial structure are important for reducing CO_2 emissions in less developed regions [62].

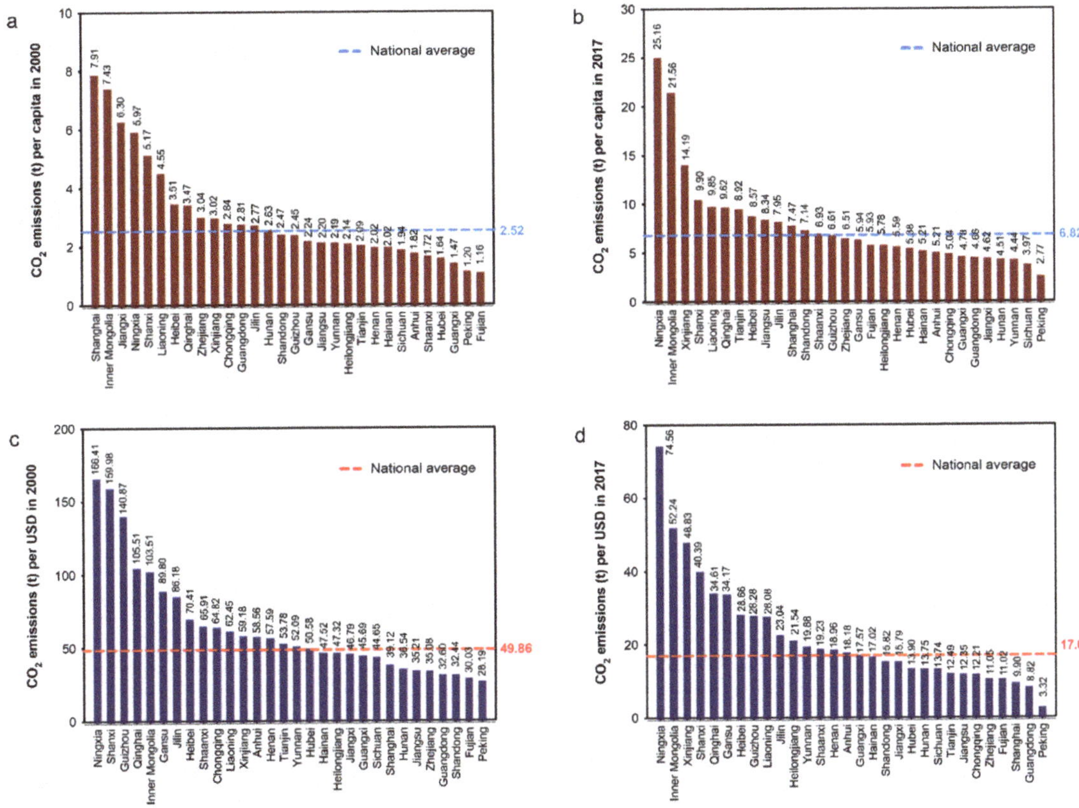

Figure 8. CO_2 emissions per capita (t) in (**a**) 2000 and (**b**) 2017 in each provincial administrative unit. CO_2 emissions per USD (g) in (**c**) 2000 and (**d**) 2017. The dotted line indicates the national average.

As increased human pressures lead to increased CO_2 emissions, spatial heterogeneity in the strength of human pressures leads to spatial heterogeneity in CO_2 emissions. Northern China is the main region contributing to increases in CO_2 emissions in response to human pressures. Thermal power generation is a major source of CO_2 emissions but an unsustainable form of energy generation in eastern and southern China. Northern China is rich in coal resources. The integrated development of coal and electricity has become widespread in China, and most thermal power generation units are concentrated in Inner Mongolia, Shanxi, and Xinjiang provinces [63], which has made northern China a major source of CO_2 emissions. Therefore, special effort is needed to control the intensity of human activities and greenhouse gas emissions in these regions. Modifications of the energy structure would be an effective approach to the reduction of CO_2 emissions in these regions [64]. Controlling the rate of increase in human pressures can also contribute to the mitigation of CO_2 emissions. There is thus a need to reduce the effect of human pressures on the environment, especially the deleterious effects of the human population, land-use structure, and urban construction.

One of the most important factors driving the reduction in energy consumption and CO_2 emissions in China is the increase in high-quality development. Photovoltaic (PV) power is considered one of the most promising low-carbon energy generation approaches in China [65]. By the end of 2015, the cumulative installed capacity of PV generation reached approximately 43 million kilowatts, which made China the largest solar power producer in the world [66]. By the end of 2030, the installed PV capacity will reach 100 million kilowatts, which is equivalent to building more than 30 fewer large coal power plants. At the Climate Ambition Summit in 2020, China's national leader proposed that the total installed capacity of wind and solar power would reach more than 1.2 billion kilowatts by 2030. Ultra-high voltage (UHV) technology makes significant contributions to the reduction of CO_2 emissions [67]. China has been committed to developing UHV technology for many years. As of 2019, more than 20 UHV transmission lines have been built in China [68]. In addition, the Chinese government has been vigorously promoting the use of new energy vehicles (NEVs). Annual sales of NEVs were 1.367 million in 2020, which is more than 160 times the number of NEVs sold in 2011 (8159) [69]. Direct policy support has promoted the rapid development of China's NEV market and has made large contributions to energy saving and emission reductions.

5. Conclusions

Along with the development of economy and the acceleration of urbanization, human pressure on the environment is increasing. An understanding of the spatiotemporal dynamics in CO_2 emissions under changing human pressures is essential for designing sustainable environmental strategies. Our findings have implications for the development of mitigation policies for CO_2 emissions by local governments. The main findings and policy implications are manifold. First, carbon emissions in China are still on the rise; there is thus a need to strengthen the implementation of CO_2 reduction measures. Second, CO_2 emissions in China are unevenly distributed spatially (generally higher in the south and east and lower in the north and west), indicating that the government needs to optimize the regional allocation of energy. Third, increased human pressures have increased the amount of CO_2 emissions, and northern China is the main region driving this pattern. Given that little can be done to alter the current economic trend, the impacts of human activities on CO_2 emissions can be reduced by optimizing land use, population density, and grazing density. Fourth, CO_2 emissions associated with anthropogenic activities are decreasing. High-quality development measures and strong national macroeconomic control instruments are needed to achieve China's goal of carbon neutrality. We believe that China is on track to meet its carbon reduction commitments on time.

Remote sensing technology has expanded peoples' abilities to understand their living environment, and has the advantages of qualitative accuracy, large observation range, high spatial resolution, simple acquisition, and strong data consistency. However, government-based datasets have certain advantages in terms of quantitative analysis. Our next goal is to conduct a bottom-up carbon footprint coupled with remote sensing data and government-based datasets for an analysis of the impact of human activities on carbon emissions.

Supplementary Materials: The following supporting information can be downloaded at: https://www.mdpi.com/article/10.3390/rs15020426/s1. Figure S1. Analytical scope of CO2 emissions including (a) spatial distribution diagram of the monitoring area at the county level and (b) spatial distribution of regions in China used for estimating CO2 emissions. Figure S2. Annual CO2 emissions by seven regions in China. Figure S3. Annual CO2 emissions (g) per USD in China during 2000–2017. National average is 1.45 g CO2 per USD based the 18-year period. Figure S4. Average CO2 emissions in China in each provincial administrative unit. Table S1. Spatio-temporal characteristics of the human footprints in China. Table S2. Spatio-temporal characteristics of CO2 emissions.

Author Contributions: Y.L. and W.M. conceived the study and wrote the paper. L.J. and Y.Z. completed some of the statistical analyses. Q.H. and Y.W. provided suggestions for manuscript revision. Y.B. provided feedback on various drafts of the manuscript. All authors have read and agreed to the published version of the manuscript.

Funding: This work was supported by the National Natural Science Foundation of China (No. 31971477; No. 42177057), the Key Laboratory of Algal Biology, Institute of Hydrobiology, Chinese Academy of Sciences (202205), Taiyuan University of Science and Technology Introduction of Talent Start-up Fund (2020051), and Incentive Fund for Outstanding Doctors Working in Shanxi (20212071).

Data Availability Statement: All the original data can be obtained from given data sources. The full dataset, including the spatio-temporal maps of CO_2 emissions and human footprint index used in this study is available. All datasets generated during this study are available from the corresponding author on reasonable request.

Conflicts of Interest: The authors declare that they have no known competing financial interest or personal relationships that could have influenced the work reported in this paper.

References

1. Corlett, R.T. The anthropocene concept in ecology and conservation. *Trends Ecol. Evol.* **2015**, *30*, 36–41. [CrossRef] [PubMed]
2. Watson, E.J.; Venter, O. Mapping the continuum of humanity's footprint on land. *One Earth* **2019**, *1*, 175–180. [CrossRef]
3. Zhang, L.; Xiong, L.; Li, J.; Huang, X. Long-term changes of nutrients and biocenoses indicating the anthropogenic influences on ecosystem in Jiaozhou Bay and Daya Bay, China. *Mar. Pollut. Bull.* **2021**, *168*, 112406. [CrossRef] [PubMed]
4. Grizzetti, B.; Pistocchi, A.; Liquete, C.; Udias, A.; Bouraoui, F.; van de Bund, W. Human pressures and ecological status of European rivers. *Sci. Rep.* **2017**, *7*, 205. [PubMed]
5. Tilman, D.; Lehman, C. Human-caused environmental change: Impacts on plant diversity and evolution. *Proc. Natl. Acad. Sci. USA* **2001**, *98*, 5433–5440. [CrossRef]
6. Marques, A.; Martins, I.S.; Kastner, T.; Plutzar, C.; Theurl, M.C.; Eisenmenger, N.; Huijbregts, M.A.J.; Wood, R.; Stadler, K.; Bruckner, M.; et al. Increasing impacts of land use on biodiversity and carbon sequestration driven by population and economic growth. *Nat. Ecol. Evol.* **2019**, *3*, 628–637.
7. Zandalinas, S.I.; Fritschi, F.B.; Mittler, R. Global warming, climate change, and environmental pollution: Recipe for a multifactorial stress combination disaster. *Trends Plant Sci.* **2021**, *26*, 588–599. [CrossRef]
8. Mazdiyasni, O.; AghaKouchak, A. Substantial increase in concurrent droughts and heatwaves in the United States. *Proc. Natl. Acad. Sci. USA* **2015**, *112*, 11484–11489.
9. Noyes, P.D.; McElwee, M.; Miller, H.D.; Clark, B.W.; Van Tiem, L.A.; Walcott, K.C.; Erwin, K.N.; Levin, E.D. The toxicology of climate change: Environmental contaminants in a warming world. *Environ. Int.* **2009**, *35*, 971–986. [CrossRef]
10. du Pont, Y.R.; Meinshausen, M. Warming assessment of the bottom-up Paris Agreement emissions pledges. *Nat. Commun.* **2018**, *9*, 4810.
11. Editorials. Accesses Net-zero carbon pledges must be meaningful to avert climate disaster. *Nature* **2021**, *592*, 8. [CrossRef]
12. FAO (Food and Agriculture Organization of the United Nations). FAOSTAT. 2019. Available online: http://faostat.fao.org/ (accessed on 1 January 2022).
13. Raupach, M.R.; Marland, G.; Ciais, P.; Le Quéré, C.; Canadell, J.G.; Klepper, G.; Field, C.B. Global and regional drivers of accelerating CO_2 emissions. *Proc. Natl. Acad. Sci. USA* **2007**, *104*, 10288–10293. [CrossRef]
14. Ibisch, P.L.; Hoffmann, M.T.; Kreft, S.; Pe'Er, G.; Kati, V.; Biber-Freudenberger, L.; DellaSala, D.A.; Vale, M.M.; Hobson, P.R.; Selva, N. A global map of roadless areas and their conservation status. *Science* **2016**, *354*, 1423–1427. [CrossRef]
15. Harris, N.L.; Brown, S.; Hagen, S.C.; Saatchi, S.S.; Petrova, S.; Salas, W.; Hansen, M.C.; Potapov, P.V.; Lotsch, A. Baseline map of carbon emissions from deforestation in tropical regions. *Science* **2012**, *336*, 1573–1576. [CrossRef]
16. Haddad, N.M.; Brudvig, L.A.; Clobert, J.; Davies, K.F.; Gonzalez, A.; Holt, R.D.; Lovejoy, T.E.; Sexton, J.O.; Austin, M.P.; Collins, C.D.; et al. Habitat fragmentation and its lasting impact on Earth's ecosystems. *Sci. Adv.* **2015**, *1*, e1500052. [CrossRef]
17. Henders, S.; Persson, U.M.; Kastner, T. Trading forests: Land-use change and carbon emissions embodied in production and exports of forest-risk commodities. *Environ. Res. Lett.* **2015**, *10*, 125012. [CrossRef]
18. Mcgowan, P.J.K. Mapping the terrestrial human footprint. *Nature* **2016**, *537*, 172–173. [CrossRef]
19. Mi, Z.; Wei, Y.-M.; Wang, B.; Meng, J.; Liu, Z.; Shan, Y.; Liu, J.; Guan, D. Socioeconomic impact assessment of China's CO_2 emissions peak prior to 2030. *J. Clean. Prod.* **2017**, *142*, 2227–2236. [CrossRef]
20. Pflugmann, F.; Blasio, N.D. The geopolitics of renewable hydrogen in low-carbon energy markets. *Geopolit. Hist. Int. Relat.* **2020**, *12*, 9–44.
21. Li, X.; Huang, C.; Zhan, S.; Wu, Y. The carbon emission reduction effect of city cluster—Evidence from the Yangtze River Economic Belt in China. *Energies* **2022**, *15*, 6210. [CrossRef]
22. Labzovskii, L.D.; Mak, H.W.L.; Kenea, S.T.; Rhee, J.S.; Lashkari, A.; Li, S.; Goo, T.Y.; Oh, Y.S.; Byun, Y.H. What can we learn about effectiveness of carbon reduction policies from interannual variability of fossil fuel CO_2 emissions in East Asia? *Environ. Sci. Policy* **2019**, *96*, 132–140. [CrossRef]
23. Sanderson, E.W.; Jaiteh, M.; Levy, M.A.; Redford, K.H.; Wannebo, A.V.; Woolmer, G. The human footprint and the last of the wild. *Bioscience* **2002**, *52*, 172–173. [CrossRef]
24. Venter, O.; Sanderson, E.W.; Magrach, A.; Allan, J.R.; Beher, J.; Jones, K.R.; Possingham, H.P.; Laurance, W.F.; Wood, P.; Fekete, B.M.; et al. Sixteen years of change in the global terrestrial human footprint and implications for biodiversity conservation. *Nat. Commun.* **2016**, *7*, 12558. [CrossRef] [PubMed]
25. Halpern, B.S.; Walbridge, S.; Selkoe, K.A.; Kappel, C.V.; Micheli, F.; D'Agrosa, C.; Bruno, J.F.; Casey, K.S.; Ebert, C.; Fox, H.E.; et al. A global map of human impact on marine ecosystems. *Science* **2008**, *319*, 948–952. [CrossRef] [PubMed]

26. Mammides, C. A global assessment of the human pressure on the world's lakes. *Global Environ. Chang.* **2020**, *63*, 102084. [CrossRef]
27. Peters, G.; Andrew, R.; Boden, T.; Canadell, J.G.; Ciais, P.; Le Quéré, C.; Marland, G.; Raupach, M.R.; Wilson, C. The challenge to keep global warming below 2 °C. *Nat. Clim. Chang.* **2012**, *2*, 2–4. [CrossRef]
28. Chiu, C.-C.; Château, P.-A.; Lin, H.-J.; Chang, Y.-C. Modeling the impacts of coastal land use changes on regional carbon balance in the Chiku coastal zone, Taiwan. *Land Use Policy* **2019**, *87*, 104079. [CrossRef]
29. Doney, S.C. The Growing Human Footprint on Coastal and Open-Ocean Biogeochemistry. *Science* **2010**, *5985*, 1512–1516. [CrossRef]
30. Findell, K.L.; Berg, A.; Gentine, P.; Krasting, J.P.; Lintner, B.R.; Malyshev, S.; Santanello, J.A., Jr.; Shevliakova, E. The impact of anthropogenic land use and land cover change on regional climate extremes. *Nat. Commun.* **2017**, *8*, 989. [CrossRef]
31. Havlík, P.; Valin, H.; Herrero, M.; Obersteiner, M.; Schmid, E.; Rufino, M.C.; Mosnier, A.; Thornton, P.K.; Böttcher, H.; Conant, R.T.; et al. Climate change mitigation through livestock system transitions. *Proc. Natl. Acad. Sci. USA* **2014**, *111*, 3709–3714. [CrossRef]
32. Hong, C.; Burney, J.A.; Pongratz, J.; Nabel, J.E.M.S.; Mueller, N.D.; Jackson, R.B.; Davis, S.J. Global and regional drivers of land-use emissions in 1961–2017. *Nature* **2021**, *589*, 554–561. [CrossRef]
33. Houghton, R.A.; House, J.I.; Pongratz, J.; Van Der Werf, G.R.; DeFries, R.S.; Hansen, M.C.; Le Quéré, C.; Ramankutty, N. Carbon emissions from land use and land-cover change. *Biogeosciences* **2012**, *9*, 5125–5142. [CrossRef]
34. Lai, L.; Huang, X.; Yang, H.; Chuai, X.; Zhang, M.; Zhong, T.; Chen, Z.; Chen, Y.; Wang, X.; Thompson, J.R. Carbon emissions from land-use change and management in China between 1990 and 2010. *Sci. Adv.* **2016**, *2*, e1601063. [CrossRef]
35. Din, S.U.; Mak, H.W.L. Retrieval of land-use/land cover change (LUCC) maps and urban expansion dynamics of Hyderabad, Pakistan via Landsat Datasets and support vector machine framework. *Remote Sens.* **2021**, *13*, 3337. [CrossRef]
36. Arshad, S.; Kazmi, J.H.; Fatima, M.; Khan, N. Change detection of land cover/land use dynamics in arid region of Bahawalpur District, Pakistan. *Appl. Geomat.* **2022**, *14*, 387–403. [CrossRef]
37. Meng, X.; Han, J. Roads, economy, population density, and CO_2: A city-scaled causality analysis. *Resour. Conserv. Recy.* **2016**, *128*, 508–515. [CrossRef]
38. Yeboah, G.; de Albuquerque, J.P.; Troilo, R.; Tregonning, G.; Perera, S.; Ahmed, S.; Ajisola, M.; Alam, O.; Aujla, N.; Azam, S.; et al. Analysis of Openstreetmap data quality at different stages of a participatory mapping process: Evidence from Slums in Africa and Asia. *ISPRS Int. J. Geo-Inf.* **2021**, *10*, 265. [CrossRef]
39. Sugar, L.; Kennedy, C.; Leman, E. Greenhouse gas emissions from Chinese cities. *J. Ind. Ecol.* **2012**, *16*, 552–563. [CrossRef]
40. Achour, H.; Belloumi, M. Investigating the causal relationship between transport infrastructure, transport energy consumption and economic growth in Tunisia. *Renew. Sustain. Energy Rev.* **2016**, *56*, 988–998. [CrossRef]
41. Wang, X.; Meng, X.; Long, Y. Projecting 1 km-grid population distributions from 2020 to 2100 globally under shared socioeconomic pathways. *Sci. Data* **2022**, *9*, 563. [CrossRef]
42. Huang, R.; Lv, G.; Chen, M.; Zhu, Z. CO_2 emissions embodied in trade: Evidence for Hong Kong SAR. *J. Clean. Prod.* **2019**, *239*, 117918. [CrossRef]
43. Da, W.; Xu, R.; Wang, Y.; Wang, Y.; Liu, Y.; Yao, T. Responses of CO_2, CH_4 and N_2O fluxes to livestock exclosure in an alpine steppe on the Tibetan Plateau, China. *Plant Soil* **2012**, *359*, 45–55.
44. Gaughan, A.E.; Oda, T.; Sorichetta, A.; Stevens, F.R.; Bondarenko, M.; Bun, R.; Krauser, L.; Yetman, G.; Nghiem, S.V. Evaluating nighttime lights and population distribution as proxies for mapping anthropogenic CO_2 emission in Vietnam, Cambodia and Laos. *Environ. Res. Commun.* **2019**, *1*, 091006. [CrossRef] [PubMed]
45. Shi, K.; Shen, J.; Wu, Y.; Liu, S.; Li, L. Carbon dioxide (CO_2) emissions from the service industry, traffic, and secondary industry as revealed by the remotely sensed nighttime light data. *Int. J. Digit. Earth* **2021**, *14*, 1514–1527. [CrossRef]
46. Bennett, M.M.; Smith, L.C. Advances in using multitemporal night-time lights satellite imagery to detect, estimate, and monitor socioeconomic dynamics. *Remote Sens. Environ.* **2017**, *192*, 176–197. [CrossRef]
47. Chen, J.; Gao, M.; Cheng, S.; Hou, W.; Song, M.; Liu, X.; Liu, Y.; Shan, Y. County-level CO_2 emissions and sequestration in China during 1997–2017. *Sci. Data* **2020**, *7*, 391. [CrossRef]
48. Zhang, Y.; Lyu, M.; Yang, P.; Lai, D.Y.; Tong, C.; Zhao, G.; Li, L.; Zhang, Y.; Yang, H. Spatial variations in CO_2 fluxes in a subtropical coastal reservoir of Southeast China were related to urbanization and land-use types. *J. Environ. Sci.* **2021**, *109*, 206–218. [CrossRef]
49. Krivoruchko, K.; Gribov, A.; Krause, E. Multivariate areal interpolation for continuous and count data. *Procedia Environ. Sci.* **2011**, *3*, 14–19. [CrossRef]
50. Pyšek, P.; Jarošík, V.; Hulme, P.E.; Kühn, I.; Wild, J.; Arianoutsou, M.; Bacher, S.; Chiron, F.; Didžiulis, V.; Essl, F.; et al. Disentangling the role of environmental and human pressures on biological invasions across Europe. *Proc. Natl. Acad. Sci. USA* **2010**, *107*, 12157–12162. [CrossRef]
51. Meng, J.; Mi, Z.; Guan, D.; Li, J.; Tao, S.; Li, Y.; Feng, K.; Liu, J.; Liu, Z.; Wang, X.; et al. The rise of south–south trade and its effect on global CO_2 emissions. *Nat. Commun.* **2018**, *9*, 1871. [CrossRef]
52. West, P.C.; Gerber, J.S.; Engstrom, P.M.; Mueller, N.D.; Brauman, K.A.; Carlson, K.M.; Cassidy, E.S.; Johnston, M.; MacDonald, G.K.; Ray, D.K.; et al. Leverage points for improving global food security and the environment. *Science* **2014**, *345*, 325–328. [CrossRef]

53. Bala, G.; Caldeira, K.; Wickett, M.; Phillips, T.J.; Lobell, D.B.; Delire, C.; Mirin, A. Combined climate and carbon cycle effects of large-scale deforestation. *Proc. Natl. Acad. Sci. USA* **2007**, *104*, 6550–6555. [CrossRef]
54. Hansis, E.; Davis, S.J.; Pongratz, J. Relevance of methodological choices for accounting of land use change carbon fluxes. *Glob. Biogeochem. Cycles* **2015**, *29*, 1230–1246. [CrossRef]
55. Shen, F.; Yang, L.; He, X.; Zhou, C.; Adams, J.M. Understanding the spatial-temporal variation of human footprint in Jiangsu Province, China, its anthropogenic and potential implications. *Sci. Rep.* **2020**, *10*, 13316. [CrossRef]
56. Liu, Z.; Geng, Y.; Lindner, S.; Guan, D. Uncovering China's greenhouse gas emission from regional and sectoral perspectives. *Energy* **2012**, *45*, 1059–1068. [CrossRef]
57. Feng, K.; Hubacek, K.; Guan, D. Lifestyles, technology and CO_2 emissions in China: A regional comparative analysis. *Ecol. Econ.* **2009**, *69*, 145–154. [CrossRef]
58. Qin, H.; Huang, Q.; Zhang, Z.; Lu, Y.; Li, M.; Xu, L.; Chen, Z. Carbon dioxide emission driving factors analysis and policy implications of Chinese cities: Combining geographically weighted regression with two-step cluster. *Sci. Total Environ.* **2019**, *684*, 413–424. [CrossRef]
59. Olivier, J.G.; Schure, K.M.; Peters, J.A.H.W. *Trends in Global CO_2 and Total Greenhouse Gas Emissions Summary of the 2017 Report*; Netherlands Environmental Assessment Agency, PBL: The Hague, The Netherlands, 2019.
60. Gregg, J.S. China: Emissions pattern of the world leader in CO_2 emissions from fossil fuel consumption and cement production. *Geophys. Res. Lett.* **2008**, *35*, L08806. [CrossRef]
61. Liu, Z.; Liang, S.; Geng, Y.; Xue, B.; Xi, F.; Pan, Y.; Zhang, T.; Fujita, T. Features, trajectories and driving forces for energy-related GHG emissions from Chinese mega cites: The case of Beijing, Tianjin, Shanghai and Chongqing. *Energy* **2012**, *37*, 245–254. [CrossRef]
62. Liu, Z.; Davis, S.J.; Feng, K.; Hubacek, K.; Liang, S.; Anadon, L.D.; Chen, B.; Liu, J.; Yan, J.; Guan, D. Targeted opportunities to address the climate–trade dilemma in China. *Nat. Clim. Chang.* **2016**, *6*, 201–206. [CrossRef]
63. CSC (China State Council). *China's 12th Five-Year Plan for Energy Development*; CSC: Beijing, China, 2012.
64. Moutinho, V.; Moreira, A.C.; Silva, P.M. The driving forces of change in energy related CO_2 emissions in Eastern, Western, Northern and Southern Europe: The LMDI approach to decomposition analysis. Renew. *Sust. Energ. Rev.* **2015**, *50*, 1485–1499. [CrossRef]
65. Adye, K.; Pearre, N.; Swan, L. Contrasting distributed and centralized photovoltaic system performance using regionally distributed pyranometers. *Sol. Energy* **2018**, *160*, 1–9. [CrossRef]
66. Liu, F.; Lv, T. Assessment of geographical distribution of photovoltaic generation in China for a low carbon electricity transition. *J. Clean. Prod.* **2019**, *212*, 655–665. [CrossRef]
67. Wei, W.; Wu, X.; Li, J.; Jiang, X.; Zhang, P.; Zhou, S.; Zhu, H.; Liu, H.; Chen, H.; Guo, J.; et al. Ultra-high voltage network induced energy cost and carbon emissions. *J. Clean. Prod.* **2018**, *178*, 276–292. [CrossRef]
68. Li, Y.; Yi, B.-W.; Wang, Y. Can ultra-high voltage power transmission bring environmental and health benefits? an assessment in China. *J. Clean. Prod.* **2020**, *178*, 124296. [CrossRef]
69. Wang, X.; Huang, L.; Daim, T.; Li, X.; Li, Z. Evaluation of China's new energy vehicle policy texts with quantitative and qualitative analysis. *Technol. Soc.* **2021**, *67*, 101770. [CrossRef]

remote sensing

Article

Comparison of Five Models for Estimating the Water Retention Service of a Typical Alpine Wetland Region in the Qinghai–Tibetan Plateau

Meiling Sun [1,2,†], Jian Hu [1,2,†], Xueling Chen [1,2], Yihe Lü [3,4] and Lixue Yang [1,2,*]

1 Sichuan Zoige Alpine Wetland Ecosystem National Observation and Research Station, Southwest Minzu University, Chengdu 610041, China
2 Institute of Qinghai-Tibetan Plateau, Southwest Minzu University, Chengdu 610041, China
3 State Key Laboratory of Urban and Regional Ecology, Research Center for Eco-Environmental Sciences, Chinese Academy of Sciences, Beijing 100085, China
4 University of Chinese Academy of Sciences, Beijing 100049, China
* Correspondence: yanglixue@swun.edu.cn; Tel.: +86-28-8552-3352
† These authors have contributed equally to this work.

Abstract: Model evaluation of water retention (WR) services has been commonly applied for national or global scientific assessment and decision making. However, evaluation results from different models are significantly uncertain, especially on a small regional scale. We compared the spatial–temporal variations and driving factors of the WR service by five models (i.e., the InVEST model (InVEST), precipitation storage model (PRS), water balance model I (WAB I), water balance model II (WAB II), and NPP-based surrogate model (NBS) based on partial correlation analysis and spatial statistics on the Ramsar international alpine wetland region of the Qinghai–Tibetan Plateau (QTP). The results showed that the wetland area continued to decrease, and built-up land increased from 2000 to 2015. The average WR volume ranged from 2.50 to 13.65 billion $m^3 \cdot yr^{-1}$, with the order from high to low being the PRS, WAB I, WAB II, and InVEST models, and the average total WR capacity was 2.21×10^9 by the NBS model. The WR service followed an increasing trend from north to south by the InVEST, PRS, WAB I, and WAB II models, while the NBS model presented a river network pattern of high values. The WR values were mainly reduced from 2000 to 2010 and increased from 2010 to 2015 in the PRS, WAB I, WAB II, and InVEST models, but the NBS model showed the opposite trend. Precipitation determined the spatial distribution of WR service in the InVEST, PRS, WAB I, and WAB II models. Still, the spatial variation was affected by climate factors, while the NPP data influenced the NBS model. In addition, the InVEST model in estimating WR values in wetlands and the PRS and WAB I models poorly estimate runoff, while the WAB II model might be the most accurate. These findings help clarify the applicability of the WR models in an alpine wetland region and provide a valuable background for improving the effectiveness of model evaluation.

Keywords: alpine wetland; water retention service; models; Qinghai–Tibetan Plateau

Citation: Sun, M.; Hu, J.; Chen, X.; Lü, Y.; Yang, L. Comparison of Five Models for Estimating the Water Retention Service of a Typical Alpine Wetland Region in the Qinghai–Tibetan Plateau. *Remote Sens.* **2022**, *14*, 6306. https://doi.org/10.3390/rs14246306

Academic Editor: Yaoming Ma

Received: 10 October 2022
Accepted: 6 December 2022
Published: 13 December 2022

Publisher's Note: MDPI stays neutral with regard to jurisdictional claims in published maps and institutional affiliations.

1. Introduction

Ecosystem services (ESs) are the benefits people obtain from ecosystems which are essential for human well-being [1]. Water retention (WR) is a critical regulative service that refers to the water retained in ecosystems within a certain period [2–5]. Although China's total water resources rank sixth in the world, it has been experiencing severe water resource shortages due to climate change and rising water demands [6,7]. The degradation of WR service has accelerated water shortages and become one of China's major ecological problems [8]. China's first ecosystem assessment showed that WR service decreased from the southeast to the northwest inland areas in 2010 [3]. Then, a similar spatial pattern was also mapped from 2000 to 2013 in another way [9,10]. However, ESs are complex due to strong scale effects, resulting in various spatial–temporal changes at small scales by models [11]. Therefore, it is necessary to compare the accuracy of WR service at a small regional scale with various models to better understand their applicability.

The Qinghai–Tibetan Plateau (QTP) is the world's largest water tower region, and poor availability and changes in its water services have been of great concern [12,13]. Researchers have mapped the spatial pattern of WR service by the Integrated Valuation of Ecosystem Services and Tradeoffs (InVEST) model, showing a decreasing trend from the southeast to the northwest of the QTP [14,15]. Several other WR models, initially developed in China, have also been applied to the QTP. The tradeoffs between carbon sequestration and WR service and the synergy between livestock production and WR service of alpine grassland were revealed by the water balance model I (WAB I) and the precipitation storage model (PRS), respectively [16,17]. The sensitivity of WR under future climate scenarios was quantified, and the ecological importance was mapped by the water balance model II (WAB II) [18,19]. In addition, the spatial–temporal characteristics of the water and nutrient retention service of the critical natural capital were displayed by the NPP-based surrogate model (NBS) [20]. However, few studies compared the results of models within different structures and characteristics of WR service, and it is necessary to reveal the WR model's performance for application in a specific region, especially for the spatial–temporal pattern and its changes, because the criteria of WR service by models are different, such as mm, m^3, or dimensionless. In addition, different results have been found in the Upper Upatoi Creek watershed and the Nansihu Lake basin between InVEST and the Soil and Water Assessment Tool (SWAT) [21,22]. Therefore, it is essential to compare various models about how well it fits applications to screen the model with strong applicability and low uncertainty to offer guidance on selecting more effective tools. The Zoige Plateau (ZP), in the eastern part of the QTP, is a critical WR region in the upper reaches of the Yellow River and is regarded as one of the most extensive alpine peatlands in the world for storing biotic carbon [23,24]. However, warming and drying climate trends and anthropogenic perturbations such as overgrazing and artificial ditch construction have resulted in the degradation of wetlands and a decrease in the runoff, threatening the stable supply of water resources [25–28]. Some researchers have evaluated the WR service of ZP [29], methane, and CO_2 emissions in the Zoige Wetland [30,31], and the ecosystem services value [32,33]. However, the accurate assessment of the spatial–temporal variation in WR service in the ZP remains a major challenge.

Researchers have evaluated the driving factors of WR service and highlighted that natural and human-induced factors are vital aspects. Additionally, these factors are independent and almost always multiple and interactive, so a one-to-one linkage between particular driving forces and particular ecosystem changes rarely exists [1]. In addition, topography, microclimate, vegetation, and hydrological processes also played vital role in soil water movement at a large scale [34]. Therefore, analyzing the combined effect of these drivers spatially to identify overlapping impacts on ESs is meaningful for spatial planning and management. Effective energy and mass transfer (EEMT), consisting of water, carbon, and energy, is a comprehensive climate indicator essential in controlling groundwater thickness and water availability [35,36]. It can be used for assessing the overlapping effects of climate-driving factors.

Therefore, our study considered ZP as the focus area and compared the WR service from 2000 to 2015 with five models, including the InVEST, PRS, WAB I, WAB II, and NBS. We used spatial statistical methods to reveal the changes in WR and the relationship with driving factors. The main objectives were to (1) quantify and compare spatial–temporal patterns and variations in WR service by five models under statistical methods, (2) reveal the relationship between natural and socioeconomic factors and WR service in different models by partial correlation analysis, and (3) discuss the applicability of five models to the alpine wetland region. This work will deepen our understanding of WR service simulation models and provide the theoretical basis for model application in an alpine wetland area of QTP.

2. Materials and Methods

2.1. Study Area

The ZP (31°50′–34°49′ N, 100°45′–103°39′ E) is located on the eastern margin of the QTP, with an area of 42,714 km^2 (Figure 1). It is a complete orbicular plateau surrounded by alpine mountains ranging from 2392 to 5059 m. Based on 39 years of data records (1980–2018), the average annual precipitation is approximately 712 mm, and the mean annual temperature is 0.87 °C. The main land types of ZP are grassland, shrubland, wetland, and forestland, accounting for 69.43%, 12.71%, 9.87%, and 5.23% of the total area in 2015, respectively. The proportion of cropland is relatively low, accounting for only 0.61% of the total area, and is mainly distributed in Aba County. The Zoige Wetland is an integral part of the Ramsar internationally important wetlands. In addition, almost all the rivers in this region belong to the Yellow River water system. Its tributaries mainly consist of the White River and Black River, providing at least 30% of the water flow into the upper Yellow River and becoming a vital water retention functional area [27]. It is crucial for China's ecological protection and high-quality development strategy in the Yellow River Basin.

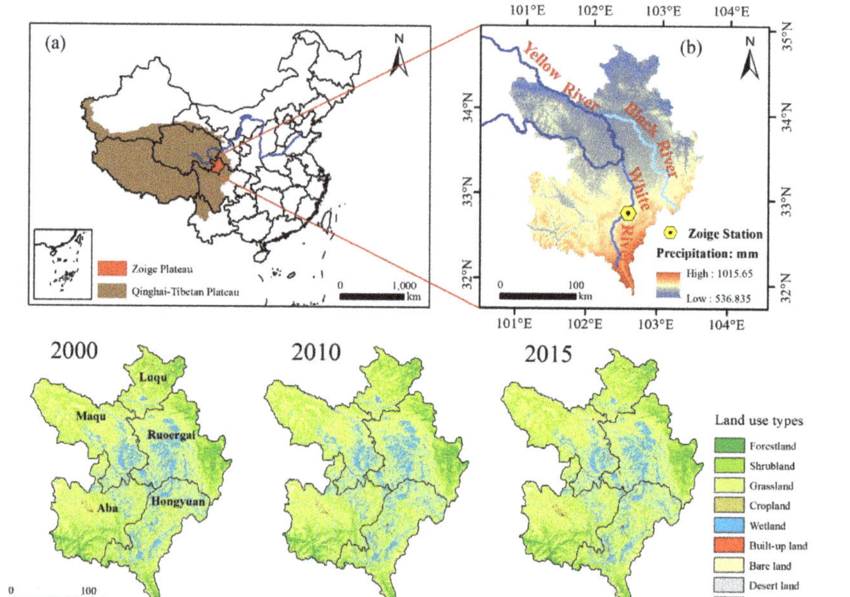

Figure 1. The location (**a**), mean annual precipitation from 1980 to 2018 (**b**), and land use type distribution from 2000 to 2015 of ZP.

2.2. Analysis Framework for Comparing the Water Retention Models

The purpose of this study was to compare the performance of WR service simulation models in an alpine wetland region, aiming to offer guidance on selecting and applying models in this area. Therefore, we selected five WR simulation models commonly used in QTP. After data preparation and preprocessing, we used the five models to map the spatial pattern of WR service from 2000 to 2015. Then, we used the spatial statistics method to compare the pattern of WR assessed by five models and its changes in the past 16 years. In addition, spatial partial correlation analysis was used to compare the relationship between WR assessed by different models and driving factors. Figure 2 shows the framework of our study.

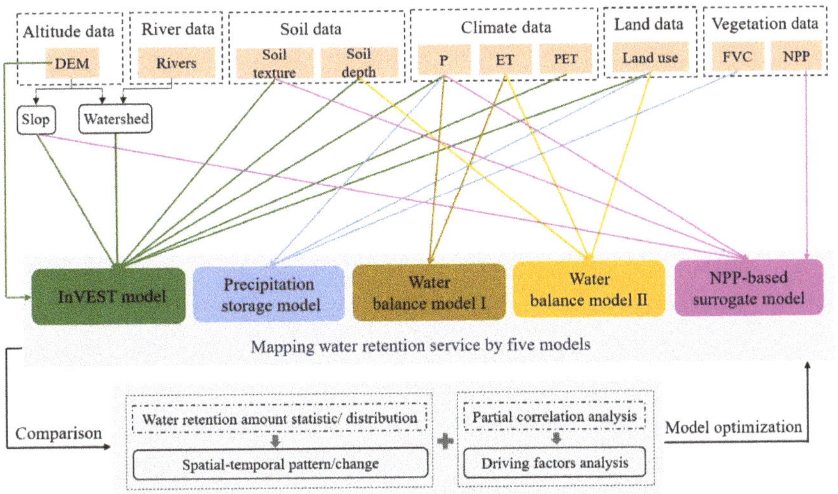

Figure 2. A framework for comparing water retention services by five models.

2.3. Data Source and Processing

The spatial data, including climate, soil, rivers, altitude, land use, vegetation, and socioeconomic data, are presented in Table 1. The land use and digital elevation model (DEM) data were obtained from the internal shared data of the Second Tibetan Plateau Scientific Expedition and Research. Origin 2022 software was used to draw the Sankey diagram of land use transfer. The gross domestic product (GDP) and population density (POP) data are raster data generated by interpolation based on GDP and population statistics data of all counties in China, considering land use types, night light brightness, residential density data, and spatial interaction with GDP. We spatially interpolated temperature and precipitation data from meteorological stations in China using Anusplin software at a 1 km resolution and then extracted them based on the study region. The resolution of all raster data was resampled to 30 m. In addition, we used the Penman–Monteith equation to calculate potential evapotranspiration (PET), and we obtained the annual PET by the sum of the monthly PET [37]. EEMT is the sum of energy input via effective precipitation and net primary production components; the calculation of EEMT (J m^{-2} s^{-1} or W m^{-2}) is given in [35] as:

$$\mathrm{EEMT} = E_{PPT} + E_{BIO} \tag{1}$$

E_{PPT} is heat energy related to effective precipitation energy and mass transfer, and E_{BIO} is NPP energy and mass transfer.

2.4. Description of Selected Models

2.4.1. InVEST Model

The InVEST model simulates and provides spatial information about ESs. The water yield module is based on the Budyko curve and water balance principles (Table 2). The following equation can calculate $WY(x)$:

$$WY(x) = \left(1 - \frac{AET(x)}{P(x)}\right) \times P(x) \tag{2}$$

where $WY(x)$ is the annual water yield for pixel x (mm), $AET(x)$ is the potential annual evapotranspiration (mm), and $P(x)$ is the annual precipitation for pixel x (mm). Taking average rainfall events from 1995–1999, 2005–2009, and 2010–2014 of four meteorological stations on the ZP as rainfall events in 2000, 2010, and 2015, the Z values were calculated as follows: 11.59, 11.09, and 10.62 (Table S1). The *WR* values (mm) of each pixel were revised by the following formula [15]:

$$WR = min\left(1,\ \frac{249}{V}\right) \times min(1,\ 0.3TI) \times min\left(1,\ \frac{K_{sat}}{300}\right) \times WY \tag{3}$$

where V is the velocity coefficient; K_{sat} is the saturated soil hydraulic conductivity (cm/d), calculated by Neuro Theta software according to the content of soil sand, silt, and clay (%); and *TI* is a topographic index, calculated from the following equation:

$$TI = \log\left(\frac{Drainage\_Area}{Soil\_Depth \times Percent\_Slop}\right) \tag{4}$$

where *Drainage_Area* is the number of catchment area grids, *Soil_Depth* is soil depth (mm), and *Percent_Slop* is the percentage slope.

2.4.2. Precipitation Storage Model (PRS Model)

The model refers to the decreased amount of water compared to bare land under the condition of rainfall generation as WR service [17]:

$$WR = A_f \times J_0 \times k \times \left(R_0 - R_f\right) \tag{5}$$

where A_f is the area of ecosystem type (km^2); J_0 is the annual precipitation (mm); k is the proportion of rainfall that can generate runoff, assigned a value of 0.6 in southern China; R_0 is the runoff ratio of bare land, assigned a value of 1; and R_f is runoff ratio of the other land use types. The runoff ratio (R_f) of grassland was obtained from the vegetation cover (f_c):

$$R_f = -0.3187 f_c + 0.36403 \tag{6}$$

2.4.3. Water Balance Model I (WAB I Model)

The WR values (mm) are considered as the balance between precipitation (mm) and actual evapotranspiration (mm) [38]:

$$WR = P - ET \tag{7}$$

Table 1. Spatial data sources and description.

Data	Spatial Resolution	Temporal Resolution	Units	Data Source
Rivers	1:1,000,000	2019	–	National Geomatics Center of China (http://www.ngcc.cn; accessed on 15 September 2021)
Soil texture/depth	1 km	–	cm	China Soil Map-Based Harmonized World Soil Database (v1.1) (http://www.ncdc.ac.cn; accessed on 15 September 2021)
Temperature (T)	1 km	1980–2018, monthly	°C	The National Meteorological Information Center of China (http://data.cma.cn/en; accessed on 20 September 2021)
Precipitation (P)	1 km	1980–2018, monthly	mm	
Potential evapotranspiration (PET)	1 km	1980–2018, monthly	mm	
Evapotranspiration (ET)	1 km	2000–2018, 10 days	mm	[39] (https://www.sciencedirect.com/; accessed on 30 October 2021)
Net primary productivity (NPP)	250 m	2000–2015, monthly	0.01 g/cm^2	Institute of Remote Sensing and Digital Earth Chinese Academy of Sciences (http://eds.ceode.ac.cn/; accessed on 15 October 2021)
Fractional vegetation cover (FVC)	250 m	2000–2015, monthly	——	
Gross Domestic Product (GDP)	1 km	2000–2015, yearly	Ten thousand yuan/km^2	Resource and Environmental Science and Data Center (http://www.resdc.cn/;accessed on 15 October 2021)
Population density (POP)	1 km	2000–2015, yearly	People/km^2	

2.4.4. Water Balance Model II (WAB II Model)

This model refers to the WR (m^3) as the difference between the precipitation (mm) and the sum of actual evapotranspiration (mm) and surface runoff (mm), which is revised from the InVEST model [3].

$$WR = \sum_{i=1}^{n} A_i \times (P_i - R_i - ET_i) \times 10^3 \quad (8)$$

where i is the ecosystem, n is the number of ecosystem types, A_i is an area of the ecosystem i (km^2), and R_i is the surface runoff of the ecosystem i, obtained by multiplying precipitation and the surface runoff coefficient. P_i and ET_i are the precipitation and actual evapotranspiration of the ecosystem i.

2.4.5. NPP-Based Surrogate Model (NBS Model)

The NPP-based surrogate model simulate a variety of services, such as water retention, soil conservation, carbon sequestration, and biodiversity conservation [9,10,20]. The ability of WR (non-dimensional) service is calculated as follows:

$$WR = NPP_{mean} \times F_{sic} \times F_{pre} \times (1 - F_{slo}) \quad (9)$$

where NPP_{mean} is the multi-year average net primary productivity, and F_{sic}, F_{pre}, and F_{slo} are normalized (0–1) treated slope factor, soil infiltration capacity factor, and multi-annual average precipitation raster, respectively.

2.5. Analysis of Climate Change Trends

The linear regression method was adopted to analyze the variation trend of climate elements, including temperature, precipitation, evapotranspiration, and effective energy and mass transfer. The linear trend was detected by the least square regression method as follows:

$$yi = a + bxi + \varepsilon \quad (10)$$

where a is the intercept, b is the slope, and ε is the residual. When b is negative, index i is a decreasing trend; when b is positive, index i is an increasing trend. MATLAB2019b and ArcGIS10.3 were used for trend analysis and spatial mapping, respectively. The change trend is significant when $p < 0.05$.

2.6. Analysis of WR Service Change and Drivers

2.6.1. Quantification of WR Values and Their Changes

After completing the WR service estimation by five models, we applied the random points tool to create 100 random points with no distance requirement to calculate the mean and standard deviation of the same model. At the same time, we removed the extreme values and retained 83 points in each model. Then, we carried out one-way ANOVA and multiple comparisons based on the LSD test to investigate the differences in model results. We used the raster calculator tool to calculate the spatial change in WR by subtracting layers of the same model from two years to obtain the number of change values on each pixel.

2.6.2. Determination of Land Use Attributes and Changes

Humans influence ecosystem structure and function by changing land characteristics and enhancing utilization to meet their demand for land supply capacity. In this study, we used land use intensity to characterize the spatial–temporal changes in land use attributes [40]:

$$L = \sum_{i=1}^{n} A_i \times C_i = \sum_{i=1}^{n} A_i \times \frac{S_i}{S} \quad (11)$$

where L is land use intensity index; A_i is the grade index of land use type i. Considering the intense grazing activities in the study area, grassland is set to 2.5, the forestland, shrubland, and wetland are all 2, cropland is 3, built-up land is 4, and other types is 1. C_i is the area percentage of land type i, S_i is the area of the land use type i, and S is the total land area of the study area.

Table 2. Basic characteristics of the five selected models.

Model	Types	Spatial Scale	Temporal Scale	Advantages and Disadvantages	Reference
InVEST model	Water balance-based	Watershed	Year	(1) Strong visualization and dynamic (2) Complicated data input (3) Ignores interaction between surface water and groundwater	[14,15,21]
PRS model	Process-based	All scales	All scales	(1) Uncertainty of runoff and rainfall parameters (2) Relative value to bare ground, not an absolute value	[17,26,41]
WAB I model	Water balance-based	All scales	All scales	(1) Easy operation (2) Ignores runoff and groundwater (3) Suitable for dry regions	[38]
WAB II model	Water balance-based	All scales	All scales	(1) Easy operation (2) Uncertainty of runoff parameters	[3,19]
NBS model	Surrogate biophysical indicators-based	Regional scale	Year	(1) Not applicable to water bodies; (2) Affected greatly by NPP data (3) Cannot give specific physical quantities	[9,10]

2.6.3. Exploration of Factors Influencing WR Service

We selected precipitation (P), evapotranspiration (ET), temperature (T), and EEMT as the natural driving factors and GDP, POP, and land use intensity (L) as the socioeconomic driving factors. Then, we used partial correlation analysis to explain the drivers related to WR service. The data were normalized first using the Z score method, and the Pearson correlation was employed to calculate the simple relationship between a dependent variable and a single independent variable. Then, the partial correlation analysis was applied when the two variables were simultaneously related to other variables (Equation (12)).

$$R_{xy,z} = \frac{r_{xy} - r_{xz}r_{yz}}{\sqrt{(1 - r_{xz}{}^2) \times (1 - r_{yz}{}^2)}} \tag{12}$$

where $R_{xy,z}$ is the partial correlation coefficient between x and y, excluding the impact of z; r_{xy} is the correlation coefficient between x and y; r_{xz} is the correlation coefficient between x and z; and r_{yz} is the correlation coefficient between y and z.

A T-test was used for the reliability of the results of the partial correlation analysis:

$$t = \frac{r_{xy,z}}{\sqrt{1 - r_{xy,z}{}^2}}\sqrt{n - m - 1} \tag{13}$$

where n is the number of years and m is the number of the independent variables. The significance level was at 0.05.

3. Results

3.1. Land Use Change from 2000 to 2015

Grassland, shrubland, forestland, wetland, and cropland were the predominant land use types, accounting for 98% of the study region (Figure 3). From 2000 to 2010, grassland increased by 109.78 km^2, while wetland and cropland decreased by 112.01 and 8.64 km^2, respectively. The shrubland and forestland increased by 3.10 and 0.65 km^2, respectively. From 2010 to 2015, grassland, shrubland, forestland, and cropland decreased by 72.49, 17.55, 3.03, 0.65, and 9.15 km^2, respectively, while built-up land, bare land, and desert land increased by 81.86, 14.62, and 6.41 km^2, respectively. Over the past 16 years, the wetland area decreased the most, reaching 129.56 km^2, and was mainly converted to grassland, bare land, and built-up land. Except for a slight increase (3.98 km^2) in Luqu County, all the other counties decreased, and Ruoergi County decreased the most, by 75.37 km^2 (Figure S1). In contrast, built-up land increased for all counties, reaching 105.26 km^2, mainly from grassland, cropland, and wetland. Grassland increased by 37.28 km^2 and was distributed primarily in Ruoergai and Maqu counties. Desert land increased by 10.11 km^2 mainly from grassland degradation. In addition, shrubland, forestland, and other land use types remained stable.

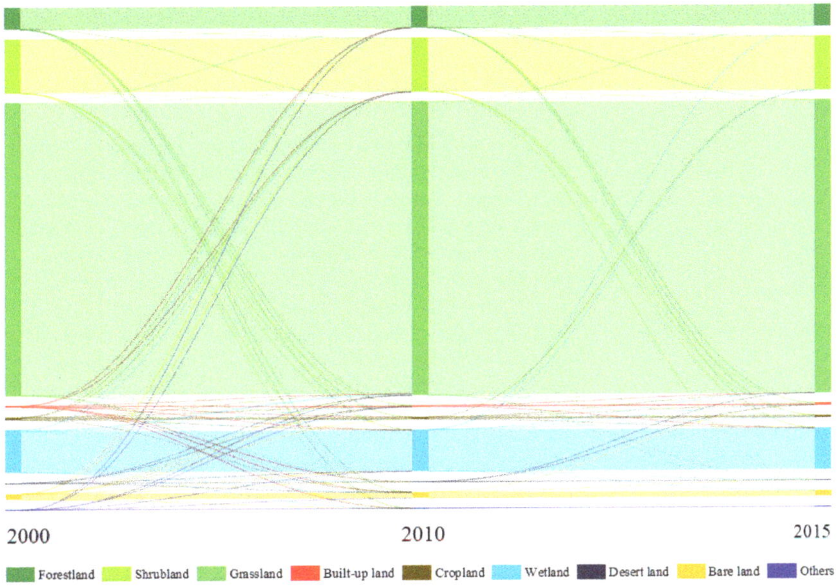

Figure 3. Land use composition and transformation from 2000 to 2015.

3.2. Climate Change Trends

From 1980 to 2018, the temperature showed a significant increasing trend ($p < 0.05$), with Luqu, Ruoergi, and Maqu counties in the north being lower than Aba and Hongyuan counties in the south and the highest increase in northwestern Maqu County (Figure 4). Precipitation decreased at approximately 42% of the regions ($p > 0.05$), mainly in northwestern Aba County, southeastern Hongyuan County, central and eastern Ruoergi County, and southern Luqu County. From 1980 to 2018, evapotranspiration showed a significant decreasing trend outside the southeast, northern, and eastern of ZP ($p < 0.05$), showing a northwest–southeast spatial distribution pattern. In addition, EEMT showed a significant increasing trend ($p < 0.05$) in other areas outside the southwest of Maqu County, parts of Hongyuan County, and Aba County.

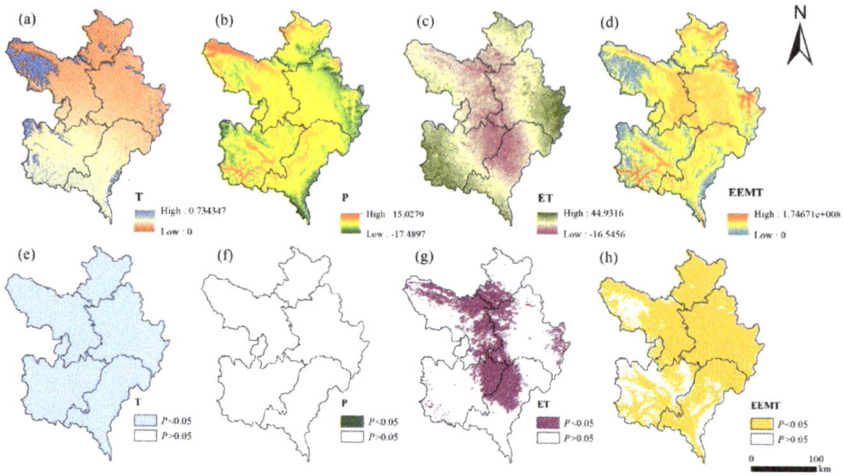

Figure 4. Spatial distribution of change trend values (**a–d**) and significance ($p < 0.05$) (**e–h**) of temperature (T), precipitation (P), evapotranspiration (ET), and effective energy and mass transfer (EEMT) of ZP.

3.3. Spatial Pattern of WR Service

The spatial pattern of the WR service showed substantial heterogeneity across the ZP. The spatial similarity was present between different models with an increasing trend from north to south in the InVEST, PRS, WAB I, and WAB II models over the past 16 years. However, the high values spread in a river network spatial pattern in the NBS model were consistent with the distribution of river channels (Figure 5). From the pixel value perspective, the maximum value in the WAB I model was 901.24 mm in 2000. The InVEST, WAB I, and WAB II models always had the largest weight of <100 mm, averaging approximately 83.61%, 43.03%, and 71.87%, respectively. The total area of pixel values ≤300 mm was approximately 93.35–99.58% of the ZP during the study period for these three models (Table S2). The PRS model had a maximum pixel area of 300–400 mm, with an area ratio of 56.04–57.21%. More than 96% of the region's WR ability by the NBS model was below 100 in the research period. In addition, all models showed that grassland was the land type with the highest WR capacity, followed by shrubland, forestland, wetland, and other land use types (Table S3).

3.4. Temporal Change in WR Service

The InVEST model had the lowest values of total WR volume, with a mean value of only 2.50 billion $m^3 \cdot yr^{-1}$ (Figure 6a). The PRS model always provided the highest WR volume, approximately 13.65 billion $m^3 \cdot yr^{-1}$. The results of the WAB I and WAB II models behaved between the above two models but were closer to the InVEST model, about 5.71 and 3.16 billion $m^3 \cdot yr^{-1}$, respectively. In addition, except for the NBS model, the WR values per grid of the PRS model were significantly higher than those of the other three models. In contrast, the InVEST model was markedly lower than PRS, WAB I, and WAB II models (Figure 6b) ($p < 0.05$). From 2000 to 2010, WR in the InVEST, WAB I, and WAB II models had a decreasing trend. The InVEST model reduced the least by 0.56 billion m^3; no decreasing regions were concentrated in the northwestern Maqu and Luqu counties, at 4493.52 km^2 (Figure 7 and Table S4). The WAB I model reduced the most, reaching 1.08 billion m^3. The unreduced area was only 2795.66 km^2, distributed in the junctional area of Ruoergai, Maqu, and Luqu counties. The WAB II model reduced 1.05 billion m^3, and the spatial variation pattern was similar to the WAB I model. However, the area unreduced was more extensive at 3111.24 km^2. The PRS model showed a slight increase of 0.11 billion m^3, and there was also an evident increase in wetland area compared to the

InVEST model, reaching 9231.54 km^2 overall. From 2010 to 2015, the above four models showed an increasing trend. Maximum and minimum increased values appeared in the PRS and InVEST models at 1.14 and 0.09 billion m^3, respectively. In addition, the WAB I and WAB II models showed 0.78 and 0.67 billion m^3 increases, respectively. It was mainly in the southwestern margin of Aba County and the eastern marginal regions of Ruoergai County for reduced areas. From 2000 to 2015, similar spatial–temporal patterns appeared between the InVEST and PRS models and the WAB I and WAB II models. The InVEST model experienced an increase of 0.03 billion m^3, with an area of 8203.94 km^2 detected to decrease in southwestern and southeastern margin areas, including Ruoergai, Hongyuan, and Aba counties. In comparison, the PRS model increased by 1.25 billion m^3 with a larger reductive area at 14,580.12 km^2, including most of Hongyuan and Aba counties and the southeastern part of Ruoergai County. The WAB I and WAB II models estimated reductions of 0.30 and 0.38 billion m^3, covering 24,613.58 and 24,970.92 km^2, respectively. A northwest–southeast distribution was found for unreduced regions. The InVEST, PRS, WAB I, and WAB II models all had more than 96.40% of the pixels whose variation in WR ranged within 200 mm from 2000 to 2015. For the NBS model, the average WR capacity per pixel was 46.93, and the total was 2.21 $\times$ 10^9. In addition, the spatial variation in the NBS model behaved differently from the other four models. There was an increasing trend from 2000 to 2010 and a decreasing trend in the central and easternmost parts. The majority decreased from 2010 to 2015 and were distributed in the region outside Luqu County for unreduced values. Over the past 16 years, the reduced pixels spread in much of the central and eastern regions, at 42,414.28 km^2. The total WR capacity changed slightly to 0.055 billion.

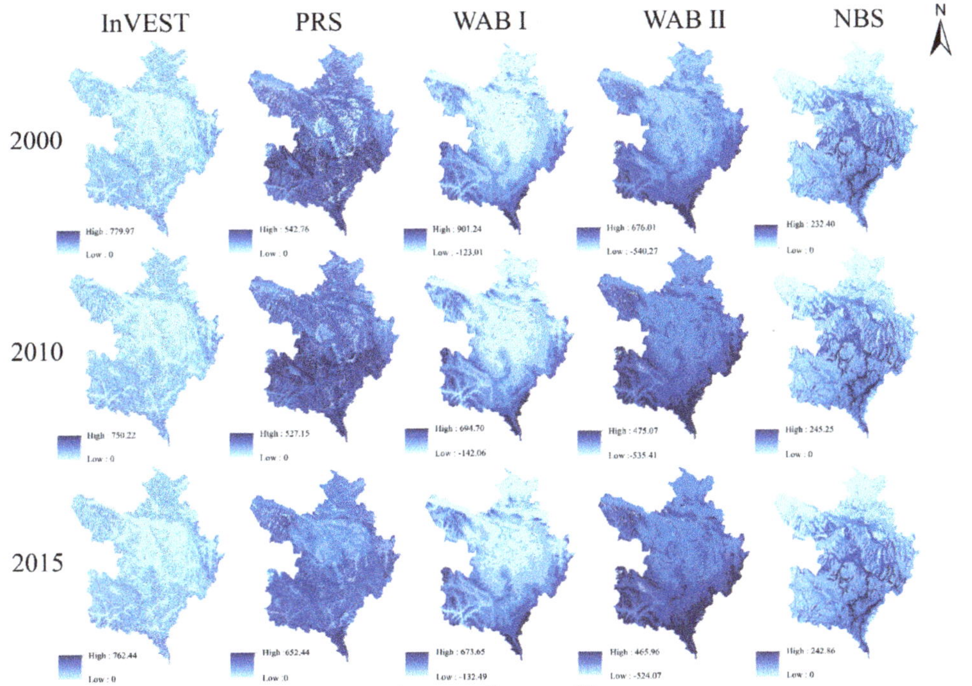

Figure 5. Spatial distribution of WR service by models from 2000 to 2015. The units for the InVEST, PRS, WAB I, and WAB II models are mm/grid and dimensionless for the NBS model.

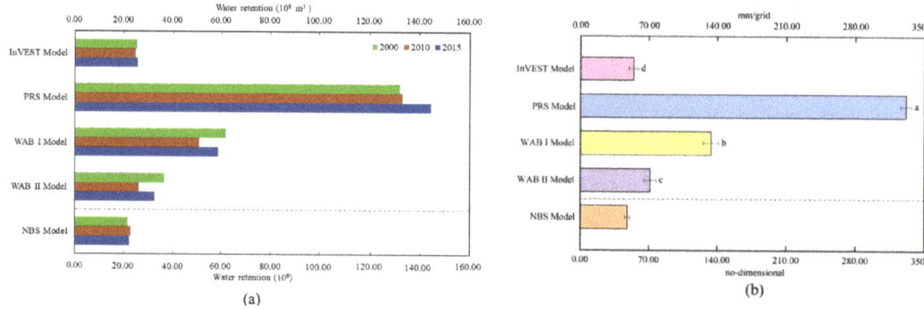

Figure 6. Characteristics of water retention volume or capability by five methods from 2000 to 2015. The units for the InVEST, PRS, WAB I, and WAB II models are m^3, and dimensionless for the NBS model (**a**). The values are the mean ± standard error, and lowercase letters indicates significant differences among the different models ($p < 0.05$) (**b**).

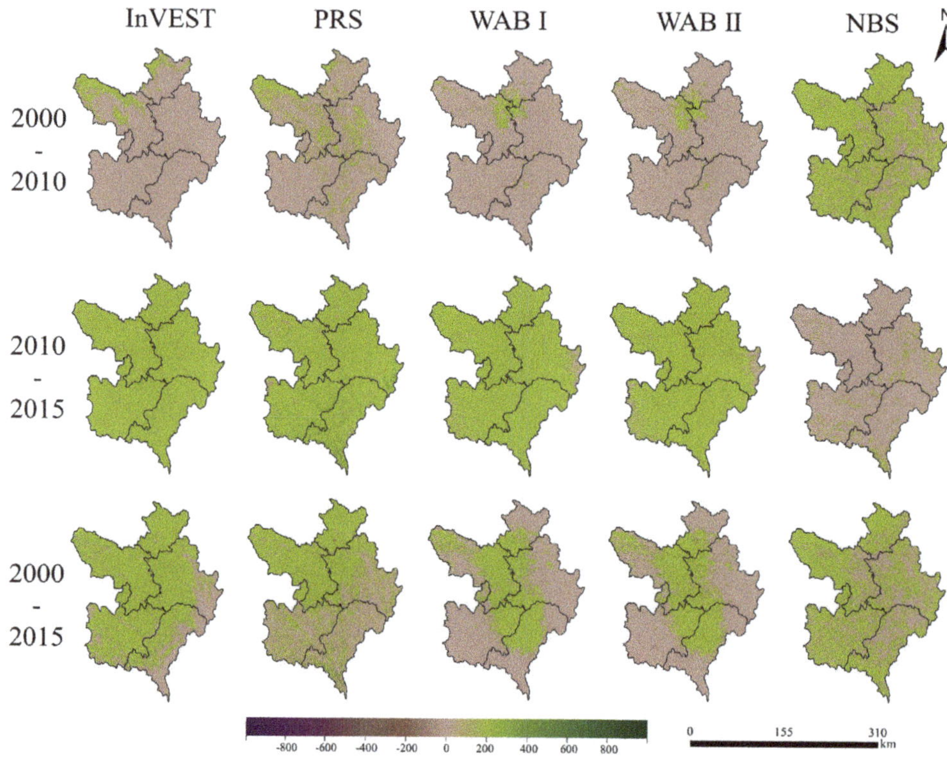

Figure 7. The spatial–temporal variations in WR service by five models in different periods. The units for models except the NBS model are mm/grid, and dimensionless for the NBS model.

3.5. The Driving Factors for WR Service Change

3.5.1. Influence of Climatic Factors on the WR Service

The WR services simulated by the five models correlated significantly with climatic factors and geographic characteristics (Figure 8 and Table 3). The results of the InVEST model showed that 56.88% of the study area had a positive relationship between WR and temperature, concentrated in Aba, Hongyuan, and Luqu counties. At the same time, the negative correlation regions were distributed in most of Maqu, Ruoergai, and the low areas in the southern mountains. In addition, the largest area with a positive relationship between WR and precipitation was found in the InVEST model, accounting for 65.90%, distributed in the vast area outside the southernmost parts of Aba and Hongyuan counties. In addition, WR was positive with evapotranspiration, mainly in central Hongyuan County, the northern margin of Luqu County, northwestern Maqu County, and northeastern Aba County. There was a significant negative correlation between WR and EEMT in the northern part of Luqu County, the eastern margin of Ruoergai County, and most of Hongyuan County by the InVEST model. The WAB I and WAB II models assumed an almost consistent spatial characteristic between WR service and four climatic factors. In detail, WR was positively correlated with temperature in the junctional region of Luqu, Maqu, and Ruoergai counties and southeastern Aba County, at approximately 41.89% and 49.65% of the total area, respectively. Likewise, the WR of these two models had a positive relationship with precipitation in northwestern Maqu County. However, the WR of the WAB I model in the western part of Aba County and the discontinuous central part of ZP showed a positive relationship with precipitation. This relationship existed in the WAB II model on the eastern margin of Ruoergai County. In addition, WR was almost negative with evapotranspiration and positive with EEMT for the WAB I and WAB II models.

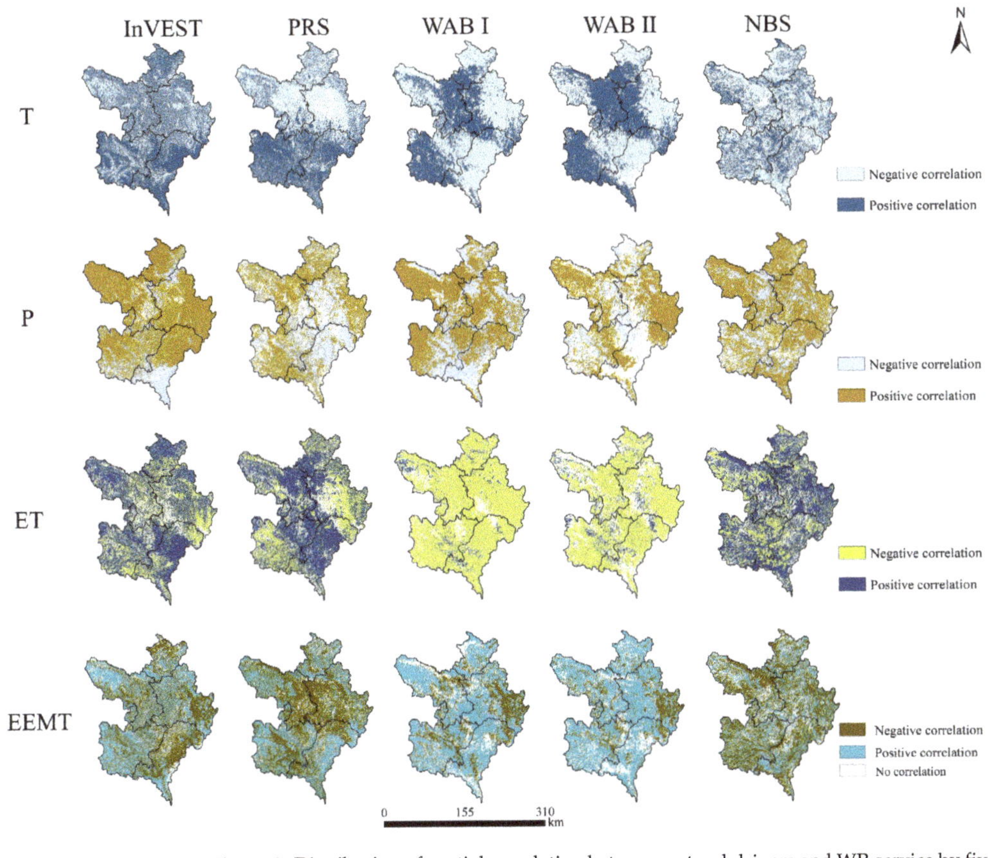

Figure 8. Distribution of partial correlation between natural drivers and WR service by five models ($p < 0.05$). T: temperature, P: precipitation, ET: evapotranspiration, EEMT: effective energy and mass transfer.

Table 3. Area and percentage of positive and negative correlations between water retention and drivers.

Drivers		Temperature		Precipitation		Evapotranspiration		EEMT		Land Use Intensity		GDP		Population Density	
Models		Area (km^2)	Percent (%)	Area (km^2)	Percent (%)	Area (km^2)	Percent (%)	Area (km^2)	Percent (%)	Area (km^2)	Percent (%)	Area (km^2)	Percent (%)	Area (km^2)	Percent (%)
InVEST model	Positive	24,294.96	56.88	28,149.75	65.90	17,523.39	41.03	18,773.75	43.95	11,741.67	27.49	15,585.16	36.49	16,831.87	39.41
	Negative	14,935.75	34.97	9061.39	21.21	17,229.58	40.34	15,292.87	35.80	10,721.12	25.10	13,114.63	30.70	12,258.14	28.70
PRS model	Positive	18,583.70	47.65	16,580.52	38.82	18,159.81	42.52	19,017.32	44.52	10,712.33	25.08	11,966.04	28.01	15,317.53	35.86
	Negative	20,415.02	52.35	15,123.32	35.41	16,899.88	39.57	16,638.75	38.95	3728.08	8.73	16,747.22	39.21	12,381.78	28.99
WAB I model	Positive	17,892.73	41.89	23,248.45	54.43	1155.51	2.71	24,399.69	57.12	9630.13	22.55	14,385.66	33.68	15,850.65	37.11
	Negative	21,615.77	50.61	15,096.95	35.34	36,965.40	86.54	8498.39	19.90	11,715.66	27.43	14,524.02	34.00	15,063.18	35.27
WAB II model	Positive	21,209.20	49.65	18,129.24	42.44	2259.73	5.29	26,194.72	61.33	10,960.71	25.66	13,685.20	32.04	16,223.72	37.98
	Negative	16,417.89	38.44	9235.47	21.62	30,617.34	71.68	6485.74	15.18	13,304.71	31.15	15,789.56	36.97	15,458.57	36.19
NBS model	Positive	14,348.02	33.59	23,972.37	56.12	18,703.74	43.79	13,826.34	32.37	13,210.07	30.93	17,309.52	40.52	15,799.46	36.99
	Negative	19,906.80	46.61	12,246.80	28.67	16,416.53	38.43	18,614.40	43.58	11,656.09	27.29	12,045.92	28.20	15,446.70	36.16

Similar to the InVEST model, WR service was positively correlated with temperature in Aba and Hongyuan counties for the PRS model. In contrast, the other three counties were opposite. There was an apparent negative relationship between WR and precipitation in the central areas of ZP and most of Hongyuan County. Evapotranspiration positively impacts the WR in a large area at the junction of Maqu, Luqu, and Ruoergai counties, northwestern Aba County, and most of Hongyuan County. In addition, the WR of the PRS model service in the central parts of ZP had a significant negative correlation with EEMT than the other four models. The NBS model was spatially different from the other four models. There was an apparent negative correlation between WR capacity and temperature. The WR capacity positively correlated with precipitation outside the northeast of Maqu County, most of Aba County, and central Hongyuan County. The NBS model also differed from the other four models between WR service and evapotranspiration, with a negative correlation mainly in southwestern Maqu County, northwestern Aba County, and Hongyuan County. The T-test performed the same as the InVEST, WAB I, and WAB II models in the eastern margin of Ruoergai County and differently from the PRS model in the central parts of ZP for the negative performance between WR service and EEMT.

3.5.2. Influence of Socioeconomic Factors on WR Service

The WR of the InVEST model showed a positive relationship with land use intensity in Hongyuan County and the eastern margin of Ruoergai County (Figure 9 and Table 3). In addition, the InVEST model showed that GDP significantly inhibited WR service in 39.49% of areas, including southeastern Maqu County and Aba and Hongyuan counties. However, the WR of the five models positively correlated with population density. The WR of the InVEST model indicated a significant negative correlation with population density at a ratio of 28.70% in southwestern Aba County and northeastern Luqu and Ruoergai counties. The WR services by the WAB I and WAB II models all showed a negative correlation with land use intensity in the central parts of ZP and a positive correlation in southern and northern Hongyuan County. Unlike the InVEST model, both models showed that WR services were positively correlated with Hongyuan County's GDP. At the same time, they were negatively correlated with GDP in the eastern parts of Ruoergai County and most of Luqu County. In addition, population density also significantly inhibited WR service in the northeastern part of Aba County and the central part of Hongyuan County.

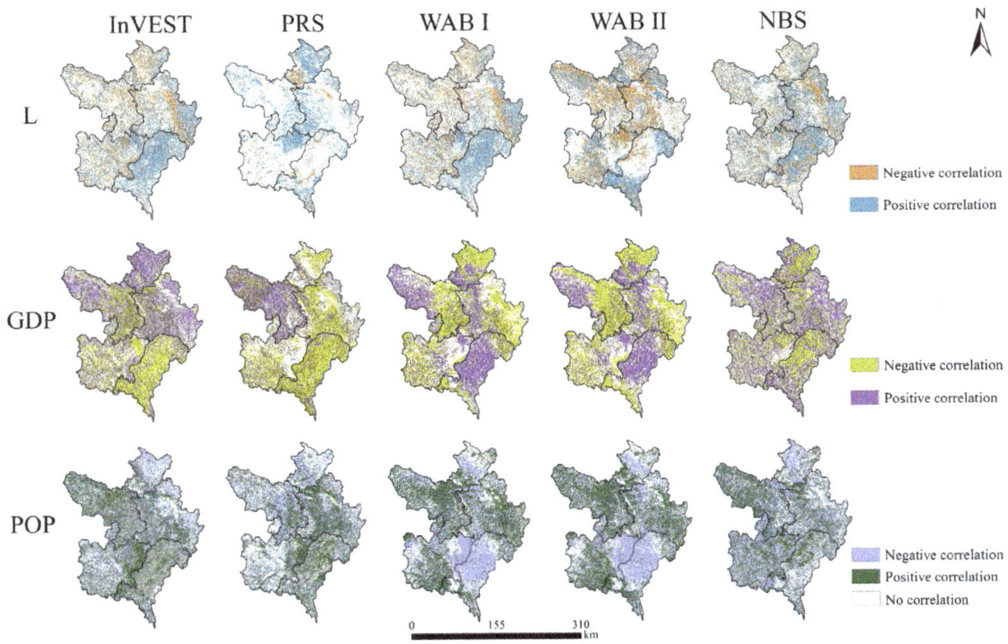

Figure 9. Distribution of partial correlation between socioeconomic factors and WR service by five models ($p < 0.05$). L: land use intensity, GDP: gross domestic product, POP: population density.

The WR of the PRS model positively correlated with land use intensity in the central part of ZP and northern Luqu County. The GDP in Maqu County and population density in the central part of ZP increased the WR of the PRS model. In addition, the WR capacity of the NBS model was equally as positive with land use intensity in Hongyuan and Ruoergai counties as of the InVEST model. In addition, WR capacity had the least negatively correlated area with the GDP, mainly in the northeastern part of Hongyuan County and the southeastern part of Luqu County. Furthermore, most central and eastern areas represented a negative relationship between WR ability and population density.

4. Discussion

4.1. Model Uncertainty and Applicability

The WAB I model takes the difference between precipitation and evapotranspiration as the WR service, which is consistent with the principle of the water yield module in the InVEST model. However, 30% of precipitation flows out as runoff in the growing season, so neither the WAB I model nor the water yield volume can be used as the actual WR volume of ZP [25,27]. The WAB I model and water yield module are more suitable for arid and semiarid areas where water budgets mainly depend on precipitation and evapotranspiration [38]. In addition, WR is the infiltration amount of the soil layer after precipitation minus evapotranspiration and surface runoff [42]. Therefore, the InVEST and WAB II models take soil infiltration over time as the WR service. However, the total WR volume of wetlands was the lowest by the InVEST model. The main reason was that many pixel values of lakes and rivers were zero by the water yield module and the InVEST model [15]. Therefore, the InVEST model may have uncertainty in assessing WR service volume in this study. In addition, the WR of the PRS model was mainly affected by precipitation and the parameters input to the model. We used the general parameter "0.6" as the proportion of precipitation that can generate runoff. However, observation data showed that evapotranspiration accounts for more than 60% of the annual precipitation, meaning that the amount of water stored in an ecosystem is less than or equal to 40% [27,41]. Therefore, the actual volume of water storage should be lower than estimated. The WAB II model may be the most accurate for WR volume despite five WR models showing a consistent spatial pattern. Furthermore, the primary indicator of the NBS model is NPP data, contributing the most to the WR values [10]. Herbaceous marsh accounted for 91.43–92.20% of wetland areas from 2000 to 2015. The high value of NPP was more extensive along rivers and wetland vegetation growth areas, where herbaceous peat bog soils are abundant [43]. However, the ZP has higher vegetation cover, prosperous plant communities, and ecosystem types than other regions, as well as a complex plateau landform landscape. Water bodies with vegetation cover were easily classified as vegetation in remote sensing data, which could cause uncertainty in the results of the NBS model [44,45]. Vegetated marshlands are probably the only spectrally unique wetland category that can be discerned from TM and ETM+ imagery [46]. Furthermore, monitoring suffered due to frequent variable weather and thick clouds, resulting in a lack of long–time series data in the QTP [45]. Therefore, high temporal and spatial resolution remote sensing products on alpine ecosystems, including land use and vegetation characteristics data, contribute to the WR service simulation.

4.2. Spatial–Temporal Patterns of WR Service

We investigated the WR service's spatial–temporal patterns from 2000 to 2015 with five WR models. The annual WR service had similar spatial patterns to the distribution characteristics of precipitation in the InVEST, PRS, WAB I, and WAB II models. The mountainous southern areas receive more rain than the plains, and the distribution of precipitation across the region increases gradually from >500 mm in the north to >1000 mm in the south. In addition, these four models also displayed higher values distributed in the high-altitude regions, with the central and northern regions having lower values, consistent with previous studies [29]. Under limited spatial data conditions, the WAB I model best shows the WR's spatial pattern since only precipitation and evapotranspiration raster data are required as inputs. From 2000 to 2015, the increased region by the WAB I and WAB II models performed a northwest–southeast distribution. Differing from the study that increased in the Aba and Ruoergai counties from 2000 to 2017, and this may be due to the different data sources and change analysis methods [29]. In addition, the WR of the InVEST and PRS models increased widely in the northwest of ZP. The results showed that the WR's changing area (increase or decrease) calculated by the InVEST model was more significant than that of the WAB II model [15,18]. High WR values are presented in river network patterns by the NBS model, mainly affected by the distribution of the NPP value of herbaceous wetlands [43]. From 2000 to 2015, wetlands' total WR ability decreased because the entire marsh NPP dropped [47]. Therefore, this model is suitable for identifying the WR capacity of herbaceous wetlands.

4.3. Driving Forces of WR Service

WR service is one of the most critical regulation services in ecosystems. It can be affected by climate change and anthropogenic perturbation. A warming and drying trend has occurred in the upper Yellow River, and increased evapotranspiration has caused water loss due to rising temperature [41]. The primary relationship between WR service and precipitation was positive for all five models and determined the spatial–temporal pattern in the InVEST, PRS, WAB I, and WAB II models. Precipitation decreased in the eastern and southern parts of the ZP, consistent with the WR service's spatial changes by InVEST from 2000 to 2015. Therefore, precipitation remains this model's critical driver of WR service changes [15]. In addition, the WAB I and WAB II models showed negative correlations with evapotranspiration over 71% area of ZP. The variation in WR values was consistent with the spatial variation in evapotranspiration. Therefore, the spatial variation in the WR service in these two models was mainly related to evapotranspiration. WR of the PRS model had a positive relationship with evapotranspiration. The evapotranspiration decreases, and the water storage of the ecosystem increases. Peatlands with carbon-rich soils have been characterized in convergent areas of greater EEMT and positively correlated with annual baseflow contributing to headwater stream runoff [48–50]. EEMT was negative in most wetland parts, mainly in the western Ruoergai County and the eastern Maqu County, by the PRS, InVEST, and NBS models, while positive in the WAB I and WAB II models. The WAB I and WAB II models may better reveal the wetland region's climate drivers. The reason may be that the PRS model evaluates the capacity of an ecosystem to regulate water, while the WAB I and WAB II models simulate the water stock of the ecosystem, both above and below ground [51]. Baseflow expressed an increasing trend with climate warming, which means water storage may become smaller [50]. Furthermore, the InVEST model needs to estimate the wetland accurately [15]. The correlation distribution between the NBS model and climate factors existed consistently, indicating that WR services were affected mainly by NPP data input and its changes [9,47].

Warming and drying trends may convert wetlands to grassland, but overgrazing, gully drainage, and peat mining are the main factors [52,53]. Since the 1960s, nearly 1000 drainage channels have been built in the Zoige wetlands, which drain groundwater, slow subsurface flow except in the rainy season, reduce the quality and area of wetlands, and lower the adjacent water table [25,47,49]. Researchers have shown that the mean annual runoff depth from the White (Black) River decreased by 28% (35%) in the human-induced period [25]. From 2000 to 2015, 129.56 km^2 of wetland areas mainly transformed into grassland, bare land, and built-up land, which altered water purification processes and weakened the capacity of soil to regulate and store water. Furthermore, this region implemented many measures and policies, including rotational and restricted grazing, filling ditches, planting grasses, and forbidding peat mining [47,54]. Afterward, the amount of wetland, grassland, and forestland areas increased, which can help to conserve water and soil [55]. Compared to natural factors, the relationship between socioeconomic factors and WR services is relatively weak, possibly due to the small spatial scale of the study area [11]. However, these human disturbances should be considered in future WR service evaluations.

In addition, precipitation, evapotranspiration, infiltration, and runoff impact the water budgets in watersheds [56]. The White River and Black River watersheds, the basic unit of the Earth's land surface system, constitute important WR functional areas in the upper Yellow River [57]. Additionally, the region is transitioning between seasonally frozen ground and permafrost. The boundaries between seasonally frozen land and permafrost are changing due to temperature increases, which impact the downstream water supply [41]. For a cold region, glaciers, snowmelt, and permafrost (frozen water components) are also essential parts of the water budget. However, the models discussed in this study may not have an adequate process to capture those budgets. Beyond that, various uncertainties exist in remote sensing data, especially in alpine regions; therefore, an integrated, distributed, and multiscale observation dataset is essential [44,45]. Comprehensive management of the Heihe River Basin in China through coupling ecohydrological and socioeconomic models undoubtedly provided scientific methods and ideas for watershed management in the Zoige Plateau.

5. Conclusions

This study compared five WR service evaluation models in terms of spatial–temporal patterns and driving factors from 2000 to 2015 based on spatial analysis and statistical methods. The InVEST, PRS, WAB I, and WAB II models showed similar spatial patterns of an increasing trend from north to south determined by the spatial distribution of precipitation. The WAB II model might be the most suitable because the InVEST model underestimates WR service due to weakening simulation in the wetland regions. The PRS model overestimates the proportion of precipitation that can generate runoff. The WAB I model lacks runoff procession but is the most convenient for showing a spatial pattern. The WAB I and WAB II models were better at revealing the overlapping effects of climate drivers in the wetland regions. However, the NBS model presented a river network pattern of high values over the past 16 years. More accurate NPP spatial data in identifying herbaceous marsh areas are essential for the assessment results of the NBS model. In addition, artificial ditch construction and ecological restoration measures should be included in evaluating WR services. The development of integrated watershed models in the alpine wetland region is needed.

Supplementary Materials: The following supporting information can be downloaded at: https://www.mdpi.com/article/10.3390/rs14246306/s1, Figure S1: Land use composition and transformation of Aba County (a), Hongyuan County (b), Luqu County (c), Maqu County (d) and Ruoergai County (e) from 2000 to 2015; Table S1: Annual rainfall events at four meteorological stations [58]; Table S2: Areas of different water retention pixel values from 2000 to 2015 based on five models; Table S3: Total water retention amount of various land types by models; Table S4: Changing areas of different pixel values in the study area from 2000 to 2015 for five models.

Author Contributions: Conceptualization, M.S., L.Y. and J.H.; Methodology, M.S., L.Y., Y.L. and J.H.; Software, M.S.; Data curation, X.C., L.Y. and J.H.; Writing—Review and Editing, M.S., X.C, Y.L. and J.H.; Visualization, J.H.; Supervision, J.H. All authors have read and agreed to the published version of the manuscript.

Funding: This study was supported by the Second Tibetan Plateau Scientific Expedition and Research of China (no. 2019QZKK0307), the Sichuan Science and Technology Program of China (no. 2022JDR0307), the National Science & Technology Fundamental Resources Investigation program of China (no. 2021FY100704), the National Natural Science Foundation of China (no. 42007057), and the Fundamental Research Funds for the Central Universities, Southwest Minzu University, China (no. 2021NYYXS14).

Data Availability Statement: Not applicable.

Acknowledgments: We thank anonymous reviewers for their original insights. We are also immensely grateful to the editor for their comments on the manuscript.

Conflicts of Interest: The authors declare no conflict of interest.

References

1. Millennium Ecosystem Assessment. *Ecosystems and Human Well-Being: Synthesis*; Island Press: Washington, DC, USA, 2005.
2. Brauman, K.A.; Daily, G.C.; Duarte, T.K.; Mooney, H.A. The nature and value of ecosystem services: An overview highlighting hydrologic services. *Annu. Rev. Environ. Resour.* **2007**, *32*, 67–98. [CrossRef]
3. Ouyang, Z.Y.; Zheng, H.; Xiao, Y.; Polasky, S.; Liu, J.G.; Xu, W.H.; Wang, Q.; Zhang, L.; Xiao, Y.; Rao, E.M.; et al. Improvements in ecosystem services from investments in natural capital. *Science* **2016**, *352*, 1455–1459. [CrossRef] [PubMed]
4. Lü, Y.H.; Hu, J.; Sun, F.X.; Zhang, L.W. Water retention and hydrological regulation: Harmony but not the same in terrestrial hydrological ecosystem services. *Acta Ecol. Sin.* **2015**, *35*, 5191–5196. (In Chinese) [CrossRef]
5. Sun, F.X.; Lyu, Y.H.; Fu, B.J.; Hu, J. Hydrological services by mountain ecosystems in Qilian Mountain of China: A review. *Chin. Geogr. Sci.* **2016**, *26*, 174–187. [CrossRef]
6. Leng, X.J.; Feng, X.M.; Fu, B.J.; Shi, Q.D.; Ye, H.P.; Zhang, Y. Asian water towers' are not a sustainable solution to the downstream water crisis. *Sci. Total Environ.* **2023**, *856*, 159237. [CrossRef]
7. Piao, S.L.; Ciais, P.; Huang, Y.; Shen, Z.H.; Peng, S.S.; Li, J.S.; Zhou, L.P.; Liu, H.Y.; Ma, Y.C.; Ding, Y.H.; et al. The impacts of climate change on water resources and agriculture in China. *Nature* **2010**, *467*, 43–51. [CrossRef]
8. Xiao, Y.; Ouyang, Z.Y. Spatial-temporal Patterns and Driving Forces of Water Retention Service in China. *Chin. Geogr. Sci.* **2019**, *29*, 100–111. [CrossRef]
9. Zhang, L.W.; Lü, Y.H.; Fu, B.J.; Dong, Z.B.; Zeng, Y.; Wu, B.F. Mapping ecosystem services for China's ecoregions with a biophysical surrogate approach. *Landsc. Urban Plan.* **2017**, *161*, 22–31. [CrossRef]
10. Zheng, H.B.; Zhang, L.W.; Wang, P.T.; Li, Y.J. The NPP-Based Composite Indicator for Assessing the Variations of Water Provision Services at the National Scale. *Water* **2019**, *11*, 1628. [CrossRef]
11. Su, C.H.; Dong, M.; Fu, B.J.; Liu, G.H. Scale effects of sediment retention, water yield, and net primary production: A case-study of the Chinese Loess Plateau. *Land Degrad. Dev.* **2020**, *31*, 1408–1421. [CrossRef]
12. Immerzeel, W.W.; Lutz, A.F.; Andrade, M.; Bahl, A.; Biemans, H.; Bolch, T.; Hyde, S.; Brumby, S.; Davies, B.J.; Elmore, A.C.; et al. Importance and vulnerability of the world's water towers. *Nature* **2020**, *577*, 364–369. [CrossRef] [PubMed]
13. Yao, T.D.; Bolch, T.; Chen, D.L.; Gao, J.; Immerzeel, W.; Piao, S.L.; Su, F.G.; Thompson, L.; Wada, Y.; Wang, L.; et al. The imbalance of the Asian water tower. *Nat. Rev. Earth Environ.* **2022**, *3*, 1–15. [CrossRef]

14. Wang, X.F.; Chu, B.Y.; Feng, X.M.; Li, Y.H.; Fu, B.J.; Liu, S.R.; Jin, J.M. Spatiotemporal variation and driving factors of water yield services on the Qingzang Plateau. *Geogr. Sustain.* **2021**, *2*, 31–39. [CrossRef]
15. Wang, Y.F.; Ye, A.Z.; Peng, D.Z.; Miao, C.Y.; Di, Z.H.; Gong, W. Spatiotemporal variations in water conservation function of the Tibetan Plateau under climate change based on InVEST model. *J. Hydrol. Reg. Stud.* **2022**, *41*, 101064. [CrossRef]
16. Pan, Y.; Wu, J.X.; Xu, Z.R. Analysis of the tradeoffs between provisioning and regulating services from the perspective of varied share of net primary production in an alpine grassland ecosystem. *Ecol. Complex.* **2014**, *17*, 79–86. [CrossRef]
17. Qian, D.W.; Du, Y.G.; Li, Q.; Guo, X.W.; Cao, G.M. Alpine grassland management based on ecosystem service relationships on the southern slopes of the Qilian Mountains, China. *J. Environ. Manag.* **2021**, *288*, 112447. [CrossRef]
18. Hua, T.; Zhao, W.W.; Cherubini, F.; Hu, X.P.; Pereira, P. Sensitivity and future exposure of ecosystem services to climate change on the Tibetan Plateau of China. *Landsc. Ecol.* **2021**, *36*, 3451–3471. [CrossRef]
19. Lin, Z.Y.; Xiao, Y.; Ouyang, Z.Y. Assessment of ecological importance of the Qinghai-Tibet Plateau based on ecosystem service flows. *J. Mt. Sci.* **2021**, *18*, 1725–1736. [CrossRef]
20. Liu, Y.X.; Lü, Y.H.; Jiang, W.; Zhao, M.Y. Mapping critical natural capital at a regional scale: Spatiotemporal variations and the effectiveness of priority conservation. *Environ. Res. Lett.* **2020**, *15*, 124025. [CrossRef]
21. Dennedy-Frank, P.J.; Muenich, R.; Chaubey, I.; Ziv, G. Comparing two tools for ecosystem service assessments regarding water resources decisions. *J. Environ. Manag.* **2016**, *177*, 331–340. [CrossRef]
22. Cong, W.C.; Sun, X.Y.; Guo, H.W.; Shan, R.F. Comparison of the SWAT and InVEST models to determine hydrological ecosystem service spatial patterns, priorities and trade-offs in a complex basin. *Ecol. Indic.* **2020**, *112*, 106089. [CrossRef]
23. Temmink, R.J.M.; Lamers, L.P.M.; Angelini, C.; Bouma, T.J.; Fritz, C.; van de Koppel, J.; Lexmond, R.; Rietkerk, M.; Silliman, B.R.; Joosten, H.; et al. Recovering wetland biogeomorphic feedbacks to restore the world's biotic carbon hotspots. *Science* **2022**, *376*, 6593. [CrossRef] [PubMed]
24. Yu, Z.C.; Joos, F.; Bauska, T.K.; Stocker, B.D.; Fischer, H.; Loisel, J.; Brovkin, V.; Hugelius, G.; Nehrbass-Ahles, C.; Kleinen, T.; et al. No support for carbon storage of >1000 GtC in northern peatlands. *Nat. Geosci.* **2021**, *14*, 465–467. [CrossRef]
25. Li, B.Q.; Yu, Z.B.; Liang, Z.M.; Song, K.C.; Li, H.X.; Wang, Y.; Zhang, W.J.; Acharya, K. Effects of Climate Variations and Human Activities on Runoff in the Zoige Alpine Wetland in the Eastern Edge of the Tibetan Plateau. *J. Hydrol. Eng.* **2014**, *19*, 1026–1035. [CrossRef]
26. Li, Z.W.; Wang, Z.Y.; Brierley, G.; Nicoll, T.; Pan, B.Z.; Li, Y.F. Shrinkage of the Ruoergai Swamp and changes to landscape connectivity, Qinghai-Tibet Plateau. *CATENA* **2015**, *126*, 155–163. [CrossRef]
27. Qin, G.H.; Li, H.X.; Zhou, Z.J.; Song, K.C.; Zhang, L. Hydrologic Variations and Stochastic Modeling of Runoff in Zoige Wetland in the Eastern Tibetan Plateau. *Adv. Meteorol.* **2015**, *2015*, 529354. [CrossRef]
28. Xi, Y.; Peng, S.S.; Ciais, P.; Chen, Y.H. Future impacts of climate change on inland Ramsar wetlands. *Nat. Clim. Chang.* **2021**, *11*, 45–51. [CrossRef]
29. Yuan, Y.; Zhang, L.; Cui, L.L. Spatiotemporal variations of water conservation capacity in Ruoergai Plateau. *Chin. J. Ecol.* **2020**, *39*, 2713–2723. (In Chinese)
30. Chen, H.; Wu, N.; Gao, Y.; Wang, Y.; Luo, P.; Tian, J. Spatial variations on methane emissions from Zoige alpine wetlands of Southwest China. *Sci. Total Environ.* **2009**, *407*, 1097–1104. [CrossRef]
31. Cui, Q.; Wang, X.; Li, C.H.; Cai, Y.P.; Liu, Q.; Li, R. Ecosystem service value analysis of CO_2 management based on land use change of Zoige alpine peat wetland, Tibetan Plateau. *Ecol. Eng.* **2015**, *76*, 158–165. [CrossRef]
32. Li, J.C.; Wang, W.L.; Hu, G.Y.; Wei, Z.H. Changes in ecosystem service values in Zoige Plateau, China. *Agric. Ecosyst. Environ.* **2010**, *139*, 766–770. [CrossRef]
33. Zhang, X.Y.; Lu, X.G. Multiple criteria evaluation of ecosystem services for the Ruoergai Plateau Marshes in southwest China. *Ecol. Econ.* **2010**, *69*, 1463–1470. [CrossRef]
34. Fu, B.J.; Wang, S.; Su, C.H.; Forsius, M. Linking ecosystem processes and ecosystem services. *Curr. Opin. Environ. Sustain.* **2013**, *5*, 4–10. [CrossRef]
35. Lü, Y.H.; Hu, J.; Fu, B.J.; Harris, P.; Wu, L.H.; Tong, X.L.; Bai, Y.F.; Comber, A.J. A framework for the regional critical zone classification: The case of the Chinese Loess Plateau. *Natl. Sci. Rev.* **2019**, *6*, 14–18. [CrossRef] [PubMed]
36. Xu, X.L.; Liu, W. The global distribution of Earth's critical zone and its controlling factors. *Geophys. Res. Lett.* **2017**, *44*, 3201–3208. [CrossRef]
37. Allen, R.G.; Pereira, L.S.; Raes, D.; Smith, M. *Crop Evapotranspiration: Guidelines for Computing Crop Water Requirements—FAO Irrigation and Drainage Paper 56*; Food and Agriculture Organization: Rome, Italy, 1998.
38. Wei, H.J.; Fan, W.G.; Ding, Z.Y.; Weng, B.Q.; Xing, K.X.; Wang, X.C.; Lu, N.C.; Ulgiati, S.; Dong, X.B. Ecosystem Services and Ecological Restoration in the Northern Shaanxi Loess Plateau, China, in Relation to Climate Fluc-tuation and Investments in Natural Capital. *Sustainability* **2017**, *9*, 199. [CrossRef]
39. Yin, L.C.; Tao, F.L.; Chen, Y.; Liu, F.S.; Hu, J. Improving terrestrial evapotranspiration estimation across China during 2000–2018 with machine learning methods. *J. Hydrol.* **2021**, *600*, 126538. [CrossRef]
40. Di, X.H.; Hou, X.Y.; Wang, Y.D.; Wu, L. Spatial-temporal characteristics of land use intensity of coastal zone in China during 2000–2010. *Chin. Geogr. Sci.* **2015**, *25*, 51–61. [CrossRef]
41. Qin, Y.; Yang, D.W.; Gao, B.; Wang, T.H.; Chen, J.S.; Chen, Y.; Wang, Y.H.; Zheng, G.H. Impacts of climate warming on the frozen ground and eco-hydrology in the Yellow River source region, China. *Sci. Total Environ.* **2017**, *605*, 830–841. [CrossRef]

42. Yu, X.X.; Zhou, B.; Lü, X.Z.; Yang, Z.G. Evaluation of water conservation function in mountain forest areas of Beijing based on InVEST Model. *Sci. Silvae. Sin.* **2012**, *48*, 1–5. (In Chinese)
43. He, L.; Li, C.Y.; He, Z.W.; Liu, X.; Qu, R. Evaluation and Validation of the Net Primary Productivity of the Zoigê Wetland Based on Grazing Coupled Remote Sensing Process Model. *IEEE. J. Sel. Top. Appl. Earth. Obs. Remote Sens.* **2022**, *15*, 440–447. [CrossRef]
44. Gao, C.; Huang, C.; Wang, J.B.; Li, Z. Modelling Dynamic Hydrological Connectivity in the Zoigê Area (China) Based on Multi-Temporal Surface Water Observation. *Remote Sens.* **2022**, *14*, 145. [CrossRef]
45. Feng, S.Y.; Li, W.L.; Xu, J.; Liang, T.G.; Ma, X.L.; Wang, W.Y.; Yu, H.Y. Land Use/Land Cover Mapping Based on GEE for the Monitoring of Changes in Ecosystem Types in the Upper Yellow River Basin over the Tibetan Plateau. *Remote Sens.* **2022**, *14*, 5361. [CrossRef]
46. Gong, P.; Wang, J.; Yu, L.; Zhao, Y.C.; Zhao, Y.Y.; Liang, L.; Niu, Z.G.; Huang, X.M.; Fu, H.H.; Liu, S.; et al. Finer resolution observation and monitoring of global land cover: First mapping results with Landsat TM and ETM+ data. *Int. J. Remote Sens.* **2012**, *34*, 2607–2654. [CrossRef]
47. Yan, W.C.; Wang, Y.Y.; Chaudhary, P.; Ju, P.J.; Zhu, Q.; Kang, X.M.; Chen, H.; He, Y.X. Effects of climate change and human activities on net primary production of wetlands on the Zoige Plateau from 1990 to 2015. *Glob. Ecol. Conserv.* **2022**, *35*, e02052. [CrossRef]
48. Zapata-Rios, X.; Brooks, P.D.; Troch, P.A.; McIntosh, J.; Rasmussen, C. Influence of climate variability on water partitioning and effective energy and mass transfer in a semi-arid critical zone. *Hydrol. Earth Syst. Sci. Discuss.* **2016**, *20*, 1103–1115. [CrossRef]
49. Li, Z.W.; Gao, P.; Lu, H.Y. Dynamic changes of groundwater storage and flows in a disturbed alpine peatland under variable climatic conditions. *J. Hydrol.* **2019**, *575*, 557–568. [CrossRef]
50. Chen, J.C.; Kuang, X.X.; Lancia, M.; Yao, Y.Y.; Zheng, C.M. Analysis of the groundwater flow system in a high-altitude headwater region under rapid climate warming: Lhasa River Basin, Tibetan Plateau. *J. Hydrol-Reg Stud.* **2021**, *36*, 100871. [CrossRef]
51. Wang, Y.F.; Ye, A.Z.; Qiao, F.; Li, Z.X.; Miao, C.Y.; Di, Z.H.; Gong, W. Review on connotation and estimation method of water conservation. *South-To-North Water Transf. Water Sci. Technol.* **2021**, *19*, 1041–1071. (In Chinese) [CrossRef]
52. Chen, X.L.; Sun, M.L.; Lü, Y.H.; Hu, J.; Yang, L.X.; Zhou, Q.P. Spatial-temporal Variability Characteristics and Its Driving Factors of Land Use in Zoige Plateau on the Eastern Edge of Qinghai-Tibetan Plateau, China. *J. Ecol. Rural. Environ.* **2022**, *38*, 11. Available online: http://www.ere.ac.cn/CN/10.19741/j.issn.1673-4831.2022.0578 (accessed on 15 October 2022). (In Chinese) [CrossRef]
53. Hu, J.; Chen, X.L.; Sun, M.L.; Zhou, Q.P. Research progresses of land-use change and its driving factors in the Zoige Plateau. *J. Southwest Minzu Univ.* **2022**, *48*, 355–361. (In Chinese) [CrossRef]
54. Hu, J.J.; Zhou, Q.P.; Cao, Q.H.; Hu, J. Effects of ecological restoration measures on vegetation and soil properties in semi-humid sandy land on the southeast Qinghai-Tibetan Plateau, China. *Glob. Ecol. Conserv.* **2022**, *33*, e02000. [CrossRef]
55. Hou, P.; Zhai, J.; Jin, D.D.; Zhou, Y.; Chen, Y.; Gao, H.F. Assessment of Changes in Key Ecosystem Factors and Water Conservation with Remote Sensing in the Zoige. *Diversity* **2022**, *14*, 552. [CrossRef]
56. Li, X.; Zhang, L.; Zheng, Y.; Yang, D.W.; Wu, F.; Tian, Y.; Han, F.; Gao, B.; Li, H.Y.; Zhang, Y.L.; et al. Novel hybrid coupling of ecohydrology and socioeconomy at river basin scale: A watershed system model for the Heihe River basin. *Environ. Model. Softw.* **2021**, *141*, 105058. [CrossRef]
57. Che, T.; Li, X.; Liu, S.M.; Li, H.Y.; Xu, Z.W.; Tan, J.L.; Zhang, Y.; Ren, Z.G.; Xiao, L.; Deng, J.; et al. Integrated hydrometeorological, snow and frozen-ground observations in the alpine region of the Heihe River Basin, China. *Earth Syst. Sci. Data.* **2019**, *11*, 1483–1499. [CrossRef]
58. Sharp, R.; Douglass, J.; Wolny, S.; Arkema, K.; Bernhardt, J.; Bierbower, W.; Chaumont, N.; Denu, D.; Fisher, D.; Glowinski, K.; et al. InVEST 3.9.0.post0+ug.gbbfa26d.d20201215 User's Guide; The Natural Capital Project, Stanford University, University of Minnesota, The Nature Conservancy, World Wildlife Fund. 2020. Available online: https://invest-userguide.readthedocs.io/_/downloads/en/3.9.0/pdf/ (accessed on 8 October 2022).

remote sensing

MDPI

Article

An Integrated Method for River Water Level Recognition from Surveillance Images Using Convolution Neural Networks

Chen Chen [1,*], Rufei Fu [1], Xiaojian Ai [1], Chengbin Huang [2], Li Cong [2], Xiaohuan Li [3], Jiange Jiang [1] and Qingqi Pei [1]

1 School of Telecommunication Engineering, Xidian University, Xi'an 710071, China
2 State Grid Jilin Province Electric Power Company Limited Information Communication Company, Changchun 130021, China
3 School of Electronics and Information Engineering, Beihang University, Beijing 100191, China
* Correspondence: cc2000@mail.xidian.edu.cn

Abstract: Water conservancy personnel usually need to know the water level by water gauge images in real-time and with an expected accuracy. However, accurately recognizing the water level from water gauge images is still a complex problem. This article proposes a composite method applied in the Wuyuan City, Jiangxi Province, in China. This method can detect water gauge areas and number areas from complex and changeable scenes, accurately detect the water level line from various water gauges, and finally, obtain the accurate water level value. Firstly, FCOS is improved by fusing a contextual adjustment module to meet the requirements of edge computing and ensure considerable detection accuracy. Secondly, to deal with scenes with indistinct water level features, we also apply the contextual adjustment module for Deeplabv3+ to segment the water gauge area above the water surface. Then, the area can be used to obtain the position of the water level line. Finally, the results of the previous two steps are combined to calculate the water level value. Detailed experiments prove that this method solves the problem of water level recognition in complex hydrological scenes. Furthermore, the recognition error of the water level by this method is less than 1 cm, proving it is capable of being applied in real river scenes.

Keywords: water level recognition; hydrological monitoring; deep learning; computer vision

Citation: Chen, C.; Fu, R.; Ai, X.; Huang, C.; Cong, L.; Li, X.; Jiang, J.; Pei, Q. An Integrated Method for River Water Level Recognition from Surveillance Images Using Convolution Neural Networks. *Remote Sens.* **2022**, *14*, 6023. https://doi.org/10.3390/rs14236023

Academic Editors: Jingzhe Wang, Zhongwen Hu, Yangyi Wu and Jie Zhang

Received: 20 October 2022
Accepted: 21 November 2022
Published: 28 November 2022

1. Introduction

Hydrological monitoring is a very important research area for many countries, especially where there are many river regions. Rivers play a critical role in human life and are also the source of floods [1]. Floods occur more frequently during the rainy season, causing substantial economic losses and disaster-induced diseases. Consequently, it is urgent and necessary to cope with flood disasters quickly. River level detection is the fundamental task of flood monitoring and needs to be accurate and fast [2].

Detecting river levels is not a trivial task [3]. With the development of surveillance cameras and modern 6G communication technologies [4], more and more hydrological stations use surveillance cameras to track water levels. Various measurement sensors are also used for automatic monitoring of water level; these are usually divided into contact or non-contact types according to their measurement methods [5]. Methods based on physical equipment include float gauge devices, pressure gauge devices, ultrasonic gauge devices, laser gauge devices, etc. [6]. However, these methods have the problem of complicated installation, great influence by environmental factors, difficult maintenance, or huge cost. In general, the most widely used measurement method is to observe a water gauge, which is an iron sheet approximately 1 meter high and 15 cm wide [7]. A water gauge is always installed in a suitable place near the monitored river, but reading of water gauges still needs to be completed manually. In remote work, water conservancy personnel read the images taken by the camera or remotely monitor the gauge [8]. In this way, water level measuring has been recognized as an important detection, classification, and tracking process. To reach closer to the real water level value, the accurate water level line position also needs to be detected. Furthermore, the existence of multiple water gauges and composite water gauges leads to negative effects on the performance of the measurement system.

For the purpose of avoiding laborious manual observation and subjective error, it is necessary to apply computer vision technology. The traditional water level detection method has three steps. The first is to locate the area containing the water gauge in the image. A typical method is to use edge segmentation to get an easier image for edge detection then get an appropriate area containing the water gauge [9]. This method is easy to implement, but the application scenarios are limited. It can only be used in scenes where the gauge is very simple to distinguish from the environment. The other locating method is to use the combination of HOG and SVM [10,11]. HOG is a classical feature extraction method, and SVM is a classical feature classification method. Usually, the sliding window method is used to generate an imprecise region as a candidate region, then, HOG is used to extract the features of the candidates, and finally, SVM is used to judge whether the region contains a water gauge. Although this combined method is partially effective, the biggest disadvantage of the method is that the feature extractor is artificially designed. Although the production standard of the water gauge is fixed, the background of the water gauge is variable. Environmental influence can also lead to situations such as defacement and obscuration of the water gauge, resulting in a human-designed feature extractor not characterizing the water gauge features well. After locating the water gauge area, the next step is to determine the number on the gauge that lies above the water surface. The area of single detection is reduced to only the area of the water gauge instead of the whole image. Water gauge numbers can be recognized using a combination of HOG and SVM methods, but due to the special features of printed numbers, optical characters are often used for identification. OCR is effective in recognizing purely digital areas of the water gauge [12], but recognition performance is often affected by many factors, such as digital distortion, water gauge defacement, and low clarity. In real hydrological scenarios, there are often many disturbances that cause the water gauge to be in a non-standard format [13], which can make the identification task very difficult. The final step is detecting the water level line. Since the water level line of the water gauge is not a straight-line feature in the conventional sense, similar shoreline detection and other types of straight-line detection methods cannot solve the problem. There is no specific solution for the water level line of the gauge scale in the current literature. However, determination of the water level line almost determines the accuracy of the water level value, so solving this problem is crucial.

Object detection is the first and most essential step in the field of computer vision [14–16] and brings a leap to many other fields, such as autonomous driving in smart transportation, intelligent security, and remote sensing [17–23]. The combined approach of HOG and SVM has poor results in object detection due to the shortcomings of the feature extractor. Deep learning has changed the situation, especially convolutional neural networks (CNNs) [24]. CNNs have exacting feature extraction capabilities and do not require human-designed feature extractors [25]. In the ImageNet 2010 challenge [26], Alex's deep convolutional neural network AlexNet achieved first place and was 10 percent ahead of the second place method in the top-five rating [27]. The superior results of AlexNet demonstrate the superiority of CNN for computer vision tasks. Since then, more and more researchers have applied CNN to target detection, semantic segmentation, and pose detection, as well as other vision tasks [28]. In the target detection task, CNN has been optimized and adjusted. YOLO [29], R-CNN [30], and SSD [31] are effective models that continuously improve the accuracy of object detection. A key step in water level recognition is the detection of the gauge and the printed numbers on the guage [32]. Therefore, it is crucial to apply such target detection models to help solve the shortcomings of traditional water level recognition methods. Another key step in the work is how to find the water level line. Conventional linear-type detection methods cannot recognize water level lines with good performance in some complex river situations [33]. The actual water gauge image has the feature that the water level line represents the split line of the water surface and the rest of the environment. Therefore, if it is possible to divide the part above the water surface, then the lower boundary of the segmented area is the water level line. Segmenting a specific region from an image belongs to the task of semantic segmentation. Similarly, CNN also has an ideal positive effect on segmentation jobs. Since the fully convolutional network (FCN) was proposed by Jonathan Long for semantic segmentation [34], CNNs have gradually become the mainstream approach. Therefore, using a semantic segmentation model to obtain the portion of the gauge above the water and calculating the water level line from the segmented polygon area is a feasible solution.

The challenge of water level recognition from water gauge images has two parts: gauge area detection and water level line detection. Existing methods focus on water gauge area detection in a single or fixed scene. They use some prior information to assist in locating the water gauge area. Therefore, they are not suitable for complex scenarios. To solve these problems efficiently, this paper introduces convolution neural networks to detect the water gauge and the printed numbers. Then, since this water level line is different from the general Riparian lines, traditional methods cannot solve this problem [35]. The semantic segmentation model is applied to segment the water gauge above the water surface to obtain the exact water level line.

Here is a summary of our work to solve the problems above. First, the fully convolutional one-stage (FCOS) object detection model was employed [36]. To efficiently detect small objects with a smaller model, FCOS was improved by fusing the context fusion model to determine the area of the gauge and the numbers above the water surface. After getting these, the rough water level value can be determined. Then, the semantic segmentation model named DeepLabv3+ was applied to segment the water gauge above the water surface [37]. After that, the water level can be determined from the result of segmentation. The above-water part and the underwater part of the water level gauge have similar image features. River water with different levels of clarity will make the underwater gauge look different. However, the clearer the water is, the more the water gauge underwater looks the same as the water gauge above the water. This phenomenon causes the model to have difficultly distinguishing between the two parts of the water gauge. We find an innovative way to solve this problem by proposing a contextual semantic fusion module for the DeepLab model. The main contributions of this paper include the following:

- In order to measure the water level in water gauge images in complex scenes, this article proposes a composite method that can accurately obtain the water level.
- In order to get the position of the water level line, this paper proposes an innovative module that divides features into different levels. This module first obtains high-level segmentation results and then gradually fuses them downward.
- Water gauge images of actual scenes and seven special scenes are used to evaluate the method proposed in this article.

2. Related Work

2.1. Physical Equipment for Water Level Recognition

Around the world, disasters caused by floods cause huge losses every year [38–40]. Detecting rising river levels is essential for flood warnings. Different physical sensor devices have been designed to apply to different environments to solve the problem by recognizing the water level. These automated water level detection devices are categorized into two types according to the measurement method: contact and non-contact. Contact devices use sensors to convert water level information into actual water level values by setting up auxiliary equipment in the water or on the shore. Non-contact devices, on the other hand, do not require fixed facilities and are usually hand-held devices that can be easily moved to multiple locations for testing. Contact-based water level measurement equipment mainly includes float-type water level meters and pressure-type water level meters, and non-contact water level measurement equipment mainly includes ultrasonic water level meters, radar-type water level meters, and laser-type water level meters. Float-type water level meters usually use floats to sense the change to the water level up or down and records the transmission record directly by mechanical means. The entirety of the equipment usually consists of floats, balance hammers, and suspension ropes, which need to be used in conjunction with water level wells. Although float-type water level meters have high measurement accuracy and a large measurement range, they have high equipment installation and maintenance difficulty and are susceptible to floods. Pressure-type water level meters [41] use underwater measurement points as the water depth and water pressure reference points; changes to the water surface height bring changes to the pressure value. According to the relationship equation between underwater pressure and water depth, the change of water surface height is calculated. Pressure-type water level meters need to be used in calm water bodies and are less stable in the field, and it is difficult to guarantee measurement accuracy. Ultrasonic-type water level meters [42] use the principle of ultrasonic reflection to measure the water level; the sensor emits ultrasonic signals to the surface of the river, the ultrasonic waves encounter the water surface and reflect to the receiving sensor, and the receiver calculates the distance by propagation time, thus calculating the water level. Ultrasonic water level meters have better performance in terms measurement accuracy, and the installation is easier in complex environments, but they are vulnerable to environmental impact. A radar-type water level meter is a special kind of water level measurement equipment; it is not affected by weather, environment, installation conditions, or other factors. The measurement principle is to use the reflection of electromagnetic waves to send a radar pulse from the radar antenna sensor to the water surface. After the pulse is reflected from the water surface, the antenna receives the reflected signal and records the time, processes the received pulse signal, and finally calculates the river level. Laser-type water level meter is a kind of water level measuring instrument using laser distance measurement. It uses the advantage of light beam propagation; the transmitter emits a high-speed laser pulse, then, the laser pulse meets the water surface and reflects, and the laser receiver receives and calculates the propagation time, thus calculating the water level height. The characteristics of these methods are shown in Table 1.

Table 1. Characteristics of different water level detection methods.

Water Level Type	Advantage	Disadvantage
Float-type	High measurement accuracy and large measurement range	Installation difficulty and bad flood performance
Pressure-type	Easy installation	Can only be used in calm water bodies
Ultrasonic-type	Easy installation, good performance in complex environments	Accuracy is impacted by environment
Radar-type	Good performance in complex environments	High cost
Laser-type	High accuracy and stability	High cost and installation difficulty

2.2. Image-Based Water Level Recognition

In recent years, most horological stations have been equipped with monitoring systems, especially video surveillance systems connected with networks [43]. More and more automatic water line recognition methods and measurements have been proposed to deal with flood-related disasters [44]. Image processing is an essential part of image-based methods in detecting water level and almost completely determines the performance of the detecting system. There are mainly two kinds of methods in image processing, which are as follows.

The first approach mimics the human vision mechanism, in which the water level is first measured by positioning the water gauge and then identifying the numbers [45,46]. Bruinink improved the segmentation method of the water gauge using a two-class random forest classifier based on a feature vector of textons [47]. Then, a Gaussian mixture model segmentation is applied to the gauge bar and numbers for reading the water gauge. However, the algorithm is relatively sensitive to the environment where the gauge is located. If the gauge itself is dirty, damaged, or lacking light, and the water surface is polluted, the performance of the algorithm is greatly affected.

The idea of the second method is from machine vision. Like the above method, the position of the water level meter is first determined. The difference is that this method converts the coordinate relationship of the water meter into a pixel histogram relation and then determines the water level [48–50]. In the recognition of two-color water gauges, the horizontal projection method is better and more popular. The projection method can be used for horizontal projection according to grayscale images [51], binary images [52], edge images [53], etc. In this method, the water level is determined by looking for points in the horizontally projected curve where the change is steep. Further, the environment can cause some noise in the desired curve. For example, refraction of the water surface affects the distribution of gray values, thereby affecting the horizontal projection curve, and the final measurement result will be discounted.

2.3. Object Detection and Semantic Segmentation

Object detection is a very basic but essential task in computer vision work, and there is a research history of nearly two decades in the academic field. Traditional object detection methods, such as HOG and DPM, rely heavily on feature extractors designed by human experts. Moreover, there is no definite paradigm to design feature extractors, which leads to the failure of traditional methods to achieve excellent results. With the development of computers, deep learning has opened up a broader path for object detection, and more and more studies have begun to explore neural networks to achieve better detection results. There are mainly two kinds of target detection algorithms: anchor-based and anchor-free. The difference is the anchor's function. In an anchor-based algorithm, such as SSD and YOLO, an anchor is used to extract candidate target frames, and preset anchors are used to obtain candidate areas and to perform predictions on these proposals. However, designing a suitable ratio is a difficult task that requires strong prior knowledge. Anchor-based methods also generate lots of redundant candidate areas, resulting in the detection effect of positive and negative samples being very different. Anchor-free methods, such as CornerNet and FCOS [54], get rid of the restriction of anchors and directly predict key points for detection and classification.

Semantic segmentation is a concept similar to target detection, but it is relatively complex. The task of image classification is to classify an image into a certain category, and semantic segmentation is a further classification for each pixel: pixels are classified into different classes based on certain rules [55]. The FCN model applies end-to-end full convolutional networks to semantic segmentation. The deconvolution layer of FCN performs upsampling interpolation operations by learning, instead of simple bilinear interpolation. The encoder–decoder structure of the Unet model effectively improves the effectiveness of training with a small number of data [56,57]. In terms of feature fusion, Unet connects the semantic information at the level of the macro information of the network with the fine-grained features at the level with more detailed information in the channel dimension. Fisher proposed dilated convolution [58]. Dilated convolution increases the corresponding perceptual field size without reducing the spatial dimensionality, which facilitates the network to obtain multi-scale contextual features. DeepLab [59] proposes atrous spatial pyramidal pooling in spatial dimensions to enhance the segmentation of multi-scale targets.

3. Methodology and Raw Data

3.1. Key Steps for Water Level Recognition

The most critical step of water level recognition based on a gauge image is focused on water gauge processing. When using the human eye observation method, it is necessary to first find the area where the water gauge exists in the image. After roughly locating the water gauge area, the individual must carefully observe the scale on the water gauge that still remains above the water surface. This is a critical step when the water is particularly clear. Then, the water level is calculated using the numbers, gauge, and position of the level. The whole process of human eye observation can be summarized in three steps: water scale area and gauge number area detection, water level line detection, and water level value calculation. The method in this paper takes a technique from deep learning to recognize the water level and also follows the above process. The first step is to detect the water gauge and the gauge numbers. This is a target detection task, and this article uses the model of target detection to detect the key targets in the water gauge images, including the numbers on the water gauges and the gauge bodies. Since there are multiple water gauges in the actual hydrological scenario, each water gauge needs to be distinguished. The water level value is calculated from the numbers present on the gauge, so the numbers on the gauge body that are above the water surface need to be detected. The second step is to perform water level line segmentation. This article uses the semantic segmentation method of deep learning to segment the part of the water gauge image that lies above the water surface, and the lower boundary of this part is the actual water level line, which needs to be extracted from the segmented region.

3.2. Water Gauge and Gauge Number Detection

The water gauge detection network in this article is modified from the FCOS network. FCOS is an excellent one-stage network that has both detection speed and detection accuracy. FCOS is a pixel-by-pixel FCN-based object detection model. This network contains three parts of backbone, feature pyramid network, and heads, which are applied for classification and regression. The model is mainly deployed with the idea of edge computing [60,61]. Edge computing is a proposed solution to solve the hydrological monitoring of a river area. In an edge computing network, many facilities involving data processing and storage can be placed closer to the data source. In this way, the preprocessing of data can ensure the real-time requirements of the application, and the security of information can also be better protected. For edge deployment, the network needs to be as small as possible while ensuring accuracy. FCOS needs improvement to detect small objects. Based on FCOS, a convolutional network for water gauge detection was proposed with the basic network ResNet-54. FCOS uses five feature maps of different scales for further detection. Since the task is single-target detection, these feature maps of different scales are not necessary, so we kept just three feature maps in this module. The heads for regression and classification are also reduced to three. After adopting the above modifications, the complexity of the model and the monitoring accuracy are taken into account at the same time. The effect of multi-scale fusion was improved based on feature pyramids [62]. The information contained in the feature maps of different levels is different. The high-level features more easily give people an intuitive feeling and contain relatively macro information, while the lower levels hide more details. Level features and underlying level feature fusion are the core of the model [63–65]. For feature maps of different levels, minimization of the loss of the original semantic information in the fusion is the focus of research. A context fusion module is used to solve this problem. This module is dedicated to discovering correlations between contexts. Contextual adjustment produces dense pixel-level contextual information while improving the efficiency of feature encoding in long connection paths. The structure diagram of the improved FCOS model is shown in Figure 1. Yellow squares indicate the CA module: each CA has a high-level feature map input, a low-level feature map input, and outputs a feature map fused with contextual information.

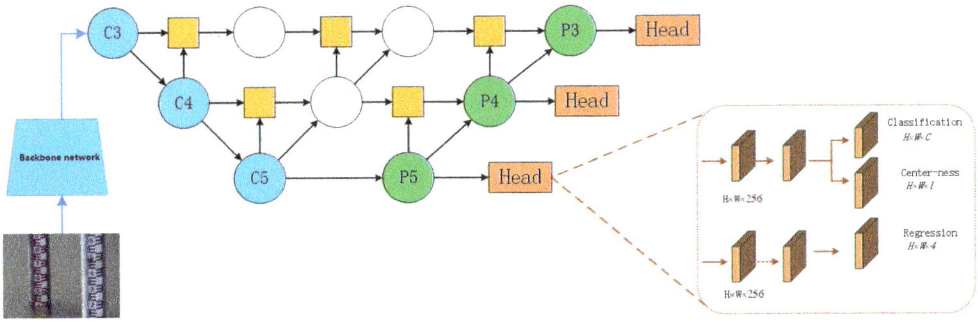

Figure 1. Structure of improved FCOS model.

3.3. Water Gauge Area Segmentation

Semantic segmentation based on CNN has reached a very impressive level [66–68]. From FPN to DeepLab series models [59,69–71], the performance of semantic segmentation is continuously improving. DeepLabv3+ employs an atrous pyramid pooling and encoder–decoder architecture to encode multi-level contextual information by processing incoming features through void convolution with multiple expansion coefficients and sensory fields for pooling operations, while the decoder network can get more informative object edges by reverting to spatial information step-by-step. DeepLabv3+ achieves state-of-the-art results on the VOC2012 dataset. However, this model still has some limitations that lead to the inability to apply the model directly to the water gauge level recognition task. The model for water level recognition needs to be run on an edge device rather than a computer in the future, so there is a limit to the size of the model [72]. Thus, ResNet-54 is the backbone network for DeepLab [73]. The reduction to the backbone network leads to a reduction in performance. To alleviate this problem, taking into consideration the segment accuracy of whether the water gauge is under the water or not, the method of feature mixing was improved by employing the contextual fusion module, which is shown in Figure 2. As a comparison, the improved DeepLab is shown in Figure 3.

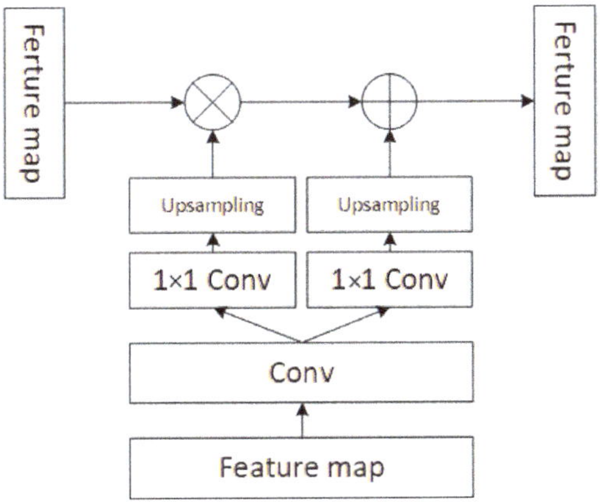

Figure 2. Contextual adjustment module.

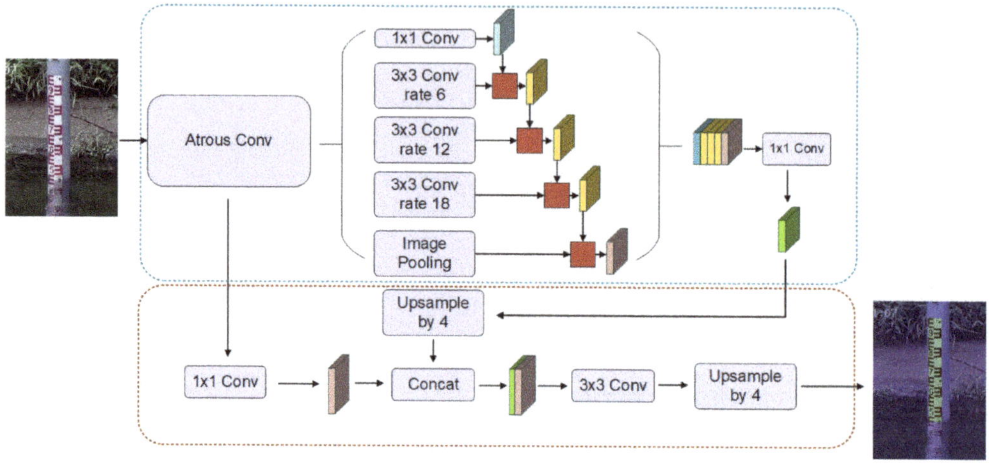

Figure 3. Structure of improved DeepLab model.

3.4. Water Level Recognition

From the above analysis of water gauge recognition, we identify the problem that must be solved in the practical application and support for our proposed CA-GAN model. In this section, the CA-GAN model is created to remove the above obstacle.

3.4.1. Water Level Line Extraction

The result of DeepLab segmentation is a polygonal area. The water level line needs to be calculated from its polygon. Firstly, this method calculates the maximum enclosing rectangle of the area. For convenience, only the vertical rectangular box is calculated. The key of the rectangular box is to calculate two coordinates: the upper left point and the lower right point. Since the segmentation result image is a black-and-white image with only two pixel values, 0 and 255, the "scan line" method can be used to get the coordinates. The water level is at the bottom of the rectangular box. The water level line extraction schematic is shown in Figure 4.

Figure 4. Water level line detection schematic.

3.4.2. Water Level Measurement

The result of FCOS object detection shows the pixel coordinates of each number, and the distance between each number can be calculated. The segmentation result of the water level line shows its pixel height, and it is then used to calculate its distance to the nearest number. The detection results of the FCOS model are rectangular boxes; ideally, a rectangular box can be given for each figure. Due to the complex background of the actual water gauges, the detection accuracy of the FCOS model cannot reach 100 percent, resulting in the existence of missed or false detection. It is necessary to use the distribution of numbers on the gauge as prior information. The numbers 0 to 9 are evenly spaced from bottom to top on the water gauge, and the detection results are processed to remove the wrong detection results.

A detection rectangle represents a region of digits, and the coordinates of the upper left point and lower right point of the rectangle are known information. The coordinates of the center of the digit rectangle are denoted as h_x, where x denotes the digit number, while the distance between two adjacent detection digits is denoted using d_{xy}. Since each digit is equally spaced, the actual distance $\bar{d}$ between two adjacent digits can be expressed by Equation (1).

$$\bar{d} = 1/n \sum d_{xy} \tag{1}$$

The value of n is determined by the actual numbering distance between the digits. If the model detects only two numbers, "8" and "2", the value of n is 6. The actual height of any one number, i.e., the physical distance, is 5 cm, and the actual physical distance between two adjacent numbers is 10 cm, i.e., the actual physical distance per unit pixel height. The actual physical distance per unit pixel height is expressed by (2).

$$d = \bar{d}/10 \tag{2}$$

The height of the center coordinate of the detected minimum number x is h_{min}, and the height of the detected level line is h_{line}; then, the actual water level value can be calculated by Equation (3).

$$f(x) = \begin{cases} x - (h_{line} - h_{min})/\bar{d}, & h_{line} \geq h_{min} \\ x + (h_{min} - h_{line})/\bar{d}, & h_{line} < h_{min} \end{cases} \tag{3}$$

The method's flowchart is shown in Figure 5: the green part shows the method's innovation.

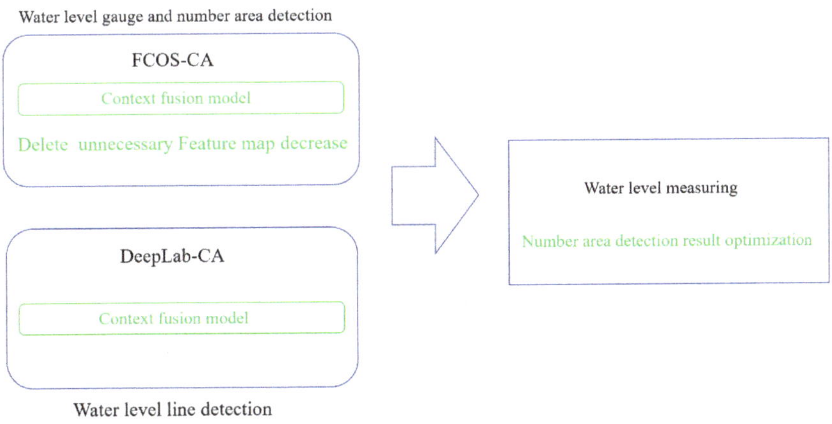

Figure 5. Flowcharts of the methods.

3.5. Dataset

At present, there are no effective relevant datasets, so the source data of the dataset were obtained by contacting the Wuyuan City Hydrological Bureau. A total of 600 basic hydrological images were collected. The source data covers 20 sites, including different environments such as water pollution, low light, etc., to meet our complex environmental needs. In addition, some pictures in the dataset that were similar to the actual shooting are from the Internet. Data augmentation was performed on the source data for a richer dataset and included annotating dials and numerals with 12 classes. Finally, the images were processed to a size suitable for the model. There are two sizes: 5.6 kb, and 0.6 kb. Samples from the dataset are shown in Figure 6.

Figure 6. Samples of the gauge dataset.

4. Experiments and Results

4.1. Evaluation Metrics

Different evaluation metrics are used for the three tasks described in this paper. For the water gauge and gauge number detection task, which is essentially an object detection problem, the more commonly used metrics were chosen, including precision, recall, and mean average accuracy. For the water gauge region segmentation task, the evaluation metrics included pixel accuracy and mean cross-merge ratio. For the water level recognition task, the metrics included relative error and absolute error.

- Precision and Recall. These two indicators consist of four base indicators, namely *TP*, *TN*, *FP*, and *FN*. '*T*' means true, '*F*' means false, and the second character means the predicted result: '*P*' and '*N*', respectively, are positive and negative. For example, *TP* represents a positive sample predicted as a positive sample. Precision and recall are calculated as in the equations below. Meanwhile, in order to consider the evaluation of these two metrics together, these two metrics can be used in order to draw a PR curve. The vertical coordinate is the accuracy of detection, and the horizontal coordinate is the recall; then, the area enclosed by the PR curve and the coordinate axis can be used as a new measurement. For a single target, this metric is called the average accuracy. For multiple targets, the average of the AP of each category is represented as mAP.

$$Precision = \frac{TP}{TP + FP} \tag{4}$$

$$Recall = \frac{TP}{TP + FN} \tag{5}$$

- Per-pixel acc. is used to indicate the accuracy of the prediction, expressed by the ratio of the count of pixels segmented correctly to all the pixels counted. For different pixel types (represented by *i*), TP_i means the count of accurate predictions of *i*-type pixels, and FP_i means the i-type pixels predicted as categories. It is easy to obtain the expression of the overall accuracy as follows:

$$per - pixelAcc = \frac{\sum_{i=0}^{n} TP_i}{\sum_{i=0}^{n}(TP_i + FP_i)} \tag{6}$$

4.2. Experiment and Analysis

4.2.1. Water Gauge Detection Experiment

The experiment used the homemade Water Gauge Dataset to train improved FCOS and test its performance; we then selected the SSD and YOLOv3 target detection models for comparison. The results of the models are shown in Table 2. The improved model in this paper is represented as FCOS-CA.

Table 2. Detection score for different methods.

Model	Precision (%)	Recall (%)	mAP (%)
SSD	77%	72%	75%
YOLOv3	78%	74%	77%
FCOS	91%	85%	87%
FCOS-CA	93%	86%	89%

From the results in Table 1, it can be seen that the FCOS-CA model is 16%, 15%, and 2% higher than SSD, YOLOv3, and FCOS, respectively, in terms of precision. For recall, compared with SSD, YOLOv3, and FCOS, FCOS-CA is increased by 14%, 12%, and 1%, respectively. Compared with SSD, YOLOv3, and FCOS, FCOS-CA increased by 14%, 12%, and 2%, respectively, in terms of mAP. The good results of FCOS-CA in the three indicators show that the model in this paper can be competent for the task of water gauge detection.

At the same time, we also tested seven difficult scenes in water level value recognition of water gauges: reflection, wind and waves, backlight, water transparency, night fill light, dirty, and sun shadow. The test results are shown in Figures 7–10.

Figure 7. Inverted water gauge image detection results.

Figure 8. Wind and wave and backlight water gauge image detection results.

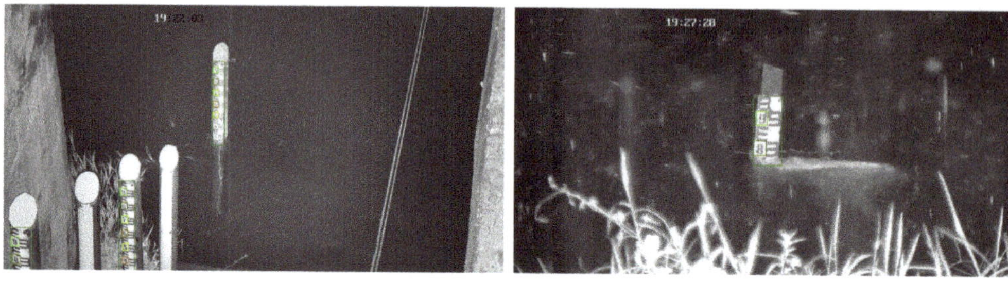

Figure 9. Nighttime fill light water gauge image detection results.

Figure 10. Dirty water gauge image inspection results.

4.2.2. Water Gauge Segmentation Experiment

In this section, the self-made Water Gauge Dataset is used to train the improved DeepLabv3+ semantic segmentation model (represented as DeepLab-CA), and then its performance is tested. The FCN and Unet segmentation models are also selected as the comparison models. The test results of the models above are listed in Table 3. In the experimental results of water gauge region segmentation, the segmentation results are represented as light green regions and are superimposed on top of the original image while reducing the brightness of the original water gauge image to obtain more considerable visualization test results. The visualization segmentation results of DeepLab-CA on the actual water gauge image are shown in Figure 11.

Table 3. Segmentation score for different methods.

Model	Pixel Acc (%)	mIOU (%)	Inference Time (s)
FCN	72%	75%	0.23
Unet++	85%	78%	0.15
DeepLabv3+	91%	82%	0.13
DeepLab-CA	93%	85%	0.17

It can be seen from the table above, in terms of processing time, although the model in this paper is slightly slower (0.02 s and 0.04 s slower than UNET and DeepLabv3+, respectively), it is improved to varying degrees in pixel ACC and mIOU. Specifically, pixel ACC was increased by 21%, 8%, and 2%, respectively, and mIOU was increased by 10%, 7%, and 3%, respectively. The good results of DeepLab-CA on the three indicators show that the model in this paper can be competent for the task of water gauge segmentation.

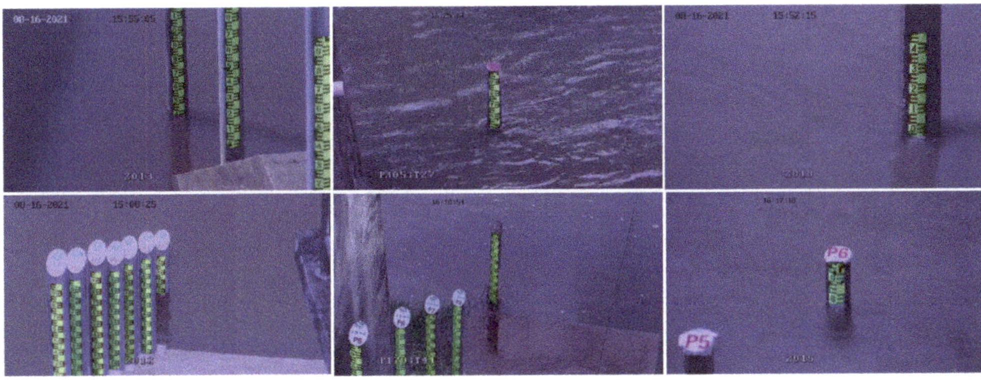

Figure 11. Water gauge image segmentation results.

From the segmentation results in Figure 11, the model in this paper shows excellent performance by accurately segmenting the part above the water surface while ignoring the submerged part of the water gauge. The segmentation effect of the model meets the needs of water level value recognition and can be applied to water gauge water level line segmentation. At the same time, this article also tested seven difficult scenes in water level value recognition of water gauges: reflection, wind and waves, backlight, water body transparency, night fill light, dirty, and sun shadow. Some segmentation results are shown in Figures 12 and 13.

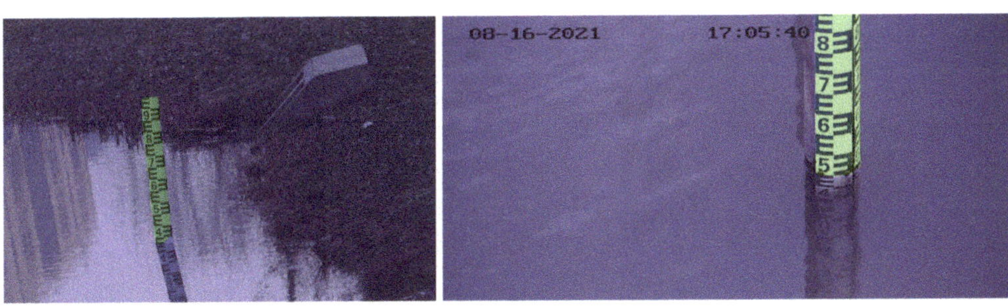

Figure 12. Reflection water gauge image segmentation results.

Figure 13. Wind–wave and backlight segmentation results.

For the tilted water gauge in the left image above, the model in this paper accurately distinguishes the two parts of the water gauge area above and under the water. In Experiment 1, the detection model of the water gauge and its body detects a number "4" on the image, which is a false detection result, and the region segmentation result of the image can be used as a mask to remove the false detection of "4". In the vertical water gauge image on the right, although it is visually difficult to distinguish whether the submerged water gauge area around the number "4" is above or under the water, the model still accurately segments the above-water portion. It can be seen that the water gauge region segmentation model in this paper is capable of segmenting the water gauge region in the reflection scene.

In terms of the segmentation index of the water gauge region, the model in this paper achieves 92 percent pixel segmentation accuracy. From the segmentation results shown above, the model in this paper not only has a significant segmentation effect in the actual complex scenes but also achieves a significant segmentation effect for the seven special scenes in the actual scenes.

4.2.3. Water Level Measurement Experiment

To verify the effectiveness of recognition in this model, the evaluation index used in this paper is the absolute difference value from the vertical height of the water level line observed by human eyes compared to that identified by the algorithm. The height here is the number on the gauge body. The schematic diagram of the water gauge reading is shown in Figure 14. The physical distance between each letter E and the flip E is 5 cm, and the physical width of each "cross" of the letter E is 1 cm. The center of each number corresponds to the position of a "cross" center in the middle of the letter E. There are 10 "crosses" between the centers of two adjacent numbers. The reading of the water level value in the chart is 5.0 or 50.0 cm, without units, which means the distance of the mark relative to the bottom of the gauge, and with units, which means the physical distance relative to the bottom of the gauge. The actual hydrological scenario of the water level value also needs to be calculated using the elevation information of the water gauge installation location, and only the relative marking distance of the gauge is used as the water level value for discussion in this paper. The water level value in the graph is 5.0.

Figure 14. Water level on water gauge reading diagram.

First of all, we selected the test set of the water gauge dataset in this experiment of water level value identification; the results are listed in Table 4, the identification results are divided into several ranges according to the absolute error, and the proportion of the sample results of each range to all samples is counted. From Table 4, it can be seen that the proportion of test samples with a difference less than 0.5 cm is 35 percent, the proportion of test samples with a difference between 0.5 cm and 1 cm is 28 percent, the proportion of test samples with errors between 1 cm and 2 cm is 27 percent, and the proportion of test samples with errors greater than 2 cm is only 10 percent. This statistical result shows that the water level identification algorithm in this paper achieves better results.

Table 4. Statistical results of water level value recognition of water gauge test dataset.

Error Range	X < 0.5	0.5 < X < 1	1 < X < 2	X > 2
Sample proportion	35%	28%	27%	10%

Then, six groups of locations with different degrees of background disturbance in different artificial lakes on campus were selected for testing; the test results are listed in Table 5. The manual readings were taken in seven places with human-eye observation readings, which were confirmed by water conservancy personnel to be the standard water level values. For each set of monitoring point water gauge images of the seven groups of water level value data, the minimum one and the maximum one were removed to reduce the error introduced by individual deviation, and the average was taken as the final water value of the gauge image. The experimental value is the water level value identified by the algorithm in this paper.

Table 5. Water level measurement results of actual monitoring points.

Error Range	A	B	C	D	E	F	H
Manual recognition (cm)	11.80	30.50	35.00	36.90	65.30	78.80	58.90
Algorithm recognition (cm)	11.21	31.77	34.53	38.53	65.87	78.58	58.38
Error (cm)	0.61	1.27	0.47	1.63	0.57	0.22	0.52

From the experiments at the actual monitoring points of the artificial campus lakes, it can be seen that the recognition accuracy of the model has excellent performance just with a tiny difference from the results of manual observation, and the absolute error is kept within 1 cm, which meets the requirements of the observation level standard. At the same time, this paper measured the recognition of the level of water gauges in seven special complex scenarios. The measurement results are shown in Table 6.

Table 6. Water level recognition results in 7 special scenarios.

Error Range	Manual (cm)	Algorithm (cm)	Error (cm)
Reversed reflection	34.50	34.46	0.04
Backlighting	47.00	46.73	0.27
Nighttime fill light	32.50	32.07	0.43
Wind and waves	30.50	29.80	0.70
Soiling	27.00	26.00	1.00
Water transparency	77.00	75.90	1.11
Sun shadow	5.00	4.70	0.30

5. Discussion

Water gauge detection and water level segmentation have been fully tested and compared with some classic models, such as YOLOv3, as well as the basic models used for this method, FCOS, and DeepLabv3+. The method proposed in this paper achieved better performance. Most images of the dataset come from Wuyuan City, Jiangxi Province, which consists of low mountain and hilly areas and is suitable for many natural river scenes. It also shows good results in local artificial lakes and has the ability to be applied to actual scenes. For this point, we should thanks to the National Key Research and Development Program of China (2020YFB1807500) and the National Natural Science Foundation of China(62072360) for the support of funding and abundant training dataset.

6. Conclusions

To solve the problem that water conservancy personnel always meet in real hydrological scenarios—detecting water levels—this article proposes a combined method for water level recognition. This combined method is CNN-based and includes a water gauge digital detection model, a water gauge area segmentation model, and a water level line extraction algorithm. The results from the campus lake experiment on our gauge dataset prove that the method effectively solves the problem of water level recognition with cameras watching water gauges. However, the model still has shortcomings in that it cannot be applied to embedded AI devices yet, and in foggy and stormy environments, the method has slightly worse performance than that of the experiment in this paper. In the future, we will conduct more experiments, enrich datasets, and improve the model to solve the problems above and expand the scope of the application.

Author Contributions: Conceptualization, C.C.; methodology, C.C., Q.P. and R.F.; software, R.F. and X.A.; validation, X.A., C.H. and J.J.; formal analysis, L.C.; investigation, X.L.; resources, X.A. and R.F.; writing—original draft preparation, X.A.; writing—review and editing, C.C. and R.F. All authors have read and agreed to the published version of the manuscript.

Funding: This work was supported by the 2022 science and technology project (2022-33) of State Grid Jilin Electric Power Company.

Data Availability Statement: Not applicable.

Acknowledgments: The authors would like to thank the Xi'an Key Laboratory of Mobile Edge Computing and Security and the Ministry of Water Resources of China for data acquisition and computation machine support.

Conflicts of Interest: The authors declare no conflict of interest.

References

1. Wu, W.; Emerton, R.; Duan, Q.; Wood, A.W.; Wetterhall, F.; Robertson, D.E. Ensemble flood forecasting: Current status and future opportunities. *Wiley Interdiscip. Rev. Water* **2020**, *7*, e1432. [CrossRef]
2. Sunkpho, J.; Ootamakorn, C. Real-time flood monitoring and warning system. *Songklanakarin J. Sci. Technol.* **2011**, *33*, 227–235.
3. Sulistyowati, R.; Sujono, H.A.; Musthofa, A.K. Design and field test equipment of river water level detection based on ultrasonic sensor and SMS gateway as flood early warning. *AIP Conf. Proc.* **2017**, *1855*, 50003.
4. Zhao, M.; Chen, C.; Liu, L.; Lan, D.; Wan, S. Orbital collaborative learning in 6G space-air-ground integrated networks. *Neurocomputing* **2022**, *497*, 94–109. [CrossRef]
5. Taylor, C.J. *Ground-Water-Level Monitoring and the Importance of Long-Term Water-Level Data*; US Geological Survey: Denver, CO, USA, 2001.
6. Hernández-Nolasco, J.A.; Ovando, M.A.W.; Acosta, F.D.; Pancardo, P. Water level meter for alerting population about floods. In Proceedings of the 2016 IEEE 30th International Conference on Advanced Information Networking and Applications (AINA), Crans-Montana, Switzerland, 23–25 March 2016; pp. 879–884.
7. Ministry of Water Resources of People's Republic of China. *Standard Stage Observation*; Ministry of Water Resources of People's Republic of China: Beijing, China, 2010.
8. Chen, C.; Ma, H.; Yao, G.; Lv, N.; Yang, H.; Li, C.; Wan, S. Remote sensing image augmentation based on text description for waterside change detection. *Remote Sens.* **2021**, *13*, 1894. [CrossRef]
9. Zhong, Z. Method of water level data capturing based on video image recognition. *Foreign Electron. Meas. Technol.* **2017**, *1*, 48–51.
10. Dalal, N.; Triggs, B. Histograms of Oriented Gradients for Human Detection. In Proceedings of the IEEE Computer Society Conference on Computer Vision & Pattern Recognition, San Diego, CA, USA, 20–25 June 2005.
11. Jakkula, V. Tutorial on support vector machine (svm). *Sch. Eecs, Wash. State Univ.* **2006**, *37*, 3.
12. Mori, S.; Suen, C.Y.; Yamamoto, K. *Historical Review of OCR Research and Development*; IEEE Computer Society Press: Washington, DC, USA, 1995,
13. Sabbatini, L.; Palma, L.; Belli, A.; Sini, F.; Pierleoni, P. A Computer Vision System for Staff Gauge in River Flood Monitoring. *Inventions* **2021**, *6*, 79. [CrossRef]
14. Viola, P.; Jones, M.J. Robust Real-time Object Detection. *Int. J. Comput. Vis.* **2001**, *57*, 87.
15. Felzenszwalb, P.F.; Girshick, R.S.; McAllester, D.; Ramanan, D. Object Detection with Discriminatively Trained Part-Based Models. *IEEE Trans. Pattern Anal. Mach. Intell.* **2010**, *32*, 1627–1645. [CrossRef]
16. Lin, T.Y.; Goyal, P.; Girshick, R.; He, K.; Dollár, P. Focal Loss for Dense Object Detection. *IEEE Trans. Pattern Anal. Mach. Intell.* **2017**, *1*, 2999–3007.

17. Bojarski, M.; Del Testa, D.; Dworakowski, D.; Firner, B.; Flepp, B.; Goyal, P.; Jackel, L.D.; Monfort, M.; Muller, U.; Zhang, J.; et al. End to end learning for self-driving cars. *arXiv* **2016**, arXiv:1604.07316.
18. Xu, Z.; Sun, Y.; Liu, M. iCurb: Imitation Learning-based Detection of Road Curbs using Aerial Images for Autonomous Driving. *IEEE Robot. Autom. Lett.* **2021**, *6*, 1097–1104. [CrossRef]
19. Wu, Y.; Feng, S.; Huang, X.; Wu, Z. L4Net: An anchor-free generic object detector with attention mechanism for autonomous driving. *IET Comput. Vis.* **2021**, *15*, 36–46. [CrossRef]
20. Schroff, F.; Kalenichenko, D.; Philbin, J. Facenet: A unified embedding for face recognition and clustering. In Proceedings of the IEEE Conference on Computer Vision and Pattern Recognition, Boston, MA, USA, 7–12 June 2015; pp. 815–823.
21. Jung, J.; Lee, S.; Oh, H.S.; Park, Y.; Park, J.; Son, S. Unified Negative Pair Generation toward Well-discriminative Feature Space for Face Recognition. *arXiv* **2022**, arXiv:2203.11593.
22. Ying, L. Design of attendance system based on face recognition. *Electron. Test* **2020**, *1*, 117–121.
23. Camps-Valls, G.; Tuia, D.; Zhu, X.X.; Reichstein, M. *Deep Learning for the Earth Sciences: A Comprehensive Approach to Remote Sensing, Climate Science and Geosciences*; John Wiley & Sons: Hoboken, NJ, USA, 2021.
24. Albawi, S.; Mohammed, T.A.; Al-Zawi, S. Understanding of a convolutional neural network. In Proceedings of the 2017 International Conference on Engineering and Technology (ICET), Antalya, Turkey, 21–23 August 2017; pp. 1–6.
25. Yu, Y.; Samali, B.; Rashidi, M.; Mohammadi, M.; Nguyen, T.N.; Zhang, G. Vision-based concrete crack detection using a hybrid framework considering noise effect. *J. Build. Eng.* **2022**, *61*, 105246. [CrossRef]
26. Krizhevsky, A.; Sutskever, I.; Hinton, G. ImageNet Classification with Deep Convolutional Neural Networks. In *Advances in Neural Information Processing Systems 25*; Curran Associates Inc.: Red Hook, NY, USA, 2012.
27. Krizhevsky, A.; Sutskever, I.; Hinton, G.E. Imagenet classification with deep convolutional neural networks. *Commun. ACM* **2017**, *60*, 84–90. [CrossRef]
28. Chauhan, R.; Ghanshala, K.K.; Joshi, R. Convolutional neural network (CNN) for image detection and recognition. In Proceedings of the 2018 First International Conference on Secure Cyber Computing and Communication (ICSCCC), Jalandhar, India, 15–17 December 2018; pp. 278–282.
29. Redmon, J.; Divvala, S.; Girshick, R.; Farhadi, A. You Only Look Once: Unified, Real-Time Object Detection. In Proceedings of the IEEE Conference on Computer Vision and Pattern Recognition (CVPR), Las Vegas, NV, USA, 27–30 June 2016; pp. 779–788.
30. Girshick, R.; Donahue, J.; Darrell, T.; Malik, J. Rich Feature Hierarchies for Accurate Object Detection and Semantic Segmentation. *arXiv* **2013**, arXiv:1311.2524.
31. Liu, W.; Anguelov, D.; Erhan, D.; Szegedy, C.; Reed, S.; Fu, C.Y.; Berg, A.C. *SSD: Single Shot MultiBox Detector*; Springer: Cham, Switzerland, 2016.
32. Xu, Z.; Feng, J.; Zhang, Z.; Duan, C. Water level estimation based on image of staff gauge in smart city. In Proceedings of the 2018 IEEE SmartWorld, Ubiquitous Intelligence & Computing, Advanced & Trusted Computing, Scalable Computing & Communications, Cloud & Big Data Computing, Internet of People and Smart City Innovation (SmartWorld/SCALCOM/UIC/ATC/CBDCom/IOP/SCI), Guangzhou, China, 8–12 October 2018; pp. 1341–1345.
33. Dou, G.; Chen, R.; Han, C.; Liu, Z.; Liu, J. Research on water-level recognition method based on image processing and convolutional neural networks. *Water* **2022**, *14*, 1890. [CrossRef]
34. Long, J.; Shelhamer, E.; Darrell, T. Fully Convolutional Networks for Semantic Segmentation. *IEEE Trans. Pattern Anal. Mach. Intell.* **2015**, *39*, 640–651.
35. Liu, Y.; Xie, Z.; Liu, H. LB-LSD: A length-based line segment detector for real-time applications. *Pattern Recognit. Lett.* **2019**, *128*. [CrossRef]
36. Tian, Z.; Shen, C.; Chen, H.; He, T. FCOS: Fully Convolutional One-Stage Object Detection. In Proceedings of the 2019 IEEE/CVF International Conference on Computer Vision (ICCV), Seoul, Republic of Korea, 27 October–2 November 2019.
37. Chen, L.C.; Zhu, Y.; Papandreou, G.; Schroff, F.; Adam, H. *Encoder-Decoder with Atrous Separable Convolution for Semantic Image Segmentation*; Springer: Cham, Switzerland, 2018.
38. Karamouz, M.; Zahmatkesh, Z.; Saad, T. Cloud Computing in Urban Flood Disaster Management. In Proceedings of the World Environmental & Water Resources Congress, Cincinnati, OH, USA, 19–23 May 2013; pp. 2747–2757.
39. Fan, Y.; He, H.; Bo, L.; Ming, L. Research on Flood Disaster Extent Dynamics Monitoring Using HJ-1 CCD—A Case Study in Fuyuan of Heilongjiang Province, Northestern China. *Remote Sens. Technol. Appl.* **2016**, *31*, 102–108.
40. Shafiai, S. Flood Disaster Management in Malaysia: A Review of Issues of Flood Disaster Relief during and Post-Disaster. In Proceedings of the ISSC 2016 International Conference on Soft Science, Kedah, Malaysia, 11–13 April 2016.
41. Abe, K. Frequency response of pressure type water level meter. *Bull. Nippon. Dent. Univ. Gen. Educ.* **2001**, *30*, 49–56.
42. Tang, X.; Liu, Y.; Shang, X. The Research On Low Power and High Accuracy Ultrasonic Water Level Meter. *Hydropower Autom. Dam Monit.* **2014**, *1*, 1.
43. Zhen, Z.; Yang, Z.; Yuchou, L.; Youjie, Y.; Xurui, L. IP camera-based LSPIV system for on-line monitoring of river flow. In Proceedings of the 2017 13th IEEE International Conference on Electronic Measurement & Instruments (ICEMI), Yangzhou, China, 20–22 October 2017; pp. 357–363.
44. Lin, Y.T.; Lin, Y.C.; Han, J.Y. Automatic water-level detection using single-camera images with varied poses. *Measurement* **2018**, *127*, 167–174. [CrossRef]

45. Huang, Z.; Xiong, H.; Zhu, M.; Cai, H. Embedded Measurement System and Interpretation Algorithm for Water Gauge Image. *Opto-Electron. Eng.* **2013**, *40*, 1–7.
46. Lin, R.F.; Hai, X.U. Automatic measurement method for canals water level based on imaging sensor. *Transducer Microsyst. Technol.* **2013**, *32*, 53–55.
47. Bruinink, M.; Chandarr, A.; Rudinac, M.; Overloop, P.; Jonker, P. Portable, automatic water level estimation using mobile phone cameras. In Proceedings of the 2015 14th IAPR International Conference on Machine Vision Applications (MVA), Tokyo, Japan, 18–22 May 2015.
48. Leduc, P.; Ashmore, P.; Sjogren, D. Technical note: Stage and water width measurement of a mountain stream using a simple time-lapse camera. *Hydrol. Earth Syst. Sci. Discuss.* **2018**, *22*, 1–11. [CrossRef]
49. Liu, Q.; Chu, B.; Peng, J.; Tang, S. A Visual Measurement of Water Content of Crude Oil Based on Image Grayscale Accumulated Value Difference. *Sensors* **2019**, *19*, 2963. [CrossRef]
50. Gilmore, T.E.; Birgand, F.; Chapman, K.W. Source and magnitude of error in an inexpensive image-based water level measurement system. *J. Hydrol.* **2013**, *496*, 178–186. [CrossRef]
51. Young, D.S.; Hart, J.K.; Martinez, K. Image analysis techniques to estimate river discharge using time-lapse cameras in remote locations. *Comput. Geosci.* **2015**, *76*, 1–10. [CrossRef]
52. Zhang, Z.; Zhou, Y.; Wang, H.; Gao, H.; Liu, H. Image-based water level measurement with standard bicolor staff gauge. *Yi Qi Yi Biao Xue Bao/Chin. J. Sci. Instrum.* **2018**, *39*, 236–245.
53. Jiang, X.Y.; Hua, Z.J. Water-Level auto reading based on image processing. *Electron. Des. Eng.* **2011**, *19*, 23–25.
54. Law, H.; Deng, J. Cornernet: Detecting objects as paired keypoints. In Proceedings of the European conference on computer vision (ECCV), Munich, Germany, 8–14 September 2018; pp. 734–750.
55. Lv, N.; Han, Z.; Chen, C.; Feng, Y.; Su, T.; Goudos, S.; Wan, S. Encoding Spectral-Spatial Features for Hyperspectral Image Classification in the Satellite Internet of Things System. *Remote Sens.* **2021**, *13*, 3561. [CrossRef]
56. Ronneberger, O.; Fischer, P.; Brox, T. U-Net: Convolutional Networks for Biomedical Image Segmentation. In *Medical Image Computing and Computer-Assisted Intervention—MICCAI 2015*; Springer International Publishing: Cham, Switzerland, 2015.
57. Lv, N.; Ma, H.; Chen, C.; Pei, Q.; Zhou, Y.; Xiao, F.; Li, J. Remote sensing data augmentation through adversarial training. *IEEE J. Sel. Top. Appl. Earth Obs. Remote. Sens.* **2021**, *14*, 9318–9333. [CrossRef]
58. Yu, F.; Koltun, V. Multi-scale context aggregation by dilated convolutions. *arXiv* **2015**, arXiv:1511.07122.
59. Chen, L.C.; Papandreou, G.; Kokkinos, I.; Murphy, K.; Yuille, A.L. DeepLab: Semantic Image Segmentation with Deep Convolutional Nets, Atrous Convolution, and Fully Connected CRFs. *IEEE Trans. Pattern Anal. Mach. Intell.* **2018**, *40*, 834–848. [CrossRef]
60. Hu, J.; Chen, C.; Cai, L.; Khosravi, M.R.; Pei, Q.; Wan, S. UAV-assisted vehicular edge computing for the 6G internet of vehicles: Architecture, intelligence, and challenges. *IEEE Commun. Stand. Mag.* **2021**, *5*, 12–18. [CrossRef]
61. Chen, C.; Zeng, Y.; Li, H.; Liu, Y.; Wan, S. A Multi-hop Task Offloading Decision Model in MEC-enabled Internet of Vehicles. *IEEE Internet Things J.* **2022**. [CrossRef]
62. Ma, X.; Li, X.; Tang, X.; Zhang, B.; Yao, R.; Lu, J. Deconvolution Feature Fusion for traffic signs detection in 5G driven unmanned vehicle. *Phys. Commun.* **2021**, *47*, 101375. [CrossRef]
63. Isola, P.; Zhu, J.Y.; Zhou, T.; Efros, A.A. Image-to-Image Translation with Conditional Adversarial Networks. In Proceedings of the IEEE Conference on Computer Vision & Pattern Recognition, Las Vegas, NV, USA, 27–30 June 2016.
64. Park, T.; Liu, M.Y.; Wang, T.C.; Zhu, J.Y. Semantic Image Synthesis With Spatially-Adaptive Normalization. In Proceedings of the 2019 IEEE/CVF Conference on Computer Vision and Pattern Recognition (CVPR), Long Beach, CA, USA, 15–20 June 2019.
65. Yu, J.; Lin, Z.; Yang, J.; Shen, X.; Lu, X.; Huang, T.S. Generative Image Inpainting with Contextual Attention. In Proceedings of the 2018 IEEE/CVF Conference on Computer Vision and Pattern Recognition, Salt Lake City, UT, USA, 18–23 June 2018.
66. Badrinarayanan, V.; Kendall, A.; Cipolla, R. SegNet: A Deep Convolutional Encoder-Decoder Architecture for Image Segmentation. *IEEE Trans. Pattern Anal. Mach. Intell.* **2017**, *39*, 2481–2495. [CrossRef]
67. Zhao, H.; Shi, J.; Qi, X.; Wang, X.; Jia, J. Pyramid Scene Parsing Network. *arXiv* **2016**, arXiv:1612.01105.
68. Lin, G.; Milan, A.; Shen, C.; Reid, I. RefineNet: Multi-path Refinement Networks for High-Resolution Semantic Segmentation. In Proceedings of the 2017 IEEE Conference on Computer Vision and Pattern Recognition (CVPR), Honolulu, HI, USA, 21–26 July 2017.
69. Lin, T.Y.; Dollar, P.; Girshick, R.; He, K.; Hariharan, B.; Belongie, S. Feature Pyramid Networks for Object Detection. In Proceedings of the 2017 IEEE Conference on Computer Vision and Pattern Recognition (CVPR), Honolulu, HI, USA, 21–26 July 2017.
70. Chen, L.C.; Papandreou, G.; Kokkinos, I.; Murphy, K.; Yuille, A.L. Semantic image segmentation with deep convolutional nets and fully connected crfs. *arXiv* **2014**, arXiv:1412.7062.
71. Chen, L.C.; Papandreou, G.; Schroff, F.; Adam, H. Rethinking atrous convolution for semantic image segmentation. *arXiv* **2017**, arXiv:1706.05587.
72. Chen, C.; Jiang, J.; Zhou, Y.; Lv, N.; Liang, X.; Wan, S. An edge intelligence empowered flooding process prediction using Internet of things in smart city. *J. Parallel Distrib. Comput.* **2022**, *165*, 66–78. [CrossRef]
73. He, K.; Zhang, X.; Ren, S.; Sun, J. Deep Residual Learning for Image Recognition. In Proceedings of the 2016 IEEE Conference on Computer Vision and Pattern Recognition (CVPR), Las Vegas, NV, USA, 27–30 June 2016.

remote sensing

MDPI

Article

The Effect of Drought on Vegetation Gross Primary Productivity under Different Vegetation Types across China from 2001 to 2020

Xiaoping Wu [1], Rongrong Zhang [1], Virgílio A. Bento [2], Song Leng [1,3], Junyu Qi [4], Jingyu Zeng [1,5] and Qianfeng Wang [1,*]

1 College of Environment & Safety Engineering, Fuzhou University, Fuzhou 350116, China
2 Instituto Dom Luiz (IDL), Faculdade de Ciências, Universidade de Lisboa, 1749-016 Lisboa, Portugal
3 School of Life Sciences, University of Technology Sydney, Sydney 2007, Australia
4 Earth System Science Interdisciplinary Center, University of Maryland, 5825 University Research Ct, College Park, MD 20740, USA
5 State Key Laboratory of Earth Surface Processes and Resource Ecology, Faculty of Geographical Science, Beijing Normal University, Beijing 100875, China
* Correspondence: wangqianfeng@fzu.edu.cn

Highlights:

- Two drought indices (SPEI and VPD) were used to characterize the degree of dryness/wetness.
- The water deficit represented by two drought indices was mostly negatively correlated with vegetation GPP, especially in summer and autumn.
- The negative impact of water deficit/drought as measured by SPEI on vegetation GPP was more severe than that revealed by VPD.
- During drought, both SPEI and VPD showed that drought had a negative impact on vegetation GPP in North China, Southwest China, and the Qinghai–Tibet Plateau.

Abstract: Climate change has exacerbated the frequency and severity of droughts worldwide. Evaluating the response of gross primary productivity (GPP) to drought is thus beneficial to improving our understanding of the impact of drought on the carbon cycle balance. Although many studies have investigated the relationship between vegetation productivity and dry/wet conditions, the capability of different drought indices of assessing the influence of water deficit is not well understood. Moreover, few studies consider the effects of drought on vegetation with a focus on periods of drought. Here, we investigated the spatial-temporal patterns of GPP, the standardized precipitation evapotranspiration index (SPEI), and the vapor pressure deficit (VPD) in China from 2001 to 2020 and examined the relationship between GPP and water deficit/drought for different vegetation types. The results revealed that SPEI and GPP were positively correlated over approximately 70.7% of the total area, and VPD was negatively correlated with GPP over about 66.2% of the domain. Furthermore, vegetation productivity was more negatively affected by water deficit in summer and autumn. During periods of drought, the greatest negative impact was on deciduous forests and croplands, and woody savannas were the least impacted. This research provides a scientific reference for developing mitigation and adaptation measures to lessen the impact of drought disasters under a changing climate.

Keywords: China; drought; SPEI; VPD; vegetation; GPP

Citation: Wu, X.; Zhang, R.; Bento, V.A.; Leng, S.; Qi, J.; Zeng, J.; Wang, Q. The Effect of Drought on Vegetation Gross Primary Productivity under Different Vegetation Types across China from 2001 to 2020. *Remote Sens.* **2022**, *14*, 4658. https://doi.org/10.3390/rs14184658

Academic Editors: Jingzhe Wang, Zhongwen Hu, Yangyi Wu and Jie Zhang

Received: 26 June 2022
Accepted: 14 September 2022
Published: 18 September 2022

1. Introduction

Droughts are among the most complex and costly natural disasters in the world. They are usually divided into meteorological, agricultural, hydrological, and economic droughts [1]. With global climate change and increased human activities, the frequent occurrence of droughts has become a serious challenge [2–5]. Under future climate change scenarios, the severity of droughts is expected to increase [6]. Since droughts profoundly impact the economy, society, ecology, and environment, they increasingly attract the attention of researchers in different fields.

China has suffered a growing number of drought events, including extreme drought episodes, since the 1990s [7,8]. Droughts in China are highly complex, since they are affected by many factors, such as climate, topography, and human activities [9]. The severity of droughts in the northern region of the country (drought-prone area) has intensified, while their frequency in the southern region is increasing [10,11]. The impacts of droughts on agriculture and ecology threaten food and ecological security in China, the most populous country in the world. Thus, assessing the temporal and spatial characteristics of droughts, as well as understanding their trends and behaviors, is a crucial task to provide a scientific basis for developing mitigation and adaptation measures to lessen the impact of drought disasters under a changing climate.

Drought indices are among the most popular methods for monitoring and evaluating droughts [12]. Many drought indices have been proposed in the literature, but the most commonly used drought indices are the standardized precipitation index (SPI) [13], Palmer drought severity index (PDSI) [14], and standardized precipitation evapotranspiration index (SPEI) [3,15]. SPI is a widely used drought index using a range of timescales that considers precipitation and ignores other climatic factors, such as temperature and evapotranspiration. On the other hand, PDSI combines the effects of precipitation and evapotranspiration, but with the handicap of having a fixed temporal scale. Vicente-Serrano et al. [15] introduced a multi-scale index based on SPI, with the inclusion of potential evapotranspiration—the SPEI. Using precipitation and evapotranspiration, SPEI not only retains PDSI's evapotranspiration sensitivity to temperature, but also includes SPI's calculation simplicity and suitability for multi-scale and multi-space comparison. Because this index is suitable for monitoring drought characteristics under climate change, it has been widely used worldwide. Moreover, SPEI may be used as a tool to assess the impacts of droughts on different systems (hydrology, agriculture, and ecology) [16].

Vapor pressure deficit (VPD) is the difference between the saturation and actual vapor pressure, and is an important indicator of the degree of air dryness considered as a key factor affecting the physiological function of vegetation [17]. High VPD is one of the main causes of vegetation drought and may inhibit photosynthesis in vegetation [18].

Vegetation is an essential component of terrestrial ecosystems, reflecting an ecosystem's production and carbon sink capacities. However, it is sensitive to climate change, and its growth is strongly affected by droughts [19,20]. Drought is a vital factor leading to the decline of vegetation productivity [21–23]. However, the response of different ecosystems to drought varies [24]. Thus, assessing the impact of drought on vegetation has become a crucial scientific issue [25]. Many studies use gross primary productivity (GPP), net primary productivity (NPP), and the normalized difference vegetation index (NDVI) to characterize ecosystem productivity and analyze the productivity response to climate change [26–28]. Among these, GPP is the number of photosynthetic products, or total organic carbon, fixed by organisms through photosynthesis in a unit of time. It is the largest component and foundation of the carbon cycle of an ecosystem [29]. Thus, changes in GPP can accurately reflect ecosystem responses to extreme events [30], whereas many studies observe that drought occurrences decrease vegetation productivity, with different regions and vegetation types responding differently [31]. As a country with a vast territory, China has undergone complicated and extensive changes in vegetation types due to rapid economic development [32]. However, there is limited research regarding the impact of large-scale drought across China on different vegetation types during different seasons. Indeed, identifying the effects of drought on the productivity of different vegetation types is a significant achievement that may help to better develop strategies to protect the environment.

The aim of this study is to investigate the effects of water deficit/drought on vegetation productivity in various vegetation types across China during different seasons from 2001 to 2020. Specifically, we used two meteorological drought indices, SPEI and VPD, to characterize water deficit and drought, along with vegetation productivity, as indicated by GOSIF-GPP. We also examined the effect patterns of water deficit/drought on various vegetation types (including evergreen forests, deciduous forests, mixed forests, woody savannas, savannas, grasslands, and croplands). This study is developed to further grasp and respond to the impact of climate change on vegetation.

2. Materials and Methods

2.1. Study Area

This study is performed across China, which covers a vast territory (Figure 1). Because of conspicuous spatial differences in temperature, precipitation, and topography, land cover types over China show evident spatial heterogeneity. China can be divided into four climate regions according to the aridity index (the ratio between evapotranspiration and precipitation): humid, semi-humid, semi-arid, and arid regions [33]. The humid regions (mainly located in the south of China) are dominated by forests and savannas; the semi-humid regions are primarily covered by croplands and grasslands; the semi-arid regions are dominated by grasslands; and the arid areas are mostly barren (Figure 1).

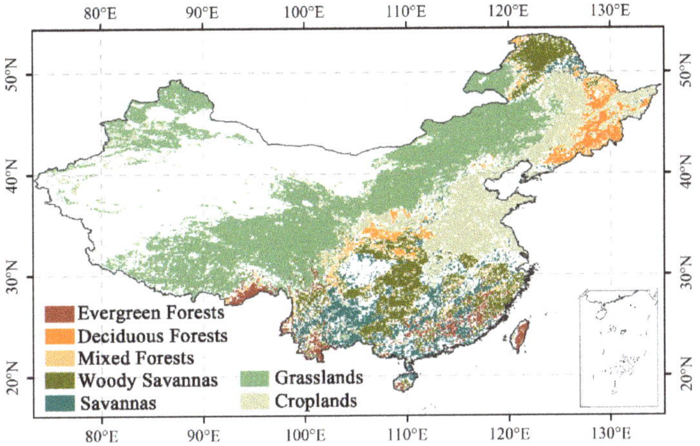

Figure 1. Spatial distribution of vegetation types in the selected study area across China (the white area represents the pixels that were excluded for further analysis, including land cover type that has changed from 2001 to 2020, non-vegetation cover, and shrublands cover).

To reduce the uncertainty of vegetation cover change, only the areas where the land cover type has not changed from 2001 to 2020 were selected for analysis. Specifically, the images of land cover in 2001 and 2020 were compared and subtracted. If the pixel value was equal to 0, it was assumed that the land cover type remained unchanged during the study period, and vice versa. In addition, the regions with non-vegetation cover (including water bodies, permanent wetlands, urban and built-up lands, permanent snow and ice, and barren) or shrublands cover (less than 100 pixels in the whole study area) were masked, i.e., excluded from this study. The final selected study area is shown in Figure 1. Within the selected study area, there are seven main vegetation types, namely evergreen forests, deciduous forests, mixed forests, woody savannas, savannas, grasslands, and croplands (Table 1). Among these, grasslands occupy the largest area, while mixed forests occupy the smallest area, accounting for 48.6% and 2.2%, respectively.

Table 1. Vegetation types and their area proportions of the selected study area.

Vegetation Types	Area Proportions (%)
Evergreen Forests	3.3
Deciduous Forests	5.2
Mixed Forests	2.2
Woody Savannas	12.7
Savannas	8.1
Grasslands	48.6
Croplands	19.9

2.2. Datasets and Preprocessing

2.2.1. Meteorological Data

In this study, we used daily precipitation, temperature, and relative humidity data from monitoring stations provided by the China Meteorological Data Network (CMDN, http://data.cma.cn/ (accessed on 20 April 2022)). Observed near-surface meteorological data at weather stations in China are routinely publicized, with a lag of about three months, a period used for data compilation and quality control. Therefore, this dataset has high accuracy and timeliness, and has been widely used in studies related to climate change [34–37]. To ensure continuous and complete data records, 786 meteorological stations were chosen for our study. The data from the meteorological stations was interpolated to 0.1° grid points using ANUSPLIN software, which was generated using the thin-plate spline algorithm [38]. In order to improve the accuracy of interpolation in complex terrain areas, it is necessary to input elevation elements in the interpolation process. In China, interpolation of temperature and relative humidity based on this method has high confidence, and error estimates of precipitation for southern China were quite low [39]. This data was used to calculate SPEI and VPD.

2.2.2. Gross Primary Productivity Data

The monthly GOSIF-GPP data, with a spatial resolution of 0.05° for 2001–2020, was used (http://globalecology.unh.edu (accessed on 22 April 2022)). The data was constructed based on meteorological data, OCO-2 SIF data, remote sensing data from the Moderate Resolution Imaging Spectroradiometer (MODIS), and meteorological reanalysis data [40]. The GOSIF-GPP dataset is highly correlated with GPP estimated from EC flux sites (R^2 = 0.73, $p < 0.001$); therefore, it is able to objectively reflect the primary productivity of vegetation in China [41].

2.2.3. Land Cover Data

The Moderate Resolution Imaging Spectroradiometer (MODIS) Land Cover Climate Modeling Grid product (MCD12C1) of IGBP (International Geosphere-Biosphere Programme) classification (https://search.earthdata.nasa.gov/ (accessed on 3 May 2022)) for 2001–2020, with a spatial resolution of 0.05°, was used. Moreover, the land cover data were reclassified into seven classes (evergreen forests, deciduous forests, mixed forests, woody savannas, savannas, grasslands, and croplands). In addition, the vegetation cover data and GOSIF-GPP data were resampled to a spatial resolution of 0.1° using a nearest neighbor interpolation method.

2.3. Methods

2.3.1. Meteorological Drought Indices Calculation

Based on temperature (T) and relative humidity (RH), monthly VPD can be obtained using the following formula:

$$VPD = 0.61078 \times e^{\left(\frac{17.27 \times T}{T+237.3}\right)} \times \left(1 - \frac{RH}{100}\right)$$

SPEI is a probability distribution function that fits and normalizes cumulative water scarcity and can characterize the wetness and dryness of a region. It can be calculated using the difference between precipitation and potential evapotranspiration. The SPEI is characterized by multiple time scales. Furthermore, to match the time scale of GPP, this study calculated the SPEI at the 1-month scale. The Hargreaves model, which has strong accuracy in China [35], was used to estimate the monthly PET. The generalized extreme value (GEV) distribution was chosen to normalize the SPEI, as it has been shown to have strong stability in calculating shorter time scales SPEI [42].

Usually, When SPEI is less than −0.5, drought is indicated [15]. Previous studies have focused on the effect of dryness/wetness on vegetation [19,43]. In order to investigate the effect of drought on vegetation GPP, all drought months (i.e., months with SPEI less than −0.5) in each year were selected. The sum of the absolute SPEI values of all drought months, along with the sum of VPD and the mean GPP values corresponding to the drought months, were calculated. The calculation equations are as follows:

$$SPEI_{drought\_month} = \left| \sum_{i=1}^{n} SPEI_i \right| (SPEI_i < -0.5)$$

$$VPD_{drought\_month} = \sum_{i}^{n} VPD_i$$

$$GPP_{drought\_month} = \frac{\sum_{i}^{n} GPP_i}{n}$$

where $SPEI_{drought\_month}$ represents the sum of the absolute values of SPEI for all drought months in a year, $VPD_{drought\_month}$ represents the sum of the values of VPD for all drought months in a year, $GPP_{drought\_month}$ represents the average GPP of all drought months in a year, i represents the ith month, and n represents the number of drought months in a year.

2.3.2. Meteorological Drought Indices Calculation

The Mann–Kendall (M–K) test was used to identify the annual trends of GPP and drought indices and their significance. This is a nonparametric statistical test method [44,45] recommended by the World Meteorological Organization. It presents advantages, since samples do not need specific distributions, and its calculation method is fairly simple. Therefore, the M–K method is widely used to analyze changing trends and qualify the significance of trends in the time series of drought, precipitation, and temperature, among others [46–49].

For the statistic Z of the M–K test, $Z > 0$ indicates an upward trend, and $Z < 0$ indicates a downward trend. When the significance levels are set to 0.01, 0.05, and 0.1, $|Za|$ is 2.58, 1.96, and 1.65, respectively. At a certain significance level, if $|Z| > |Za|$, the statistics pass the corresponding significance test.

2.3.3. Correlation Analysis

The Pearson correlation analysis is usually used to investigate the degree of correlation between variables [50]. This method is frequently used in meteorology and ecology [31]. We selected this method to examine the response mechanism of vegetation to drought. An F test was performed, and it indicated that the correlation is significant when $p < 0.05$. Positive and negative correlations were represented by $R > 0$ and $R < 0$, respectively.

Moreover, the closer $|R|$ is to 1, the stronger the correlation. The correlation coefficient R was calculated as follows:

$$R = \frac{\sum_{i=1}^{n}(x_i - \overline{x})(y_i - \overline{y})}{\sqrt{\sum_{i=1}^{n}(x_i - \overline{x})^2}\sqrt{\sum_{i=1}^{n}(y_i - \overline{x})^2}}$$

where R ranges from -1 to 1; x_i and y_i represent the value of factors in x and y in period i, respectively; $\overline{x}$ and $\overline{y}$ represent the average value of factors; and n represents the time series length.

3. Results

3.1. Variation Trends of Meteorological Drought Indices and Vegetation GPP

We used the M–K test method to analyze the annual trend, seasonal trend, and significance of SPEI and VPD trends from 2001 to 2020 (Figure 2). From the perspective of the annual average change trend (Figure 2a), the SPEI, accounting for 68.0% of the total area, showed an upward trend. The areas showing a decreasing trend were mainly distributed in southwest and northern China.

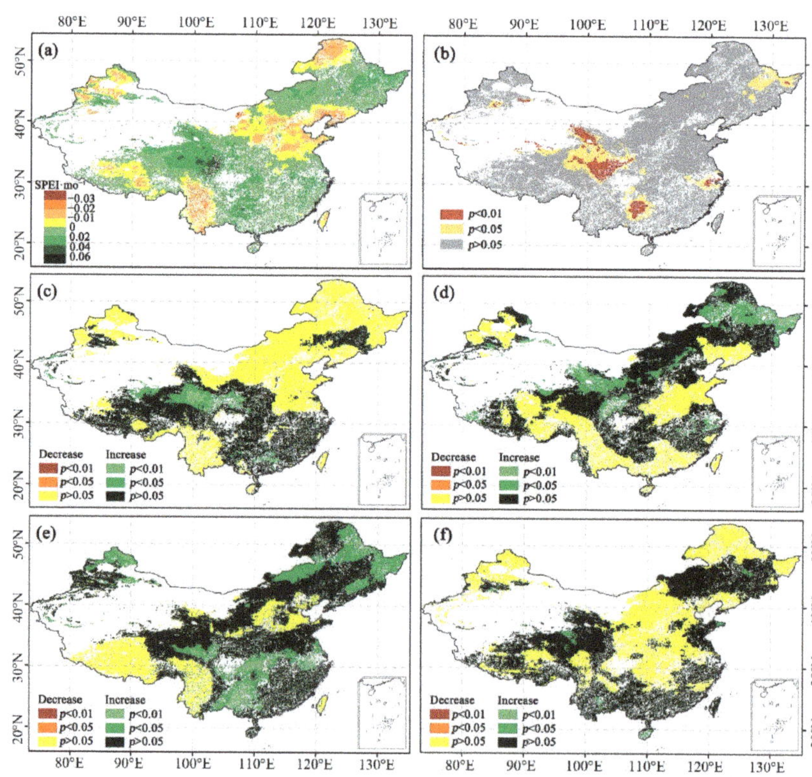

Figure 2. Spatial distribution of annual (**a**) changing trend, (**b**) *p*-value, and seasonal changing trend in (**c**) spring, (**d**) summer, (**e**) autumn, and (**f**) winter across China from 2001 to 2020 using the standardized precipitation evapotranspiration index (SPEI) based on the Mann–Kendall (M–K) trend test.

The analysis of the seasonal change trend in SPEI showed large differences in the changing trend when different seasons were considered (Figure 2c–f). Among the different seasons, the largest proportion of area (~54.3%) with a decreasing trend according to the SPEI was found in spring, while the smallest proportion (~19.9%) was found in autumn. However, it is worth noting that the regions with decreasing trends were not significant ($p > 0.05$) during the four seasons. In terms of vegetation types, for the annually averaged SPEI trends, evergreen forests presented the largest proportion (~46.4%) of decreasing trends, while mixed forests showed the smallest (~15.1%) (Table 2).

Table 2. The area proportions of the annual decreasing trend of SPEI, increasing trend of VPD, and increasing trend of GPP from 2001 to 2020 across China, under different vegetation types (unit: %).

Vegetation Types	SPEI	VPD	GPP
Evergreen Forests	46.4	49.7	80.1
Deciduous Forests	16.2	27.6	97.5
Mixed Forests	15.1	25.2	90.4
Woody Savannas	27.1	31.2	96.8
Savannas	20.7	37.1	96.1
Grasslands	32.6	65.1	87.2
Croplands	41.6	54.7	93.2

Similarly, the spatial distribution of the annual average and seasonal trends of VPD are shown in Figure 3. The trend of dryness and wetness changes, characterized by VPD and SPEI, are generally consistent. The area with an increasing trend of annual average VPD accounted for 53.1% of the total area. Furthermore, the regions with significant trends showed upward tendencies. As with SPEI, the changing trend of VPD in different seasons also varied. Among the different seasons, the largest proportion (~71.5%) of areas with an increasing trend of VPD was found in spring, while the smallest (~41.6%) was found in autumn. In all four seasons, a significant upward trend was observed. As for different vegetation types, grasslands had the largest proportion (~65.1%) of increasing trends, while mixed forests had the smallest (~25.2%) (Table 2).

Finally, the same results are presented for GPP in Figure 4. Annual GPP showed an overall upward trend, with 90.7% of the regions showing an increasing trend and 9.3% showing a decreasing trend from 2001 to 2020. Except for the decline in GPP in some areas in western China, most of the other regions presented positive tendencies, with highly significant increasing trends ($p < 0.01$). On the contrary, the values for the areas with a downward trend were not significant ($p > 0.05$). Among the different seasons, the largest proportion (~87.8%) of areas with an increasing trend of GPP was found in autumn, while the smallest (~65.9%) was found in winter. From the perspective of different vegetation types, deciduous forests had the largest proportion (~97.5%) of increasing trends, while evergreen forests had the smallest (80.1%) (Table 2).

3.2. The Relationship between Meteorological Drought Indices and Vegetation GPP

To characterize the influence of water deficit on vegetation, the Pearson correlation was applied to evaluate the correlations between GPP and meteorological drought indices. As shown in Figure 5, 70.7% of the total area showed that SPEI and GPP were positively correlated. In addition, 9.8% of the area showed a significant positive correlation ($p < 0.05$), and 6.1% showed a highly significant positive correlation ($p < 0.01$). The regions that showed a significant positive correlation were mainly concentrated in northern China. For different seasons, the proportion of positively correlated GPP and SPEI is larger in summer and autumn, accounting for 64.6% (with 27.9% having a significant positive correlation, namely, $p < 0.05$ and $p < 0.01$) and 65.3% (with 11.1% having a significant positive correlation, namely, $p < 0.05$ and $p < 0.01$), respectively. On the contrary, the proportion of positive correlation in spring and winter is smaller, accounting for 37.6% and 43.2%, respectively.

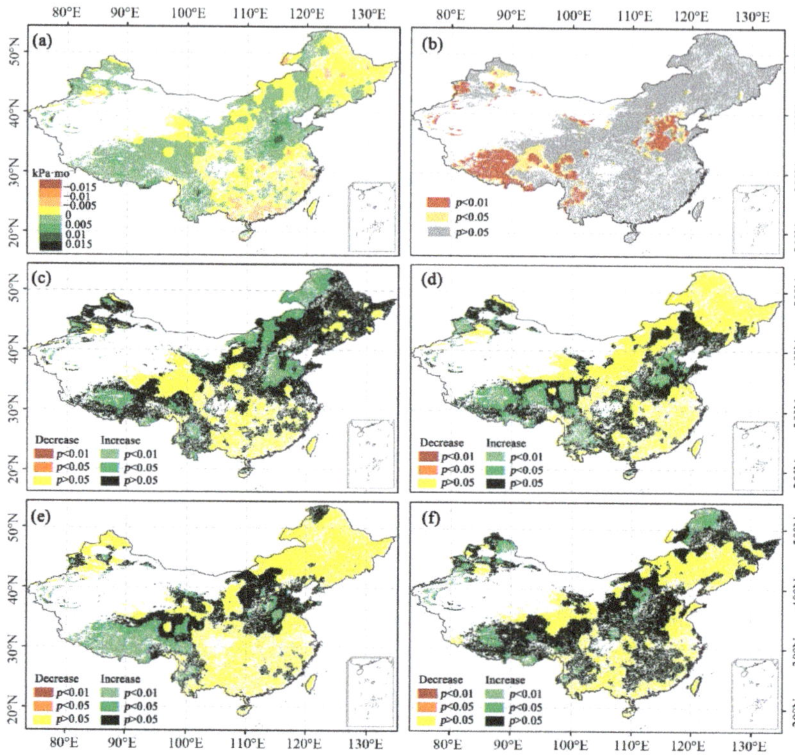

Figure 3. Spatial distribution of annual (**a**) changing trend, (**b**) p-value, and seasonal changing trend in (**c**) spring, (**d**) summer, (**e**) autumn, and (**f**) winter of vapor pressure deficit (VPD) across China from 2001 to 2020 based on the Mann–Kendall (M–K) trend test.

Figure 6 shows the frequency distribution (%) of the areas with a positive correlation ($R > 0$) between GPP and SPEI among different seasons and vegetation types. At the annual scale, water deficit had the greatest negative impact ($R > 0$) on GPP in savannas (80.0%) and the least impact on evergreen forests (54.8%). The sensitivity of vegetation to water deficit varied in different seasons. Except for grasslands, all vegetation types showed the strongest sensitivity in autumn. In spring, savannas were the most sensitive to water deficit. Grasslands were the most sensitive to water deficit in both summer and winter, while in autumn, croplands were the most sensitive to water deficit.

We further examined the correlation between GPP and VPD. As shown in Figure 7, 66.2% of the total area showed that GPP and VPD were negatively correlated, indicating that water deficit has a particular inhibitory effect on GPP. In addition, 9.9% of the area showed a significant negative correlation ($p < 0.05$), and 11.6% showed a highly significant negative correlation ($p < 0.01$). Similar to SPEI, the regions that showed significant negative correlations were mainly concentrated in northern China. For different seasons, the proportion of negatively correlated GPP and VPD was larger in summer and autumn, accounting for 64.9% (with 28.8% having a significant positive correlation, namely, $p < 0.05$ and $p < 0.01$) and 64.2% (with 13.6% having a significant positive correlation, namely, $p < 0.05$ and $p < 0.01$), respectively. On the contrary, the proportion of negative correlation in spring and winter was smaller, 34.0% and 43.7%, respectively.

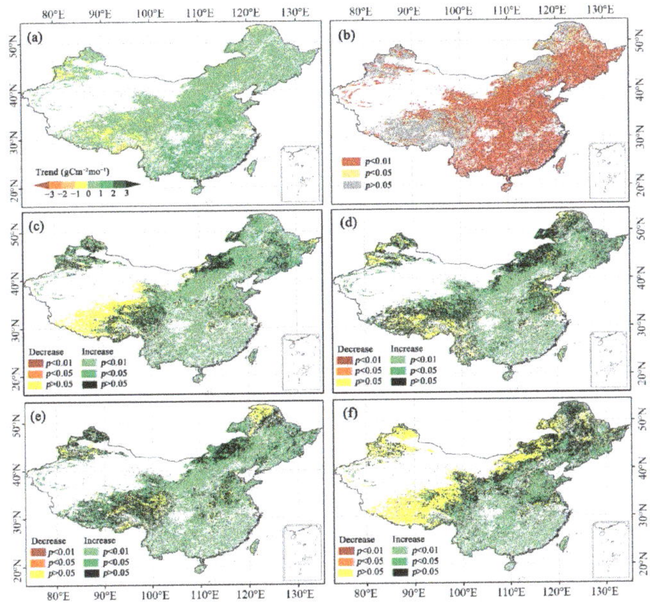

Figure 4. Spatial distribution of annual (**a**) changing trend, (**b**) *p*-value, and seasonal changing trend in (**c**) spring, (**d**) summer, (**e**) autumn, and (**f**) winter of gross primary productivity (GPP) across China from 2001 to 2020 based on the Mann–Kendall (M–K) trend test.

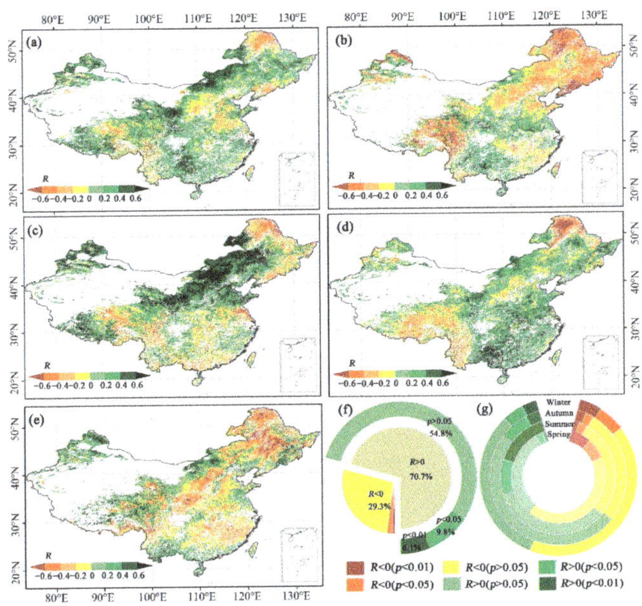

Figure 5. Spatial distribution of the correlation coefficients (R) between gross primary productivity (GPP) and standardized precipitation evapotranspiration index (SPEI) across China, under (**a**) annual and seasonal temporal scales including (**b**) spring, (**c**) summer, (**d**) autumn, and (**e**) winter from 2001 to 2020, and (**f**) annual and (**g**) seasonal frequency distribution of correlation coefficients (R) between GPP and SPEI.

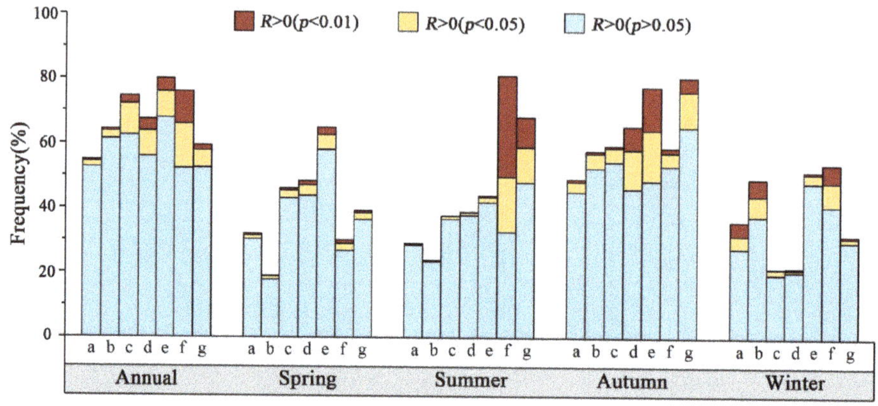

Figure 6. Frequency distribution of the areas with positive correlation ($R > 0$) between gross primary productivity (GPP) and standardized precipitation evapotranspiration index (SPEI) for different vegetation types, including (a) evergreen forests, (b) deciduous forests, (c) mixed forests, (d) woody savannas, (e) savannas, (f) grasslands, and (g) croplands under different temporal scales (annual and seasonal) across China from 2001 to 2020.

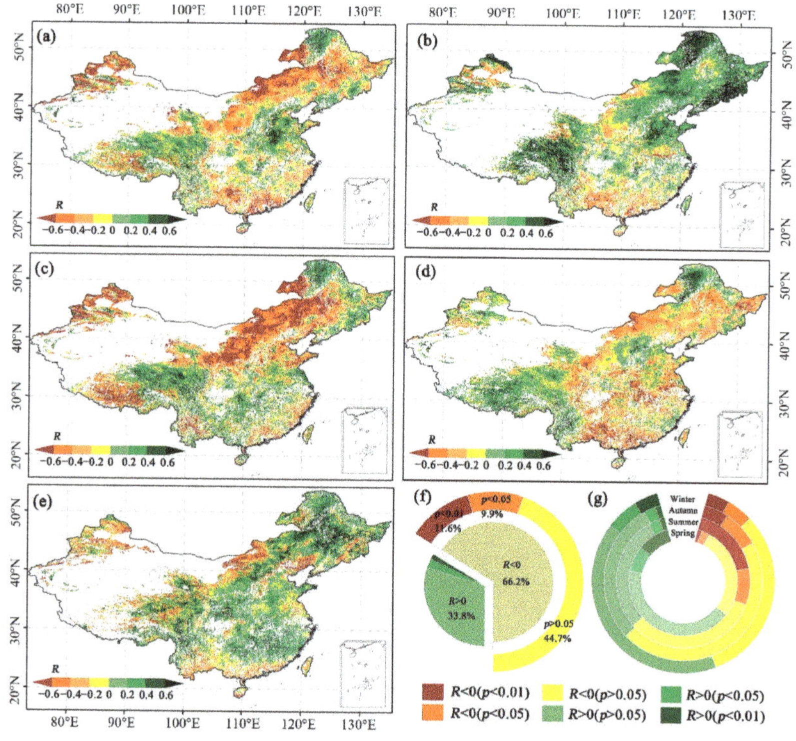

Figure 7. Spatial distribution of the correlation coefficients (R) between gross primary productivity (GPP) and vapor pressure deficit (VPD) across China under (**a**) annual and seasonal temporal scales including (**b**) spring, (**c**) summer, (**d**) autumn, and (**e**) winter from 2001 to 2020, and (**f**) annual and (**g**) seasonal frequency distribution of correlation coefficients (R) between GPP and VPD.

Figure 8 shows the frequency distribution (%) of the areas with a negative correlation ($R < 0$) between GPP and VPD for different seasons and vegetation types in the study area. At the annual scale, water deficit had the greatest negative impact ($R < 0$) on GPP in savannas (73.0%) and the least impact on croplands (56.3%). The sensitivity of vegetation to water deficit varied in different seasons. Except for grasslands, all vegetation types showed the strongest sensitivity in autumn. In spring, savannas were the most sensitive to water deficit. Grasslands were the most sensitive to water deficit in both summer and winter. In autumn, croplands were the most sensitive to water deficit. This result is consistent with the relationship between GPP and SPEI.

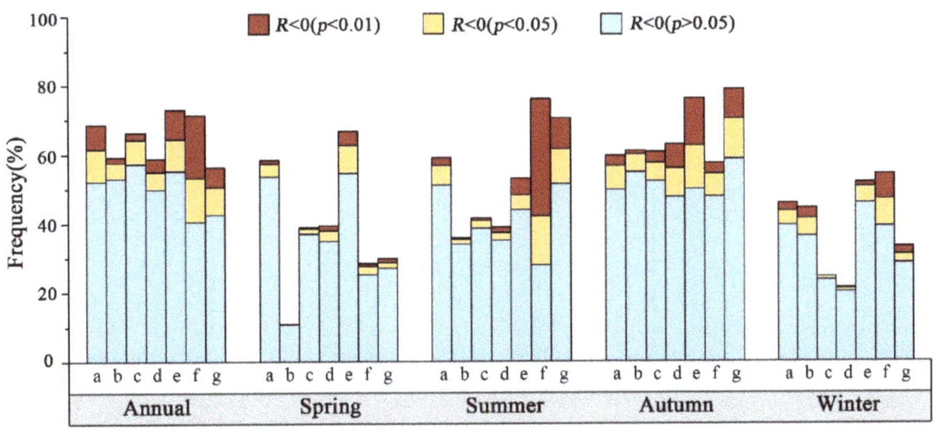

Figure 8. Frequency distribution of the areas with negative correlation ($R < 0$) between gross primary productivity (GPP) and vapor pressure deficit (VPD) for different vegetation types (a) evergreen forests, (b) deciduous forests, (c) mixed forests, (d) woody savannas, (e) savannas, (f) grasslands, and (g) croplands under different temporal scales (annual and seasonal) across China from 2001 to 2020.

3.3. The Impact of Drought on Vegetation GPP

The spatial distribution of the correlation coefficients between $GPP_{drought\_month}$ and $SPEI_{drought\_month}/VPD_{drought\_month}$ and the p-value are shown in Figure 9. We found that the $GPP_{drought\_month}$ was negatively correlated with $SPEI_{drought\_month}$ over 66.3% of the study area (Figure 9a). Among these, 5.1% showed a significant negative correlation ($p < 0.05$), which indicated an inhibitory effect of drought on GPP. The area in which $GPP_{drought\_month}$ was negatively correlated with $VPD_{drought\_month}$ was 48.7%, with only 1.1% showing a significant negative correlation ($p < 0.05$). Relative to VPD, the drought indicated by SPEI had a more severe negative impact on vegetation GPP. Furthermore, the spatial distribution showed that the correlation coefficients between the two drought indices and GPP are roughly the same during drought periods, and the regions with larger differences are mainly concentrated in northeast and northwestern China. Both drought indices showed that vegetation growth was inhibited during the drought period in northern and southwestern China, as well as in the Qinghai–Tibet Plateau (situated in southwestern China).

The correlation between $GPP_{drought\_month}$ and $SPEI_{drought\_month}/VPD_{drought\_month}$ for the seven vegetation types in the study area is shown in Figure 10. The correlation coefficients between $GPP_{drought\_month}$ and $SPEI_{drought\_month}$ were mostly negative. Among them, deciduous forests were the most negatively affected by drought, and woody savannas were the least affected. However, the relationship between $GPP_{drought\_month}$ and $VPD_{drought\_month}$ of different types of vegetation showed that the proportion of positive correlation was less than that of $SPEI_{drought\_month}$. Croplands were the most negatively affected by drought, while woody savannas were the least.

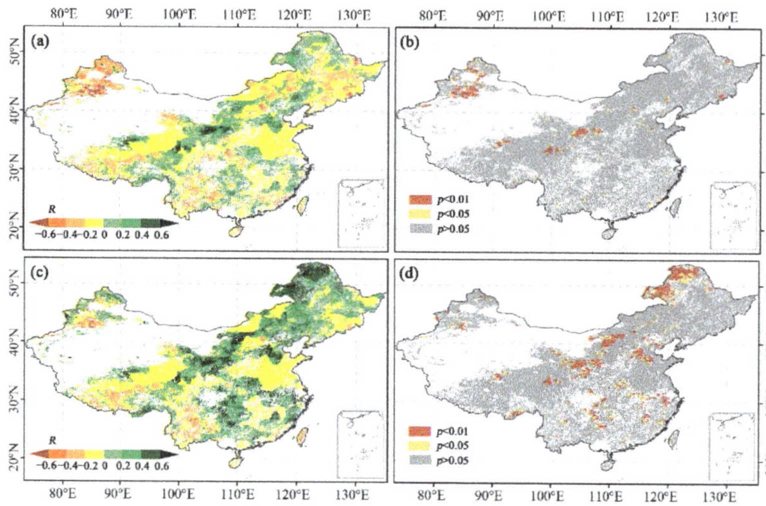

Figure 9. Spatial distribution of (**a**) the correlation coefficients (R) and (**b**) statistical significance (p-value) between $GPP_{drought\_month}$ (the average GPP of all drought months in a year) and $SPEI_{drought\_month}$ (the sum of the absolute values of SPEI for all drought months in a year), and the spatial distribution of (**c**) the correlation coefficients (R) and (**d**) statistical significance (p-value) between $GPP_{drought\_month}$ and $VPD_{drought\_month}$ (the sum of the values of VPD for all drought months in a year).

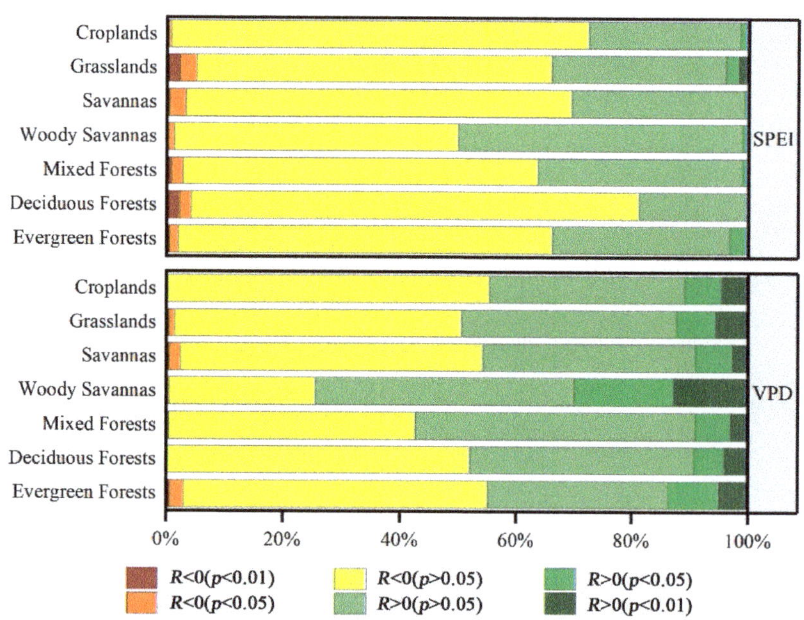

Figure 10. Frequency distribution of the areas with correlation coefficients (R) between $GPP_{drought\_month}$ (the average GPP of all drought months in a year) and $SPEI_{drought\_month}$ (the sum of the absolute values of SPEI for all drought months in a year; upper panel)/$VPD_{drought\_month}$ (the sum of the values of VPD for all drought months in a year; lower panel) for different vegetation types.

4. Discussion

4.1. Validation of Meteorological Interpolation Data

The accuracy of meteorological data seriously affects the results of drought event assessment. We compared the interpolation data with the European Centre for Medium-Range Weather Forecasts Reanalysis 5 (ERA5, https://cds.climate.copernicus.eu/#!/home (accessed on 30 July 2022)) at monthly time-steps for the same period for verification. The datasets from the ERA5 potentially supersede other reanalysis products due to their high spatial and temporal resolution, and they have been widely used in meteorology-related research [51,52]. We analyzed the performance of the dataset across different climate zones according to new and improved current (1980–2016) Köppen–Geiger classifications (Figure 11). The classification is based on the threshold values and seasonality of monthly air temperature and precipitation, which can be used for studies based on differences in climatic regimes [53]. We selected climate classifications with an area greater than 8% of the total area (BWk, ET, BSk, Cfa, Dwa, Cwa, and Dwb) for analysis. Together, these classifications account for more than 90% of the total area. The selected Köppen–Geiger classifications as listed in Table 3.

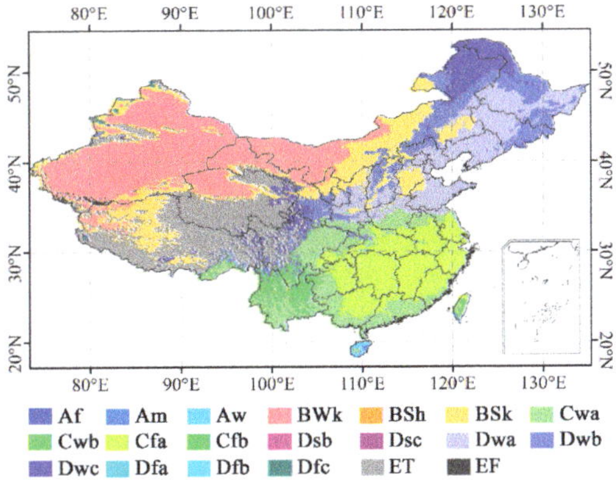

Figure 11. New and improved current (1980–2016) Köppen-Geiger classifications of China (for explanations of the abbreviations, please see Table 3).

The verification results of temperature, precipitation, and relative humidity data are shown in Figures 12–14, respectively. The interpolation data were in good agreement with the ERA5 data, with most R^2 greater than 0.90. The R^2 of temperature data for climate classifications in China were all equal to 0.99, while precipitation data were all ≥ 0.94, and relative humidity data were all ≥ 0.72. In general, the interpolation data met the data accuracy requirements.

4.2. Spatiotemporal Variation and Characteristics of Drought and GPP

By analyzing the annual average trends of SPEI and VPD, we observed that both indicated a drying trend in northern China, southwest China (especially Yunnan Province), and the Qinghai–Tibet Plateau. Since the 21st century, southwest China has become one of the regions with the largest frequency of droughts [54]. This region is mainly influenced by the southwest monsoon, having the multi-seasonal and multi-level drought states superimposed, giving rise to the complexity of the drought situation. Therefore, the characteristics of drought events and their impacts in this region have been widely studied in recent years [24,55,56]. Another area frequently affected by drought is northern China, a region that is an important, but ecologically fragile, grain-producing hotspot; thus, its drying trend requires attention from the authorities. Due to the sensitivity of the plateau region to climate change, many researchers have analyzed the temporal and spatial characteristics of drought in the Qinghai–Tibet Plateau [57,58]. The results using SPEI showed a large area of the Qinghai–Tibet Plateau region with a drying trend in summer and autumn, while the VPD showed a more significant drying trend in autumn. Notably, in the future, projections point to the increasing risk of drought in southwest China and the Qinghai–Tibet Plateau, with the risk being almost twice as high as that for the rest of China [59].

Table 3. Köppen–Geiger classifications and their area proportions of China (classifications with area proportions greater than 8% were selected for further analysis).

Köppen–Geiger Classifications	Abbreviations	Area Proportions (%)
Arid, desert, cold	BWk	21.806
Polar, tundra	ET	15.098
Arid, steppe, cold	BSk	12.792
Temperate, no dry season, hot summer	Cfa	11.570
Cold, dry winter, hot summer	Dwa	11.344
Temperate, dry winter, hot summer	Cwa	9.271
Cold, dry winter, warm summer	Dwb	8.545
Cold, dry winter, cold summer	Dwc	5.602
Temperate, dry winter, warm summer	Cwb	2.937
Tropical, savannah	Aw	0.293
Polar, frost	EF	0.264
Cold, no dry season, cold summer	Dfc	0.202
Temperate, no dry season, warm summer	Cfb	0.082
Tropical, monsoon	Am	0.073
Cold, no dry season, warm summer	Dfb	0.033
Cold, no dry season, hot summer	Dfa	0.028
Cold, dry summer, cold summer	Dsc	0.021
Arid, steppe, hot	BSh	0.014
Tropical, rainforest	Af	0.013
Cold, dry summer, warm summer	Dsb	0.012

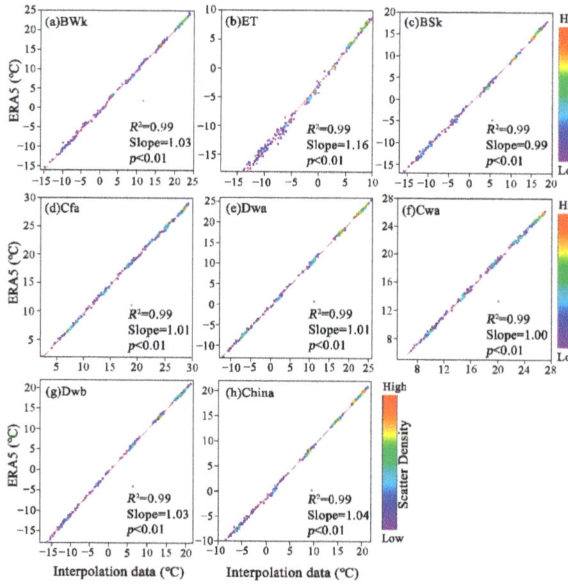

Figure 12. Comparison and verification of interpolation data and the European Centre for Medium-Range Weather Forecasts Reanalysis 5 (ERA5) data of temperature for selected Köppen–Geiger classifications (**a**) BWk, (**b**) ET, (**c**) BSk, (**d**) Cfa, (**e**) Dwa, (**f**) Cwa, (**g**) Dwb, and (**h**) across China (see Table 3 for explanations of the abbreviations).

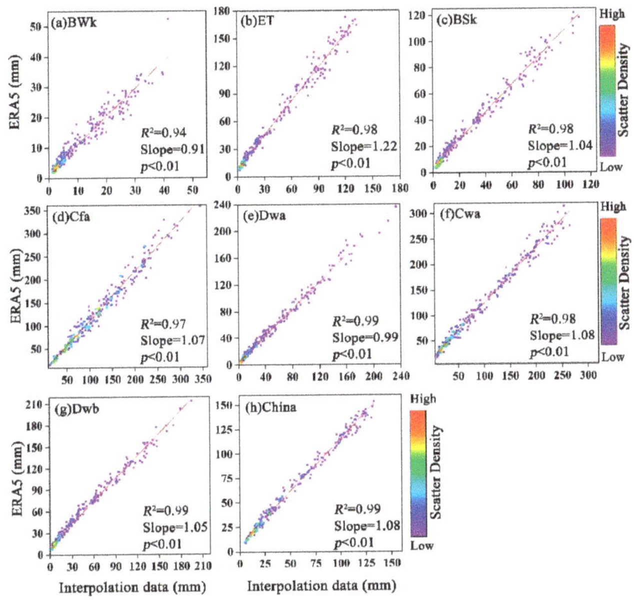

Figure 13. Comparison and verification of interpolation data and the European Centre for Medium-Range Weather Forecasts Reanalysis 5 (ERA5) data of precipitation for selected Köppen–Geiger classifications (**a**) BWk, (**b**) ET, (**c**) BSk, (**d**) Cfa, (**e**) Dwa, (**f**) Cwa, (**g**) Dwb, and (**h**) across China (see Table 3 for explanations of the abbreviations).

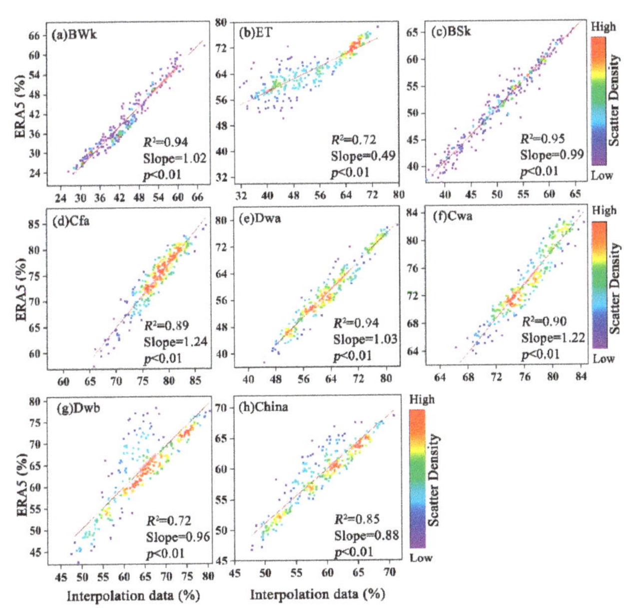

Figure 14. Comparison and verification of interpolation data and the European Centre for Medium-Range Weather Forecasts Reanalysis 5 (ERA5) data of relative humidity for selected Köppen–Geiger classifications (**a**) BWk, (**b**) ET, (**c**) BSk, (**d**) Cfa, (**e**) Dwa, (**f**) Cwa, (**g**) Dwb, and (**h**) across China (see Table 3 for explanations of the abbreviations).

As for the changing trend of vegetation GPP, we observed a significant upward trend in most parts of China. This phenomenon may be caused by several factors, such as atmospheric-CO_2 fertilization effects, the increase in temperature, and solar radiation [60,61]. In addition, ecological restoration measures, such as the implementation of the Grain for Green Project, also play a vital role [62]. However, studies have shown that the increase in forest cover may lead to a decrease in soil moisture, which in turn may have negative impacts on plant growth and water resources [63]. The regions with a decreasing trend of GPP are mainly concentrated in the Qinghai–Tibet Plateau, with the trend being more obvious in spring and winter. This may be due to drought events caused by low rainfall. Water availability is the most important factor influencing GPP changes in this region [64].

4.3. Effects of Drought on Vegetation GPP

Few studies have analyzed the relationship between vegetation GPP and water deficit in different seasons at the national level. In China, we observed significant differences in the effects of water deficit on vegetation in different seasons. For different seasons, GPP correlated (in both positive and negative directions) almost identically with VPD and SPEI (Figures 7 and 9). The relationship between SPEI and GPP and between VPD and GPP both indicated that the proportion of areas where water deficit negatively affected vegetation was greater in summer and autumn than in spring and winter. This result is consistent with previous studies in which vegetation activity was found to be more sensitive to drought events during the growing season [65]. In spring, vegetation grows well as long as there is sufficient soil moisture and atmospheric vapor pressure [66]. In addition, related studies have confirmed that spring vegetation greening may exacerbate summer drought [67,68]. Summer drought limits vegetation transpiration and directly reduces photosynthesis [69]. Similarly, autumn drought will also promote stomatal closure of leaves to decrease transpiration [70]. Moreover, the correlation between meteorological drought and hydrological drought is stronger in summer and autumn compared to spring and winter [71], which may be another crucial reason for the findings of our study. It was found that different vegetation had distinct sensitivity to water deficit in various seasons. All vegetation types were the most negatively affected by water deficit in the fall, except for grasslands, which were the most negatively affected in the summer. Previous studies only considered the relationship between drought index and vegetation in all periods of time, but did not pay attention to the relationship between drought index and GPP only during drought periods. This study found that there were some differences between the two results; in terms of spatial characteristics, drought had negative effects on vegetation GPP in northern China, southwest China, and the Qinghai–Tibet Plateau. It is worth noting that these areas are highly overlapped with areas where drought severity and frequency have been increasing in recent decades.

The serious adverse effects of water deficit on vegetation GPP were mainly concentrated in Inner Mongolia, of which the main vegetation type is grassland. Grasslands are among the most fragile ecosystems, where both climate change and human activities are significantly impacting their productivity [72]. Grasslands consist mainly of herbaceous plants whose roots can store a limited amount of water, resulting in a weak resistance to water deficit [73]. Furthermore, there are differences in the response of various grassland types. This study also investigated savannas and concluded that these were some of the more significantly negatively affected by water deficit among the different vegetation types. Indeed, more studies have focused on the response of forests and grasslands to drought, and relatively few studies have been conducted on the impact of drought on savannas [74,75]. Severe and persistent drought can significantly reduce savanna productivity, leading to structural and compositional changes and even widespread vegetation mortality [75]. As for croplands, this is the vegetation type that receives the most human intervention (e.g., irrigation), but it is still vulnerable to drought. Additional comparison of the differences in the response of crops, including detailed species, to drought would be better conducted at a regional scale. Environment monitoring by using remote sensing is an import measurement [76–82].

4.4. Potential Impact of Drought on the Ecosystem Regarding Future Climate Change

There is a common recognition that climate change poses an adverse effect on the natural ecosystem and human well-being [83]. In past decades, drought has led to severe effects on the terrestrial ecosystem in China [84]. Although we have investigate the impact of drought on vegetation GPP across different land cover types, excluding the land cover changed regions, the land use and land cover at the local scales may exacerbate or alleviate extreme climate events (such as drought) under climate change conditions [85]. It is necessary to consider land-use changes and drought patterns when exploring the impact of drought on ecosystems in the future by using ecological process models. The intensification of droughts, along with storm rainfall projected in future climate change scenarios, will increase soil carbon loss due to soil erosion [83]. However, the increasing CO_2 may compensated for the negative effects of drought on net carbon uptake in the future [86]. Therefore, we should try to consider more environmental factors to analyze the relationships between drought and vegetation GPP in future research.

4.5. Impacts, Limitations, and Future Work

In this research, we used the M–K test to analyze changing trends, and Pearson correlation analysis was used to study the correlation between drought indices and vegetation GPP. This method is fairly simple to calculate and can be applied to assess and predict the impact of droughts and other extreme climate events on vegetation. Furthermore, we innovatively used two indices, SPEI and VPD, to characterize drought. Although both indices are able to assess the degree of drought, they have different theoretical foundations that result in different conditions characterized by different drought indices. In order to measure the effects of drought on vegetation, the combination of these two indices can comprehensively depict the drought dynamics and effectively reduce the uncertainty induced by each individual index. In this study, the trends and the effects of water deficit on vegetation assessed using the two drought indices are generally consistent, which improves the reliability of the results and strengthens our findings. The results also revealed that drought characterized by SPEI was negatively correlated with vegetation during drought in more areas compared to VPD. The main reason for this phenomenon may be the significant spatial and temporal differences in the effects of temperature and precipitation on drought [87]. SPEI is based on both precipitation and potential evapotranspiration (PET). PET is strongly influenced by temperature and is closely related to VPD. By contrast, VPD is computed without including precipitation. Therefore, further research could be conducted to explore the contribution of temperature and precipitation to drought in different regions of China.

Nevertheless, this study has some limitations. The first is data accuracy. There were still differences between the GOSIF data and the data measured on site [88]. Because of the influence of soil moisture on light energy use efficiency, GPP, obtained using remote sensing, tends to underestimate the impact of droughts on vegetation [89]. In addition, this study only considered the impact of drought on GPP. Vegetation growth results from a combination of several factors, and may also be affected by other extreme weather events, such as high temperatures, low temperatures, and floods [90]. Thus, the complexity of the causes of drought and the influencing factors on vegetation growth bring certain uncertainty to the results. Despite these shortcomings, the methods and conclusions of this study can be used as a reference for determining the impact of droughts on vegetation. These can be used to take timely and effective measures to prevent and mitigate the adverse effects of droughts on vegetation in relevant areas.

5. Conclusions

In the context of global climate change, enhancing our understanding of the effects of droughts on vegetation is of great significance for maintaining the balance of the carbon cycle in the ecosystem. In this study, we used SPEI and VPD to analyze spatial-temporal patterns of drought in China from 2001 to 2020. The vegetation parameter GPP was used to evaluate vegetation productivity and analyze the annual and seasonal trends of drought and GPP under different vegetation types. The response of vegetation GPP to water deficit/drought under different vegetation types was assessed using Pearson correlation analysis. The conclusions of this study are as follows:

In terms of annual trends, China's SPEI and GPP showed an upward trend in most regions, whereas VPD showed an overall downward trend. Among the different seasons, both SPEI and VPD showed the most areas with an increasing trend of drought in spring and the least in autumn. GPP shows an upward trend, with the largest share of areas showing this trend in autumn.

About 70.7% of the pixels of the SPEI and GPP were positively correlated, which was greater than that of the negative correlation. About 66.2% of the total area showed that GPP and VPD were negatively correlated. The relationships of SPEI, VPD, and GPP indicated that the negative effects of water deficit on vegetation were more pronounced in summer and autumn. In spring, savannas were the most sensitive to water deficit. Grasslands were most sensitive to water deficit in both summer and winter. By contrast, croplands were the most sensitive to water deficit in autumn. During drought periods, vegetation growth was inhibited, especially in northern China, southwest China, and the Qinghai–Tibet Plateau.

Author Contributions: Conceptualization, X.W., R.Z. and Q.W.; data curation, R.Z.; formal analysis, X.W. and R.Z.; funding acquisition, Q.W.; investigation, X.W., R.Z., V.A.B., S.L., J.Q. and J.Z.; methodology, X.W. and Q.W.; project administration, Q.W.; software, X.W.; supervision, Q.W.; validation, V.A.B., S.L., J.Q., J.Z. and Q.W.; visualization, X.W.; writing—original draft, X.W.; writing—review and editing, R.Z., V.A.B., S.L., J.Q., J.Z. and Q.W. All authors have read and agreed to the published version of the manuscript.

Funding: This research was funded by the Natural Science Foundation of Fujian Province (2021J01627) and the National Natural Science Foundation of China (41601562).

Data Availability Statement: The SPEI and VPD data in this study are available from the corresponding author upon reasonable request.

Acknowledgments: Thanks to the Natural Science Foundation of Fujian Province (2021J01627) and the National Natural Science Foundation of China (No. 41601562) for funding this work. Special thanks to Wenping Yuan's team for providing the interpolated meteorological data.

Conflicts of Interest: The authors declare no conflict of interest.

References

1. Heim, R.R. A review of twentieth-century drought indices used in the United States. *Bull. Am. Meteorol. Soc.* **2002**, *83*, 1149–1165. [CrossRef]
2. Mukherjee, S.; Mishra, A.; Trenberth, K.E. Climate Change and Drought: A Perspective on Drought Indices. *Curr. Clim. Change Rep.* **2018**, *4*, 145–163. [CrossRef]
3. Wang, Q.; Shi, P.; Lei, T.; Geng, G.; Liu, J.; Mo, X.; Li, X.; Zhou, H.; Wu, J. The alleviating trend of drought in the Huang-Huai-Hai Plain of China based on the daily SPEI. *Int. J. Climatol.* **2015**, *35*, 3760–3769. [CrossRef]
4. Wang, Q.; Qi, J.; Wu, H.; Zeng, Y.; Shui, W.; Zeng, J.; Zhang, X. Freeze-Thaw cycle representation alters response of watershed hydrology to future climate change. *Catena* **2020**, *195*, 104767. [CrossRef]
5. Zhang, R.; Wu, X.; Zhou, X.; Ren, B.; Zeng, J.; Wang, Q. Investigating the effect of improved drought events extraction method on spatiotemporal characteristics of drought. *Theor. Appl. Climatol.* **2022**, *147*, 395–408. [CrossRef]
6. Vicente-Serrano, S.M.; Quiring, S.M.; Pena-Gallardo, M.; Yuan, S.; Dominguez-Castro, F. A review of environmental droughts: Increased risk under global warming? *Earth-Sci. Rev.* **2020**, *201*, 102953. [CrossRef]
7. Chen, H.; Sun, J. Changes in Drought Characteristics over China Using the Standardized Precipitation Evapotranspiration Index. *J. Clim.* **2015**, *28*, 5430–5447. [CrossRef]
8. Yu, M.; Li, Q.; Hayes, M.J.; Svoboda, M.D.; Heim, R.R. Are droughts becoming more frequent or severe in China based on the Standardized Precipitation Evapotranspiration Index: 1951–2010? *Int. J. Climatol.* **2014**, *34*, 545–558. [CrossRef]
9. Zhang, Q.; Yao, Y.; Li, Y.; Huang, J.; Ma, Z.; Wang, Z.; Wang, S.; Wang, Y.; Zhang, Y. Causes and Changes of Drought in China: Research Progress and Prospects. *J. Meteorol. Res.* **2020**, *34*, 460–481. [CrossRef]

10. Begueria, S.; Vicente-Serrano, S.M.; Reig, F.; Latorre, B. Standardized precipitation evapotranspiration index (SPEI) revisited: Parameter fitting, evapotranspiration models, tools, datasets and drought monitoring. *Int. J. Climatol.* **2014**, *34*, 3001–3023. [CrossRef]
11. Su, B.; Huang, J.; Fischer, T.; Wang, Y.; Kundzewicz, Z.W.; Zhai, J.; Sun, H.; Wang, A.; Zeng, X.; Wang, G.; et al. Drought losses in China might double between the 1.5 degrees C and 2.0 degrees C warming. *Proc. Natl. Acad. Sci. USA* **2018**, *115*, 10600–10605. [CrossRef] [PubMed]
12. Zeng, J.; Zhang, R.; Qu, Y.; Bentod, V.A.; Zhou, T.; Lin, Y.; Wu, X.; Qi, J.; Shui, W.; Wang, Q. Improving the drought monitoring capability of VHI at the global scale via ensemble indices for various vegetation types from 2001 to 2018. *Weather Clim. Extrem.* **2022**, *35*, 100412. [CrossRef]
13. McKee, T.B.; Doesken, N.J.; Kleist, J.R. The Relationship of Drought Frequency and Duration to Time Scales. In Proceedings of the 8th Conference on Applied Climatology, Anaheim CA, USA, 17–22 January 1993; American Meteorological Society: Boston, MA, USA, 1993.
14. Palmer, W.C. *Meteorological Drought*; US Department of Commerce, Weather Bureau: Washington, DC, USA, 1965; Volume 30.
15. Vicente-Serrano, S.M.; Begueria, S.; Lopez-Moreno, J.I. A Multiscalar Drought Index Sensitive to Global Warming: The Standardized Precipitation Evapotranspiration Index. *J. Clim.* **2010**, *23*, 1696–1718. [CrossRef]
16. Vicente-Serrano, S.M.; Begueria, S.; Lorenzo-Lacruz, J.; Julio Camarero, J.; Lopez-Moreno, J.I.; Azorin-Molina, C.; Revuelto, J.; Moran-Tejeda, E.; Sanchez-Lorenzo, A. Performance of Drought Indices for Ecological, Agricultural, and Hydrological Applications. *Earth Interact.* **2012**, *16*, 1–27. [CrossRef]
17. Grossiord, C.; Buckley, T.N.; Cernusak, L.A.; Novick, K.A.; Poulter, B.; Siegwolf, R.T.W.; Sperry, J.S.; McDowell, N.G. Plant responses to rising vapor pressure deficit. *New Phytol.* **2020**, *226*, 1550–1566. [CrossRef]
18. Liu, L.; Gudmundsson, L.; Hauser, M.; Qin, D.; Li, S.; Seneviratne, S.I. Soil moisture dominates dryness stress on ecosystem production globally. *Nat. Commun.* **2020**, *11*, 4892. [CrossRef]
19. Ding, Y.; Xu, J.; Wang, X.; Peng, X.; Cai, H. Spatial and temporal effects of drought on Chinese vegetation under different coverage levels. *Sci. Total Environ.* **2020**, *716*, 137166. [CrossRef]
20. Zhang, R.; Qi, J.; Leng, S.; Wang, Q. Long-Term Vegetation Phenology Changes and Responses to Preseason Temperature and Precipitation in Northern China. *Remote Sens.* **2022**, *14*, 1396. [CrossRef]
21. Sperry, J.S.; Love, D.M. What plant hydraulics can tell us about responses to climate-change droughts. *New Phytol.* **2015**, *207*, 14–27. [CrossRef]
22. Leng, S.; Huete, A.; Cleverly, J.; Yu, Q.; Zhang, R.; Wang, Q. Spatiotemporal Variations of Dryland Vegetation Phenology Revealed by Satellite-Observed Fluorescence and Greenness across the North Australian Tropical Transect. *Remote Sens.* **2022**, *14*, 2985. [CrossRef]
23. Leng, S.; Huete, A.; Cleverly, J.; Gao, S.; Yu, Q.; Meng, X.; Qi, J.; Zhang, R.; Wang, Q. Assessing the Impact of Extreme Droughts on Dryland Vegetation by Multi-Satellite Solar-Induced Chlorophyll Fluorescence. *Remote Sens.* **2022**, *14*, 1581. [CrossRef]
24. Li, X.; Li, Y.; Chen, A.; Gao, M.; Slette, I.J.; Piao, S. The impact of the 2009/2010 drought on vegetation growth and terrestrial carbon balance in Southwest China. *Agric. For. Meteorol.* **2019**, *269*, 239–248. [CrossRef]
25. Zhang, Q.; Kong, D.; Singh, V.P.; Shi, P. Response of vegetation to different time-scales drought across China: Spatiotemporal patterns, causes and implications. *Glob. Planet. Change* **2017**, *152*, 1–11. [CrossRef]
26. Wu, D.; Zhao, X.; Liang, S.; Zhou, T.; Huang, K.; Tang, B.; Zhao, W. Time-lag effects of global vegetation responses to climate change. *Glob. Change Biol.* **2015**, *21*, 3520–3531. [CrossRef] [PubMed]
27. Zheng, K.; Wei, J.-Z.; Pei, J.-Y.; Cheng, H.; Zhang, X.-L.; Huang, F.-Q.; Li, F.-M.; Ye, J.-S. Impacts of climate change and human activities on grassland vegetation variation in the Chinese Loess Plateau. *Sci. Total Environ.* **2019**, *660*, 236–244. [CrossRef]
28. Campbell, J.E.; Berry, J.A.; Seibt, U.; Smith, S.J.; Montzka, S.A.; Launois, T.; Belviso, S.; Bopp, L.; Laine, M. Large historical growth in global terrestrial gross primary production. *Nature* **2017**, *544*, 84–87. [CrossRef]
29. Verma, M.; Schimel, D.; Evans, B.; Frankenberg, C.; Beringer, J.; Drewry, D.T.; Magney, T.; Marang, I.; Hutley, L.; Moore, C.; et al. Effect of environmental conditions on the relationship between solar-induced fluorescence and gross primary productivity at an OzFlux grassland site. *J. Geophys. Res. Biogeosci.* **2017**, *122*, 716–733. [CrossRef]
30. Zhang, Y.; Xiao, X.; Wu, X.; Zhou, S.; Zhang, G.; Qin, Y.; Dong, J. Data Descriptor: A global moderate resolution dataset of gross primary production of vegetation for 2000–2016. *Sci. Data* **2017**, *4*, 170165. [CrossRef]
31. Chu, H.; Venevsky, S.; Wu, C.; Wang, M. NDVI-based vegetation dynamics and its response to climate changes at Amur-Heilongjiang River Basin from 1982 to 2015. *Sci. Total Environ.* **2019**, *650*, 2051–2062. [CrossRef]
32. Liang, W.; Yang, Y.; Fan, D.; Guan, H.; Zhang, T.; Long, D.; Zhou, Y.; Bai, D. Analysis of spatial and temporal patterns of net primary production and their climate controls in China from 1982 to 2010. *Agric. For. Meteorol.* **2015**, *204*, 22–36. [CrossRef]
33. Ma, D.; Yin, Y.; Wu, S.; Zheng, D. Sensitivity of arid/humid patterns in China to future climate change under high emission scenario. *Acta Geogr. Sin.* **2019**, *74*, 857–874. [CrossRef]
34. He, J.; Yang, K.; Tang, W.; Lu, H.; Qin, J.; Chen, Y.; Li, X. The first high-resolution meteorological forcing dataset for land process studies over China. *Sci. Data* **2020**, *7*, 25. [CrossRef] [PubMed]
35. Wang, Q.; Zeng, J.; Qi, J.; Zhang, X.; Zeng, Y.; Shui, W.; Xu, Z.; Zhang, R.; Wu, X.; Cong, J. A multi-scale daily SPEI dataset for drought characterization at observation stations over mainland China from 1961 to 2018. *Earth Syst. Sci. Data* **2021**, *13*, 331–341. [CrossRef]

36. Wang, Q.; Zhang, R.; Qi, J.; Zeng, J.; Wu, J.; Shui, W.; Wu, X.; Li, J. An improved daily standardized precipitation index dataset for mainland China from 1961 to 2018. *Sci. Data* **2022**, *9*, 124. [CrossRef]
37. He, B.; Wang, H.L.; Wang, Q.F.; Di, Z.H. A quantitative assessment of the relationship between precipitation deficits and air temperature variations. *J. Geophys. Res. Atmos.* **2015**, *120*, 5951–5961. [CrossRef]
38. Hutchinson, M.F. Interpolating mean rainfall using thin-plate smoothing splines. *Int. J. Geogr. Inf. Syst.* **1995**, *9*, 385–403. [CrossRef]
39. Yuan, W.; Xu, B.; Chen, Z.; Xia, J.; Xu, W.; Chen, Y.; Wu, X.; Fu, Y. Validation of China-wide interpolated daily climate variables from 1960 to 2011. *Theor. Appl. Climatol.* **2015**, *119*, 689–700. [CrossRef]
40. Li, X.; Xiao, J. A global, 0.05-degree product of solar-induced chlorophyll fluorescence derived from OCO-2, MODIS, and reanalysis data. *Remote Sens.* **2019**, *11*, 517. [CrossRef]
41. Zhang, X.; Wang, H.; Yan, H.; Ai, J. Analysis of spatio-temporal changes of gross primary productivity in China from 2001 to 2018 based on Romote Sensing. *Acta Ecol. Sin.* **2021**, *41*, 6351–6362.
42. Khadka, D.; Babel, M.S.; Shrestha, S.; Virdis, S.G.P.; Collins, M. Multivariate and multi-temporal analysis of meteorological drought in the northeast of Thailand. *Weather Clim. Extrem.* **2021**, *34*, 100399. [CrossRef]
43. Zhao, A.; Zhang, A.; Cao, S.; Liu, X.; Liu, J.; Cheng, D. Responses of vegetation productivity to multi-scale drought in Loess Plateau, China. *Catena* **2018**, *163*, 165–171. [CrossRef]
44. Kendall, M.G. *Kendall Rank Correlation Methods*; Griffin: London, UK, 1975.
45. Mann, H.B. Nonparametric Tests Against Trend. *Econometrica* **1945**, *13*, 245–259. [CrossRef]
46. Gocic, M.; Trajkovic, S. Analysis of changes in meteorological variables using Mann-Kendall and Sen's slope estimator statistical tests in Serbia. *Glob. Planet. Change* **2013**, *100*, 172–182. [CrossRef]
47. Tosunoglu, F.; Kisi, O. Trend Analysis of Maximum Hydrologic Drought Variables Using Mann-Kendall and Sen's Innovative Trend Method. *River Res. Appl.* **2017**, *33*, 597–610. [CrossRef]
48. Wang, Q.; Tang, J.; Zeng, J.; Qu, Y.; Zhang, Q.; Shui, W.; Wang, W.; Yi, L.; Leng, S. Spatial-temporal evolution of vegetation evapotranspiration in Hebei Province, China. *J. Integr. Agric.* **2018**, *17*, 2107–2117. [CrossRef]
49. Ghosh, K.G. Analysis of rainfall trends and its spatial patterns during the last century over the Gangetic West Bengal, Eastern India. *J. Geovis. Spat. Anal.* **2018**, *2*, 15. [CrossRef]
50. Schober, P.; Boer, C.; Schwarte, L.A. Correlation Coefficients: Appropriate Use and Interpretation. *Anesth. Analg.* **2018**, *126*, 1763–1768. [CrossRef]
51. Mukherjee, S.; Mishra, A.K. Increase in Compound Drought and Heatwaves in a Warming World. *Geophys. Res. Lett.* **2021**, *48*, e2020GL090617. [CrossRef]
52. Hersbach, H.; Bell, B.; Berrisford, P.; Hirahara, S.; Horanyi, A.; Munoz-Sabater, J.; Nicolas, J.; Peubey, C.; Radu, R.; Schepers, D.; et al. The ERA5 global reanalysis. *Q. J. R. Meteorol. Soc.* **2020**, *146*, 1999–2049. [CrossRef]
53. Beck, H.E.; Zimmermann, N.E.; McVicar, T.R.; Vergopolan, N.; Berg, A.; Wood, E.F. Present and future Koppen-Geiger climate classification maps at 1-km resolution. *Sci. Data* **2018**, *5*, 180214. [CrossRef]
54. Zhang, Q.; Yao, Y.; Li, Y.; Huang, J.; Ma, Z.; Wang, Z.; Wang, S.; Wang, Y.; Zhang, Y. Progress and prospect on the study of causes and variation regularity of droughts in China. *Acta Meteorol. Sin.* **2020**, *78*, 500–521. [CrossRef]
55. Sun, X.; Lai, P.; Wang, S.; Song, L.; Ma, M.; Han, X. Monitoring of Extreme Agricultural Drought of the Past 20 Years in Southwest China Using GLDAS Soil Moisture. *Remote Sens.* **2022**, *14*, 1323. [CrossRef]
56. Mokhtar, A.; He, H.; Alsafadi, K.; Mohammed, S.; Ayantobo, O.O.; Elbeltagi, A.; Abdelwahab, O.M.M.; Zhao, H.; Quan, Y.; Abdo, H.G.; et al. Assessment of the effects of spatiotemporal characteristics of drought on crop yields in southwest China. *Int. J. Climatol.* **2022**, *42*, 3056–3075. [CrossRef]
57. Gao, Y.; Li, X.; Leung, L.R.; Chen, D.; Xu, J. Aridity changes in the Tibetan Plateau in a warming climate. *Environ. Res. Lett.* **2015**, *10*, 034013. [CrossRef]
58. Feng, W.; Lu, H.; Yao, T.; Yu, Q. Drought characteristics and its elevation dependence in the Qinghai-Tibet plateau during the last half-century. *Sci. Rep.* **2020**, *10*, 14323. [CrossRef]
59. Wang, L.; Chen, W. A CMIP5 multimodel projection of future temperature, precipitation, and climatological drought in China. *Int. J. Climatol.* **2014**, *34*, 2059–2078. [CrossRef]
60. Zhao, M.; Running, S.W. Drought-Induced Reduction in Global Terrestrial Net Primary Production from 2000 Through 2009. *Science* **2010**, *329*, 940–943. [CrossRef] [PubMed]
61. Yan, H.; Wang, S.-Q.; Cao, Y.; Xu, L.-L.; Wu, M.-x.; Cheng, L.; Mao, L.-X.; Zhang, X.-Z.; Liu, Y.-F.; Wang, Y.-F.; et al. Multi-model analysis of climate impacts on plant photosynthesis in China during 2000–2015. *Int. J. Climatol.* **2019**, *39*, 5539–5555. [CrossRef]
62. Qiu, L.; Chen, Y.; Wu, Y.; Xue, Q.; Shi, Z.; Lei, X.; Liao, W.; Zhao, F.; Wang, W. The Water Availability on the Chinese Loess Plateau since the Implementation of the Grain for Green Project as Indicated by the Evaporative Stress Index. *Remote Sens.* **2021**, *13*, 3302. [CrossRef]
63. Qiu, L.; Wu, Y.; Shi, Z.; Yu, M.; Zhao, F.; Guan, Y. Quantifying spatiotemporal variations in soil moisture driven by vegetation restoration on the Loess Plateau of China. *J. Hydrol.* **2021**, *600*, 126580. [CrossRef]
64. Zhang, T.; Zhang, Y.; Xu, M.; Zhu, J.; Chen, N.; Jiang, Y.; Huang, K.; Zu, J.; Liu, Y.; Yu, G. Water availability is more important than temperature in driving the carbon fluxes of an alpine meadow on the Tibetan Plateau. *Agric. For. Meteorol.* **2018**, *256*, 22–31. [CrossRef]

65. Hua, T.; Wang, X.; Zhang, C.; Lang, L.; Li, H. Response of vegetation activity to drought in Northern China. *Land Degrad. Dev.* **2017**, *28*, 1913–1921. [CrossRef]
66. Wu, X.; Liu, H.; Li, X.; Liang, E.; Beck, P.S.A.; Huang, Y. Seasonal divergence in the interannual responses of Northern Hemisphere vegetation activity to variations in diurnal climate. *Sci. Rep.* **2016**, *6*, 19000. [CrossRef] [PubMed]
67. Buermann, W.; Bikash, P.R.; Jung, M.; Burn, D.H.; Reichstein, M. Earlier springs decrease peak summer productivity in North American boreal forests. *Environ. Res. Lett.* **2013**, *8*, 024027. [CrossRef]
68. Lian, X.; Piao, S.; Li, L.Z.X.; Li, Y.; Huntingford, C.; Ciais, P.; Cescatti, A.; Janssens, I.A.; Penuelas, J.; Buermann, W.; et al. Summer soil drying exacerbated by earlier spring greening of northern vegetation. *Sci. Adv.* **2020**, *6*, eaax0255. [CrossRef]
69. Zhang, L.; Xiao, J.; Zhou, Y.; Zheng, Y.; Li, J.; Xiao, H. Drought events and their effects on vegetation productivity in China. *Ecosphere* **2016**, *7*, e01591. [CrossRef]
70. Kang, W.; Wang, T.; Liu, S. The Response of Vegetation Phenology and Productivity to Drought in Semi-Arid Regions of Northern China. *Remote Sens.* **2018**, *10*, 727. [CrossRef]
71. Ding, Y.; Xu, J.; Wang, X.; Cai, H.; Zhou, Z.; Sun, Y.; Shi, H. Propagation of meteorological to hydrological drought for different climate regions in China. *J. Environ. Manag.* **2021**, *283*, 111980. [CrossRef]
72. Wang, X.; Pan, S.; Pan, N.; Pan, P. Grassland productivity response to droughts in northern China monitored by satellite-based solar-induced chlorophyll fluorescence. *Sci. Total Environ.* **2022**, *830*, 154550. [CrossRef]
73. Xu, H.; Wang, X.; Zhao, C. Drought sensitivity of vegetation photosynthesis along the aridity gradient in northern China. *Int. J. Appl. Earth Obs. Geoinf.* **2021**, *102*, 102418. [CrossRef]
74. Case, M.F.; Wigley-Coetsee, C.; Nzima, N.; Scogings, P.F.; Staver, A.C. Severe drought limits trees in a semi-arid savanna. *Ecology* **2019**, *100*, e02842. [CrossRef] [PubMed]
75. Sankaran, M.; Staver, C. Droughts and the ecological future of tropical savanna vegetation. *J. Ecol.* **2019**, *107*, 1531–1549. [CrossRef]
76. Wang, J.; Ding, J.; Li, G.; Liang, J.; Yu, D.; Aishan, T.; Zhang, F.; Yang, J.; Abuldmiti, A.; Liu, J. Dynamic detection of water surface area of Ebinur Lake using multi-source satellite data (Landsat and Sentinel-1A) and its responses to changing environment. *Catena* **2019**, *177*, 189–201. [CrossRef]
77. Wang, J.; Ding, J.; Yu, D.; Ma, X.; Zhang, Z.; Ge, X.; Teng, D.; Li, X.; Liang, J.; Lizag, I.; et al. Capability of Sentinel-2 MSI data for monitoring and mapping of soil salinity in dry and wet seasons in the Ebinur Lake region, Xinjiang, China. *Geoderma* **2019**, *353*, 172–187. [CrossRef]
78. Wang, J.; Ding, J.; Yu, D.; Teng, D.; He, B.; Chen, X.; Ge, X.; Zhang, Z.; Wang, Y.; Yang, X.; et al. Machine learning-based detection of soil salinity in an arid desert region, Northwest China: A comparison between Landsat-8 OLI and Sentinel-2 MSI. *Sci. Total Environ.* **2020**, *707*, 136092. [CrossRef] [PubMed]
79. Wang, J.; Shi, T.; Yu, D.; Teng, D.; Ge, X.; Zhang, Z.; Yang, X.; Wang, H.; Wu, G. Ensemble machine-learning-based framework for estimating total nitrogen concentration in water using drone-borne hyperspectral imagery of emergent plants: A case study in an arid oasis, NW China. *Environ. Pollut.* **2020**, *266*, 115412. [CrossRef] [PubMed]
80. Wang, J.; Hu, X.; Shi, T.; He, L.; Hu, W.; Wu, G. Assessing toxic metal chromium in the soil in coal mining areas via proximal sensing: Prerequisites for land rehabilitation and sustainable development. *Geoderma* **2022**, *405*, 115399. [CrossRef]
81. Yang, X.; Chen, Y.; Wang, J. Combined use of Sentinel-2 and Landsat 8 to monitor water surface area dynamics using Google Earth Engine. *Remote Sens. Lett.* **2020**, *11*, 687–696. [CrossRef]
82. Ghasemloo, N.; Matkan, A.A.; Alimohammadi, A.; Aghighi, H.; Mirbagheri, B. Estimating the Agricultural Farm Soil Moisture Using Spectral Indices of Landsat 8, and Sentinel-1, and Artificial Neural Networks. *J. Geovis. Spat. Anal.* **2022**, *6*, 19. [CrossRef]
83. Frank, D.; Reichstein, M.; Bahn, M.; Thonicke, K.; Frank, D.; Mahecha, M.D.; Smith, P.; Van der Velde, M.; Vicca, S.; Babst, F.; et al. Effects of climate extremes on the terrestrial carbon cycle: Concepts, processes and potential future impacts. *Glob. Change Biol.* **2015**, *21*, 2861–2880. [CrossRef]
84. Deng, Y.; Wu, D.; Wang, X.; Xie, Z. Responding time scales of vegetation production to extreme droughts over China. *Ecol. Indic.* **2022**, *136*, 108630. [CrossRef]
85. Kim, J.; Choi, J.; Choi, C.; Park, S. Impacts of changes in climate and land use/land cover under IPCC RCP scenarios on streamflow in the Hoeya River Basin, Korea. *Sci. Total Environ.* **2013**, *452*, 181–195. [CrossRef] [PubMed]
86. Roy, J.; Picon-Cochard, C.; Augusti, A.; Benot, M.-L.; Thiery, L.; Darsonville, O.; Landais, D.; Piel, C.; Defossez, M.; Devidal, S.; et al. Elevated CO_2 maintains grassland net carbon uptake under a future heat and drought extreme. *Proc. Natl. Acad. Sci. USA* **2016**, *113*, 6224–6229. [CrossRef] [PubMed]
87. Sun, C.; Zhu, L.; Liu, Y.; Hao, Z.; Zhang, J. Changes in the drought condition over northern East Asia and the connections with extreme temperature and precipitation indices. *Glob. Planet. Change* **2021**, *207*, 103645. [CrossRef]
88. Gao, H.; Liu, S.; Lu, W.; Smith, A.R.; Valbuena, R.; Yan, W.; Wang, Z.; Li, X.; Peng, X.; Li, Q.; et al. Global Analysis of the Relationship between Reconstructed Solar-Induced Chlorophyll Fluorescence (SIF) and Gross Primary Production (GPP). *Remote Sens.* **2021**, *13*, 2824. [CrossRef]
89. Flach, M.; Brenning, A.; Gans, F.; Reichstein, M.; Sippel, S.; Mahecha, M.D. Vegetation modulates the impact of climate extremes on gross primary production. *Biogeosciences* **2021**, *18*, 39–53. [CrossRef]
90. Zhu, X.; Zhang, S.; Liu, T.; Liu, Y. Impacts of Heat and Drought on Gross Primary Productivity in China. *Remote Sens.* **2021**, *13*, 378. [CrossRef]

remote sensing

MDPI

Article

Spatiotemporal Variations of Dryland Vegetation Phenology Revealed by Satellite-Observed Fluorescence and Greenness across the North Australian Tropical Transect

Song Leng [1,2], Alfredo Huete [2], Jamie Cleverly [3], Qiang Yu [4], Rongrong Zhang [1] and Qianfeng Wang [1,5,*]

1. Fujian Provincial Key Laboratory of Remote Sensing of Soil Erosion, College of Environment & Safety Engineering, Fuzhou University, Fuzhou 350116, China; s_leng2022@126.com (S.L.); n190620018@fzu.edu.cn (R.Z.)
2. Faculty of Science, University of Technology Sydney, Ultimo, NSW 2007, Australia; alfredo.huete@uts.edu.au
3. College of Science and Engineering, James Cook University, Cairns, QLD 4878, Australia; dr.jr.cleverly@gmail.com
4. State Key Laboratory of Soil Erosion and Dryland Farming on the Loess Plateau, Northwest A&F University, Xianyang 712100, China; yuq@nwafu.edu.cn
5. Key Lab of Spatial Data Mining & Information Sharing, Ministry of Education of China, Fuzhou 350116, China

* Correspondence: wangqianfeng@fzu.edu.cn

Citation: Leng, S.; Huete, A.; Cleverly, J.; Yu, Q.; Zhang, R.; Wang, Q. Spatiotemporal Variations of Dryland Vegetation Phenology Revealed by Satellite-Observed Fluorescence and Greenness across the North Australian Tropical Transect. *Remote Sens.* **2022**, *14*, 2985. https://doi.org/10.3390/rs14132985

Academic Editors: Jingzhe Wang, Zhongwen Hu, Yangyi Wu and Jie Zhang

Received: 19 May 2022
Accepted: 19 June 2022
Published: 22 June 2022

Publisher's Note: MDPI stays neutral with regard to jurisdictional claims in published maps and institutional affiliations.

Abstract: Accurate characterization of spatial patterns and temporal variations in dryland vegetation is of great importance for improving our understanding of terrestrial ecosystem functioning under changing climates. Here, we explored the spatiotemporal variability of dryland vegetation phenology using satellite-observed Solar-Induced chlorophyll Fluorescence (SIF) and the Enhanced Vegetation Index (EVI) along the North Australian Tropical Transect (NATT). Substantial impacts of extreme drought and intense wetness on the phenology and productivity of dryland vegetation are observed by both SIF and EVI, especially in the arid/semiarid interior of Australia without detectable seasonality in the dry year of 2018–2019. The greenness-based vegetation index (EVI) can more accurately capture the seasonal and interannual variation in vegetation production than SIF (EVI r^2: 0.47~0.86, SIF r^2: 0.47~0.78). However, during the brown-down periods, the rate of decline in EVI is evidently slower than that in SIF and in situ measurement of gross primary productivity (GPP), due partially to the advanced seasonality of absorbed photosynthetically active radiation. Over 70% of the variability of EVI (except for Hummock grasslands) and 40% of the variability of SIF (except for shrublands) can be explained by the water-related drivers (rainfall and soil moisture). By contrast, air temperature contributed to 25~40% of the variability of the effective fluorescence yield (SIF_{yield}) across all biomes. In spite of high retrieval noises and variable accuracy in phenological metrics (MAE: 8~60 days), spaceborne SIF observations, offsetting the drawbacks of greenness-based phenology products with a potentially lagged end of the season, have the promising capability of mapping and characterizing the spatiotemporal dynamics of dryland vegetation phenology.

Keywords: SIF; EVI; phenology; NATT

1. Introduction

Vegetation phenology, the study of the periodic biological life cycle events of plants, is a critical regulator of carbon and water cycling in terrestrial ecosystems [1]. The trend of global warming has aroused great interest in understanding and monitoring the dynamics of vegetation phenology under the changing climate [2]. As a valuable indicator of climate variability and ecosystem responses [1,3], accurate measurement of land surface phenology (LSP) is crucial for better explicating the land–atmosphere–energy exchange and its representation in terrestrial biosphere models [4–7].

Phenological studies are usually conducted at the species- or ecosystem-level through ground-based field techniques [4], such as visual inspection, eddy covariance flux towers, near-surface spectral radiometers, and digital cameras [8]. Satellite-based observation, providing a unique and feasible way for examining broader-scale phenomena [4], has notably expanded the horizon of traditional phenology observation [2]. Vegetation indices, which are combinations of surface reflectance at two or more wavelengths designed to highlight a particular property of vegetation [9–11], can depict spatiotemporal patterns of the timing of plant growth, senescence, and dormancy at seasonal and inter-annual time scales [12]. The enhanced vegetation index [9], normalized difference vegetation index [13], and leaf area index (LAI) derived from optical satellite sensors are widely applied for tracking large-scale vegetation seasonality.

Dryland vegetation, covering approximately 40% of the global land surface [14], is exceedingly susceptible to climate conditions, mainly owing to chronic physiological stress [10,15–17]. Numerous phenology studies based on satellite remote sensing have examined temperature-driven ecosystems of the Northern Hemisphere [4,18,19]; however, the phenological dynamics of dryland vegetation under climate variability and extreme events have not been thoroughly investigated as a result of the intrinsic sensitivity and complexity [15]. Given that the mixed and heterogeneous arrangement of water-limited ecosystems confounds the extraction of the phenophase, Walker et al. (2014) assessed dryland vegetation phenology across an elevation gradient in Arizona, USA, by fusing synthetic Moderate Resolution Imaging Spectroradiometer (MODIS) and Landsat imagery [15]. The significant spatiotemporal variations in savanna biomes with diverse tree-grass ratios implies that grass-containing xeric savanna is especially sensitive and vulnerable to hydro-climatic variability [20]. Besides, previous studies found that growing season retrieved from the greenness-based approach overestimates the photosynthetically active duration, indicating a divergence between greenness and function [21,22]. On the basis of the complex composition of plant functional types, as well as the highly dynamic climate in Australia, accurately assessing the phenological characteristics of dryland vegetation remains challenging [23].

In contrast to the reflectance-based approaches, fluorescence measurements directly from the core of photosynthetic machinery present a fresh manner of remotely sensing vegetation growth and response. Recent studies found that satellite-based observations of SIF can detect interannual and seasonal variations in gross primary production of vegetation in North America [24–26]. Wu et al. (2018) found strong spatiotemporal consistency between satellite-based SIF and GPP products across the Contiguous United States [27]. Although photosynthesis and greenness exhibit an analogous seasonality in deciduous forests, satellite chlorophyll fluorescence measurements reveal large-scale decoupling of photosynthesis and greenness dynamics in boreal evergreen forests [28,29]. Continuous observations of satellite-based SIF and vegetation indices contribute to improved understanding of large-scale variations in vegetation phenology and function [21].

By comparing EVI and the photochemical reflectivity index, Smith et al. (2018) demonstrated that chlorophyll fluorescence derived from GOME-2 more accurately captures seasonal and interannual variations of gross primary productivity across dryland ecosystems of southwestern North America [30]. Through evaluating the SIF time series derived from GOME-2 with tower-based GPP in Australia, Sanders et al. (2016) found that natural biome types, such as savanna and open woodland, showed weaker correlations relative to agricultural biomes [31]. Taking into account the sparse spatial resolution of the existing spaceborne SIF product, as well as the varying mixtures of trees, shrubs, and grasses in Australia's dryland ecosystems, a high-resolution contiguous SIF product (SIF_{oco2_005}) was utilized to explore the phenology dynamics of dryland vegetation along the North Australian Tropical Transect [23]. They found that these up-scaled SIF data at a 0.05° spatial resolution outperformed EVI for characterizing seasonal onset and senescence of dryland vegetation and suggested it had the potential for large-scale mapping of phenology dynamics as opposed to traditional reflectance-based vegetation indices. Nevertheless, SIF_{oco2_005} was generated by the OCO-2 native SIF along with MODIS reflectance using machine learning [32], and it therefore showed an enhanced accuracy in determining the phenological transition of GPP derived from flux tower measurement as a consequence of it containing both structural and physiological information. However, whether the original SIF observation can accurately capture the phenology dynamics of Australia's dryland vegetation under climate variability remains unclear.

In arid and semiarid ecosystems, rainfall strongly affects the strength and occurrence of photosynthetic and respiratory activities and is the dominant driver for vegetation phenology and productivity [33–35]. The seasonality of Acacia-dominated woodlands (as measured by EVI) was found to show substantial responsiveness to hydroclimatic variability [4]. Additionally, approximately 80% of the variations in the length of the growing season of major dryland biomes along the NATT could be attributed to the variability of annual precipitation [4]. Nevertheless, there is a knowledge gap regarding how and to what extent major environmental and climatic drivers determine the phenological dynamics of fluorescence (as a surrogate of photosynthesis).

The objectives of this study were: (1) to evaluate the spatial patterns and seasonal variations of dryland vegetation phenology across NATT under a dynamic climate; (2) to contrast the differences in phenological metrics derived from satellite-observed SIF and EVI; (3) to determine the dominant factor that drives the interannual and seasonal variability of each vegetation variable among major biome types.

2. Materials and Methods

2.1. Study Area

This study was conducted at a regional scale between 12°S and 23°S and 128°E and 138°E, which is known as the Northern Australian Tropical Transect (Figure 1). This region, particularly the northern NATT, has a classic monsoon climate pattern, which receives more than 80% of its annual precipitation during November to April [4]. From the northern mesic tropics to xeric central Australia, mean annual rainfall ranges from 1700 mm to approximately 300 mm [4].

Correspondingly, the vegetation follows a wet–dry gradient that shifts from Eucalyptus-dominated forests, open forests, and woodlands in the coastal northern areas to Acacia-dominated open woodlands, scattered shrubs, and Hummock grassland into the vast inland [4]. More detailed descriptions with respect to ecosystems, climate, and soils of the entire study region can be found in Ma et al. (2013) and Hutley et al. (2011) [4,36].

To contrast satellite observation with ground-based evidence, we selected five representative flux tower sites across the extensive study area: Howard Springs (AU-How (131.15°E, 12.495°S)), Dry River (AU-Dry (132.371°E, 15.259°S)), Sturt Plains (AU-Stp (133.3502°E, 17.1507°S)), Alice Springs Mulga (AU-ASM, (133.2493°E, 22.2828°S)), and Ti Tree East (AU-TTE (133.64°E, 22.287°S)) [37]. The major vegetation types over the five sites, respectively, are Eucalyptus woodland (AU-How), Eucalyptus open forest (AU-Dry), Tussock grassland (AU-Stp), Mulga open woodland (AU-ASM), and Corymbia savanna (AU-TTE) (http://www.ozflux.org.au/, accessed on 1 October 2020).

2.2. Satellite Data

In this study, we utilised satellite-based SIF records obtained from the Global Ozone Monitoring Experiment-2 (GOME-2) onboard the MetOp-B platform. This dataset is the retrieval of the far-red chlorophyll fluorescence peaking at 740 nm, based on a simplified radiative transfer model in the company of a principal component analysis [22]. Monthly global coverage of SIF data at a 0.5° × 0.5° spatial resolution (level 3, Version 28) from March 2013 to March 2019 was obtained from NASA Goddard Space Flight Centre (https://avdc.gsfc.nasa.gov/, accessed on 1 July 2020). The daily orbital data (level 2, Version) were also used to aggregate 16-day interval records for higher temporal resolution analysis. Given that the data provider of GOME-2 SIF suspended updating in March 2019, we introduced a novel SIF product (from July 2018 to June 2019) derived from TROPOMI onboard the Sentinel-5 Precursor satellite as a supplementary in this study. A data-driven method was employed to retrieve the SIF signal using spectral measurements ranging from

743 nm to 758 nm. The daily orbital TROPOMI SIF at a 0.05° spatial resolution (obtained from ftp://fluo.gps.caltech.edu/data/tropomi/, accessed on 1 August 2020) was likewise aggregated to 16-day series by the mean value.

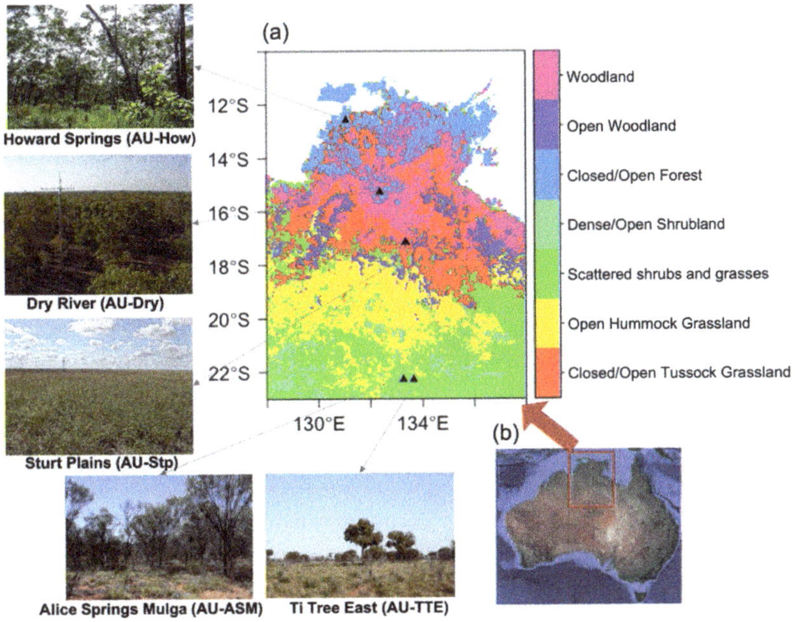

Figure 1. (**a**) Land cover map of the Northern Australian Tropical Transect (NATT) study area (data source: Dynamic Land Cover Dataset). Black triangles refer to the five flux tower sites. Photographs show the ground-view of each flux tower site linked with the black arrow (image source: www.ozflux.org.au, accessed on 1 October 2020). (**b**) The locations of the study area over the Australian continent (image source: Google Earth).

We used the Moderate Resolution Imaging Spectroradiometer (onboard Aqua, Collection 6) MYD13C1 (0.05°, 16-day) and MYD13C2 (0.05°, monthly) Vegetation Indices' products from January 2007 to June 2019 downloaded from NASA Earth Observation data (https://search.earthdata.nasa.gov/search, accessed on 1 August 2020). EVI is an optimized version of vegetation indices that effectively reduces soil background influences and is widely used as a proxy of canopy greenness. The equation of EVI is:

$$EVI = 2.5\frac{\rho_{NIR} - \rho_{red}}{\rho_{NIR} + 6\rho_{red} - 7.5\rho_{blue} + 1} \quad (1)$$

where ρ_{blue}, ρ_{red}, and ρ_{NIR} are reflectance in the blue, red, and near-infrared bands, respectively. To reduce noise and uncertainties, only the best-quality data were retained in this study by removing pixels for which the quality control flag of the first 2 bits was neither 00 nor 01, and pixelwise EVI time series data were smoothed using the Savitzky–Golay filter.

MODIS daytime Land Surface Temperature (LST, MYD11C3, Version 6) at a monthly scale and a 0.05° spatial resolution was included in this study, collected from NASA Earth Observation data (https://search.earthdata.nasa.gov/search, accessed on 1 August 2020). Similarly, bad-quality data were removed by eliminating pixels with a quality control flag.

To examine the impact of solar radiation on vegetation seasonality, monthly photosynthetic active radiation (PAR) at a 1° spatial resolution was downloaded from the NASA Langley Research Centre, Cloud and Earth's Radiant Energy System (CERES, Ed4.1), including adjusted surface PAR, both direct and diffuse fluxes under all-sky conditions (http://ceres.larc.nasa.gov, accessed on 1 August 2020). The total PAR was computed as the sum of both direct and diffuse PAR [38]. As recent studies suggested EVI outperformed the MODIS fPAR (fraction of absorbed PAR) products in estimating the APAR [39], we refer to EVI × PAR as an alternative estimate of APAR:

$$APAR \approx EVI \times PAR \quad (2)$$

2.3. Climate Data and Land Cover Map

To assess the interaction of environmental drivers and vegetation, monthly air temperature (at a 2 m height) and soil moisture content (surface 0–7 cm depth, root zone 28–100 cm depth) based on ERA-5 reanalysis data were downloaded from Copernicus Climate Change Service (https://cds.climate.copernicus.eu/, accessed on 1 July 2020).

Global monthly precipitation based on Integrated Multi-Satellite Retrievals for Global Precipitation Measurement (IMERG, Version 6, Final run, 2007–2019) at a monthly scale and a 0.1° spatial resolution was collected from the NASA Precipitation Processing System (https://pps.gsfc.nasa.gov/, accessed on 1 August 2020).

The National Dynamic Land Cover Dataset (DLCD) was used in this research, obtained from Geoscience Australia and the Bureau of Agricultural and Resource Economics and Sciences (http://www.ga.gov.au/scientific-topics/earth-obs/landcover, accessed on 1 July 2020). Given that some biome types were only covered by a few pixels over the study region, closed Tussock grassland, dense shrubland, and closed forest were, respectively, re-grouped into open/closed Tussock grassland, dense/open shrubland, and closed/open forest (Figure 1). This dataset, validated with abundant field sites, was aggregated to a 0.5° spatial resolution by the most frequent values.

2.4. Eddy Covariance Data

The original level 3 (AU-How, AU-Dry, AU-Stp) and level 6 (AU-ASM, AU-TTE) flux data provided by the OzFlux network (http://www.ozflux.org.au/, accessed on 1 October 2020) were used to pre-process, including quality control assessment, removal of outliers, and gap-filling [33]. In order to estimate daily mean GPP with hourly eddy covariance and meteorological data, flux partitioning for level 3 data was conducted in the open-source R scientific computation environment (Version 3.5.1) associated with the REddyProc package (Version 1.2) [40]. This tool used the gap-filling and flux partitioning algorithms to partition level 3 data into GPP and field ecosystem respiration [41]. The daily-estimated flux data were, respectively, aggregated into monthly and 16-day GPP to match with satellite-based observations.

2.5. Phenological Metrics

Owing to the fact that Ma et al. (2013) demonstrated the capability of singular spectrum analysis (SSA) in the analysis of nonlinear dynamics in NATT [4], we also employed the same method to smooth and reduce the noise in satellite-based SIF, EVI time series. Correspondingly, following Ma et al. (2013) [4], 37 composite periods of the window length and four leading components were selected to configure the parameters in the SSA implementation (Figure 2). After interpolating to daily time series from SSA-reconstructed SIF, EVI series, we used the PhenoDeriv function from the "GreenBrown" package to derive the key phenological metrics. Five metrics were extracted:

1. The start of the growing season (SOS), defined as the date halfway between the minimum value and the fastest greening rate;
2. The peak of the growing season (POS), the date of the maximum value;
3. The end of the growing season (EOS), the date halfway between the fastest brown-down rate and minimum value;
4. The rate of spring green-up (RSP), the amplitude of EVI or SIF between POS and SOS divided by the periods (days) between POS and SOS;

5. The rate of autumn senescence (RAU), the amplitude of EVI or SIF between POS and EOS divided by the periods (days) between POS and EOS

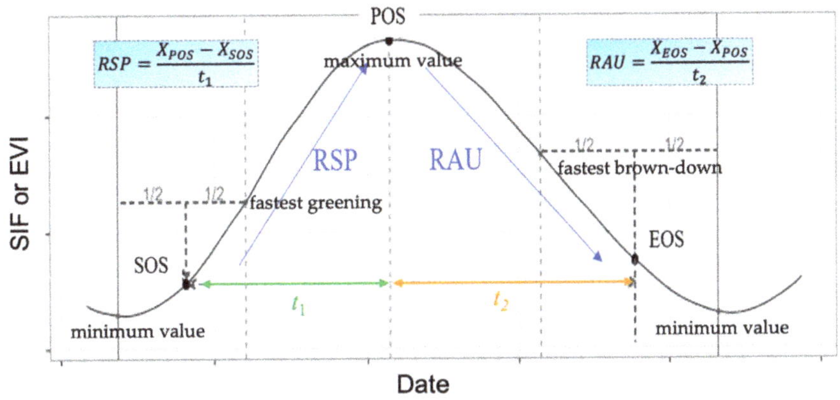

Figure 2. Schematic diagram of phenological metrics' retrieval. The curve refers to the seasonal SSA-re-reconstructed EVI or SIF profile at a daily scale. SOS: the start of the growing season; POS: the peak of the growing season; EOS: the end of the growing season; RSP: the rate of spring green-up; RAU: the rate of autumn senescence.

To examine the interaction of environmental drivers and vegetation variables, the coefficient of determination (r^2) was calculated across four major biome types. A t-test was utilized to examine the statistical significance level of the relationships (p-value). To further explore the difference between SIF and EVI, we interpreted SIF with fPAR and SIF_{yield}, expressed as:

$$SIF_{PAR} = \frac{SIF}{PAR} \tag{3}$$

$$SIF_{yield} = \frac{SIF}{EVI \times PAR} \tag{4}$$

Data processing, analysis, and visualization were conducted in the open-source R scientific computation environment (Version 3.6.2) and the associated packages contributed by the R user community (http://cran.r-project.org, accessed on 1 August 2020).

3. Results

3.1. Seasonal and Inter-Annual Variations over Local Sites

Figure 3 shows the inter-annual variations in eddy-covariance-estimated GPP and satellite-based SIF, EVI over five selected flux sites during 2014–2019. Generally, both satellite variables, especially EVI, exhibited the capacity for capturing the seasonal and inter-annual dynamics of dryland vegetation as indicated by tower-based GPP (r^2 ranging from 0.47 to 0.86). Compared with northern mesic sites, the two southern semi-arid sites (AU-ASM, AU-TTE) displayed much higher inter-annual variability, of which there was no seasonality detected by satellite observation nor field measurement in 2018–2019 (Figure 3d,e). Likewise, there was significant hysteretic senescence in EVI as opposed to GPP and SIF over two southern sites, in particular in the wet year of 2016–2017 (Figure 3d,e). The POS of SIF was mostly advanced relative to those of EVI. Furthermore, we found that multi-year series of GOME-2 SIF were more erratic and deviated from the fitted SSA-reconstructed curves than GPP and EVI across northern humid sites, as well as southern arid sites.

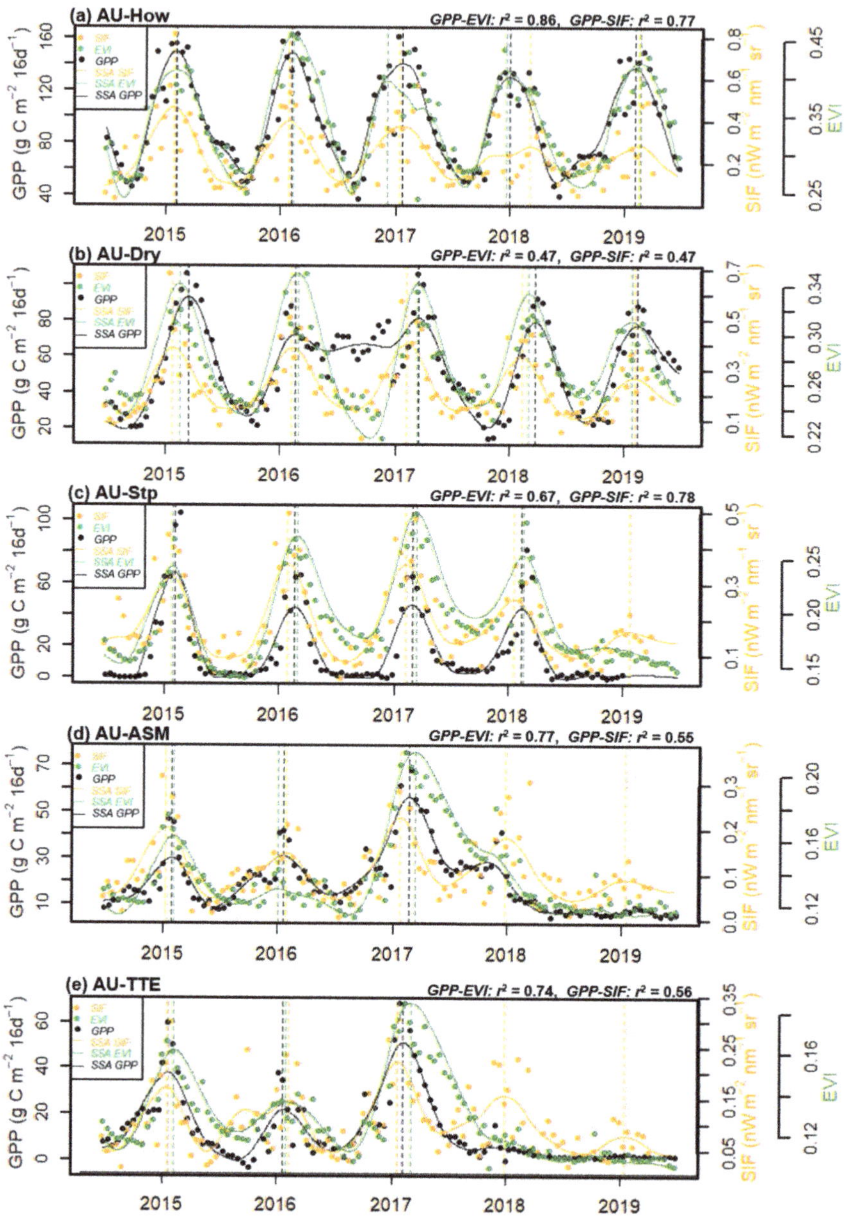

Figure 3. Time series of SSA-reconstructed tower-based GPP, GOME-2 SIF, MODIS EVI over (**a**) AU-How, (**b**) AU-Dry, (**c**) AU-Stp, (**d**) AU-ASM, and (**e**) AU-TTE. R-squared (r^2) refers to the relationship between GPP and EVI (green) and GPP and SIF (orange), respectively. Vertical dashed lines refer to the peak of the growing season in each hydrological year.

To assess the ability of SIF and EVI to track the seasonal dynamics of dryland vegetation as delineated by tower-based GPP, mean seasonal cycles of vegetation variables (normalized to unity at the maxima) over five selected tower sites during 2014–2019 are shown in Figure 4. For northern mesic site AU-How, GOME-2 SIF indicates a consistent growing season with tower data, in contrast to the 1–2-month lagged seasonal profile of EVI during both the green-up and senescent periods (Figure 4a). Conversely, a notably advanced springtime increase and earlier autumn drop are observed in SIF relative to tower-based GPP and EVI for the AU-Dry site (Figure 4b). For three semi-arid and arid sites (AU-Stp, AU-ASM, AU-TTE), both GPP and SIF show a slightly narrower mean seasonal profile as opposed to two northern sites (Figure 4c–e). Besides, there is substantial later senescence in EVI as compared with GPP, SIF, and APAR (≈EVI× PAR) over southern water-limited sites. Likewise, the seasonal profiles of APAR significantly shifted 1~2 months earlier relative to those of EVI over all selected sites.

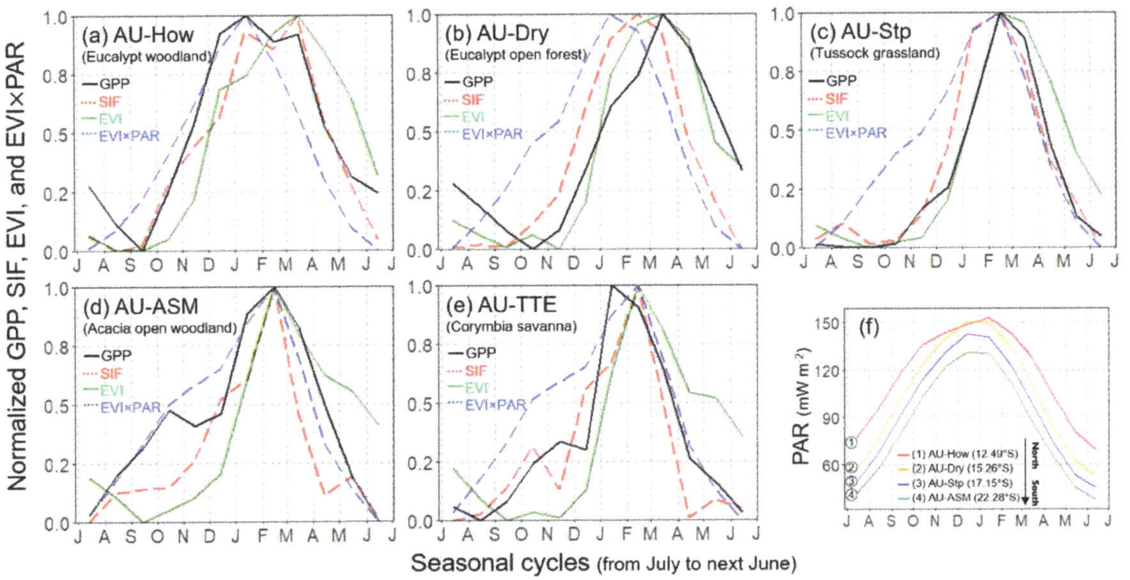

Figure 4. (**a–e**) Seasonal cycle (mean) of tower-based GPP, GOME-2 SIF, MODIS EVI, and EVI × PAR over 5 selected local sites during 2014–2019. Curves are normalized with respect to unity at the maximum annual value. (**f**) Seasonal cycle (mean) of PAR over 5 selected local sites during 2014–2019. Satellite observations were extracted within a 3 × 3 window centred at each flux tower site (SIF: 1.5° × 1.5°; EVI and EVI × PAR: 0.15° × 0.15°).

The differences of key phenological metrics (SOS and EOS) derived from satellite-observed SIF and EVI as compared with tower-based GPP over five flux tower sites during 2014–2019 are presented in Table 1. For SOS over three northern sites (AU-How, AU-Dry, and AU-Stp), the mean absolute errors (MAE) of both SIF and EVI ranging from 7 to 32 days were generally less than those over two southern arid sites (AU-ASM and AU-TTE, ranging from 31 to 54 days). On the contrary, there was a larger discrepancy in EOS at AU-How and AU-Dry (MAE ranges from 39 to 60 days). Besides, for two southern sites (AU-ASM and AU-TTE), the differences of EOS between EVI and GPP were notably larger than those between SIF and GPP.

Table 1. Summary of differences (days) in SOS and EOS derived from SIF, EVI compared with that derived from tower-based GPP. Negative values represent an earlier green-up (SOS) or senescence (EOS) of SIF, EVI relative to that of GPP.

		SOS						EOS					
Site	**Data**	**2014–2015**	**2015–2016**	**2016–2017**	**2017–2018**	**2018–2019**	**MAE [1]**	**2014–2015**	**2015–2016**	**2016–2017**	**2017–2018**	**2018–2019**	**MAE**
AU-How	SIF-GPP	6	24	−4	−21	0	11	−80	20	7	44	−42	39
	EVI-GPP	−17	−12	10	3	−6	10	9	74	83	24	15	41
AU-Dry	SIF-GPP	−56	3	−53	−38	−10	32	−63	−71	−42	−47	−75	60
	EVI-GPP	−6	5	−27	−16	−8	12	−39	−55	−5	−23	−78	40
AU-Stp	SIF-GPP	2	−21	−33	−21		19	−5	−14	−14	0		8
	EVI-GPP	−3	0	6	17		7	3	12	22	10		12
AU-ASM	SIF-GPP	22	108	−25			52	−11	9	−31			17
	EVI-GPP	37	86	−10			44	14	84	34			44
AU-TTE	SIF-GPP	15	−134	−13			54	−2	9	−27			13
	EVI-GPP	67	−25	1			31	20	84	42			49

[1] MAE: mean absolute error.

3.2. Biogeographic Patterns of Vegetation Phenology

To contrast the difference in the timing of seasonal greening derived from two satellite-based vegetation indicators and assess spatial variability over the NATT, the region-wide phenology maps based on EVI (Figure 5) and SIF (Figure 6) were generated. Besides, three representative years, 2014–2015 (normal year), 2016–2017 (wet year), and 2018–2019 (dry year), were selected to evaluate the impact of inter-annual precipitation variability on vegetation phenology. We found large spatial and inter-annual variations in the timing of key phenology transition dates retrieved from EVI (Figure 5), particularly in the southern arid/semi-arid NATT (18°S~23°S). In the extremely dry year of 2018–2019, EVI exhibited no seasonality over nearly the entire southern NATT as compared with detectable phenology in normal and wet years (Figure 5). There is a remarkable lagging trend in key phenological metrics (especially POS and EOS) in 2014–2015 (normal year) and 2016–2017 (wet year) from north to south across the study area, of which Eucalyptus-dominated woodlands distributed in the humid northern NATT started to green-up earlier (September to October). By contrast, the SOS of Acacia-dominated woodlands and Hummock-dominated grasslands distributed in the arid inland was generally 1–2 months behind (November to December). Similarly, the peak and end of the growing season in the south were 2–3 months delayed compared to those in the north during the normal/wet years.

The spatial patterns of vegetation phenology derived from SIF across NATT are shown in Figure 6. Consistent with the in situ comparison, the key phenological metrics of most pixels derived from SIF were generally earlier than those derived from EVI. There was a contrary spatial pattern in the timing of transition dates between EVI and SIF, in which arid/semiarid ecosystems over the southern NATT exhibited advanced POS (December to January) and EOS (March to April) as opposed to POS (February to March) and EOS (April to May) over the northern humid area (Figure 6d–h). With reference to the normal/wet years, larger latitudinal changes in the peak of season between the north (11°S–17°S) and south (17°S–23°S) NATT under the extremely dry condition of 2018–2019 were observed, and fractional pixels in the south were likewise without detectable phenology (Figure 6c,f).

With the purpose of further contrasting the differences in the seasonal profile derived from SIF and EVI, the relationship between green-up rate and brown-down rate of two satellite-based indicators among four major vegetation types are shown in Figure 7. For the semiarid/arid biomes (Hummock grasslands and shrublands), the majority of pixels displayed considerably higher green-up rates than senescence rates ($RSP_{EVI} > |RAU_{EVI}|$), of which the absolute senescence rate ($|RAU_{EVI}|$) of Hummock grasslands was mostly less than 0.002 d^{-1} (Figure 7a,b). By contrast, for northern sub-humid/humid biomes (Eucalyptus-dominated forests and woodlands and Tussock grasslands), there were notably rapid and comparable rates of green-up and senescence (0.0005~0.004 d^{-1}) (Figure 7c,d). On the other hand, RSP_{SIF} and RAU_{SIF} ranging from 0.002~0.010 nW m^{-2} nm^{-1} sr^{-1} d^{-1} had fewer discrepancies in spite of diverse biomes (Figure 7e–h).

3.3. Interaction between Environmental Drivers and Vegetation Variables

To investigate the dominant factors controlling the seasonal and inter-annual variations in vegetation, the biome-specific relationships of the time series of principle environmental drivers and vegetation variables (SIF, SIF_{PAR}, SIF_{yield}, and EVI) during 2014–2019 are shown in Figure 8. Apart from forests and woodlands, root zone soil moisture was most relevant for EVI (r^2: 0.42~0.79) relative to other drivers; among those, temperature-related drivers were poorly correlated ($r^2 < 0.16$) (Figure 8a2–a4). By contrast, both moisture-related drivers, as well as land surface temperature highly corresponded with EVI over northern humid forests and woodlands (Figure 8a1). There were strong correlations ($r^2 > 0.8$) between SIF and EVI over forests and woodlands and Tussock grasslands as compared with moderate correlations (r^2: 0.21, 0.38) over water-limited ecosystems (Figure 8b1–b4). Similarly, compared with temperature-related drivers (r^2: 0~0.3), water-related factors (especially soil moisture) were more associated with SIF among all biomes. After removing the impact of solar radiation, there was an enhanced agreement between EVI and PAR-normalized SIF (SIF_{PAR}) over Tussock grassland, shrublands, and Hummock grasslands (r^2: 0.51~0.86) (Figure 8c2–c4). Likewise, the correlation between root zone soil moisture and SIF_{PAR} was strengthened relative to those with SIF regardless of different vegetation types. Precipitation and surface soil moisture agreed well with the effective fluorescence yield (SIF_{yield}) over forests and woodlands, as well as Tussock grasslands (r^2: 0.38~0.62) (Figure 8d1–d2). In addition, there was a moderately increased correlation between air temperature and SIF_{yield} (r^2: 0.22~0.4) in comparison to those with SIF or SIF_{PAR} (r^2: 0~0.2).

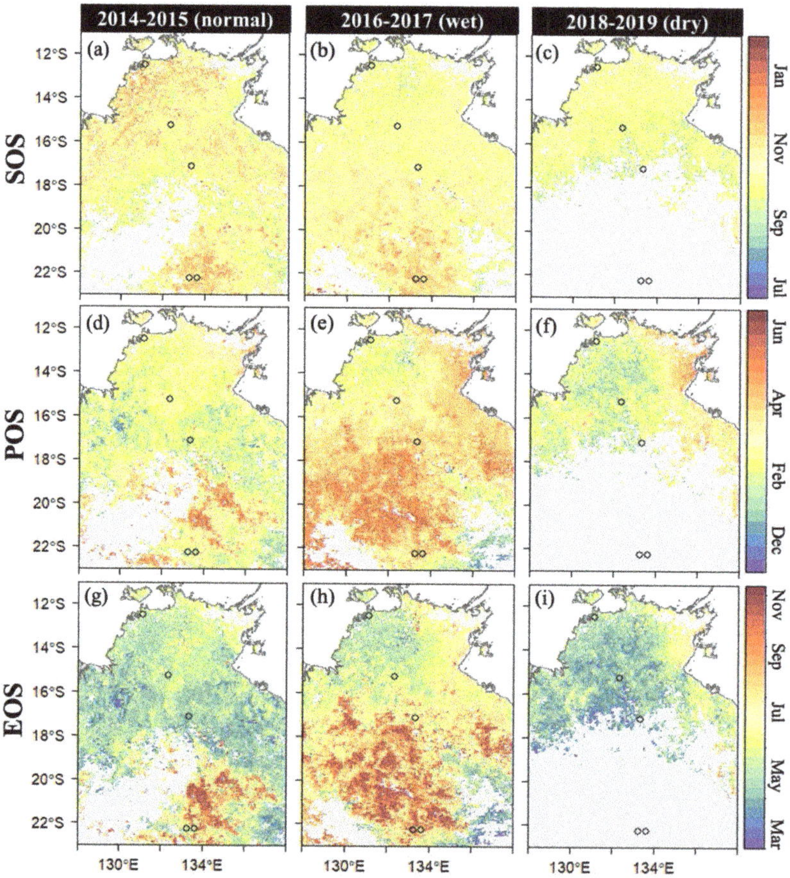

Figure 5. Spatial patterns of vegetation phenology based on EVI over the NATT study area across three representative hydrological years, (**a**,**d**,**g**) 2014–2015 (normal year), (**b**,**e**,**h**) 2016–2017 (wet year), and (**c**,**f**,**i**) 2018–2019 (dry year). The filled pixels (grey shaded areas) are either water body or without detectable phenology. SOS: the start of growing season; POS: the peak of growing season; EOS: the end of growing season. Blank circles represent 5 selected eddy covariance flux tower sites.

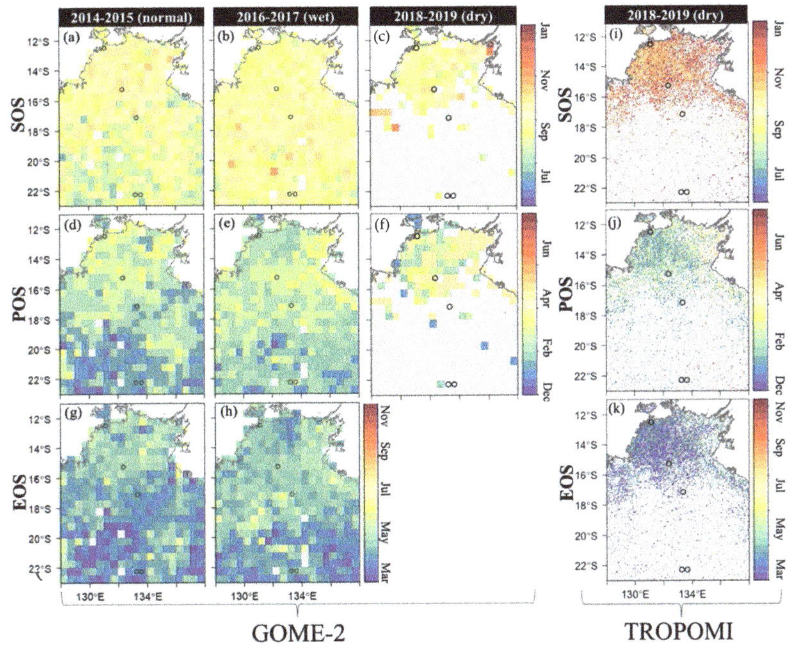

Figure 6. Spatial patterns of vegetation phenology based on (**a**–**h**) GOME-2 SIF and (**i**–**k**) TROPOMI SIF over the NATT study area across three representative hydrological years, 2014–2015 (normal year), 2016–2017 (wet year), and 2018–2019 (dry year). The filled pixels (grey shaded areas) are either water body or without detectable phenology. SOS: the start of growing season; POS: the peak of growing season; EOS: the end of growing season. Blank circles represent 5 selected eddy covariance flux tower sites.

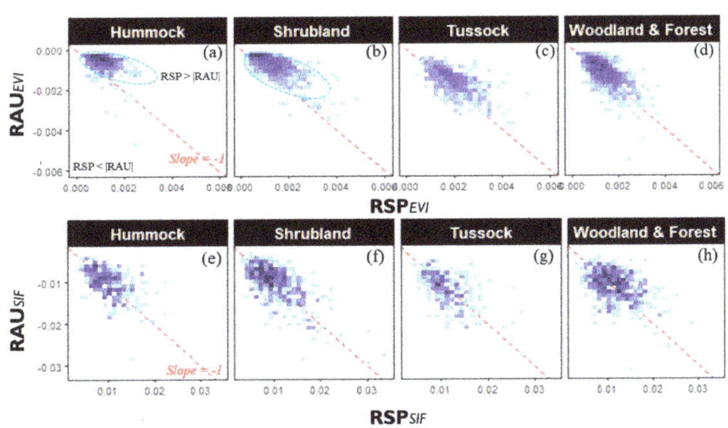

Figure 7. Scatter plot between spring green-up rates (RSP) and autumn senescence rates (RAU) of (**a**–**d**) EVI and (**e**–**h**) SIF among four major biomes across the entire study region. Red dashed lines refer to a −1:1 diagonal.

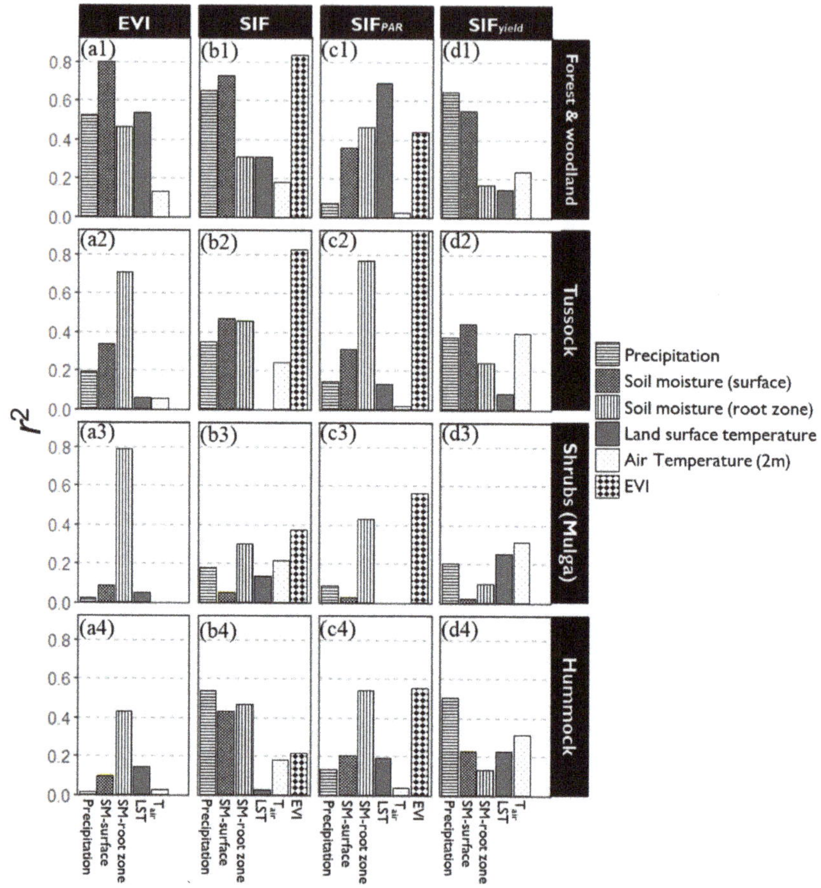

Figure 8. Relationships between climatic–environmental drivers and vegetation variables of (**a1–a4**) EVI, (**b1–b4**) SIF, (**c1–c4**) SIF_{PAR}, and (**d1–d4**) SIF_{yield} among four major biome types across the NATT during 2014–2019. (p-value < 0.001).

4. Discussion

4.1. Ground Interpretations of the Satellite-Observed Vegetation Phenology

Although the five selected sites exhibited distinct seasonality revealed by eddy covariance flux measurements of vegetation production (GPP), both satellite-based SIF and EVI generally captured seasonal dynamics and inter-annual variations over a variety of biomes (Figures 3 and 4). In comparison with the greenness-based vegetation index, GOME-2 SIF displayed more consistently a seasonal profile with tower-based GPP (Figure 4). Especially, there was a significant "hysteresis effect" during the senescent period of EVI relative to that of GPP over water-limited ecosystems (Hummock grasslands and shrublands), consistent with previous findings [4]. In the wet year of 2016–2017, there was a considerably delayed senescence of EVI over two southern sites (AU-ASM and AU-TTE), relative to those of GPP and SIF (Figure 3). Besides, we found that the autumn senescence rates of EVI were considerably slower than the spring green-up rates of EVI over these arid/semiarid biomes (Figure 7); however, RAU and RSP of SIF were generally comparable. Aside from the slow chlorophyll degradation leading to a gradual decrease of EVI [42], another possible reason causing the discrepancies between EVI and SIF during the brown-down phase is that the rapid decline in solar radiation in the arid southern NATT gave rise to the swiftly dropping SIF signals (Figure 4f). After removing the impact of the PAR on SIF, there was a remarkably enhanced correlation between EVI and SIF_{PAR} across southern water-limited biomes (Figure 8b3–b4).

In contrast to GOME-2 SIF, EVI had an improved capability (higher r^2) of tracking the interannual and seasonal variations in GPP over most sites (except AU-Stp, Figure 3), probably owing to the substantial footprint mismatch between flux tower measurement and satellite observations (EVI: ~5 km, SIF: ~50 km). Moreover, as compared with the reflectance-based vegetation index, the high retrieval noises over the low-productivity region resulted in more erratic SIF signals [43], giving rise to a relatively weaker correlation with tower-based GPP. Given the sparse spatial resolution of satellite-based SIF data, Wang et al. (2019) utilized a high-resolution contiguous SIF product (SIF_{OCO2_005}) over the NATT and found that SIF_{OCO2_005} outperformed EVI at AU-ASM with a stronger temporal consistency with tower-based GPP [23]. However, this up-scaled dataset was generated by native OCO-2 SIF signals along with MODIS reflectance through a machine learning method [32], and it therefore contains information of canopy structure and chlorophyll content, like greenness-based indices. Even though EVI showed a tighter correlation with the GPP (higher r^2) than SIF_{OCO2_005} at AU-Dry and AU-Stp in Wang et al.'s (2019) studies, congruent with our results [23], nevertheless, the high spatiotemporal SIF data derived from TROPOMI onboard the Sentinel-5 Precursor launched in 2017 exhibiting striking consistency with field measurement has great potential of characterizing the phenological dynamics of dryland vegetation in the future [44].

4.2. Spatial Patterns of Vegetation Phenology

Distinct biogeographic patterns in the timing of transition dates, especially POS and EOS, derived from SIF and EVI were observed (Figures 5 and 6), of which there was a considerable latitudinal shift in vegetation phenology (EVI) with a gradually delayed trend from north to south. This is in accordance with the in situ comparison that EVI displayed significantly late senescence at southern xeric sites as compared with GPP and SIF, particularly in the wet year of 2016–2017. In addition, the spatial phenomenon was in line with Ma et al.'s (2013) findings [4], that the majority of EOS based on EVI in wet years (2005–2006, 2010–2011, and 2016–2017), especially in the southern NATT, was generally later than those in normal (2001–2002, 2014–2015) and drought years (2007–2008, 2018–2019) (Figure 5), probably as a consequence of the aforementioned "hysteresis effect". On the contrary, there was less difference in the spatial patterns of EOS based on SIF between the north and south NATT, as well as in normal/wet years (Figure 6), of which both showed apparently earlier EOS, presumably due to the rapidly decreasing solar radiation over the southern inland. However, as an example of moderately inconsistent phenological metrics compared with field measurement GPP (Table 1), the coarse spatial and temporal resolution of GOME-2 SIF, as well as high retrieval noises, impeded the full potential of capturing the seasonal and interannual variations in vegetation. Considering the drawbacks of existing greenness-based phenology products, as well as the highly heterogeneous composition of dryland ecosystems, our findings imply that new spaceborne SIF with improved spatiotemporal resolution, such as TROPOMI, has great capability for advancing our understanding of phenological characterization in Australia.

Furthermore, SIF and EVI displayed a stronger temporal consistency over northern mesic biomes (forests and woodlands and Tussock grasslands) relative to southern xeric biomes (Hummock grasslands and shrublands) (Figure 8). Soil moisture can explain more than 60% of the seasonal and interannual variability in EVI over most biomes (except Hummock grasslands), suggesting it is the dominant factor controlling the dynamics of vegetation greenness across the NATT. By contrast, the temperature-related drivers could barely explain less than 5% of the variability (except forests and woodlands). Despite the fact that solar radiation has substantial impacts on the SIF signal, temperature- and moisture-related factors almost equally contributed to the process of light use efficiency for fluorescence (SIF_{yield}, except forests and woodlands). This could partially explain the temporal discrepancies between SIF and EVI across diverse biomes.

5. Conclusions

In summary, we utilized satellite-based SIF and EVI observations to investigate the spatial patterns and seasonal dynamics of vegetation phenology across wet and dry years along the North Australian Tropical Transect. Considerable impacts of drought and wet extremes on the phenology and production of dryland vegetation were revealed by both SIF and EVI, especially in the arid/semiarid interior of Australia. Although EVI exhibited a considerably delayed senescence relative to SIF and GPP, phenological metrics derived from SIF had more variable accuracy in contrast to those derived from the greenness-based vegetation index. In addition, EVI exhibiting stronger correlation with tower-based GPP (r^2: 0.47~0.86) can be a superior indicator to track the seasonal and interannual variation in dryland vegetation production than the spatially coarse SIF dataset (r^2: 0.47~0.78). In spite of the sparse sampling and high retrieval noises of GOME-2 SIF products, which offset the drawbacks of greenness-based phenology products with a potentially lagged end of the season, spaceborne SIF retrieved from state-of-the-art instruments (such as TROPOMI, OCO-2) has the promising potential of assessing the characterization of the phenology dynamics of dryland ecosystems.

Author Contributions: Conceptualization, S.L. and A.H.; methodology, S.L.; validation, S.L.; formal analysis, S.L.; investigation, S.L.; data curation, S.L. and J.C.; writing—original draft preparation, S.L.; writing—review and editing, A.H., Q.Y., R.Z., and Q.W.; visualization, S.L. and Q.W.; supervision, A.H. and Q.Y.; funding acquisition, Q.W. All authors have read and agreed to the published version of the manuscript.

Funding: This research was funded by the Natural Science Foundation of Fujian Province (2021J01627) and the National Natural Science Foundation of China (41601562).

Data Availability Statement: All satellite-based data, as well as other datasets in this study are available from the corresponding author upon reasonable request.

Acknowledgments: The first author acknowledges the financial support from Chinese Scholarship Council, the Ministry of Education, China. We also would like to thank the opening of flux tower data for public access by the Australian Terrestrial Ecosystem Research Network (TERN, Available online: www.tern.org.au)—OzFlux facility (Available online: www.ozflux.org.au, accessed on 1 October 2020).

Conflicts of Interest: The authors declare no conflict of interest.

References

1. Peng, D.; Wu, C.; Zhang, X.; Yu, L.; Huete, A.R.; Wang, F.; Luo, S.; Liu, X.; Zhang, H. Scaling up Spring Phenology Derived from Remote Sensing Images. *Agric. For. Meteorol.* **2018**, *256–257*, 207–219. [CrossRef]
2. Piao, S.; Liu, Q.; Chen, A.; Janssens, I.A.; Fu, Y.; Dai, J.; Liu, L.; Lian, X.; Shen, M.; Zhu, X. Plant Phenology and Global Climate Change: Current Progresses and Challenges. *Glob. Change Biol.* **2019**, *25*, 1922–1940. [CrossRef] [PubMed]
3. Zhang, R.; Qi, J.; Leng, S.; Wang, Q. Long-Term Vegetation Phenology Changes and Responses to Preseason Temperature and Precipitation in Northern China. *Remote Sens.* **2022**, *14*, 1396. [CrossRef]
4. Ma, X.; Huete, A.; Yu, Q.; Coupe, N.R.; Davies, K.; Broich, M.; Ratana, P.; Beringer, J.; Hutley, L.B.; Cleverly, J. Spatial Patterns and Temporal Dynamics in Savanna Vegetation Phenology across the North Australian Tropical Transect. *Remote Sens. Environ.* **2013**, *139*, 97–115. [CrossRef]
5. Zhang, Q.; Kong, D.; Shi, P.; Singh, V.P.; Sun, P. Vegetation Phenology on the Qinghai-Tibetan Plateau and Its Response to Climate Change (1982–2013). *Agric. For. Meteorol.* **2018**, *248*, 408–417. [CrossRef]
6. Wang, Q.; Qi, J.; Wu, H.; Zeng, Y.; Shui, W.; Zeng, J.; Zhang, X. Freeze-Thaw Cycle Representation Alters Response of Watershed Hydrology to Future Climate Change. *CATENA* **2020**, *195*, 104767. [CrossRef]
7. Wang, Q.; Qi, J.; Qiu, H.; Li, J.; Cole, J.; Waldhoff, S.; Zhang, X. Pronounced Increases in Future Soil Erosion and Sediment Deposition as Influenced by Freeze–Thaw Cycles in the Upper Mississippi River Basin. *Environ. Sci. Technol.* **2021**, *55*, 9905–9915. [CrossRef]
8. Verger, A.; Filella, I.; Baret, F.; Peñuelas, J. Vegetation Baseline Phenology from Kilometric Global LAI Satellite Products. *Remote Sens. Environ.* **2016**, *178*, 1–14. [CrossRef]

9. Huete, A.; Didan, K.; Miura, T.; Rodriguez, E.P.; Gao, X.; Ferreira, L.G. Overview of the Radiometric and Biophysical Performance of the MODIS Vegetation Indices. *Remote Sens. Environ.* **2002**, *83*, 195–213. [CrossRef]
10. Wang, J.; Shi, T.; Yu, D.; Teng, D.; Ge, X.; Zhang, Z.; Yang, X.; Wang, H.; Wu, G. Ensemble Machine-Learning-Based Framework for Estimating Total Nitrogen Concentration in Water Using Drone-Borne Hyperspectral Imagery of Emergent Plants: A Case Study in an Arid Oasis, NW China. *Environ. Pollut.* **2020**, *266*, 115412. [CrossRef]
11. Wang, J.; Ding, J.; Li, G.; Liang, J.; Yu, D.; Aishan, T.; Zhang, F.; Yang, J.; Abulimiti, A.; Liu, J. Dynamic Detection of Water Surface Area of Ebinur Lake Using Multi-Source Satellite Data (Landsat and Sentinel-1A) and Its Responses to Changing Environment. *CATENA* **2019**, *177*, 189–201. [CrossRef]
12. Broich, M.; Tulbure, M.G.; Verbesselt, J.; Xin, Q.; Wearne, J. Quantifying Australia's Dryland Vegetation Response to Flooding and Drought at Sub-Continental Scale. *Remote Sens. Environ.* **2018**, *212*, 60–78. [CrossRef]
13. Tucker, C.J. Red and Photographic Infrared Linear Combinations for Monitoring Vegetation. *Remote Sens. Environ.* **1979**, *8*, 127–150. [CrossRef]
14. Reynolds, J.F.; Smith, D.M.S.; Lambin, E.F.; Turner, B.L.; Mortimore, M.; Batterbury, S.P.J.; Downing, T.E.; Dowlatabadi, H.; Fernández, R.J.; Herrick, J.E. Global Desertification: Building a Science for Dryland Development. *Science* **2007**, *316*, 847–851. [CrossRef]
15. Walker, J.J.; de Beurs, K.M.; Wynne, R.H. Dryland Vegetation Phenology across an Elevation Gradient in Arizona, USA, Investigated with Fused MODIS and Landsat Data. *Remote Sens. Environ.* **2014**, *144*, 85–97. [CrossRef]
16. Zeng, J.; Zhang, R.; Qu, Y.; Bento, V.A.; Zhou, T.; Lin, Y.; Wu, X.; Qi, J.; Shui, W.; Wang, Q. Improving the Drought Monitoring Capability of VHI at the Global Scale via Ensemble Indices for Various Vegetation Types from 2001 to 2018. *Weather. Clim. Extrem.* **2022**, *35*, 100412. [CrossRef]
17. Zhang, J.; Ding, J.; Wu, P.; Tan, J.; Huang, S.; Teng, D.; Cao, X.; Wang, J.; Chen, W. Assessing Arid Inland Lake Watershed Area and Vegetation Response to Multiple Temporal Scales of Drought Across the Ebinur Lake Watershed. *Sci. Rep.* **2020**, *10*, 1354. [CrossRef]
18. Wang, J.; Ding, J.; Yu, D.; Ma, X.; Zhang, Z.; Ge, X.; Teng, D.; Li, X.; Liang, J.; Lizaga, I.; et al. Capability of Sentinel-2 MSI Data for Monitoring and Mapping of Soil Salinity in Dry and Wet Seasons in the Ebinur Lake Region, Xinjiang, China. *Geoderma* **2019**, *353*, 172–187. [CrossRef]
19. Wang, J.; Hu, X.; Shi, T.; He, L.; Hu, W.; Wu, G. Assessing Toxic Metal Chromium in the Soil in Coal Mining Areas via Proximal Sensing: Prerequisites for Land Rehabilitation and Sustainable Development. *Geoderma* **2022**, *405*, 115399. [CrossRef]
20. Ma, X.; Huete, A.; Moore, C.E.; Cleverly, J.; Hutley, L.B.; Beringer, J.; Leng, S.; Xie, Z.; Yu, Q.; Eamus, D. Spatiotemporal Partitioning of Savanna Plant Functional Type Productivity along NATT. *Remote Sens. Environ.* **2020**, *246*, 111855. [CrossRef]
21. Jeong, S.-J.; Schimel, D.; Frankenberg, C.; Drewry, D.T.; Fisher, J.B.; Verma, M.; Berry, J.A.; Lee, J.-E.; Joiner, J. Application of Satellite Solar-Induced Chlorophyll Fluorescence to Understanding Large-Scale Variations in Vegetation Phenology and Function over Northern High Latitude Forests. *Remote Sens. Environ.* **2017**, *190*, 178–187. [CrossRef]
22. Joiner, J.; Yoshida, Y.; Vasilkov, A.P.; Schaefer, K.; Jung, M.; Guanter, L.; Zhang, Y.; Garrity, S.; Middleton, E.M.; Huemmrich, K.F.; et al. The Seasonal Cycle of Satellite Chlorophyll Fluorescence Observations and Its Relationship to Vegetation Phenology and Ecosystem Atmosphere Carbon Exchange. *Remote Sens. Environ.* **2014**, *152*, 375–391. [CrossRef]
23. Wang, C.; Beringer, J.; Hutley, L.B.; Cleverly, J.; Li, J.; Liu, Q.; Sun, Y. Phenology Dynamics of Dryland Ecosystems along the North Australian Tropical Transect Revealed by Satellite Solar-induced Chlorophyll Fluorescence. *Geophys. Res. Lett.* **2019**, *46*, 5294–5302. [CrossRef]
24. Yang, X.; Tang, J.; Mustard, J.F.; Lee, J.-E.; Rossini, M.; Joiner, J.; Munger, J.W.; Kornfeld, A.; Richardson, A.D. Solar-Induced Chlorophyll Fluorescence That Correlates with Canopy Photosynthesis on Diurnal and Seasonal Scales in a Temperate Deciduous Forest. *Geophys. Res. Lett.* **2015**, *42*, 2977–2987. [CrossRef]
25. Zhang, Y.; Xiao, X.; Jin, C.; Dong, J.; Zhou, S.; Wagle, P.; Joiner, J.; Guanter, L.; Zhang, Y.; Zhang, G.; et al. Consistency between Sun-Induced Chlorophyll Fluorescence and Gross Primary Production of Vegetation in North America. *Remote Sens. Environ.* **2016**, *183*, 154–169. [CrossRef]
26. Zuromski, L.M.; Bowling, D.R.; Köhler, P.; Frankenberg, C.; Goulden, M.L.; Blanken, P.D.; Lin, J.C. Solar-Induced Fluorescence Detects Interannual Variation in Gross Primary Production of Coniferous Forests in the Western United States. *Geophys. Res. Lett.* **2018**, *45*, 7184–7193. [CrossRef]
27. Wu, X.; Xiao, X.; Zhang, Y.; He, W.; Wolf, S.; Chen, J.; He, M.; Gough, C.M.; Qin, Y.; Zhou, Y. Spatiotemporal Consistency of Four Gross Primary Production Products and Solar-Induced Chlorophyll Fluorescence in Response to Climate Extremes Across CONUS in 2012. *J. Geophys. Res. Biogeosci.* **2018**, *123*, 3140–3161. [CrossRef]
28. Chang, Q.; Xiao, X.; Jiao, W.; Wu, X.; Doughty, R.; Wang, J.; Du, L.; Zou, Z.; Qin, Y. Assessing Consistency of Spring Phenology of Snow-Covered Forests as Estimated by Vegetation Indices, Gross Primary Production, and Solar-Induced Chlorophyll Fluorescence. *Agric. For. Meteorol.* **2019**, *275*, 305–316. [CrossRef]
29. Walther, S.; Voigt, M.; Thum, T.; Gonsamo, A.; Zhang, Y.; Kohler, P.; Jung, M.; Varlagin, A.; Guanter, L. Satellite Chlorophyll Fluorescence Measurements Reveal Large-Scale Decoupling of Photosynthesis and Greenness Dynamics in Boreal Evergreen Forests. *Glob. Chang. Biol.* **2016**, *22*, 2979–2996. [CrossRef]

30. Smith, W.K.; Biederman, J.A.; Scott, R.L.; Moore, D.J.P.; He, M.; Kimball, J.S.; Yan, D.; Hudson, A.; Barnes, M.L.; MacBean, N.; et al. Chlorophyll Fluorescence Better Captures Seasonal and Interannual Gross Primary Productivity Dynamics Across Dryland Ecosystems of Southwestern North America. *Geophys. Res. Lett.* **2018**, *45*, 748–757. [CrossRef]
31. Sanders, A.; Verstraeten, W.; Kooreman, M.; van Leth, T.; Beringer, J.; Joiner, J. Spaceborne Sun-Induced Vegetation Fluorescence Time Series from 2007 to 2015 Evaluated with Australian Flux Tower Measurements. *Remote Sens.* **2016**, *8*, 895. [CrossRef]
32. Yu, L.; Wen, J.; Chang, C.Y.; Frankenberg, C.; Sun, Y. High-Resolution Global Contiguous SIF of OCO-2. *Geophys. Res. Lett.* **2019**, *46*, 1449–1458. [CrossRef]
33. Cleverly, J.; Chen, C.; Boulain, N.; Villalobos-Vega, R.; Faux, R.; Grant, N.; Yu, Q.; Eamus, D. Aerodynamic Resistance and Penman–Monteith Evapotranspiration over a Seasonally Two-Layered Canopy in Semiarid Central Australia. *J. Hydrometeorol.* **2013**, *14*, 1562–1570. [CrossRef]
34. Wang, Q.; Zhang, R.; Qi, J.; Zeng, J.; Wu, J.; Shui, W.; Wu, X.; Li, J. An Improved Daily Standardized Precipitation Index Dataset for Mainland China from 1961 to 2018. *Scientific Data* **2022**, *9*, 124. [CrossRef] [PubMed]
35. Wang, Q.; Zeng, J.; Leng, S.; Fan, B.; Tang, J.; Jiang, C.; Huang, Y.; Zhang, Q.; Qu, Y.; Wang, W.; et al. The Effects of Air Temperature and Precipitation on the Net Primary Productivity in China during the Early 21st Century. *Front. Earth Sci.* **2018**, *12*, 818–833. [CrossRef]
36. Hutley, L.B.; Beringer, J.; Isaac, P.R.; Hacker, J.M.; Cernusak, L.A. A Sub-Continental Scale Living Laboratory: Spatial Patterns of Savanna Vegetation over a Rainfall Gradient in Northern Australia. *Agric. For. Meteorol.* **2011**, *151*, 1417–1428. [CrossRef]
37. Beringer, J.; Hutley, L.B.; McHugh, I.; Arndt, S.K.; Campbell, D.; Cleugh, H.A.; Cleverly, J.; de Dios, V.R.; Eamus, D.; Evans, B. An Introduction to the Australian and New Zealand Flux Tower Network-OzFlux. *Biogeosciences* **2016**, *13*, 5895–5916. [CrossRef]
38. Wang, X.; Qiu, B.; Li, W.; Zhang, Q. Impacts of Drought and Heatwave on the Terrestrial Ecosystem in China as Revealed by Satellite Solar-Induced Chlorophyll Fluorescence. *Sci. Total Environ.* **2019**, *693*, 133627. [CrossRef]
39. Lu, X.; Cheng, X.; Li, X.; Chen, J.; Sun, M.; Ji, M.; He, H.; Wang, S.; Li, S.; Tang, J. Seasonal Patterns of Canopy Photosynthesis Captured by Remotely Sensed Sun-Induced Fluorescence and Vegetation Indexes in Mid-to-High Latitude Forests: A Cross-Platform Comparison. *Sci. Total Environ.* **2018**, *644*, 439–451. [CrossRef]
40. Wutzler, T.; Lucas-Moffat, A.; Migliavacca, M.; Knauer, J.; Sickel, K.; Šigut, L.; Menzer, O.; Reichstein, M. Basic and Extensible Post-Processing of Eddy Covariance Flux Data with REddyProc. *Biogeosciences* **2018**, *15*, 5015–5030. [CrossRef]
41. Reichstein, M.; Falge, E.; Baldocchi, D.; Papale, D.; Aubinet, M.; Berbigier, P.; Bernhofer, C.; Buchmann, N.; Gilmanov, T.; Granier, A. On the Separation of Net Ecosystem Exchange into Assimilation and Ecosystem Respiration: Review and Improved Algorithm. *Glob. Chang. Biol.* **2005**, *11*, 1424–1439. [CrossRef]
42. Jenkins, J.P.; Richardson, A.D.; Braswell, B.H.; Ollinger, S.v.; Hollinger, D.Y.; Smith, M.-L. Refining Light-Use Efficiency Calculations for a Deciduous Forest Canopy Using Simultaneous Tower-Based Carbon Flux and Radiometric Measurements. *Agric. For. Meteorol.* **2007**, *143*, 64–79. [CrossRef]
43. Geruo, A.; Velicogna, I.; Kimball, J.S.; Du, J.; Kim, Y.; Colliander, A.; Njoku, E. Satellite-Observed Changes in Vegetation Sensitivities to Surface Soil Moisture and Total Water Storage Variations since the 2011 Texas Drought. *Environ. Res. Lett.* **2017**, *12*, 054006.
44. Leng, S.; Huete, A.; Cleverly, J.; Gao, S.; Yu, Q.; Meng, X.; Qi, J.; Zhang, R.; Wang, Q. Assessing the Impact of Extreme Droughts on Dryland Vegetation by Multi-Satellite Solar-Induced Chlorophyll Fluorescence. *Remote Sens.* **2022**, *14*, 1581. [CrossRef]

remote sensing

MDPI

Article

Assessing the Impact of Extreme Droughts on Dryland Vegetation by Multi-Satellite Solar-Induced Chlorophyll Fluorescence

Song Leng [1,2], Alfredo Huete [2], Jamie Cleverly [3], Sicong Gao [4], Qiang Yu [2,5], Xianyong Meng [6], Junyu Qi [7], Rongrong Zhang [1] and Qianfeng Wang [1,8,*]

1 Fujian Provincial Key Laboratory of Remote Sensing of Soil Erosion, College of Environment & Safety Engineering, Fuzhou University, Fuzhou 350116, China; song.leng@student.uts.edu.au (S.L.); n190620018@fzu.edu.cn (R.Z.)
2 School of Life Sciences, University of Technology Sydney, Ultimo, NSW 2007, Australia; alfredo.huete@uts.edu.au (A.H.); qiang.yu@uts.edu.au (Q.Y.)
3 College of Science and Engineering, James Cook University, Cairns, QLD 4878, Australia; dr.jr.cleverly@gmail.com
4 Land and Water, Commonwealth Scientific and Industrial Research Organization, Adelaide, SA 5064, Australia; steve.gao@csiro.au
5 State Key Laboratory of Soil Erosion and Dryland Farming on the Loess Plateau, Northwest A&F University, Xianyang 712100, China
6 School of Atmospheric Physics, Nanjing University of Information Science & Technology, Nanjing 210044, China; xymeng@nuist.edu.cn
7 Earth System Science Interdisciplinary Center, University of Maryland, College Park, MD 20740, USA; junyuqi@umd.edu
8 Key Lab of Spatial Data Mining & Information Sharing, Ministry of Education of China, Fuzhou 350116, China
* Correspondence: wangqianfeng@fzu.edu.cn

Citation: Leng, S.; Huete, A.; Cleverly, J.; Gao, S.; Yu, Q.; Meng, X.; Qi, J.; Zhang, R.; Wang, Q. Assessing the Impact of Extreme Droughts on Dryland Vegetation by Multi-Satellite Solar-Induced Chlorophyll Fluorescence. *Remote Sens.* **2022**, *14*, 1581. https://doi.org/10.3390/rs14071581

Academic Editors: Jingzhe Wang, Zhongwen Hu, Yangyi Wu and Jie Zhang

Received: 5 March 2022
Accepted: 22 March 2022
Published: 25 March 2022

Abstract: Satellite-estimated solar-induced chlorophyll fluorescence (SIF) is proven to be an effective indicator for dynamic drought monitoring, while the capability of SIF to assess the variability of dryland vegetation under water and heat stress remains challenging. This study presents an analysis of the responses of dryland vegetation to the worst extreme drought over the past two decades in Australia, using multi-source spaceborne SIF derived from the Global Ozone Monitoring Experiment-2 (GOME-2) and TROPOspheric Monitoring Instrument (TROPOMI). Vegetation functioning was substantially constrained by this extreme event, especially in the interior of Australia, in which there was hardly seasonal growth detected by neither satellite-based observations nor tower-based flux measurements. At a 16-day interval, both SIF and enhanced vegetation index (EVI) can timely capture the reduction at the onset of drought over dryland ecosystems. The results demonstrate that satellite-observed SIF has the potential for characterizing and monitoring the spatiotemporal dynamics of drought over water-limited ecosystems, despite coarse spatial resolution coupled with high-retrieval noise as compared with EVI. Furthermore, our study highlights that SIF retrieved from TROPOMI featuring substantially enhanced spatiotemporal resolution has the promising capability for accurately tracking the drought-induced variation of heterogeneous dryland vegetation.

Keywords: dryland; SIF; EVI; TROPOMI; extreme drought

1. Introduction

Large-scale drought, as one of most costly and pressing natural hazards, has profound impacts on terrestrial ecosystems, water resource, agriculture, and social economics [1]. In addition, global warming increases the exposure frequency to extreme heatwaves, which raises the risk of water deficit and leads to substantial losses of vegetation production [2,3]. As a consequence of highly variable climate, Australia is particularly vulnerable to extreme drought, as evidenced by Millennium Drought (2001–2009) in the Murray–Darling basin as well as the 2012–2013 drought over arid inland [4,5]. Australia tends to be a global hot spot for variability, with semiarid vegetation in that country exhibiting increased carbon uptake [6]. By persisting through the 'big dry' and responding favorably in a following 'big wet' [7], Australia's ecosystems play a significant role in regulating the global water and carbon cycles [4]. Enhancing our understanding with respect to dryland ecosystem in response to climate extreme is paramount for effectively managing the environment and global change research [8–11].

Satellite-based observation offers a method for monitoring and characterizing the spatiotemporal dynamics of vegetation under changing climates [12], which is especially valuable for remote areas such as most Australia's interior with very sparse monitoring sites. Traditional reflectance-based vegetation indices are widely applied to assess the effects of extreme drought on ecosystem functioning and vegetation productivity at a regional, continental, or global scale [8,13–17]. Dramatic impacts of climate extremes on vegetation dynamics (as measured by EVI) with abrupt changes in phenology and productivity over southeast Australia demonstrates that semiarid ecosystems exhibit the largest sensitivity to hydro-climatic variations [18,19]. Although vegetation-index-based approaches are essential for evaluating vegetation variation under large-scale drought events, the observations they provide are not directly associated with vegetation functioning [20].

During recent years, satellite retrievals of SIF based on energy reemitted directly from the core of photosynthetic machinery present a fresh manner to observe vegetation growth and response [21–23]. On account of energy conservation, variations in SIF signal include information regarding vegetation physiological, biochemical, and metabolic functions in addition to the amount of absorbed photosynthetically active radiation [20]. Numerous studies have examined the relationship between GPP and SIF for each biome at various scales retrieved from different satellite sensors [24–27], implying that SIF can be unbiased in monitoring of vegetation productivity and further contribute to the understanding of the global terrestrial carbon cycle [28,29]. Both SIF and Moderate Resolution Imaging Spectroradiometer (MODIS) GPP respond to water availability in much the same direction beyond their previously determined correlations, although biome-specific distinction remains evident [30].

Relative to traditional vegetation indices (e.g., normalized difference vegetation index, enhanced vegetation index, leaf area index, etc.), spaceborne SIF or fluorescence yield was found to show an earlier and more significant response to rainfall deficits and heat stress over tropical vegetated region, forest, and cropland at monthly scale [31,32]. It highlights that SIF is susceptible to both the structural and physiological variations of vegetation and can be beneficial to timely monitor and characterize the development of drought and heatwave [33–35]. During the extreme drought of 2015–2016 driven by a strong El Nino event, large-scale decoupling of greenness and photosynthesis over Amazon forests was observed by satellite, of which SIF showed a pronounced reduction as opposed to a slight increase of canopy greenness [36]. The magnitude of divergence in greenness and photosynthesis that are related to times of fluctuant soil moisture differ in sign along a tree cover gradient, and those of forested ecosystems display contrasting average responses [37]. Conversely, both greenness and photosynthesis of non-woody semiarid ecosystems exhibiting a strong response to the availability of soil moisture usually alter contemporaneously [37]. To assess the effect of drought-related stress, there is a need for remotely sensed indicators that measure not only how green the plants are, but also to what extent photosynthetic activity is.

By investigating the response of vegetation growth to the 2019 extreme heatwave in Australia, Qiu et al. [38] concluded that SIF observed by both GOME-2 and Orbiting Carbon Observatory-2 (OCO-2) are more sensitive to water and heat stress than the greenness-based vegetation index (EVI). In addition, OCO-2 SIF exhibits a more pronounced decrease and one-month earlier than GOME-2 SIF, primarily owing to the diverse timing of observations (overpass timing: GOME-2 at near 09:30 a.m. and OCO-2 at 1:36 p.m. local solar time). Nonetheless, this study was conducted with limited baseline years (2015–2019), and it remains uncertain at a relatively coarse scale (1° spatial resolution and monthly interval) for heterogeneous Australia's dryland vegetation under highly dynamic climate. The newly launched TROPOMI facilitates a considerable enhancement in SIF observation as a result of unprecedented spatial resolution (up to 7 km × 3.5 km at nadir) and near-daily global coverage as contrasted with GOME-2 and OCO-2 [26,39], which enables the assessment of the relationship of SIF-GPP at an ecologically meaningful scale.

Characteristic drought time-scale can play a critical role in determining the sensitivity of biome-specific vegetation to drought [40]. Arid and humid biomes mostly respond to short time-scales, while semiarid and subhumid biomes respond to long time scales [40]. The response of the annual crop yield in five main dryland cultivations in the United States to different time-scales of drought is highly spatial variable, of which some crops (e.g., winter wheat) responded to drought at medium to long SPEI time-scales, while other crops (e.g., soybean and corn) responded to short or long drought time-scales [41]. It remains unclear how major biome types of natural dryland vegetation respond to drought at different intensities and time scales.

Here, we utilize the large-scale 2018–2019 drought across Australia as a natural experiment to investigate and characterize the spatiotemporal response of dryland vegetation to this extreme event, using multiple-source satellite observations of solar-induced chlorophyll fluorescence and vegetation greenness. Specifically, our objectives are: (1) to investigate the spatiotemporal variability of photosynthesis and greenness of Australia's dryland vegetation under 2018–2019 extreme drought; (2) to examine whether SIF could earlier capture drought stress than EVI at monthly/semi-monthly scale; (3) to assess the capability of SIF retrieved from TROPOMI for dynamic drought monitoring over water-limited ecosystems.

2. Materials and Methods

2.1. Study Area

This study was conducted at a sub-continental scale between 10°S to 26°S and 113°E to 138°E, which encompassed northern and central Australia by a relatively constant decrease in rainfall with distance inland (Figure 1). This region, particularly for northern Australia, has a classic monsoon climate pattern, which receives more than 80% of annual precipitation during November to April [42]. From northern mesic tropics to the xeric central Australia, mean annual rainfall ranges from 1700 mm to approximately 300 mm (Bureau of Meteorology, Available online: https://www.bom.gov.au (accessed on 20 January 2020)), in line with the aridity index (AI) decreasing from 0.8 to 0.1 (Figure 1b). Correspondingly, the vegetation follows a wet–dry gradient that shifts from *Eucalyptus* dominated forests, open forests, and woodlands in the coastal northern areas to *Acacia*-dominated open woodlands, scattered shrubs, and hummock grassland into the vast inland [42] (Figure 1a).

To assess satellite observation with ground-based evidence, we selected five representative flux tower sites across the extensive study area: Howard Springs (AU-How), Dry River (AU-Dry), Sturt Plains (AU-Stp), Alice Spring Mulga (AU-ASM), and Ti Tree East (AU-TTE) [43]. The detailed description of five sites is shown in Table 1 (Available online: www.ozflux.org.au (accessed on 15 July 2019)). Apart from two northern sites (AU-How and AU-Dry), the remaining sites in the interior of Australia experienced extreme drought in 2018–2019.

Table 1. Summary of five eddy covariance flux tower sites.

Site	Abbreviation	Longitude (°E)	Latitude (°S)	Vegetation Type	Data Coverage
Howard Springs	AU-How	131.15	12.495	Eucalypt woodlands	2001–2019
Dry River	AU-Dry	132.371	15.259	Open forest savanna	2008–2019
Sturt Plains	AU-Stp	133.3502	17.1507	Tussock grasslands	2008–2018
Alice Spring Mulga	AU-ASM	133.2493	22.2828	*Acacia* woodlands	2010–2019
Ti Tree East	AU-TTE	133.64	22.287	*Corymbia* savanna	2012–2019

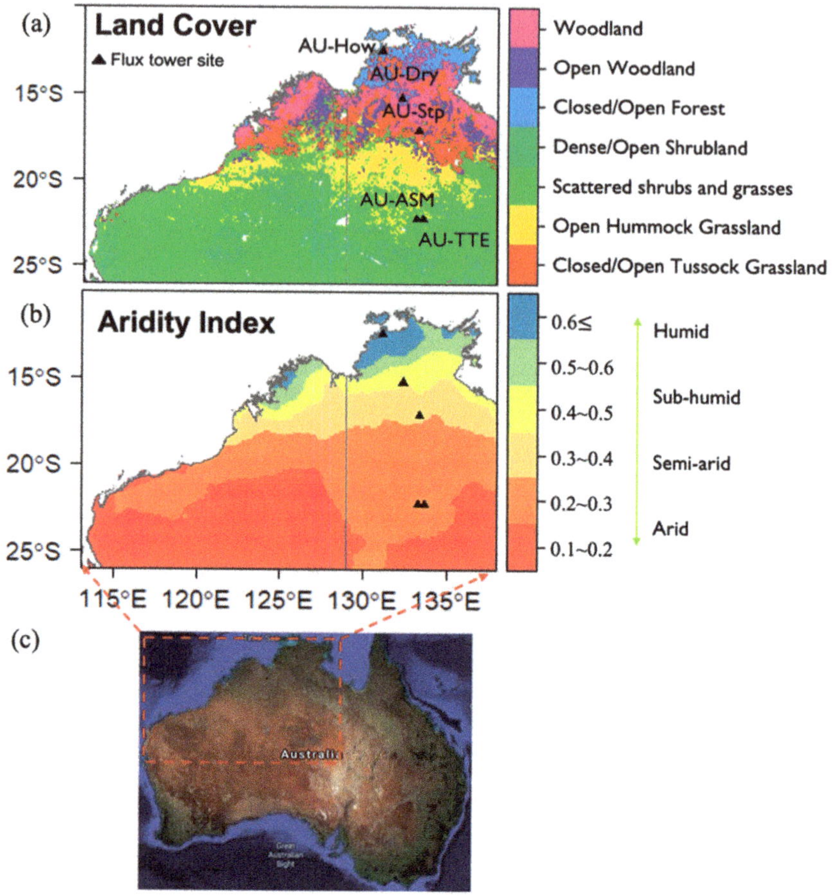

Figure 1. (**a**) Land cover map based on national Dynamic Land Cover Dataset. (**b**) Spatial pattern of aridity index. (**c**) The locations of study area over Australian continent (image source: Google Earth). Black triangles refer to five flux tower sites.

2.2. Satellite Data

A summary of satellite-based datasets used in this study is shown in Table 2. We utilized two set of satellite-based SIF records. (1) The first dataset was obtained from Global Ozone Monitoring Experiment-2 onboard EUMETSAT's MetOp-A as well as MetOp-B platform (indicated as GOME-2A and GOME-2B, respectively, in this study). SIF is the retrieval of the far-red chlorophyll fluorescence peaking at 740 nm, based on a simplified radiative transfer model in the company of a principal component analysis [44]. Monthly global coverage of SIF data based on GOME-2A at 0.5° spatial resolution (Level 3, Version 28) from February 2007 to March 2019 was obtained from NASA Goddard Space Flight Centre (Available online: https://avdc.gsfc.nasa.gov/ (accessed on 10 May 2020)). Given the degradation of the GOME-2A instrument during its lifetime, SIF observed from GOME-2B was also downloaded, which spans from March 2013 to March 2019. The daily orbital data (Level 2, Version 28) was also used to aggregate 16-day interval records for higher temporal-resolution analysis. (2) We used a newly released SIF dataset based on TROPOMI onboard the Sentinel-5 Precursor satellite. A data-driven method was employed to retrieve the SIF signal using spectral measurements ranging from 743 nm to 758 nm [26]. Daily corrected ungirded TROPOMI SIF data at a 0.05° spatial resolution (2018–2019) was available at (Available online: ftp://fluo.gps.caltech.edu/data/tropomi/ (accessed on 1 December 2020)). The daily orbital TROPOMI SIF were likewise aggregated to monthly and 16-day series by mean value, consistent with temporal resolution of GOME-2 SIF records.

Table 2. Summary of satellite-based datasets.

Dataset	Source	Resolution	Data Coverage
solar-induced chlorophyll fluorescence (SIF)	GOME-2 (MetOp-A)	Monthly/daily, 0.5°	2007–2019
solar-induced chlorophyll fluorescence (SIF)	GOME-2 (MetOp-B)	Monthly/daily, 0.5°	2013–2019
solar-induced chlorophyll fluorescence (SIF)	TROPOMI (Sentinel-5 Precursor)	Monthly/daily, 0.05°	2018–2019
enhanced vegetation index (EVI)	MODIS (Aqua)	Monthly/16-day, 0.05°	2007–2019
gross primary productivity (GPP)	FluxSat (MODIS and FLUXNET 2015)	Monthly, 0.5°	2007–2019
photosynthetic active radiation (PAR)	CERES (Aqua)	Monthly, 1°	2007–2019
land surface temperature (LST)	MODIS (Aqua)	Monthly, 0.05°	2007–2019
Precipitation	IMERG (GPM)	Monthly, 0.1°	2007–2019

We used Moderate Resolution Imaging Spectroradiometer (MODIS on board Aqua, Collection 6) MYD13C1 (0.05°, 16-day) and MYD13C2 (0.05°, monthly) Vegetation Indices products from January 2007 to June 2019 downloaded from NASA Earth Observation data (Available online: https://search.earthdata.nasa.gov/search (accessed on 1 August 2020)). EVI is an optimized version of vegetation indices that effectively reduces soil background influences and is widely used as a proxy of canopy greenness [45–47]. The equation (1) of EVI is:

$$EVI = 2.5\frac{\rho_{NIR} - \rho_{red}}{\rho_{NIR} + 6\rho_{red} - 7.5\rho_{blue} + 1} \tag{1}$$

where ρ_{blue}, ρ_{red}, ρ_{NIR} are reflectance in the blue, red, and near infrared bands, respectively. To reduce noise and uncertainties, only best quality data remained in this study through removing pixels of which quality control flag of the first 2 bits neither 00 nor 01, and pixel-wise EVI time series data was smoothed using the Savitzky–Golay filter.

To contrast with satellite-observed greenness and fluorescence, data-driven gross primary productivity (FluxSat_GPP, version 1.1, 2007–2019) derived primarily from MODIS reflectance product was used as the spatial reference of vegetation activity, obtained from NASA Goddard Space Flight Centre (Available online: https://avdc.gsfc.nasa.gov/ (accessed on 5 August 2020)). This is a monthly dataset with 0.5° spatial resolution, calibrated by a set of the FLUXNET 2015 eddy covariance data [48].

Monthly photosynthetic active radiation (PAR) at 1° resolution grid was downloaded from the NASA Langley Research Centre, Cloud and Earth's Radiant Energy System (CERES, Ed4.1), including adjusted surface PAR both direct and diffuse fluxes under all sky conditions. The total PAR was computed as the sum of both direct and diffuse PAR, and then used to remove the effects of PAR on SIF by normalizing after aggregating into the SIF grid resolution:

$$SIF_{PAR} = \frac{SIF}{PAR} \tag{2}$$

To assess the relationships between vegetation variables and meteor-environmental drivers under drought stress, MODIS daytime land surface temperature (LST, MYD11C3, Collection 6) at monthly scale was included in this study, collected from NASA Earth Observation data (Available online: https://search.earthdata.nasa.gov (accessed on 12 August 2020)). Similarly, best quality data remained through eliminating pixels with quality control flag. Global monthly precipitation at 0.1° resolution grid was based on Integrated Multi-Satellite Retrievals for Global Precipitation Measurement (IMERG, Version 6, Final run, 2007–2019) was collected from NASA Precipitation Processing System (Available online: https://pps.gsfc.nasa.gov/ (accessed on 12 November 2020)).

2.3. Climate DATA and Land Cover Map

In this study, we utilized monthly the Standardized Precipitation Evapotranspiration Index (SPEI) to characterize the spatial and temporal dynamics of the 2018–2019 extreme drought event across Australia [49]. This dataset at 1° spatial resolution with 1-month, 6-month timescale (indicated as $SPEI_{1\text{-month}}$, $SPEI_{6\text{-month}}$), representing the cumulative water status (deficit or surplus) over the preceding 1 or 6 months respectively, was downloaded from SPEI Global Drought Monitor (Available online: https://spei.csic.es/ (accessed on 20 July 2020)).

Monthly air temperature (at 2-m height) and soil moisture content (surface 0–7 cm depth, root zone 28–100 cm depth) based on ERA-5 reanalysis data were downloaded from Copernicus Climate Change Service (Available online: https://cds.climate.copernicus.eu/ (accessed on 28 September 2020)) [50]. Gridded annual potential evapotranspiration (PET) and monthly mean decadal precipitation were collected from the Bureau of Meteorology (Available online: http://www.bom.gov.au/ (accessed on 5 April 2020)). Firstly, we converted monthly precipitation into annual scale, and then aggregated into 0.1° spatial grid consistent with *PET*. Secondly, aridity index was computed as annual precipitation normalized by annual *PET*. The Equation (3) is:

$$AI = \frac{P}{PET} \tag{3}$$

Finally, pixels within the entire study area were binned by AI (every 0.1 increment) into 6 groups from humid to arid region, as shown in Figure 1b.

National Dynamic Land Cover Dataset (DLCD) was used in this research, obtained from Geoscience Australia and Bureau of Agricultural and Resource Economics and Sciences (Available online: http://www.ga.gov.au/scientific-topics/earth-obs/landcover (accessed on 10 May 2020)). This dataset validated with abundant field sites was aggregated to 0.5° spatial resolution by most frequent values.

2.4. Eddy Covariance Data

We used five selected flux towers to interpret the satellite-observed dynamic of vegetation under extreme drought. The original level 3 (AU-How, AU-Dry, AU-Stp) and level 6 (AU-ASM, AU-TTE) flux data provided by the OzFlux network (Available online: http://www.ozflux.org.au/ (accessed on 15 July 2020)) were used to pre-process, including quality control assessment, removal of outliers, and gap-filling [51]. The R package, REddyProc, was implemented for Level 3 data to estimate daily mean GPP with hourly eddy covariance and meteorological data [52]. This tool used the gap-filling and flux partitioning algorithms to partition Level 3 data into GPP and field ecosystem respiration [53], conducted in open-source R scientific computation environment (Version 3.5.1). The estimated daily GPP were aggregated into monthly and 16-day GPP to match with satellite-based observations.

2.5. Analysis

To investigate the responses of dryland vegetation to different drought intensities, we defined two categories of dry conditions based on $SPEI_{6\text{-month}}$, extreme dry ($SPEI \leq -2$), and medium dry ($-2 < SPEI \leq -0.5$). The R package, GeoRange, was implemented for monthly SPEI data to estimate the total area under extreme dry during the recent two decades (2000–2020). The Mann–Kendall trend test was utilized for time series of the total area and region-wide mean SPEI within the study area. Besides, the statistic ($SPEI_{1\text{-month}} \leq -2$) was conducted to generate the map of the pixel-wise extreme-dry month during November 2018 to April 2019.

To further examine the response of dryland biomes to extreme drought, we selected four-pair regions of interest (labelled as ROI-1, ROI-2, ROI-3, and ROI-4) distributed from mesic northern to xeric southern areas. The selection criteria are: (1) same or highly similar biome types in each ROI; (2) the sub-regions within ROIs underwent different drought conditions (extreme dry, medium dry, respectively). ROI-1 are mainly covered by woodlands and open woodlands. ROI-2 and ROI-3 are covered by scattered shrubs and grasses. Vegetation in ROI-4 is dominated by Hummock grasses mixed with scattered shrubs. In addition, ROI-2 and ROI-4 were also under extreme dry conditions in the monsoon season of 2012–2013 and 2015–2016, respectively (Figure S1).

With the purpose of detecting drought-related signals by eliminating seasonality, monthly anomalies (ΔX) and standardized anomalies (X_{SA}) of all the aforementioned variables were calculated as a deviation from their corresponding multiyear (2007–2018) mean of each month. The Equations (4) and (5) are:

$$\Delta X = X_m - \overline{X_m} \tag{4}$$

$$X_{SA} = \frac{X_{nm} - \overline{X_m}}{\sigma_m} \tag{5}$$

where n is the yearly temporal coverage from 2007 to 2018, X_{nm} is the monthly ranging from July to June, $\overline{X_m}$ and σ_m are the mean and standard deviation of time series X at month m.

To quantify the relationship between satellite-based observation and tower-based measurement, the coefficient of determination (r^2) was calculated across five selected tower sites at different spatial (0.5°, 0.05°) and temporal (monthly, 16-day) resolutions. All satellite datasets were extracted from a 3 × 3 window centered on each flux tower site respectively. A t-test was utilized to examine the statistically significant level of the relationships (p-value).

Data processing, analysis, and visualization were conducted in open-source R scientific computation environment (Version 3.6.2) and associated packages contributed by the R user community (Available online: http://cran.r-project.org (accessed on 20 August 2020)).

3. Results

3.1. Spatiotemporal Dynamics of the 2018–2019 Extreme Drought

From the start of the new century, there was a significant increase trend in the total area influenced by extreme dry condition and an opposite trend in the region-wide mean SPEI (p-value < 0.0001). The extreme, 4-month drought spanning from December 2018 to March 2019 was the worst drought event during the last two decades over central and northwest Australia (Figure 2), of which almost the entire region was influenced by this event (Figure 3a). During the 2018–2019 monsoon season, a total area of 1.75 million km^2 was under extreme dry condition extending from north to xeric central Australia (15°–26°S and 118°–138°E), particularly in December 2018 and January 2019. In addition, the xeric inland experienced a longer extreme dry period (more than 2 months) relative to the mesic northern coastal regions (Figure 3b).

Temporal variations in satellite-based SIF, SIF_{PAR}, EVI, and GPP averaged within the extent of extreme and medium dry, respectively, are displayed in Figure 4. Compared with approximately one standard deviation (SD) decline in all vegetation variables under medium dry, those exhibited 2–3 times SD decrease relative to climatology under extreme situation. Correspondingly, both SIFPAR and EVI showed the largest reduction from January 2019 as a consequence of persistent rainfall deficit as well as increased temperature since November 2018.

Spatial patterns of standardized anomalies of precipitation, surface/root zone soil moisture, LST, SIF, SIF_{PAR}, EVI, and GPP during the 2018–2019 monsoon season are shown in Figure 5, along with the corresponding pixel density classified by extreme and medium dry. Meteorological data show that 87%, 25% of study region appeared rainfall anomalies below than −1 SD, −2 SD, respectively, and 91%, 60% of the area exhibited air temperature anomalies larger than 1 SD and 2 SD, revealing it was a large-scale drought event coupled with an extreme heatwave (Figure 5a,d). The persistently reduced precipitation along with higher temperature results in depletion of both surface and root zone soil water content over most areas, of which 70%, 50% showed less than −1 SD, respectively (Figure 5b,c).

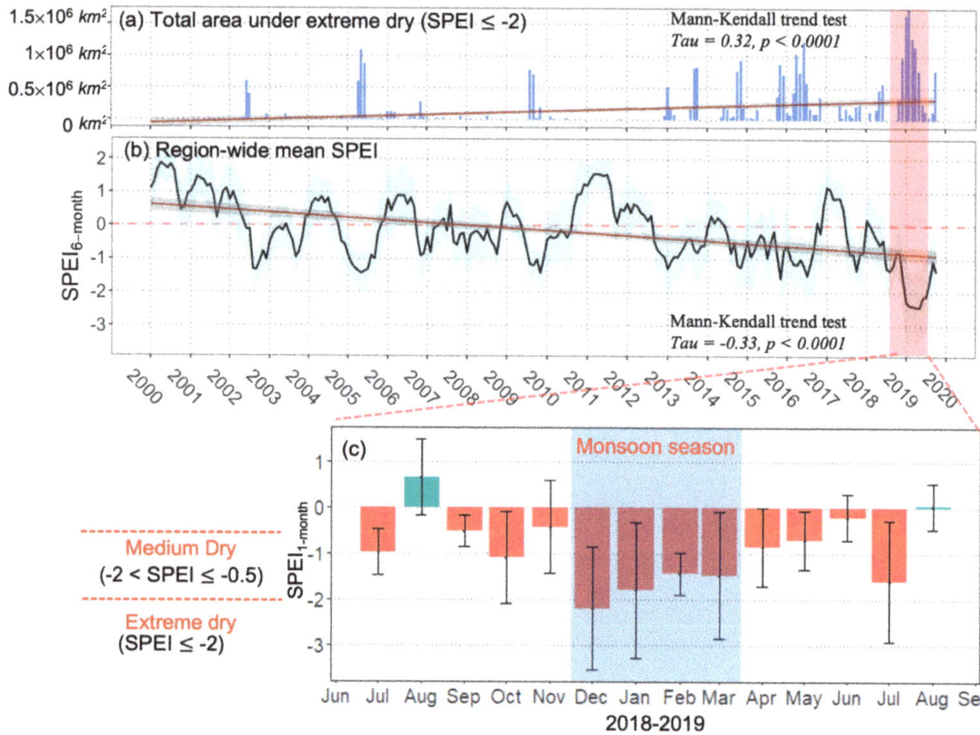

Figure 2. (**a**) Time series of the total area under extreme dry within the extent of study region during 2000–2019. Red line refers to the linear regression line. (**b**) Region-wide mean monthly $SPEI_{6\text{-month}}$ during 2000–2019. (**c**) Region-wide mean monthly $SPEI_{1\text{-month}}$ during 2018–2019.

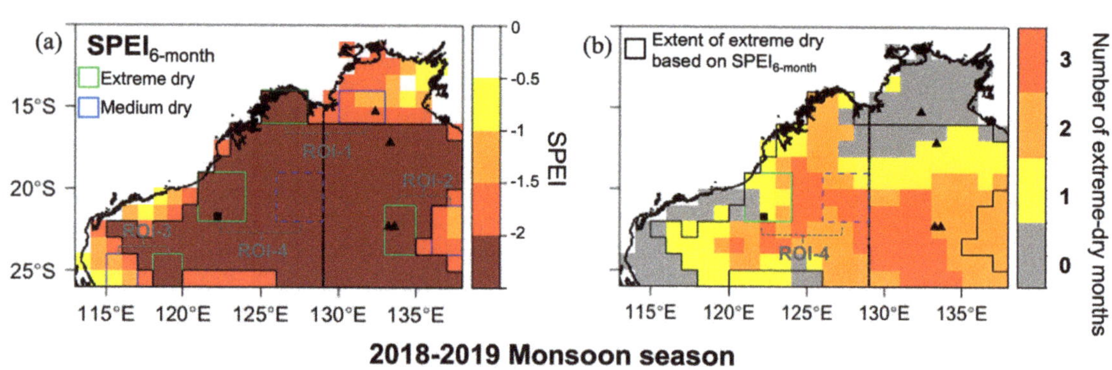

Figure 3. (**a**) Spatial pattern of drought intensity based on $SPEI_{6\text{-month}}$ during 2018–2019 monsoon season. (**b**) Pixel-wise number of extreme-dry months based on $SPEI_{1\text{-month}}$. Four pairs of ROIs are shown in Figure 3a. The black boundary represents the extent of extreme (within) and medium (outside) dry.

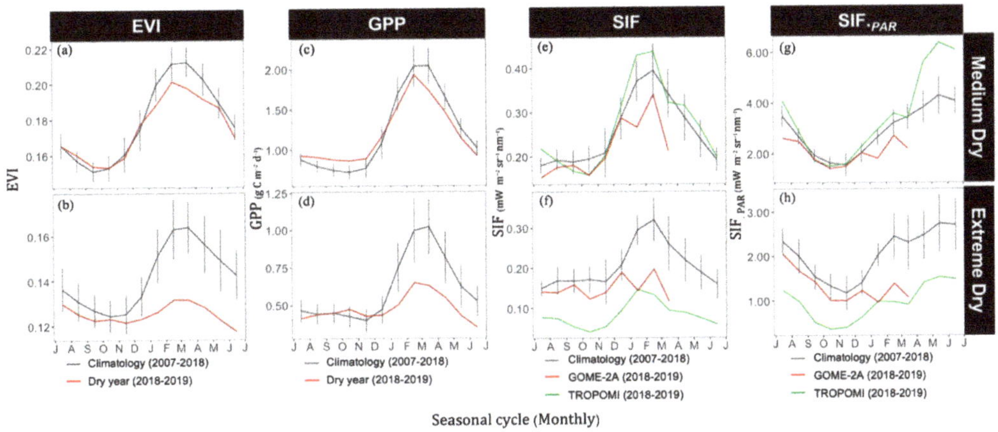

Figure 4. The spatially averaged seasonal cycles of monthly EVI, GPP, SIF, and SIF_{PAR} within (**a**,**c**,**e**,**g**) medium and (**b**,**d**,**f**,**h**) extreme dry extent from July 2018 to June 2019. The grey curves represent the monthly multiyear mean (2007–2018). Vertical lines refer to ±1 standard deviation.

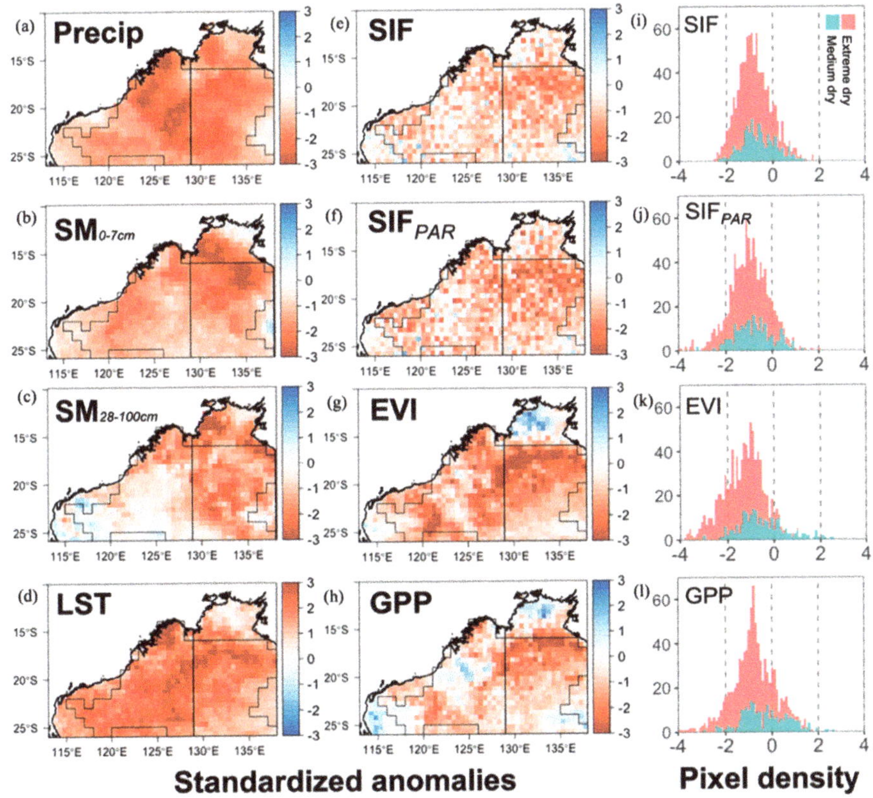

Figure 5. (**a**–**h**) Spatial distributions of the standardized anomalies of precipitation, surface/root zone soil moisture, LST, SIF, SIF_{PAR}, EVI, and GPP during 2018–2019 monsoon season. (**i**–**l**) Pixel density of four vegetation variables within the extent of extreme and medium dry.

Vegetation growth is largely affected by the extreme drought, and the majority of the area showed negative anomalies for all vegetation variables relative to the multiyear mean (Figure 5e–h). SIF_{PAR} and EVI exhibited a similar spatial pattern, where those were remarkably negative anomalies (<−2 SD) between 17°–20°S and 129°–138°E (Figure 5f,g). SIF and SIFPAR showed proportional pixels distribution under extreme dry and medium dry, while there was an obvious distinction for EVI (Figure 5i–k). This discrepancy between SIF and EVI was mainly caused by the divergent responses over northern humid northern Australia (11°–15°S), where EVI exhibited evidently positive anomalies in comparison to slightly depressed SIF, SIF_{PAR}.

3.2. Responses of Dryland Vegetation to Diverse Drought Intensity

To investigate the response of major biome types of dryland vegetation to different drought intensity, four pairs of ROIs were inter- and intra-compared during 2018–2019 extreme drought as well as associated 2012–2013 and 2015–2016 drought events. Temporal variation in SIF, EVI averaged within ROIs are presented, along with seasonal anomalies of SIF, EVI (Figure 6). There were prominent differences among ROI-1 and ROI-3, of which both SIF and EVI showed larger negative anomalies under extreme dry relative to medium dry (Figure 6a,b,g,h). Within semi-arid ROI-2, both vegetation variables exhibited equivalent magnitude of negative anomalies regardless of drought severity under two contrasting drought events of 2012–2013 and 2018–2019 (Figure 6c–f). Similarly, semi-arid region partially covered with hummock grass (ROI-4), SIF also showed comparably negative anomalies under both extreme and medium dry of 2015–2016 (Figure 6l), while EVI displayed distinct responses (Figure 6k). In addition to the drought intensity, the response of vegetation was also affected by time scales of extreme drought, as evident by ROI-4 in 2018–2019 (Figure 6i,j). We found both sub-regions of ROI-4 were under extreme dry conditions, but with different time scales of extreme-dry period (Figure 3a,b), which leads to the larger reduction of SIF and EVI under longer periods of extreme dry conditions.

3.3. Comparison of SIF and EVI in Response to Drought

To further explore the difference in response of biome-specific vegetation of SIF and EVI to extreme drought, anomalies of SIF and EVI among the entire region under extreme and medium dry are shown in Figure 7. In general, the magnitude of anomalies of SIF and EVI was proportional to the mean SPEI among each category (Figure 7a,d). Tussock grasslands under the worst dry situation in 2018–2019 also exhibited the largest reduction of both SIF and EVI compared with other major biomes (Figure 7b,c). Apart from this, the largest divergent magnitude of anomalies under extreme and medium dry conditions was also shown. Relative to arid/semiarid biomes (e.g., hummock grassland, open shrubland, and scattered shrubs and grasses), mesic ecosystems (e.g., woodland, open woodland) exhibited larger differences under two drought intensities (Figure 7b,c). There was an evident trend for EVI, with the amplified difference between extreme and medium dry from arid to humid climate regimes (Figure 7f), especially for positively anomalous EVI under medium dry (AI > 0.5), mainly distributed over northern coastal areas (Figure 5). Conversely, the majority of SIF across different biome types or climate regimes showed negative anomalies with comparable magnitude, regardless of diverse drought intensity (Figure 7b,e).

Temporal percentage of drought-related vegetation decline indicated by EVI and SIF at a monthly and 16-day scale is shown in Figure 8. At the beginning of this drought event (November 2018), approximately 38%, 20% of the area was affected by moderate ($<-0.5\sigma$) and severe ($<-1\sigma$) losses, respectively, indicated by EVI, and increased to over 80%, 50% with the development of extreme drought (Figure 8a). The percentage of moderate and severe reduction observed by SIF, SIF_{PAR} in November of 2018 was around 36%, 18%. In addition, the percentages of affected area revealed by SIF slightly decreased after reaching the peak in February of 2019, while the impacted region indicated by EVI remained growing (Figure 8a–c).

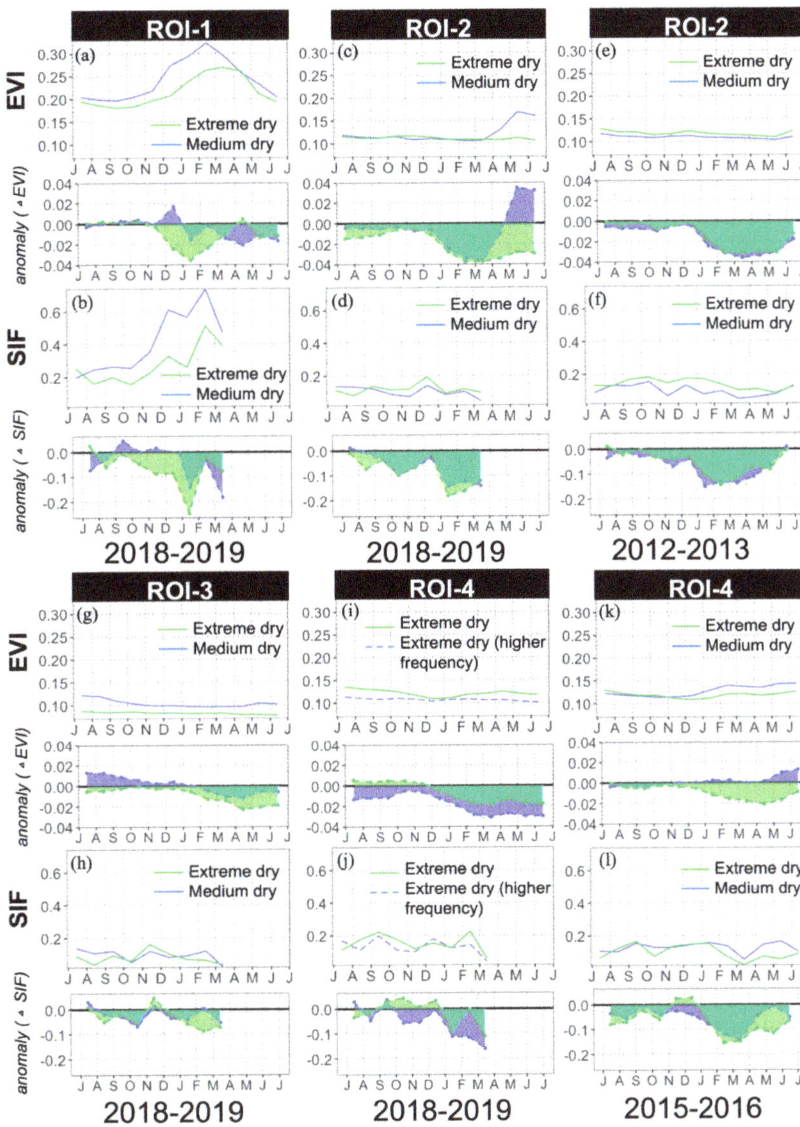

Figure 6. Region-wide seasonal cycles and anomalies of (**a**,**c**,**g**,**i**) EVI and (**b**,**d**,**h**,**j**) SIF within four pairs of ROIs in 2018–2019. Seasonal profiles of SIF and EVI within (**e**,**f**) ROI–2 in 2012–2013 and (**k**,**l**) ROI–4 in 2015–2016. Bottom panels of each figure refer to the anomalies of SIF or EVI relative to climatology (2007–2018).

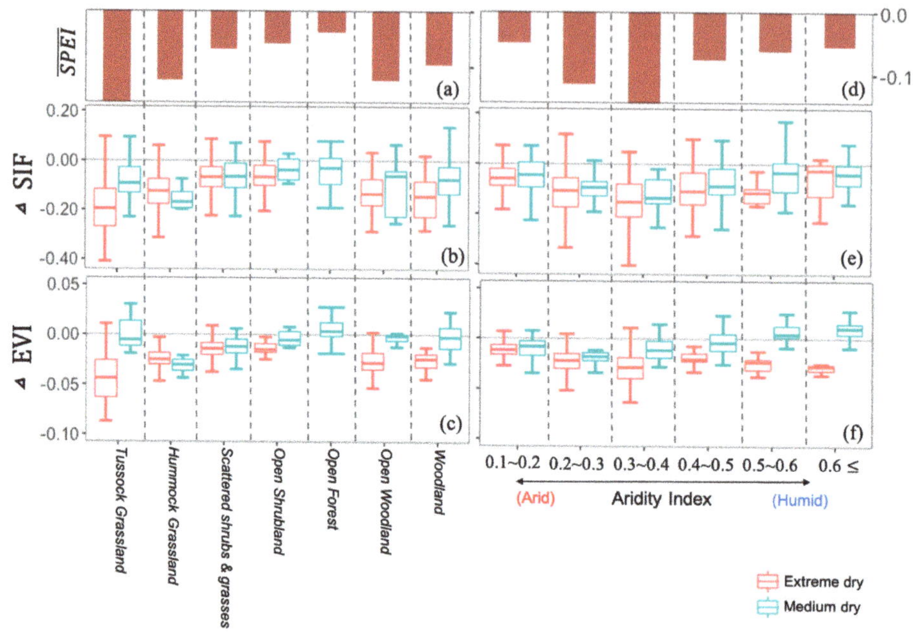

Figure 7. Boxplot of anomalies of EVI and SIF under extreme and medium dry in 2018–2019 grouped by (**b**,**c**) major biome types and (**e**,**f**) aridity index. (**a**,**d**) The top panel refers to the spatially mean SPEI of each category.

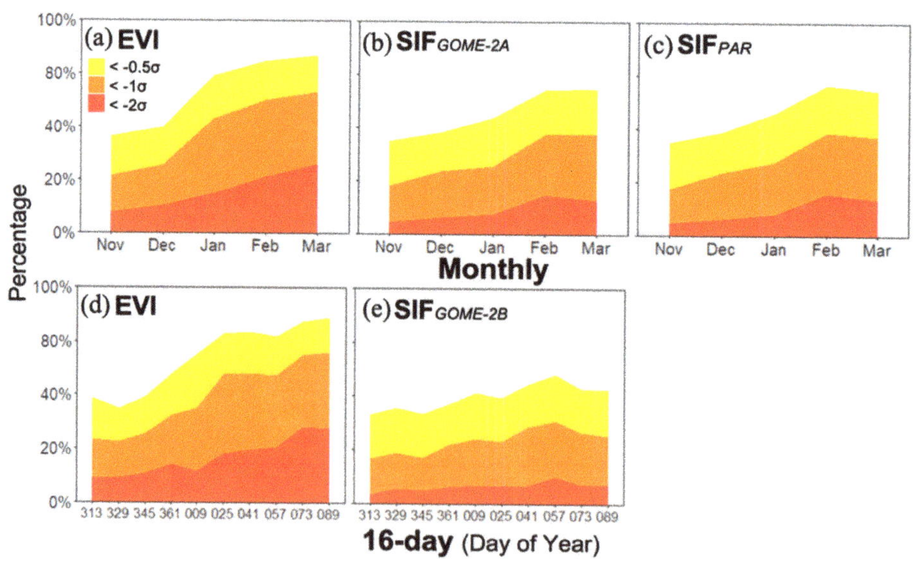

Figure 8. (**a**,**b**,**c**) Monthly and (**d**,**e**) 16-day time series of the percentage of drought-induced vegetation reduction indicated by EVI, SIF, and SIF_{PAR} under moderately ($<-0.5\sigma$), severely ($<-1\sigma$), and extremely ($<-2\sigma$) negative anomalies. σ indicates the standard deviation of the monthly SIF, EVI during 2007–2018 as well as 16-day SIF, EVI during 2013–2018. The text of the x-axis in Figure 9d,e refer to the day of year from November 2018 to March 2019.

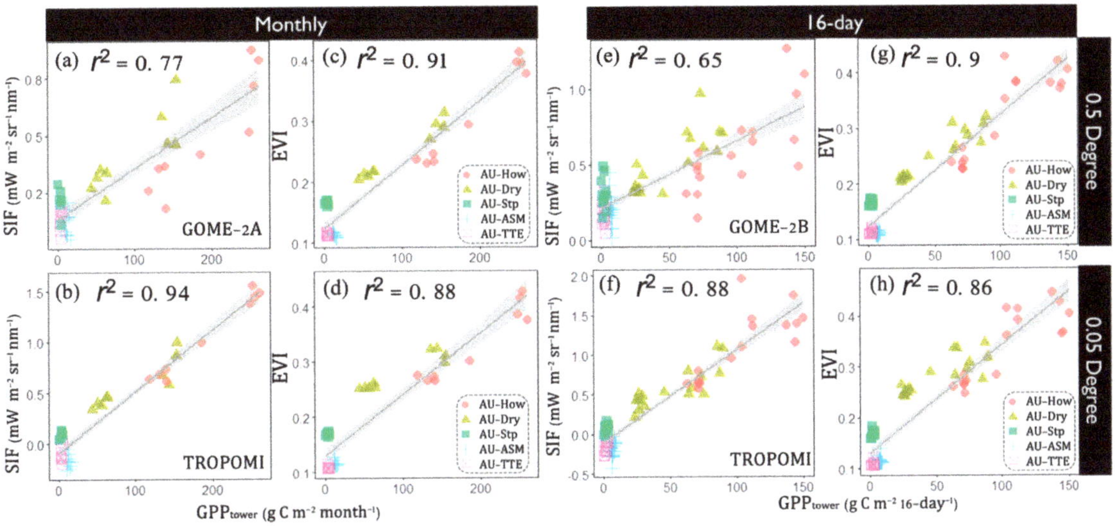

Figure 9. (**a**–**h**) Relationship between tower-based GPP and satellite-based SIF, EVI in 2018–2019 across five flux tower sites at different spatial (0.5°, 0.05° grid) and temporal (monthly, 16-day) resolution. The bottom-right embedded box shows the legend of the five flux tower sites. (p value < 0.001).

To further contrast the responses of SIF and EVI at a higher temporal scale, 16-day time series of percentage of moderate/severe/extreme anomalies indicated by EVI and SIF was presented in Figure 8d,e. There was apparently a larger proportion of vegetation decline indicated by EVI than SIF at each category of anomalies, especially over 20% of the area in the middle of March (DOY = 73) which was affected by extreme ($<-2\sigma$) reduction observed by EVI relative to the maximum 7%, indicated by SIF at the end of February (DOY = 57). Besides, the percent of pixels with no seasonality detected by EVI was also higher than that observed by SIF.

3.4. Cross-Comparison with In-Situ Measurements

With the purpose of validating satellite-based observation with field measurement, Figure 9 presents the relationship between tower-based GPP and satellite-based SIF, EVI in 2018–2019 across five selected sites at different spatial and temporal resolutions. Despite diverse spatiotemporal resolution, both satellite-observed SIF, EVI were overall significantly (p-value < 0.001) associated with in-situ measurement GPP (r^2 ranging from 0.65 to 0.94) in the extremely dry year of 2018–2019. There was an enhanced correspondence between GPP and SIF with increasing spatial resolution (from 0.5° to 0.05° grid) for both monthly and 16-day temporal scales (Figure 9a,b,e,f). Besides, we found a stronger relationship between field measurement and satellite-based observation at monthly series (r^2 = 0.77, 0.94) relative to 16-day series (r^2 = 0.65, 0.88). The close correlation between GPP and EVI remained across different spatial and temporal resolution (r^2 ranging from 0.86 to 0.91, Figure 9c,d,g,h). For three southern sites (AU-Stp, AU-ASM, AU-TTE) that suffered from extreme dryness in 2018–2019 there was large reduction, and seasonal amplitude was barely detected by either in-situ measurement or satellite observation (EVI).

4. Discussion

4.1. Potential of Spaceborne SIF for Drought Monitoring over Water-Limited Ecosystems

Satellite-based SIF observations capture the spatial and temporal variations in dryland vegetation under 2018–2019 extreme drought in Australia (Figures 4 and 5). Relative to EVI and GPP, the fluctuated SIF series and notable spatial speckling were probably due to high retrieval noise in sparsely vegetated regions [54], especially for lower signal levels under extreme drought. Overall, SIF and EVI exhibited resemblance to spatial and temporal responses of dryland vegetation to this extreme event, wherein the central study area (between 17–20°S and 129–138°E) showed the most remarkable reduction (Figure 5). The primary difference in spatial dynamics of two variables occurred over the northern humid region (between 12–15°S and 129–138°E) under medium drought, where SIF showed marginally negative anomalies in comparison to positive EVI anomalies (Figure 5). This is consistent with previous studies [32,34,36,55], revealing that there was a considerable decoupling of photosynthesis and greenness dynamics under moderate dry conditions over mesic ecosystems (e.g., forests and woodlands).

Nonetheless, there was concurrently substantial loss in both SIF and EVI over arid and semiarid biome types, and those suffered from extreme rainfall deficit as well as severe heat wave (Figure 5). In particular, a higher percentage of the area indicated by EVI was affected by extreme ($<-2\sigma$) losses than that by SIF throughout the entire 2018–2019 monsoon season (Figures 5 and 8). This finding was against a similar study [38], which concludes satellite SIF observations are more sensitive to water and heat stress than EVI over arid central Australia in the 2019 heatwave. However, the divergent results were primarily ascribed to the different methods to define the anomaly, and they utilized the relative anomaly (a departure from the climatology and divided by the multiyear mean) instead of the standardized anomaly used in this study. The larger variation in magnitude of SIF than that of greenness-based VIs gave rise to the sharper reduction in SIF, indicated by relative anomaly relative to EVI. In addition, insufficient baseline years (2015–2018) in their study may induce more uncertainties as a consequence of highly dynamic climate regime over inland Australia.

On the contrary, we contrasted the responses of two sources of SIF data derived from GOME-2A and GOME-2B with MODIS EVI under two reference climatologies (2007–2018 and 2014–2018), respectively. We found that the percentage of the drought-affected area captured by EVI was consistently higher than that of SIF regardless of diverse temporal scales (Figure 8). The reason for the more significantly negative standardized anomalies of EVI relative to those of SIF was possibly owing to the fact that the dramatically drought-induced reduction in vegetation leads to barely increased magnitude of seasonal cycles over the vast inland area (Figure 9). Apart from this, high-retrieval noise provokes the larger variability of SIF signal over arid/semiarid ecosystems.

Compared with coarse SIF data derived from GOME-2, the state-of-the-art TROPOMI SIF observation with substantially improved spatiotemporal resolution shows an enhanced correlation with tower-based GPP in the extremely dry year of 2018–2019 (Figure 9), implying that this advanced dataset has promising potential for drought monitoring over heterogeneous arid and semiarid ecosystems. As opposed to the fluctuated temporal series and spatial speckling of GOME-2 SIF signal (Figures 4–6 and 9), SIF retrieved from remarkable radiometric performance of TROPOMI exhibits more spatial and temporal consistency, representing a step change in SIF remote sensing capabilities [26,56]. It is worth noting that an abundance of negative values of TROPOMI SIF was observed over central Australia in 2018–2019, probably owing to the retrieval noise. Köhler et al. [26] suggested that negative SIF values mainly occur for single TROPOMI measurements, which should not be over-interpreted. A comprehensive assessment of the capability of the novel dataset over dryland ecosystem needs to be conducted in the future.

4.2. Dynamics of Dryland Vegetation under Different Drought Scenarios

Varying responses of major biome types of dryland vegetation to different drought severity (extreme dry vs medium dry) were observed (Figures 6 and 7). We found enhanced magnitude of reduction of both SIF and EVI in conjunction with increasing drought severity over humid/sub-humid biomes (e.g., forest, woodland, and tussock grasslands). By contrast, the two vegetation variables of arid/semiarid vegetation (e.g., hummock grasslands, open shrublands) showed comparable losses under extreme and medium dry condition, indicating the largest sensitivity and vulnerability of semiarid ecosystems to the hydroclimatic variations [18,57]. It is not only evident by the 2018–2019 extreme drought, but also the 2012–2013, 2015–2016 drought events in central Australia (Figure 6e–l). As prolonged drought is projected to increase both its frequency and intensity [58,59], drylands will have a more critical role in regulating the global carbon and water cycle [60,61].

Given that the coarse spatial grid (0.5°) induced uncertainty of biome-specific analysis over heterogeneous drylands, re-grouped pixels based on aridity index were used to investigate the responses to the extreme drought at different time scales (Figure 7). As opposed to notably depressed SIF over mesic ecosystem (AI > 0.6), EVI remained positive anomalies even under one-month extreme dryness, in line with reported studies concerning dry-season greening of forests [62,63]. Despite that, canopy greenness of humid ecosystems exhibited a significant reduction when the extreme-dry period exceeded one month ($\geq$2). Likewise, we found larger magnitudes of decline in arid regions (AI < 0.3) resulting from the increasing extreme-dry months, although both sub-regions of ROI-4 suffered from extreme dryness in 2018–2019 (Figure 3a,b, Figures 6l–n and 7). In contrast, both SIF and EVI in semiarid regions (AI: 0.3–0.4) exhibited the largest reduction under short timescales (extreme-dry month $\leq$ 1), confirming semiarid biomes are the most susceptible to extreme drought.

5. Conclusions

In summary, we investigated the spatial and temporal responses of SIF and EVI of dryland vegetation to the 2018–2019 extreme drought over Australia using multi-source satellite-based SIF observations. In contrast to the enhanced magnitude of reduction of both SIF and EVI in conjunction with increasing drought severity over humid/sub-humid biomes, the two vegetation variables of arid/semiarid vegetation showed comparable negative anomalies regardless of extreme or medium dry conditions, indicating the largest sensitivity and vulnerability of semiarid ecosystems to the hydroclimatic variations. At a 16-day interval, both satellite-based indicators (SIF and EVI) can synchronously detect the impact of extreme drought on dryland vegetation growth. Besides, we find that space-borne SIF is capable of characterizing the spatiotemporal dynamics of drought over water-limited ecosystems despite high retrieval noises, as compared with EVI. The unprecedented SIF derived from TROPOMI shows remarkably enhanced agreement with tower-based GPP in the dry year of 2018–2019, demonstrating the great potential of the advanced dataset to track the dynamics of dryland vegetation under future changing climates. In addition, semiarid ecosystems exhibiting the largest reduction, regardless of drought severity and time scales, are the most sensitive to climatic extremes.

Supplementary Materials: The following supporting information can be downloaded at: https://www.mdpi.com/article/10.3390/rs14071581/s1, Figure S1: Spatial pattern of drought intensity based on $SPEI_{6\text{-month}}$ during 2018–2019, 2015–2016, and 2012–2013 monsoon season. Green and blue rectangles refer to the region of interest (ROI). Black triangles refer to five flux tower sites.

Author Contributions: Conceptualization, S.L. and A.H.; methodology, S.L.; validation, S.L.; formal analysis, S.L.; investigation, S.L.; data curation, S.L., S.G. and J.C.; writing—original draft preparation, S.L.; writing—review and editing, A.H., Q.Y., X.M., J.Q., R.Z. and Q.W.; visualization, S.L. and Q.W.; supervision, A.H. and Q.Y.; funding acquisition, Q.W. All authors have read and agreed to the published version of the manuscript.

Funding: This research was funded by the Natural Science Foundation of Fujian Province (2021J01627) and the National Natural Science Foundation of China (41601562).

Institutional Review Board Statement: Not applicable.

Informed Consent Statement: Not applicable.

Data Availability Statement: All satellite-based data well as other datasets in this study are available from the corresponding author upon reasonable request.

Acknowledgments: This work was supported by the Natural Science Foundation of Fujian Province (2021J01627) and the National Natural Science Foundation of China (No. 41601562) for funding this work. The first author acknowledges the financial support from Chinese Scholarship Council, Ministry of Education, China. We also would like to thank the opening of flux tower data for public access by the Australian Terrestrial Ecosystem Research Network (TERN, Available online: www.tern.org.au (accessed on 30 June 2020))—OzFlux facility (Available online: www.ozflux.org.au (accessed on 30 June 2020)).

Conflicts of Interest: The authors declare no conflict of interest.

References

1. Jiao, W.; Chang, Q.; Wang, L. The Sensitivity of Satellite Solar-Induced Chlorophyll Fluorescence to Meteorological Drought. *Earth's Future* **2019**, *7*, 558–573.
2. Cataldo, E.; Salvi, L.; Mattii, G.B. Effects of Irrigation on Ecophysiology, Sugar Content Andthiol Precursors (3-S-Cysteinylhexan-1-Ol and 3-S-Glutathionylhexan-1-Ol) on Vitis Vinifera Cv. Sauvignon Blanc. *Plant Physiol. Biochem.* **2021**, *164*, 247–259. [CrossRef] [PubMed]
3. Cataldo, E.; Fucile, M.; Mattii, G.B. A Review: Soil Management, Sustainable Strategies and Approaches to Improve the Quality of Modern Viticulture. *Agronomy* **2021**, *11*, 2359. [CrossRef]
4. Ma, X.; Huete, A.; Cleverly, J.; Eamus, D.; Chevallier, F.; Joiner, J.; Poulter, B.; Zhang, Y.; Guanter, L.; Meyer, W.; et al. Drought Rapidly Diminishes the Large Net CO_2 Uptake in 2011 over Semi-Arid Australia. *Sci. Rep.* **2016**, *6*, 37747. [CrossRef]
5. van Dijk, A.I.J.M.; Beck, H.E.; Crosbie, R.S.; de Jeu, R.A.M.; Liu, Y.Y.; Podger, G.M.; Timbal, B.; Viney, N.R. The Millennium Drought in Southeast Australia (2001–2009): Natural and Human Causes and Implications for Water Resources, Ecosystems, Economy, and Society. *Water Resour. Res.* **2013**, *49*, 1040–1057.
6. Haverd, V.; Ahlstrom, A.; Smith, B.; Canadell, J.G. Carbon Cycle Responses of Semi-Arid Ecosystems to Positive Asymmetry in Rainfall. *Glob. Chang. Biol.* **2017**, *23*, 793–800. [CrossRef]
7. Xie, Z.; Huete, A.; Cleverly, J.; Phinn, S.; McDonald-Madden, E.; Cao, Y.; Qin, F. Multi-Climate Mode Interactions Drive Hydrological and Vegetation Responses to Hydroclimatic Extremes in Australia. *Remote Sens. Environ.* **2019**, *231*, 111270. [CrossRef]
8. Broich, M.; Tulbure, M.G.; Verbesselt, J.; Xin, Q.; Wearne, J. Quantifying Australia's Dryland Vegetation Response to Flooding and Drought at Sub-Continental Scale. *Remote Sens. Environ.* **2018**, *212*, 60–78. [CrossRef]
9. Wang, J.; Ding, J.; Yu, D.; Teng, D.; He, B.; Chen, X.; Ge, X.; Zhang, Z.; Wang, Y.; Yang, X.; et al. Machine Learning-Based Detection of Soil Salinity in an Arid Desert Region, Northwest China: A Comparison between Landsat-8 OLI and Sentinel-2 MSI. *Sci. Total Environ.* **2020**, *707*, 136092. [CrossRef]
10. Wang, J.; Ding, J.; Li, G.; Liang, J.; Yu, D.; Aishan, T.; Zhang, F.; Yang, J.; Abulimiti, A.; Liu, J. Dynamic Detection of Water Surface Area of Ebinur Lake Using Multi-Source Satellite Data (Landsat and Sentinel-1A) and Its Responses to Changing Environment. *CATENA* **2019**, *177*, 189–201. [CrossRef]
11. Zhang, J.; Ding, J.; Wu, P.; Tan, J.; Huang, S.; Teng, D.; Cao, X.; Wang, J.; Chen, W. Assessing Arid Inland Lake Watershed Area and Vegetation Response to Multiple Temporal Scales of Drought Across the Ebinur Lake Watershed. *Sci. Rep.* **2020**, *10*, 1354. [CrossRef] [PubMed]
12. Huete, A.R.; Restrepo-Coupe, N.; Ratana, P.; Didan, K.; Saleska, S.R.; Ichii, K.; Panuthai, S.; Gamo, M. Multiple Site Tower Flux and Remote Sensing Comparisons of Tropical Forest Dynamics in Monsoon Asia. *Agric. For. Meteorol.* **2008**, *148*, 748–760. [CrossRef]
13. Kath, J.; le Brocque, A.F.; Reardon-Smith, K.; Apan, A. Remotely Sensed Agricultural Grassland Productivity Responses to Land Use and Hydro-Climatic Drivers under Extreme Drought and Rainfall. *Agric. For. Meteorol.* **2019**, *268*, 11–22. [CrossRef]
14. Wang, Q.; Yang, Y.; Liu, Y.; Tong, L.; Zhang, Q.; Li, J. Assessing the Impacts of Drought on Grassland Net Primary Production at the Global Scale. *Sci. Rep.* **2019**, *9*, 14041. [PubMed]
15. Xu, H.; Wang, X.; Zhao, C.; Yang, X. Diverse Responses of Vegetation Growth to Meteorological Drought across Climate Zones and Land Biomes in Northern China from 1981 to 2014. *Agric. For. Meteorol.* **2018**, *262*, 1–13. [CrossRef]
16. Zeng, J.; Zhang, R.; Qu, Y.; Bento, V.A.; Zhou, T.; Lin, Y.; Wu, X.; Qi, J.; Shui, W.; Wang, Q. Improving the Drought Monitoring Capability of VHI at the Global Scale via Ensemble Indices for Various Vegetation Types from 2001 to 2018. *Weather Clim. Extrem.* **2022**, *35*, 100412. [CrossRef]
17. Zhang, R.; Wu, X.; Zhou, X.; Ren, B.; Zeng, J.; Wang, Q. Investigating the Effect of Improved Drought Events Extraction Method on Spatiotemporal Characteristics of Drought. *Theor. Appl. Climatol.* **2022**, *147*, 395–408. [CrossRef]
18. Ma, X.; Huete, A.; Moran, S.; Ponce-Campos, G.; Eamus, D. Abrupt Shifts in Phenology and Vegetation Productivity under Climate Extremes. *J. Geophys. Res. Biogeosci.* **2015**, *120*, 2036–2052. [CrossRef]
19. Wang, Q.; Qi, J.; Qiu, H.; Li, J.; Cole, J.; Waldhoff, S.; Zhang, X. Pronounced Increases in Future Soil Erosion and Sediment Deposition as Influenced by Freeze–Thaw Cycles in the Upper Mississippi River Basin. *Environ. Sci. Technol.* **2021**, *55*, 9905–9915. [CrossRef]

20. Sun, Y.; Fu, R.; Dickinson, R.; Joiner, J.; Frankenberg, C.; Gu, L.; Xia, Y.; Fernando, N. Drought Onset Mechanisms Revealed by Satellite Solar-Induced Chlorophyll Fluorescence: Insights from Two Contrasting Extreme Events. *J. Geophys. Res. Biogeosci.* **2015**, *120*, 2427–2440. [CrossRef]
21. Frankenberg, C.; Fisher, J.B.; Worden, J.; Badgley, G.; Saatchi, S.S.; Lee, J.; Toon, G.C.; Butz, A.; Jung, M.; Kuze, A. New Global Observations of the Terrestrial Carbon Cycle from GOSAT: Patterns of Plant Fluorescence with Gross Primary Productivity. *Geophys. Res. Lett.* **2011**, *38*, L17706.
22. Guan, K.; Pan, M.; Li, H.; Wolf, A.; Wu, J.; Medvigy, D.; Caylor, K.K.; Sheffield, J.; Wood, E.F.; Malhi, Y.; et al. Photosynthetic Seasonality of Global Tropical Forests Constrained by Hydroclimate. *Nat. Geosci.* **2015**, *8*, 284–289. [CrossRef]
23. Sun, Y.; Frankenberg, C.; Wood, J.D.; Schimel, D.S.; Jung, M.; Guanter, L.; Drewry, D.T.; Verma, M.; Porcar-Castell, A.; Griffis, T.J. OCO-2 Advances Photosynthesis Observation from Space via Solar-Induced Chlorophyll Fluorescence. *Science* **2017**, *358*, 713–728.
24. Guan, K.; Berry, J.A.; Zhang, Y.; Joiner, J.; Guanter, L.; Badgley, G.; Lobell, D.B. Improving the Monitoring of Crop Productivity Using Spaceborne Solar-Induced Fluorescence. *Glob. Chang. Biol.* **2016**, *22*, 716–726. [CrossRef]
25. Kira, O.; Sun, Y. Extraction of Sub-Pixel C3/C4 Emissions of Solar-Induced Chlorophyll Fluorescence (SIF) Using Artificial Neural Network. *ISPRS J. Photogramm. Remote Sens.* **2020**, *161*, 135–146.
26. Köhler, P.; Frankenberg, C.; Magney, T.S.; Guanter, L.; Joiner, J.; Landgraf, J. Global Retrievals of Solar-induced Chlorophyll Fluorescence with TROPOMI: First Results and Intersensor Comparison to OCO-2. *Geophys. Res. Lett.* **2018**, *45*, 10–456.
27. Yang, X.; Tang, J.; Mustard, J.F.; Lee, J.-E.; Rossini, M.; Joiner, J.; Munger, J.W.; Kornfeld, A.; Richardson, A.D. Solar-Induced Chlorophyll Fluorescence That Correlates with Canopy Photosynthesis on Diurnal and Seasonal Scales in a Temperate Deciduous Forest. *Geophys. Res. Lett.* **2015**, *42*, 2977–2987. [CrossRef]
28. Biederman, J.A.; Scott, R.L.; Bell, T.W.; Bowling, D.R.; Dore, S.; Garatuza-Payan, J.; Kolb, T.E.; Krishnan, P.; Krofcheck, D.J.; Litvak, M.E. CO 2 Exchange and Evapotranspiration across Dryland Ecosystems of Southwestern North America. *Glob. Chang. Biol.* **2017**, *23*, 4204–4221.
29. Guanter, L.; Zhang, Y.; Jung, M.; Joiner, J.; Voigt, M.; Berry, J.A.; Frankenberg, C.; Huete, A.R.; Zarco-Tejada, P.; Lee, J.-E. Global and Time-Resolved Monitoring of Crop Photosynthesis with Chlorophyll Fluorescence. *Proc. Natl. Acad. Sci. USA* **2014**, *111*, E1327–E1333.
30. Short Gianotti, D.J.; Rigden, A.J.; Salvucci, G.D.; Entekhabi, D. Satellite and Station Observations Demonstrate Water Availability's Effect on Continental-scale Evaporative and Photosynthetic Land Surface Dynamics. *Water Resourc. Res.* **2019**, *55*, 540–554.
31. Song, L.; Guanter, L.; Guan, K.; You, L.; Huete, A.; Ju, W.; Zhang, Y. Satellite Sun-induced Chlorophyll Fluorescence Detects Early Response of Winter Wheat to Heat Stress in the Indian Indo-Gangetic Plains. *Glob. Chang. Biol.* **2018**, *24*, 4023–4037. [PubMed]
32. Zhang, L.; Qiao, N.; Huang, C.; Wang, S. Monitoring Drought Effects on Vegetation Productivity Using Satellite Solar-Induced Chlorophyll Fluorescence. *Remote Sens.* **2019**, *11*, 378.
33. Qian, X.; Qiu, B.; Zhang, Y. Widespread Decline in Vegetation Photosynthesis in Southeast Asia Due to the Prolonged Drought During the 2015/2016 El Niño. *Remote Sens.* **2019**, *11*, 910.
34. Wang, X.; Wang, T.; Liu, D.; Zhang, T.; Xu, J.; Cui, G.; Lv, G.; Huang, H. Multisatellite Analyses of Spatiotemporal Variability in Photosynthetic Activity Over the Tibetan Plateau. *J. Geophys. Res. Biogeosci.* **2019**, *124*, 3778–3797.
35. Wang, X.; Qiu, B.; Li, W.; Zhang, Q. Impacts of Drought and Heatwave on the Terrestrial Ecosystem in China as Revealed by Satellite Solar-Induced Chlorophyll Fluorescence. *Sci. Total Environ.* **2019**, *693*, 133627.
36. Yang, J.; Tian, H.; Pan, S.; Chen, G.; Zhang, B.; Dangal, S. Amazon Drought and Forest Response: Largely Reduced Forest Photosynthesis but Slightly Increased Canopy Greenness during the Extreme Drought of 2015/2016. *Glob. Chang. Biol.* **2018**, *24*, 1919–1934. [CrossRef]
37. Walther, S.; Duveiller, G.; Jung, M.; Guanter, L.; Cescatti, A.; Camps-Valls, G. Satellite Observations of the Contrasting Response of Trees and Grasses to Variations in Water Availability. Geophys. *Res. Lett.* **2019**, *46*, 1429–1440.
38. Qiu, B.; Ge, J.; Guo, W.; Pitman, A.J.; Mu, M. Responses of Australian Dryland Vegetation to the 2019 Heatwave at a Sub-daily Scale. *Geophys. Res. Lett.* **2020**, *47*, e2019GL086569.
39. Doughty, R.; Köhler, P.; Frankenberg, C.; Magney, T.S.; Xiao, X.; Qin, Y.; Wu, X.; Moore, B. TROPOMI Reveals Dry-Season Increase of Solar-Induced Chlorophyll Fluorescence in the Amazon Forest. *Proc. Natl. Acad. Sci. USA* **2019**, *116*, 22393–22398.
40. Vicente-Serrano, S.M.; Gouveia, C.; Camarero, J.J.; Beguería, S.; Trigo, R.; López-Moreno, J.I.; Azorín-Molina, C.; Pasho, E.; Lorenzo-Lacruz, J.; Revuelto, J. Response of Vegetation to Drought Time-Scales across Global Land Biomes. *Proc. Natl. Acad. Sci. USA* **2013**, *110*, 52–57.
41. Peña-Gallardo, M.; Vicente-Serrano, S.M.; Quiring, S.; Svoboda, M.; Hannaford, J.; Tomas-Burguera, M.; Martín-Hernández, N.; Domínguez-Castro, F.; el Kenawy, A. Response of Crop Yield to Different Time-Scales of Drought in the United States: Spatio-Temporal Patterns and Climatic and Environmental Drivers. *Agric. For. Meteorol.* **2019**, *264*, 40–55. [CrossRef]
42. Ma, X.; Huete, A.; Yu, Q.; Coupe, N.R.; Davies, K.; Broich, M.; Ratana, P.; Beringer, J.; Hutley, L.B.; Cleverly, J. Spatial Patterns and Temporal Dynamics in Savanna Vegetation Phenology across the North Australian Tropical Transect. *Remote Sen. Environ.* **2013**, *139*, 97–115.
43. Beringer, J.; Hutley, L.B.; McHugh, I.; Arndt, S.K.; Campbell, D.; Cleugh, H.A.; Cleverly, J.; de Dios, V.R.; Eamus, D.; Evans, B. An Introduction to the Australian and New Zealand Flux Tower Network-OzFlux. *Biogeosciences* **2016**, *13*, 5895–5916.

44. Joiner, J.; Yoshida, Y.; Vasilkov, A.P.; Schaefer, K.; Jung, M.; Guanter, L.; Zhang, Y.; Garrity, S.; Middleton, E.M.; Huemmrich, K.F.; et al. The Seasonal Cycle of Satellite Chlorophyll Fluorescence Observations and Its Relationship to Vegetation Phenology and Ecosystem Atmosphere Carbon Exchange. *Remote Sens. Environ.* **2014**, *152*, 375–391. [CrossRef]
45. Huete, A.; Didan, K.; Miura, T.; Rodriguez, E.P.; Gao, X.; Ferreira, L.G. Overview of the Radiometric and Biophysical Performance of the MODIS Vegetation Indices. *Remote Sens. Environ.* **2002**, *83*, 195–213.
46. Wang, J.; Hu, X.; Shi, T.; He, L.; Hu, W.; Wu, G. Assessing Toxic Metal Chromium in the Soil in Coal Mining Areas via Proximal Sensing: Prerequisites for Land Rehabilitation and Sustainable Development. *Geoderma* **2022**, *405*, 115399. [CrossRef]
47. Wang, J.; Ding, J.; Yu, D.; Ma, X.; Zhang, Z.; Ge, X.; Teng, D.; Li, X.; Liang, J.; Lizaga, I.; et al. Capability of Sentinel-2 MSI Data for Monitoring and Mapping of Soil Salinity in Dry and Wet Seasons in the Ebinur Lake Region, Xinjiang, China. *Geoderma* **2019**, *353*, 172–187. [CrossRef]
48. Joiner, J.; Yoshida, Y.; Zhang, Y.; Duveiller, G.; Jung, M.; Lyapustin, A.; Wang, Y.; Tucker, C.J. Estimation of Terrestrial Global Gross Primary Production (GPP) with Satellite Data-Driven Models and Eddy Covariance Flux Data. *Remote Sens.* **2018**, *10*, 1346.
49. Wang, Q.; Zeng, J.; Qi, J.; Zhang, X.; Zeng, Y.; Shui, W.; Xu, Z.; Zhang, R.; Wu, X.; Cong, J. A Multi-Scale Daily SPEI Dataset for Drought Characterization at Observation Stations over Mainland China from 1961 to 2018. *Earth Syst. Sci. Data* **2021**, *13*, 331–341. [CrossRef]
50. Zhang, W.; Zhang, B.; Zhu, W.; Tang, X.; Li, F.; Liu, X.; Yu, Q. Comprehensive Assessment of MODIS-Derived near-Surface Air Temperature Using Wide Elevation-Spanned Measurements in China. *Sci. Total Environ.* **2021**, *800*, 149535. [CrossRef]
51. Cleverly, J.; Chen, C.; Boulain, N.; Villalobos-Vega, R.; Faux, R.; Grant, N.; Yu, Q.; Eamus, D. Aerodynamic Resistance and Penman–Monteith Evapotranspiration over a Seasonally Two-Layered Canopy in Semiarid Central Australia. *J. Hydrometeorol.* **2013**, *14*, 1562–1570. [CrossRef]
52. Wutzler, T.; Lucas-Moffat, A.; Migliavacca, M.; Knauer, J.; Sickel, K.; Šigut, L.; Menzer, O.; Reichstein, M. Basic and Extensible Post-Processing of Eddy Covariance Flux Data with REddyProc. *Biogeosciences* **2018**, *15*, 5015–5030.
53. Reichstein, M.; Falge, E.; Baldocchi, D.; Papale, D.; Aubinet, M.; Berbigier, P.; Bernhofer, C.; Buchmann, N.; Gilmanov, T.; Granier, A. On the Separation of Net Ecosystem Exchange into Assimilation and Ecosystem Respiration: Review and Improved Algorithm. *Glob. Chang. Biol.* **2005**, *11*, 1424–1439.
54. Geruo, A.; Velicogna, I.; Kimball, J.S.; Du, J.; Kim, Y.; Colliander, A.; Njoku, E. Satellite-Observed Changes in Vegetation Sensitivities to Surface Soil Moisture and Total Water Storage Variations since the 2011 Texas Drought. *Environ. Res. Lett.* **2017**, *12*, 054006. [CrossRef]
55. Walther, S.; Voigt, M.; Thum, T.; Gonsamo, A.; Zhang, Y.; Kohler, P.; Jung, M.; Varlagin, A.; Guanter, L. Satellite Chlorophyll Fluorescence Measurements Reveal Large-Scale Decoupling of Photosynthesis and Greenness Dynamics in Boreal Evergreen Forests. *Glob. Chang. Biol.* **2016**, *22*, 2979–2996. [CrossRef] [PubMed]
56. Wang, C.; Guan, K.; Peng, B.; Chen, M.; Jiang, C.; Zeng, Y.; Wu, G.; Wang, S.; Wu, J.; Yang, X. Satellite Footprint Data from OCO-2 and TROPOMI Reveal Significant Spatio-Temporal and Inter-Vegetation Type Variabilities of Solar-Induced Fluorescence Yield in the US Midwest. *Remote Sens. Environ.* **2020**, *241*, 111728.
57. Felton, A.J.; Zavislan-Pullaro, S.; Smith, M.D. Semiarid Ecosystem Sensitivity to Precipitation Extremes: Weak Evidence for Vegetation Constraints. *Ecology* **2019**, *100*, e02572.
58. Huang, J.; Yu, H.; Guan, X.; Wang, G.; Guo, R. Accelerated Dryland Expansion under Climate Change. *Nat. Clim. Chang.* **2016**, *6*, 166.
59. Min, S.-K.; Zhang, X.; Zwiers, F.W.; Hegerl, G.C. Human Contribution to More-Intense Precipitation Extremes. *Nature* **2011**, *470*, 378.
60. Ahlström, A.; Raupach, M.R.; Schurgers, G.; Smith, B.; Arneth, A.; Jung, M.; Reichstein, M.; Canadell, J.G.; Friedlingstein, P.; Jain, A.K. The Dominant Role of Semi-Arid Ecosystems in the Trend and Variability of the Land CO2 Sink. *Science* **2015**, *348*, 895–899.
61. Poulter, B.; Frank, D.; Ciais, P.; Myneni, R.B.; Andela, N.; Bi, J.; Broquet, G.; Canadell, J.G.; Chevallier, F.; Liu, Y.Y.; et al. Contribution of Semi-Arid Ecosystems to Interannual Variability of the Global Carbon Cycle. *Nature* **2014**, *509*, 600–603. [CrossRef] [PubMed]
62. Morton, D.C.; Nagol, J.; Carabajal, C.C.; Rosette, J.; Palace, M.; Cook, B.D.; Vermote, E.F.; Harding, D.J.; North, P.R. Amazon Forests Maintain Consistent Canopy Structure and Greenness during the Dry Season. *Nature* **2014**, *506*, 221–224. [CrossRef] [PubMed]
63. Saleska, S.R.; Wu, J.; Guan, K.; Araujo, A.C.; Huete, A.; Nobre, A.D.; Restrepo-Coupe, N. Dry-Season Greening of Amazon Forests. *Nature* **2016**, *531*, E4–E5. [CrossRef] [PubMed]

remote sensing

Article

Field Model-Based Cultural Diffusion Patterns and GIS Spatial Analysis Study on the Spatial Diffusion Patterns of Qijia Culture in China

Yuanyuan Wang [1], Naiang Wang [1,*], Xuepeng Zhao [2], Xueran Liang [1], Jiang Liu [1], Ping Yang [1], Yipeng Wang [1] and Yixin Wang [1]

1 Center for Glacier and Desert Research, College of Earth and Environmental Sciences, Lanzhou University, No. 222 South Tianshui Road, Lanzhou 730000, China; yywang19@lzu.edu.cn (Y.W.); liangxr19@lzu.edu.cn (X.L.); liujiang2020@lzu.edu.cn (J.L.); yangp2018@lzu.edu.cn (P.Y.); wangyp2019@lzu.edu.cn (Y.W.); wangyx17@lzu.edu.cn (Y.W.)

2 Chinese Department, School of Humanities and Communication, Changji University, Changji 831100, China; zxp19@cjc.edu.cn

* Correspondence: wangna@lzu.edu.cn

Abstract: Cultural diffusion is one of the core issues among researchers in the field of cultural geography. This study aimed to examine the spatial diffusion patterns of the Qijia culture (QJC) to clarify the origin and formation process of Chinese field model-based cultural diffusion patterns (FM-CDP) and geographic information system (GIS) spatial analysis methods. It used the point data of Qijia cultural sites without time information and combined them with the relevant records of Qijia cultural and historical documents, as well as archaeological excavation materials. Starting with the spatial location information of cultural distribution, it comprehensively analysed the cultural hearth, regions, diffusion patterns, and diffusion paths. The results indicated the following. (1) The QJC's heart is in the southeast of Gansu Province, where the Shizhaocun and Xishanping sites are distributed. (2) Five different levels of cultural regions were formed, which demonstrated different diffusion patterns at different regional scales. On a large regional scale, many cultural regions belong to relocation diffusion patterns. Meanwhile, at the small regional scale (in the Gansu–Qinghai region), there are two patterns of diffusion: expansion diffusion and relocation diffusion; however, the expansion diffusion pattern is the main one. (3) Based on the relationship between the QJC, altitude, and the water system, the culture also has the characteristics of diffusion to low altitude areas and a pattern of diffusion along water systems. (4) There is a circular structure of the core, periphery, and fringe regions of the QJC. Finally, (5) the dry and cold climate around 4000a B.P., the cultural exchange between Europe and the Asian continent (the introduction of barley, wheat, livestock and sheep, and copper smelting technology), and the war in the late Neolithic period were important factors affecting the diffusion of the QJC.

Keywords: Qijia culture (QJC); spatial diffusion patterns; cultural hearth; circle structure

Citation: Wang, Y.; Wang, N.; Zhao, X.; Liang, X.; Liu, J.; Yang, P.; Wang, Y.; Wang, Y. Field Model-Based Cultural Diffusion Patterns and GIS Spatial Analysis Study on the Spatial Diffusion Patterns of Qijia Culture in China. *Remote Sens.* **2022**, *14*, 1422. https://doi.org/10.3390/rs14061422

Academic Editors: Jingzhe Wang, Zhongwen Hu, Yangyi Wu and Jie Zhang

Received: 6 January 2022
Accepted: 13 March 2022
Published: 15 March 2022

1. Introduction

Research on the origin of Chinese civilisation has attracted extensive attention since the 20th century [1]. Since the 21st century, with the launch of the 'source exploration project of Chinese civilisation', the exploration of the origin of Chinese civilisation has entered a new stage [2]. The Qijia culture (QJC) was formed in the late Neolithic period, in 4300–3500a B.P. This is roughly equivalent to the history of the Xia Dynasty. It laid the tone for Chinese civilisation, and has become the focus of exploring the origin of Chinese civilisation [3]. The spatial diffusion of the QJC has provided valuable contributions to the extensive scientific projects of the origin exploration project of Chinese civilisation [1,4]. It is also helpful to explore the development track of prehistoric culture from region to continent more clearly. Many studies show that the prehistoric social process, especially the rise and fall of ancient civilisations, is related to the spatial expansion of ancient culture [5]. For example, in the development of the Chinese civilisation, cultural expansion played a role in promoting the evolution from Neolithic culture to Bronze Age culture [6]. Therefore, exploring the spatial diffusion patterns of the QJC can help us understand the origin of Chinese civilization.

Since the 1980s, owing to the rapid growth of China's economy, the number of archaeological field investigations and excavations has exponentially increased [7]; this work has given rise to a wealth of data for future generations [8,9]. Regionally, research on the QJC is limited to certain regions. These include the eastern or north-eastern Tibetan Plateau, the Loess Plateau, and the Gansu and Qinghai regions [10]. Therefore, the current research on the QJC is limited to the small area listed above, and there are few studies on the QJC in the Yellow River Basin and the whole of China. In terms of research content, existing studies have mainly focused on the classification of the QJC using stages and types [11], agricultural and economic forms [12], and features of tombs [13]. For example, regarding the division of stages and types of the QJC, Xie Duanju divides the QJC into the eastern region (the eastern region of Gansu, including the Jingshui River, Weihe River, and the upper reaches of the Western Han River), the central region (the central region of Gansu, including the upper reaches of the Yellow River and its tributaries, the Taohe River basin, and the Daxia River basin), and the western region (western Gansu province and eastern Qinghai Province, including the upstream of the Yellow River in Qinghai province and its tributaries, the HuangShui and Hexi corridor) [11]. Ye Maolin et al. discussed the agricultural and economic forms of the QJC period based on millets, animal remains, bone needles, jade and stone tools, and other relics unearthed from the QJC site [12]. As for the study of tombs, Xie Duanju et al. made a comprehensive preliminary summary of the shape and structure of the tombs, burial tools, and burial products based on the burial materials unearthed in the Qijia cultural site [11]. The above research provides rich information for an in-depth discussion of related QJC issues. However, it only takes the excavated relics as the main research object from the perspective of history and archaeology, and uses the qualitative description method to summarize the macro research on related content. The research content is relatively simple. There is a lack of research using quantitative and qualitative methods for specific issues of QJC. For example, the cultural hearth of the QJC has not been adequately researched. There are few relevant historical records, and most of the available materials are mixed with myths and legends, which have low credibility. Therefore, Han Jianye and other relevant scholars, based on archaeological excavation data and the types and forms of unearthed artefacts, believe that the southeast Gansu province, representing the remains of the seventh period of Shi Zhaocun of Tianshui, is the hearth of the QJC [14]; however, this study lacks verification based on quantitative methods. Most studies on the spatial diffusion of the QJC focus on the expansion scope of the QJC. Li Shuicheng and Dong Guanghui have outlined the spatial scope of QJC in Ganqing based on the existing cultural relics survey results and archaeological survey reports [15]. How did the QJC spread out from its cultural hearth? At present, there is a lack of research on the diffusion patterns of the cultural region formed by cultural diffusion.

Geographic information system (GIS) technology was initially applied to archaeological research, regional systematic investigation and site planning, prehistoric settlement and environmental archaeology, and prehistoric cultural sites [16]. For example, Gaffney and others used GIS to study the territorial boundary, land use, and site selection factors of the Hvar region [17]. Van Leusen discussed the relationship between the site vision difference and its functional difference of South Limburg in the south of the Netherlands through cluster analysis [18]. In the application of GIS in heritage protection planning, Nanjing Normal University first developed the GIS system for archaeological sites in the lower reaches of the Yangtze River, and discussed the role of GIS in archaeological information management systems [19]. The application of GIS in settlement and environmental archaeology mainly focuses on social complexity and population research. For example, Liu Jianguo used

GIS to study the archaeological and social complexity of settlements in the Linfen basin, Luoyang basin, Huanhe, Qixing River, and Meiyang River [20]. The application of GIS spatial analysis in the field of prehistoric culture research mainly focuses on the study of the spatial distribution and pattern of prehistoric cultural sites. For example, WuLi et al. compared and analysed the spatial and temporal distribution characteristics of prehistoric cultures in Zhejiang since the Holocene based on GIS spatial analysis [21]. Dominic Hosner et al. studied the spatial and temporal distribution pattern of 51,074 archaeological sites from the Neolithic to the Iron age in China based on GIS spatial analysis [22]. In general, people pay increasingly more attention to the research on prehistoric culture supported by GIS. However, few studies have applied GIS spatial analysis and cultural diffusion patterns to prehistoric culture.

The main purpose of this paper is to assess the diffusion patterns of the QJC in space. However, the time information of most Qijia cultural sites is missing. The field model-based cultural diffusion pattern (FM-CDP) model studies the spatial diffusion mode of the QJC without the time information, mainly generating the density surface of the distribution of cultural sites according to the dispersion degree of the QJC sites. By calculating the density surface, the distance between the QJC sites is skilfully transformed into the mutual relationship of QJC. To complete the transformation of the QJC from "point" diffusion to the "surface" of cultural region, we extract the cultural hearth, cultural region, diffusion patterns, and internal classification structure of the QJC space diffusion, and quantitatively describe QJC space diffusion [23–26]. Therefore, the FM-CDP model plays an important role in the analysis of QJC diffusion. It is a method of cultural diffusion based on the field model, that is, by using the point data of a cultural event, it detects cultural areas at different levels, automatically identifies their spatial diffusion mode, and calculates the spatial diffusion relationship between cultural areas. Zhang Haiping [25,26] verified the effectiveness of the FM-CDP by analysing the spatial diffusion patterns of rural places in China. The results showed that the FM-CDP can effectively identify the existing cultural diffusion patterns using the point data of specific cultural characteristics. This method can also be applied to analyse other cultural diffusion patterns that represent certain cultural phenomena in the form of point features. Because the QJC site data lack information on time, it is difficult to study the spatial diffusion of the QJC. Therefore, this article embarks from the space perspective of geography, with the strong objectivity of archaeological reports and the basis of historical documents to confirm each other, on the basis of the theory of cultural geography spatial diffusion, based on the cultural diffusion model (FM-CDP), from the spatial location information of the distribution of cultural sites, to explore its cultural hearth and its diffusion patterns. This provides a new idea for the study of prehistoric culture without time information.

Regional Environments

According to currently known archaeological excavation data and historical documents, there are more than 3000 sites with QJC characteristics in mainland China. These sites are mainly distributed in the Yellow River Basin and its tributaries [11]. The distribution ranges from the eastern foot of the Qinghai–Tibet Plateau in the west to the Guanzhong Plain in the east [27], and from the Inner Mongolia Plateau in the northwest to the Sichuan Plateau in the south [28,29]. Regarding the administrative division, it extends west to Qinghai Province [30,31], east to Henan Province [32], north to Inner Mongolia Alxa [27], and south to Sichuan Province [33]. However, the core region of its distribution is in Gansu Province [34].

Qijia culture is mainly distributed in the intersection of the Qinghai–Tibet Alpine region, the eastern monsoon region, and the arid northwest region (Figure 1). The regional climate difference is obvious, and the annual average temperature and annual precipitation are also different. The annual average temperature is as high as 12 °C and as low as −6 °C [35]. The highest annual precipitation is 800 mm, and the lowest is less than 50 mm. The climate gradually becomes drier from the southeast to the northwest, and the vegetation gradually changes from grassland to desert, which is the transition zone from semi-humid to semi-arid to arid. Climatic conditions and the natural environment are relatively harsh, with strong sensitivity and instability [36]. Therefore, human activities in this region are greatly influenced by the natural environment more than other regions in the same latitude. The main distribution area of the culture is the intersection of the Qinghai–Tibet plateau, Loess Plateau, and Inner Mongolia Plateau, with complex and diverse landforms. The lowest altitude is less than 600 m, and the highest is more than 6000 m. The terrain characteristics are obviously different, mainly including mountains, plateaus, plains, and basins. The region includes the Qilian Mountains, Hexi Corridor, Loess Plateau, and other topographic areas. Such complex topographic features promote or hinder the diffusion of the Qijia culture [35]. At lower altitudes, the hydrothermal conditions are better, the loess accumulation is thicker, and the cultivation conditions are better. Therefore, most of the QJC is distributed at lower altitudes. The QJC was mainly distributed in the middle and lower reaches of the Yellow River Basin, including Weishui, Taohe, Daxia, and Huangshui [36]. Rivers are affected by terrain, and most of their flow direction is consistent with the direction of mountains and terrain. The sites are generally located on loess platforms on both sides of rivers, and are close to water sources.

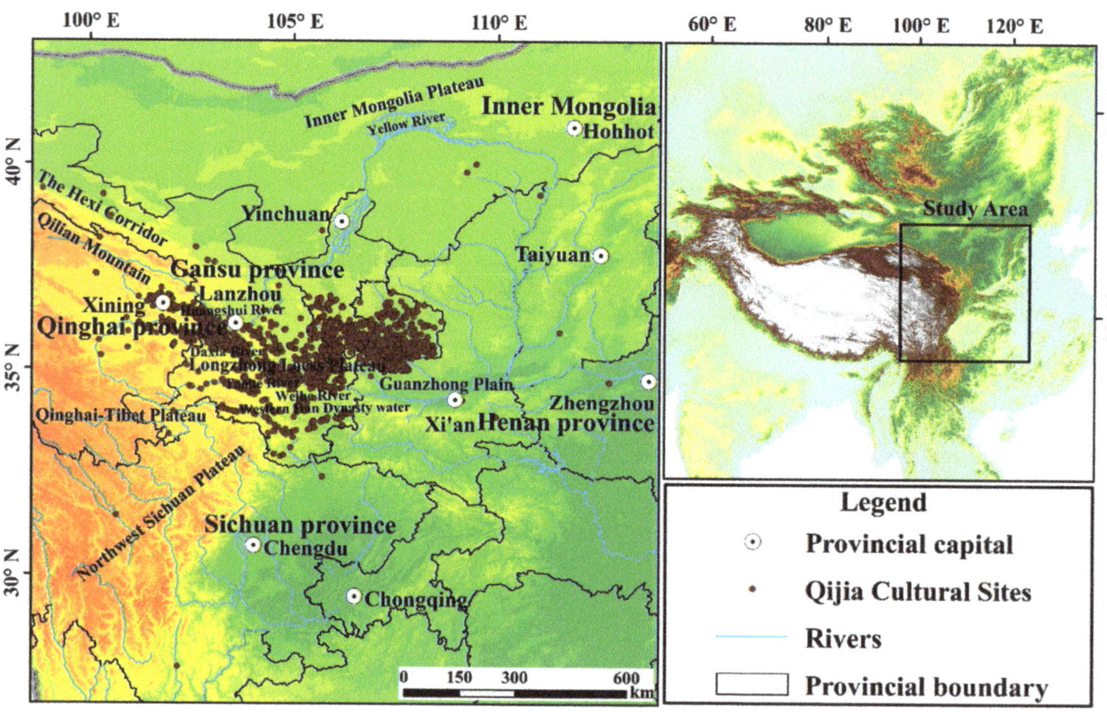

Figure 1. The major provinces where Qijia cultural sites are distributed, along with their rivers and topographies. Also depicted are the archaeological site data of Chinese provinces and autonomous regions published in the *Atlas of Chinese Cultural Relics* (brown points) series.

2. Materials and Methods

2.1. Sources of Archaeological Information

The first national survey of cultural relics in 1956 left no statistical data; therefore, the data of the Qijia cultural sites used in this study have been taken from the results of the second National Survey of Cultural Relics conducted by the State Administration of Cultural Heritage from 1981 to 1985: the *Atlas of Chinese Cultural Relics Gansu Volume*, the

Atlas of Chinese Cultural Relics Qinghai Volume, the *Atlas of Chinese Cultural Relics Ningxia Volume*, the *Atlas of Chinese Cultural Relics Inner Mongolia Volume*, the *Atlas of Chinese Cultural Relics Shaanxi Volume*, the *Atlas of Chinese Cultural Relics Shanxi Volume*, the *Atlas of Chinese Cultural Relics Sichuan Volume*, the *Atlas of Chinese Cultural Relics Henan Volume*, *Thirty Years of Cultural Relics and Archaeology*, *Fifty Years of Archaeology in New China*, *Sixty Years of Chinese Archaeology*, *Archaeological Discoveries and Research in New China*, *Chinese Archaeology—Neolithic Age Volume*, and an excavation briefing of the relevant sites. Among them, most of the data were collected from the results of the second national cultural relics census hosted by the State Administration of Cultural Relics from 1981–1985 [37–44]. According to the published archaeological excavation briefs and historical documents, no distribution of QJC was found in other Chinese provinces. Therefore, it was not included in this study's data collection scope. The *Atlas of Chinese Cultural Relics* in the provinces used in this study consists of a site location map and a concise description. The atlas provides relatively comprehensive coverage of all known sites obtained through rescue, archaeological excavation, systematic investigation, and accidental discovery. However, the specific geographical coordinates of each site and the specific activity age data of the site point in the map set were missing. Therefore, it was impossible to use the conventional mathematical statistical model such as quantitative analysis based on statistics and behavioural geography to study the related culture [45–47]. The DEM data used in this study primarily came from the GDEMV3 30M resolution digital elevation data jointly released by NASA and METI on 5 August 2019. The data format is Geo TIFF (signed 16 bits). Users can download it from the Geospatial data cloud platform of the computer network information centre of the Chinese Academy of Sciences (http://www.gscloud.cn, accessed on 5 August 2019). In addition, China's administrative boundaries, rivers, provincial capitals, and other data were derived from the 1:250,000 topographic element data of the State Bureau of Geographic Surveying and Mapping Information.

2.2. Data Processing

In this study, the QJC data are represented by QJC sites or sites with Qijia cultural elements; however, the scales and areas of these sites are inconsistent. To unify the data format, points are used in this study to represent the QJC; that is, a point represents a QJC site (see Figure 2a). Before analysing the diffusion patterns of the QJC, it was necessary to process the cultural sites' data. First, an Excel table of basic information such as the number, specific site name, administrative affiliation, longitude, latitude, site area, altitude, and other basic information of each Qijia cultural site was developed. Since the specific geographical coordinates of most sites were unknown, it was necessary to locate the coordinates based on the site location, administrative division, and text descriptions of geographical locations within the map. This work, combined with the use of Google Earth to obtain the data with decimal coordinate values according to the geographical location and administrative division of the site points in the relevant cultural relics' atlas, facilitated the direct use of the ArcGIS 10.3 software. After obtaining the corresponding coordinate values, we used the 'coordinate inverse check' function in Google Maps to check the sites to ensure the accuracy of their positioning. Second, the point layer was created using ArcGIS 10.3 based on the Excel table data. We selected the GCS_ WGS_ 1984 coordinate system, used Add X Y Data in the file to load the created Excel table, and selected the corresponding longitude and latitude values in the X and Y drop-down menus. After the point layer was generated, the output data were the vector point data in the Shapefile format. Finally, in the geographic coordinate systems (GCS) and projected coordinate systems (PCS), by taking 'm' as the unit of site point data, the decimal latitude and longitude data were added. We selected the WGS_1984 coordinate system in GCS; in order to perform the data calculation, we changed the data unit to m and selected WGS_1984_UTM_Zone_48N from the PCS. We ensured the correct coordinate system during data analysis.

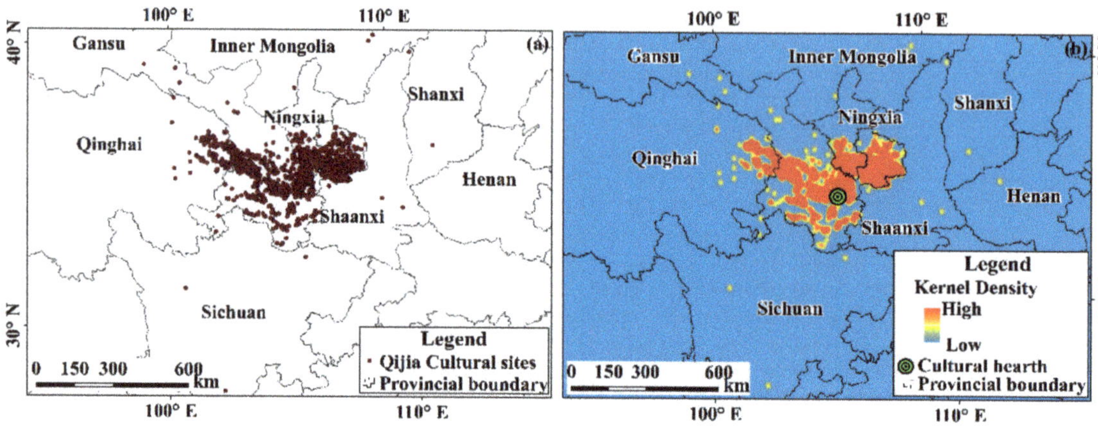

Figure 2. Qijia cultural site and density surface field. (**a**) Points represent Qijia cultural elements and (**b**) Kernel density map of the Qijia culture.

2.3. FM-CDP Diffusion Model

The FM-CDP model is a quantitative model based on the theory of cultural diffusion. It explores the mode of cultural space diffusion based on the GIS field model method, and visually expresses the cultural spatial diffusion model [22–26]. This model was originally used to investigate the spatial diffusion pattern of the place name cluster. This is because the data sources of the Qijia cultural site and the place name cluster do not have time attributes, and merely possess point elements that show the status. However, this type of prehistoric culture has continuity. Therefore, this model can effectively explore the spatial diffusion patterns of the QJC. The FM-CDP model is composed of three sub-models: DCR (detection of cultural region), RDT (recognition for diffusion types of cultural region), and ISCD (inferring for internal structure of cultural diffusion). The DCR model is mainly used to identify cultural hearths and detect cultural regions. The RDT method recognises the diffusion patterns of cultural regions, and the ISCD model infers the internal structure of cultural diffusion.

The FM-CDP model is a method used by Zhang Haiping et al. that uses the results of the second national census of place names to solve the problem that some elements of place names have time attributes, but most of them do not. GIS is used to quantify the analytical method; the cultural diffusion model is proposed around the question of "how to use quantitative methods to mine the cultural spatial diffusion patterns of place names and cultural hearth" [22–26]. Cultural diffusion is regarded as a continuous process because its boundary is fuzzy. The field model in GIS can efficiently solve the problem of boundary fuzziness and continuous change in space. As a result, the model, based on the GIS field model and the use of distribution on the space of a cultural phenomenon or event of point data, tests the different levels of cultural regions, automatically identifies the spatial diffusion types of cultures and the spatial diffusion relations between cultural regions, and effectively determines the cultural diffusion patterns on the basis of point elements [22–26]. In addition, this common method can be applied to the analysis of other cultural diffusion patterns that are represented by point elements. This model emphasises the expansion and relocation diffusions based on the relationship between distance and neighbours in geospatial space.

The GIS field model is one of the most basic GIS data models [22–26]. The GIS field model emphasises the continuous features of geographic objects. Based on raster data, the model can construct a smooth and continuously changing density surface on the basis of a certain cultural feature point in space, such that high-density regions can be recognised well and the spatial relationship between these regions can be calculated to a certain extent. On the one hand, a field model is better for modelling continuous phenomena and geographical boundaries with fuzzy borders, such as the analysis of mountains, temperature, and pollutant diffusion. Therefore, a field model is generally used to solve these problems. On the other hand, a field model can be converted into a data structure, such as contour lines, which can reflect hierarchy. The density surface of an area of cultural phenomena and elements is located in the region with the locally highest value. We can extract the local maxima of the density surface of cultural phenomena, such as extracting the peaks of mountains, by using a digital elevation model (DEM), providing a reference for detecting cultural regions [26]. In identifying cultural region types, the cultural phenomenon itself exhibits blurred boundaries, but a field model is good at extracting spatial relationships from continuous surface models, and thus identifying cultural diffusion types in cultural regions becomes possible. Moreover, a field model can efficiently extract the diffusion pattern from the characteristics of GIS spatial analysis, because the density surface constructed by the point features is hierarchical. Obtaining a high-density cultural region or calculating the type of diffusion in a cultural region is relatively easy.

3. Results

3.1. Qijia Cultural Hearth and Cultural Regions

The cultural hearth was the place where specific cultural aspects, phenomena, or systems originated. Based on the FM-CDP cultural diffusion model and the GIS database of Qijia cultural site points, the points were converted into a density-based grid surface. The local maximum of the new grid was obtained using the GIS field model, and the hearth of QJC was obtained according to the difference between the density-based surface field, Figure 2b, and the local maximum of the new grid. Through the archaeological excavation data related to the QJC, the literature, and the distribution range of cultural site clusters [37–44], it was verified that the eastern part of Gansu Province was dictated by the Shizhaocun and Xishanping sites as the cultural hearth of the QJC [14]. Therefore, this study took its distribution area as the hearth of the QJC for follow-up analysis.

Through the FM-CDP sub model DCR and the spatial distribution characteristics of the QJC, several cultural regions were obtained (Figure 3). Based on the kernel density analysis, the sub model transforms the distance between cultural sites into the interaction relationship of cultural sites through the density calculation, and completes the transformation of culture diffusion from "points" to "surfaces" of cultural regions. Figure 3 shows that a continuous and smooth density surface field was obtained according to the Qijia cultural points with different weights in the search area. This indicates that the hot spots with a high density and a certain scale were Qijia cultural areas. According to the scale and size of the cultural regions, they were divided into five different levels. Level 1 indicates that the distribution scope and density of the QJC are the largest, and Level 5, the smallest. Based on the spatial distribution of the cultural regions, several major paths that diffused from the cultural hearth of the QJC were manually drawn to obtain several main diffusion directions. These included the Hexi Corridor, Inner Mongolia Plateau, Northwest Sichuan Plateau, and Guanzhong Basin. The Northwest Sichuan Plateau and Guanzhong region formed a far-reaching diffusion path.

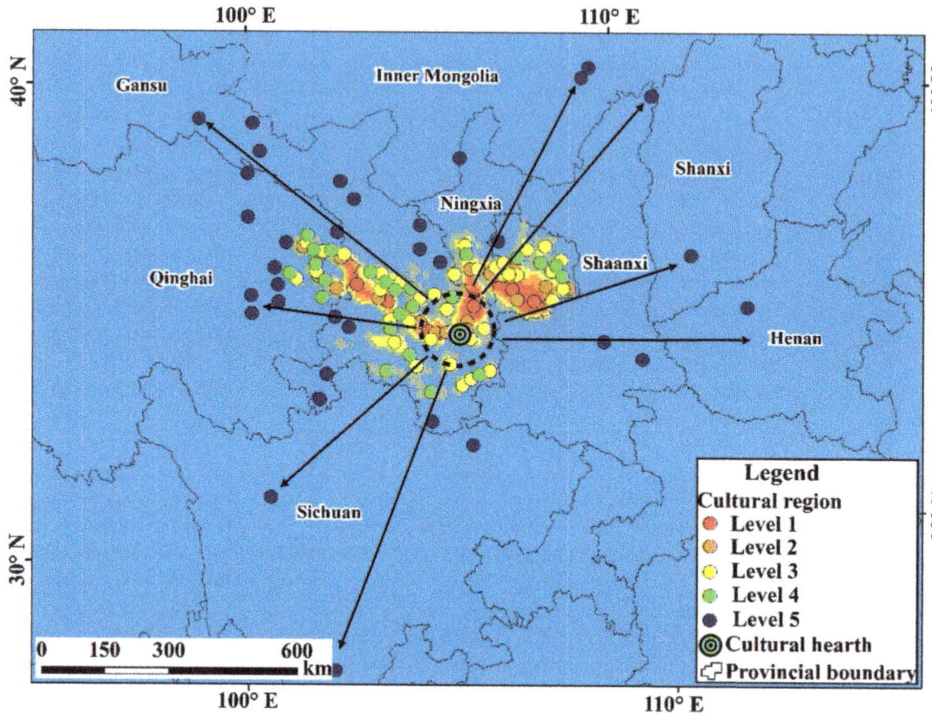

Figure 3. Spatial diffusion results of the QJC (cultural regions are divided into five grades based on scale and size).

3.2. Spatial Diffusion Patterns of QJC

We used the FM-CDP sub-models, RDT and ISCD, to quantitatively identify the diffusion mode of the QJC and its diffusion patterns at the large regional scale (Figure 4a). We also used the small regional scale in the Gansu–Qinghai region (Figure 4b). Among them, the blue cultural region is the result of relocation and diffusion, and the relocation diffusion pattern is represented by a broken line. The red cultural region is the result of expansion diffusion, which is represented by a solid line. The results show that the expansion diffusion pattern at a large regional scale changes into the coexistence of expansion and relocation at a small regional scale in the Gansu–Qing region. At the small regional scale, the QJC is a diffusion pattern with expansion as the primary and relocation as the auxiliary pattern. Th expansion diffusion was the result of diffusion near the cultural hearth, whereas the result of relocation diffusion was far away from the cultural hearth.

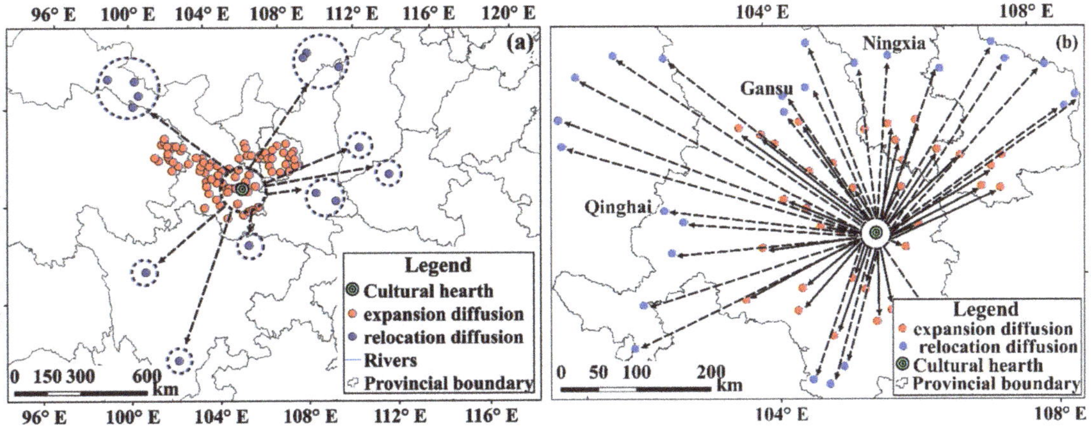

Figure 4. Patterns of Qijia cultural diffusion (cultural regions indicated by red and blue points, respectively, belong to expansion and relocation diffusions). (**a**) Patterns of Qijia cultural diffusion on a large regional scale and (**b**) patterns of Qijia cultural diffusion on a small regional scale in the Gansu–Qinghai region.

Regarding the specific regions, the QJC mainly diffused westward to the Hexi Corridor, and even to the terraces of the Hehuang (water) River Valley and its tributaries in the eastern agricultural area of Qinghai on the south-eastern edge of the Tibetan Plateau. It diffused eastward to the Guanzhong Plain and southward to the Xianshui River, a tributary of the Yalong River. It also diffused northward to the Hetao area of the Inner Mongolia Plateau. However, the main diffusion area was in the Yellow River Basin.

Based on the diffusion model, the diffusion path map of the QJC is obtained as shown in Figure 5. The density of the points representing cultural phenomena varies by cultural regions; additionally, their distribution ranges are also different. Therefore, as combined with the density value and distribution range, the cultural region is divided into five levels. At a large regional level, the cultural region directly or indirectly connected with the cultural hearth belongs to the expansion diffusion pattern. Conversely, it is the relocation diffusion pattern. Figure 5 indicates that relocation diffusion occurs mainly in smaller cultural regions. Observing the diffusion pattern around the cultural hearth, we noted that several high-level cultural regions formed near the cultural hearth. The closer a region is to the cultural hearth, that is, the closer it is to the southeast of Gansu province, the higher its level of Qijia culture. This indicates that the characteristics of the Qijia cultural elements are clear: the further away from the cultural hearth—that is, the closer to the Hexi Corridor Area, Hetao area, Guanzhong Plain, Northwest Sichuan Plateau, and other areas—a region is, the lower its level of Qijia culture (i.e., there are fewer and less obvious Qijia cultural elements).

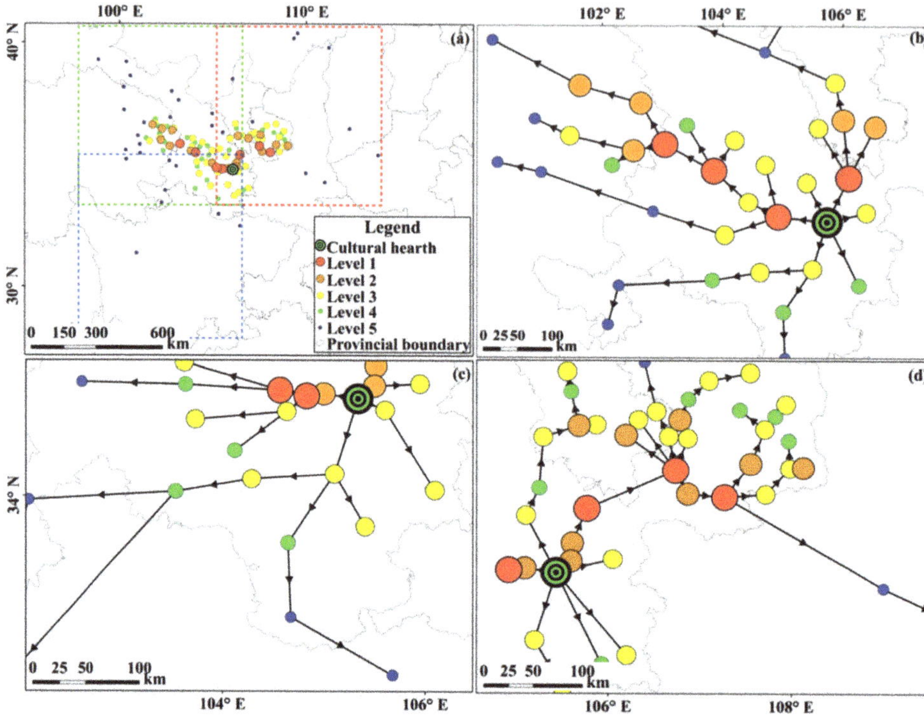

Figure 5. The diffusion path map of QJC. (**a**) The spatial distribution of the cultural hearth, the distribution of different levels of cultural regions, and the internal structure of cultural diffusion. (**b–d**) Local enlarged maps representing the paths of Qijia cultural diffusion in different regions. The green, blue and red rectangles represent the areas of the (**b–d**) diagram respectively.

3.3. *Characteristics of QJC Diffusion to Low Altitude*

Majiayao culture and QJC are both Neolithic cultures with roughly the same spatial distribution [31,36]. In order to better explore the diffusion characteristics of the QJC, Majiayao culture is selected for comparative analysis. Due to the influence of natural geographical factors such as climate, water sources, and topography in the distribution area of Qijia cultural sites, the QJC has the spatial distribution characteristics of relative dispersion at a large regional scale and relative concentration on a small regional scale [48]. Compared with the Majiayao culture, the QJC has an expanded spatial distribution range and obvious south diffusion characteristics. The centre of gravity retreats to the south of 36°N. In low altitude areas with dense rivers, the culture significantly increased. Simultaneously, the QJC has an obvious diffusion trend from the upper reaches of the Yellow River to the middle and lower reaches of the river.

The relationship between the spatial distribution and altitude of the QJC and the Majiayao culture was obtained from the ArcGIS 13.0 spatial analysis tool (Figure 6a). By taking the distribution frequency of cultural sites as the index and considering the selection of cultural sites at different elevations in the same cultural era, there was a change in the number of cultural sites of the Majiayao culture and the QJC at an elevation of 2000 m; therefore, 2000 m was selected as the elevation critical value. As can be seen from the bar chart in Figure 6, the Majiayao culture accounts for 56.34% of the areas above 2000 m, showing the characteristics of concentrated distribution at high altitude. The proportion of QJC and Majiayao culture sites below 2000 m is 74.39% and 43.66%, respectively. It shows that the QJC has the characteristics of spreading more to low altitude areas compared to the Majiayao culture.

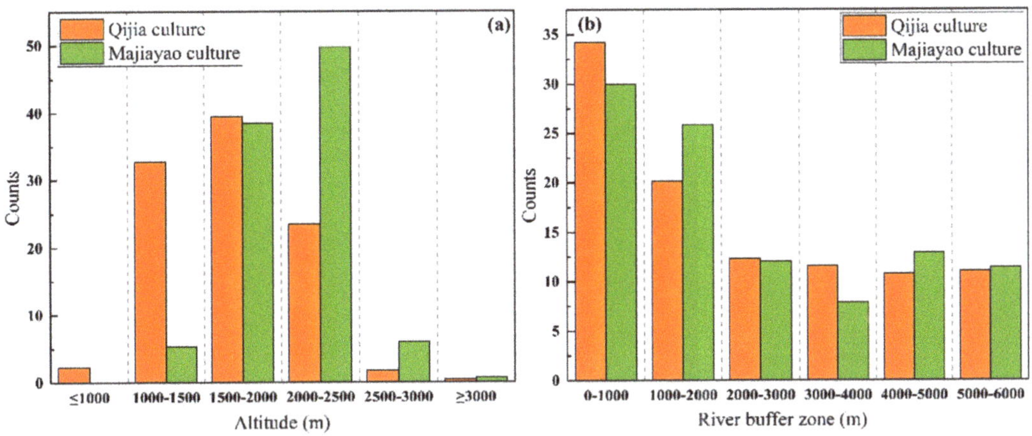

Figure 6. The relationship between the spatial distribution of culture, altitude, and water. (**a**) Counts of QJC and Majiayao culture at different elevations and (**b**) counts of QJC and Majiayao culture at different distances to rivers.

3.4. 'Water Diffusion' of QJC

The QJC period involved a low-level stage of social development. The natural environment had a significant restrictive effect on human production and life. The ancestors of the Qijia people mainly lived near water to adapt to and make use of their natural environment. Finally, the water system facilitated cultural diffusion.

The ArcGIS 13.0 spatial analysis tool was used to analyse the buffer zone of the water system in the distribution area of the QJC and the Majiayao culture at an interval of 1000 m (Figure 6b). This was where the cultural points with 0 m that were not in the buffer zone were deleted without statistical analysis to obtain the relationship between the spatial distribution of the QJC, the Majiayao culture, and the water system. The histogram indicates that the QJC and the Majiayao culture are mainly distributed within 2000 m of the river, accounting for 54.5% and 55.8%, respectively. This indicated that both QJC and prehistoric culture have the characteristics of a concentrated distribution near water. Within 3000 m of the river buffer zone, the QJC accounts for 66.6% of the total culture, and the Majiayao culture accounts for 67.8% of the total culture. Compared with the QJC, the Majiayao culture was concentrated within 3000 m of the river, indicating that, compared with the Majiayao culture, the cultural diffusion of the QJC could not be separated from the support of the water system, and the diffusion characteristics along the water system were obvious.

The QJC takes the cultural hearth as the centre and mainly diffuses to the surrounding areas along the main stream of the Yellow River and its tributaries. The Yellow River connected the QJC in the east, west, north, and south regions in a series to form a point axis structure. It was then extended to the north and south, showing the characteristics of scattering from the east, west, north, and south.

3.5. Circular Structure of Qijia Culture

The circular structure of the QJC occurred as a result of its spatial diffusion. After nearly 800 years of continuous diffusion, the QJC diffused along the Yellow River and its tributaries to the southeast edge of the Tibetan Plateau, northwest Sichuan Plateau, Inner Mongolia Plateau, Guanzhong Plain, and other regions. This is where it formed a circular structure (Figure 7). According to the theory of cultural geography, the cultural characteristics of the cultural hearth area are the most obvious and typical. With an increase in the distance from the cultural hearth area, the cultural characteristics gradually weaken, conforming to the principle of distance attenuation, and finally disappear [49]. Given the lack of time information of most Qijia cultural sites, the Qijia cultural region has been detected based on the ArcGIS 13.0 buffer analysis function. After several tests, the results are the most ideal when the search radius is 260 km, without considering topography and other factors. Therefore, taking the cultural hearth area of Qijia as the core, 260 km as the radius parameter, and province as the basic unit to establish the buffer zone, the core area of the spatial diffusion of the QJC can be obtained. When the radius parameters are 520 km and 770 km, the sphere and domain regions of culture are formed.

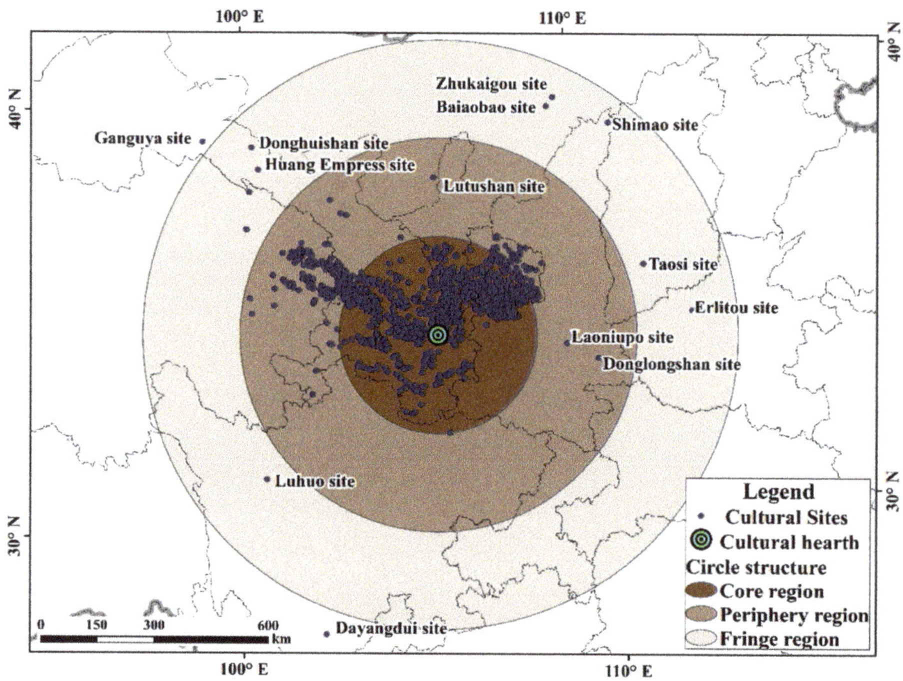

Figure 7. The circle structure of QJC space diffusion.

The cultural core area is close to the cultural hearth, and the number of cultures account for 83.0% of the total cultures. According to the theory of cultural geography, the spatial diffusion and influence of cultural phenomena are consistent with the attenuation law of spatial distance [50]. The Qijia population was the core of the group with cultural characteristics and the development of cultural inheritance, as well as an important foundation for cultural diffusion. The closer the spatial distance from the cultural hearth, the more likely the Qijia group was to accept the culture; in turn, the stronger the popularity of the culture and the more significant the cultural phenomenon. Therefore, a cultural core region dominated by an expansion diffusion pattern was formed. The number of QJCs in the sphere area of culture accounted for 16.7%, which was a common result of expansion and relocation diffusion. With the increase in spatial distance, the understanding and identity of the Qijia people in this culture decreased compared with that of the cultural core region. The popularity of the culture decreased, and the cultural phenomenon became increasingly unclear. It thus formed a spherical cultural region with less clear Qijia cultural elements. The marginal area was formed by relocation diffusion, and the number of QJCs in this area was only 0.3%. This indicated that with the increase in spatial distance, the resistance to cultural diffusion gradually increased, and was greatly affected by the culture of the adjacent regions. This reduced the influence and audience of this culture compared with other cultures in the same period to form a dominant cultural region with no obvious Qijia cultural elements.

4. Discussion

4.1. Applicability of FM-CDP Cultural Diffusion Model

Under the external driving force and the internal power of its own survival and development, cultural diffusion occurs in space [51]. Similarly, as a kind of culture, the QJC diffused around under the external driving force of the prehistoric environment and the internal driving force of culture. In the field of cultural diffusion research, there are four main methods to study cultural diffusion at present: the semiquantitative analysis method based on maps [52–54], quantitative analysis methods based on statistics [55–57], quantitative analysis methods based on behaviour geography [58,59], and FM-CDP [22–26]. Semiquantitative analysis methods based on maps distinguish the diffusion of a cultural phenomenon through observation [60]. Quantitative analysis methods based on statistics help determine cultural diffusion primarily through observation maps or simple statistical analyses [61]. Quantitative analysis methods based on behaviour geography mainly apply to the study of cultural diffusion in time and space [58]. For complex cultural phenomena, it is difficult to use the first three semiquantitative methods to describe the spatial diffusion patterns of a culture. In addition, from the perspective of structuralism, the first three methods lack reliable quantitative analysis models, so their analysis results are not rigorous. In addition, most quantitative methods based on spatial statistics usually disregard the paradigm of cultural diffusion in traditional cultural geography; furthermore, a complete system that illustrates cultural origins, cultural regions, and types of cultural diffusion and their internal structures is lacking. Methods based on traditional spatial analysis also exhibit the same defects. The diffusion of the QJC is a long process that took hundreds of years, and it is not easy for researchers to discover it. Therefore, the absence of relevant information such as the time of diffusion leads to the absence of long-term sequence data in the data source. Therefore, it is difficult to infer cultural diffusion using the first three methods. However, the cultural diffusion pattern analysis method (FM-CDP) based on a field model in GIS can detect different levels of cultural regions, and identify the diffusion patterns of these cultural regions.

The QJC is a cultural distribution with space and region as the carrier created by the ancestors of Qijia through practice in a certain geographical environment. The spatial diffusion of its culture is a relatively long process that takes hundreds of years to complete. It is a cultural phenomenon with a significant geographical proximity effect and continuous spatial processes. Cultural diffusion is mainly restricted by traditional geographical distances, and the boundary of cultural diffusion is ambiguous. Since it is a prehistoric culture, it is impossible to obtain specific temporal or other information on cultural diffusion. Therefore, the data are different from the general point data of modern times. Existing time information on most Qijia cultural sites is missing, and quantitative analysis of QJC based on time has become a challenge. However, the cultural diffusion model analysis method using the GIS field model (FM-CDP), based on the powerful spatial analysis function of GIS, can overcome the shortcomings of QJC, such as the non-computability of a lack of time information.

4.2. Influencing Factors of Spatial Diffusion of QJC

Based on the above research results, we can further analyse the reasons for the spatial diffusion of the QJC, which are mainly climatic factors and non-climatic factors.

4.2.1. Climatic Factors of Spatial Diffusion of QJC

The QJC is a late Holocene culture from around 4000a B.P. [62,63]. The temperate continental climate and temperate monsoon climate zones of the area played an important role in the spatial diffusion of local Holocene culture [64–66]. The reconstruction of climate history in the late Holocene by relevant scholars, especially from around 4000a B.P., provides a favourable foundation for exploring the relationship between QJC and climate change [8,9].

Combined with related research results on climate change in the Gansu and Qinghai areas and their surroundings, the climatic factors of the spatial diffusion of the QJC were systematically analysed. Through the summary of many climatic and environmental proxy indicators, it was found that during the late Holocene, especially around 4000a B.P., there were cooler and drier climatic events [67–69] and the climate accordingly cooled and dried.

From the perspective of the core region of the distribution of the QJC, it has been found that the climate in Gansu was in a state of low temperature and drought during the QJC period (Figure 8). For example, climate proxies, such as magnetic susceptibility, organic matter, and pollination of the loess profile in Gansu indicate that the monsoon was generally strong and humid before 4000a B.P. After 4000a B.P., the monsoon strongly retreated and the climate rapidly dried [70].

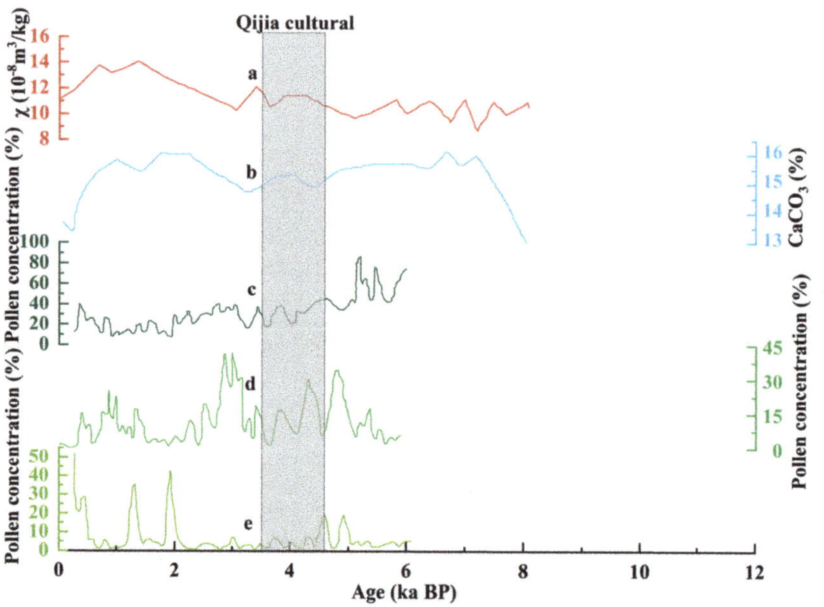

Figure 8. The study area's Holocene climate change records. a: Magnetic susceptibility of Dadiwan section; b: $CaCO_3$ content in Dadiwan profile; c–e: pollen in Dadiwan (changes in trees and shrubs, Artemisia, Chenopodiaceae) [71]. During this period, the content of arbour in the pollen of Dadiwan decreased, the proportion of terrestrial herb pollen increased, and the content of Artemisia pollen resistant to cold and drought significantly increased. The $CaCO_3$ content in this profile increased [34]. These observations are similar to others from 4000a B.P. regarding the Holocene temperature change in China and the high sea precipitation change [72]. The climate characteristics of the temperature drop and monsoon precipitation fluctuations are the same.

From the perspective of the surrounding area where the QJC spread (Figure 9), Sichuan has a southern latitude and numerous river systems compared to Gansu. During the QJC period, the total organic carbon content in the Ruoergai Red plain peat was low, indicating that the area was warmer and wetter than Gansu during the QJC period [73]. Shaanxi, located in the Guanzhong region, was significantly affected by the East Asian monsoon. Compared to Gansu in the same period, the climate was warmer and wetter, and water was abundant. According to the oxygen isotope records of the Jiuxian cave, there was more precipitation in the region during the Qijia period [74,75]. The conditions of animal husbandry in Inner Mongolia and Ningxia were better. The paleoclimate records of the Daihai basin show that the precipitation range in Inner Mongolia changed little, which provided favourable growth conditions for the emergence of large grasslands in central and southern Inner Mongolia; these met the needs of animal husbandry development [76]. It can be seen from the oxygen isotope records of the Dunde ice core in the Qilian Mountains that during the QJC period, the temperature dropped and the vegetation changed significantly, from the former forest or forest steppe to the steppe environment, and the prehistoric ancestors adopted the mixed mode of animal husbandry and planting. Henan, located in the Central Plains, is a semi-humid continental climate zone with abundant precipitation, and the basin plain can conveniently use the water resources of the Yi-Luo River and other rivers. Precipitation fluctuation has little influence on regional cultural development. After the Holocene drying and cooling event at about 4000a B.P., the climate in this region was still sub-humid. Although the annual precipitation decreased, it was still higher than the minimum precipitation required for dryland agriculture, thus providing favourable conditions for the diffusion of the QJC to this region [77–79]. The above shows that when the climate in the Gansu–Qinghai region tends to be dry and cool, the vegetation belt moves to the north and south, which in most cases will drive people with different economic modes to move from north to south, thus leading to the diffusion of the QJC to the surrounding areas that are more conducive to survival [80].

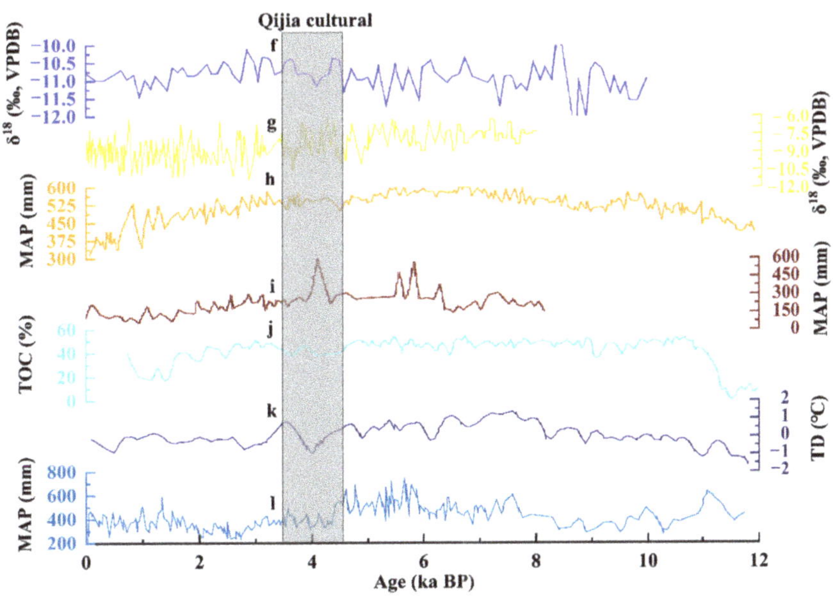

Figure 9. The study area's Holocene climate change records. f: the δ^{18}O record from Dunde ice core [79]; g: the δ^{18}O record from Jiuxian Cave [74,75]; h: the precipitation in Daihai [81]; i: the precipitation in Qinghai Lake—Darien Sea [82]; j: Ruo'ergai peat [77]; k: Holocene temperature integration sequence in China [81]; l: high seas precipitation [82].

In addition, the climate changed from relatively warm and wet to cool and dry around 4000a B.P. In the warm and humid stage, the river was fully replenished, and the ancestors lived at a moderate distance from the river [75]. After entering the cool and dry stage, the continuous cool and dry climate led to changes in the industrial structure during the Qijia period. Millet and primitive millet agriculture that relied on hydrothermal conditions declined, and animal husbandry increased and became dominant [83,84]. To solve the shortage of food resources and maintain stable productivity and life, ancestors with primitive agriculture as their main mode of production migrated to low altitudes and river valleys with better hydrothermal conditions [85]. In addition, studies have shown that water resources were more dependent on rivers [5] due to the low productivity of prehistoric society and the fact that living near water was the main mode of production of ancient humans. Therefore, the diffusion patterns of QJC to low altitudes and water systems were formed.

In conclusion, climate change caused by the change of temperature, precipitation, and other factors led to a low level of productivity for the ancestors who lived in this environment. The ability of the people to respond to the changes in the natural environment was low. This forced them to lower elevations, with a combination of drainage and hot conditions, as it was more suitable for the development of farming and animal husbandry [86].

4.2.2. Non-Climatic Factors of Spatial Diffusion of QJC

The spatial diffusion of the QJC was not only affected by climate change, but also by cultural exchanges, wars, and other factors in Eurasia, such as the emergence of wheat crops, livestock and sheep, and bronze smelting technology, to name but a few.

Exchange of Cultures between Europe and Asia

Around 4000a B.P., the introduction of barley and wheat [87,88], sheep [89], and metallurgical technology from the European continent to China occurred [90,91], along with the spread of Chinese millet to Europe [92]. The 'prehistoric food globalisation' promoted communication and exchange between cross-continental cultures [93]. This transcontinental cultural element was introduced into China in the form of the 'cultural package' [94]. It improved the ability of the prehistoric ancestors to adapt to the environment [95]. It also accelerated the diffusion of the QJC to the surrounding areas to a certain extent in the dry and cold environments.

War and Chaos

In addition, the warm and humid environment in the early stage of the QJC led to the continuous expansion of the Qijia population and the gradual prominence of the contradiction between man and land. This caused existing levels of productivity to fail to keep up with the pressures of large-scale populations and the continuous war and conflict between different cultural groups. These factors promoted the diffusion of culture to a certain extent [96].

For example, the human bones buried in the martyrdom pit found in the Xishanping site are all men aged 20–40 years [96]. Among the five random burial tombs—M35, M36, M57, M58, and M62—found in the fourth excavation of the Huang Niangniangtai site in Wuwei City, there are no fixed burial types. Some have sullied skeletons, separated heads, incomplete limbs, or no heads, and there are almost no funerary objects [96]. The phenomenon of martyred human burial was found in the Mogou Qijia cultural cemetery. The martyred humans included men and women, both old and young. Although women and children were the main burials, they did not completely follow the practice of male superiority, which was the result of the war. Perhaps in the course of the war, most men on the defeated side were killed and eventually martyred, while women and children were captured, so women and children accounted for a large proportion of the surviving captives [97]. The above shows that these people did not die normally, and that the dead were victims of increasingly frequent tribal conflict or tribe wars [98]. This may reflect that the diffusion of the QJC was affected by war. In addition, in the QJC period, war was more common. As a form of war defence, it is generally believed that the city site appeared to meet the needs of war. Now, the Huanhao settlements of the QJC period have been found. For example, in the Qijia cultural site in Qinghai Province, three large moats with a width of more than 10 m and a depth of 5–6 m have been found, and Huanhao settlements also exist in the Shenna site [99,100].

In conclusion, the influence of natural environmental factors such as climate on cultural diffusion is greater than that of social factors such as cultural exchange and war (Table 1). During the QJC period, productivity was low, and the prehistoric ancestors could not well understand and master some natural laws; therefore, they could only change their way of life, such as using subsistence strategies, according to the climate and other natural environmental factors, so as to conform to nature. Some researchers believe that ancient civilizations would have been favoured when the climate deteriorated to the limit of human endurance. Resource shortage leads to increased social complexity, which leads to an ecological adjustment of human cultural behaviour [101]. The national migration during the cooling period may also adjust the social structure, leading to the outbreak of large-scale wars, and eventually to the diffusion of culture [102].

Table 1. Influencing factors of spatial diffusion of QJC.

Influencing Factors	Time	Region	Financial Situation
Climate Factors	Around 4000a B.P.	Gansu Region	Around 4000a B.P., the climatic events of cooling and drying led to low temperatures and drought in Gansu, the decline of primitive agriculture, and the shortage of food resources, which forced the Qijia ancestors to migrate to the surrounding regions with better hydrothermal conditions [62–66].
		Sichuan Region	Sichuan has a southern latitude and numerous river systems compared to Gansu. During the QJC period, the area was warmer and wetter than Gansu, which attracted the diffusion of the QJC to this region [73].
		Shaanxi Region	Shaanxi was significantly affected by the East Asian monsoon. Compared to Gansu in the same period, the climate was warmer and wetter, and water was abundant, which provided favourable conditions for the diffusion of culture [74,75].
		Mongolia and Ningxia	The emergence of large grasslands in south–central Inner Mongolia and southern Ningxia during the Qijia period met the needs of the Qijia culture's livelihood patterns and stimulated the diffusion of the Qijia culture to the region [79].
		Hexi Corridor Region	During the Qijia period, the temperature dropped and the vegetation changed significantly, from the forest or forest steppe to the steppe environment, which promoted the development of the livelihood patterns of semi-farming and semi-grazing [81].
		Henan Region	After the Holocene drying and cooling event at about 4000a B.P., the climate in the region was still sub-humid. Although the annual precipitation decreased, it was still higher than the minimum precipitation required for dryland agriculture, thus providing favourable conditions for the diffusion of the QJC to this region [83–85].
Eurasian Cultural Exchange	Around 4000a B.P.	Yellow River Basin	Around 4000a B.P., the introduction of barley and wheat, sheep, and smelting technology from the European continent to China occurred, along with the spread of Chinese millet to Europe. The 'prehistoric food globalisation' promoted communication and exchange between cross-continental cultures [87,88].
	4000–2000a B.P.	Northern China	About 4500–4000a B.P., the introduction, domestication, and breeding of cattle and sheep to northern China accelerated the long-distance diffusion of culture [89].
	4000–3000a B.P.	Yellow River Basin	Around 4000a B.P., the transcontinental cultural element was introduced into China in the form of the cultural package. It improved the ability of the prehistoric ancestors to adapt to the environment. It also accelerated the diffusion of the QJC [90–95].

Table 1. *Cont.*

Influencing Factors	Time	Region	Financial Situation
War and Chaos	4300–3400a B.P.	Gansu Region	The age of human bones, the proportion of women and children, and the messy burial styles in the Xishanping, Huangniangniangtai, and Mogou sites indicate that the warm and humid environment in the early stage of the QJC led to the continuous expansion of the Qijia population and the gradual prominence of the contradiction between man and land. This caused existing levels of productivity to fail to keep up with the pressures of large-scale populations and the continuous war and conflict between different cultural groups. These factors promoted the diffusion of culture to a certain extent [96–100].

5. Conclusions

Based on the data from Qijia cultural sites with a lack of time information, this study explored the diffusion patterns of the QJC by using the spatial diffusion model (FM-CDP) based on the GIS field model and obtained the following conclusions:

(1) The QJC has different diffusion patterns at different regional scales, and the spatial diffusion pattern of the QJC has the characteristic trend of near-geographical unit elements at the same regional scale. This shows that the spatial diffusion patterns of the QJC vary, not only with different spatial scales, but also with changes in different geographical units. From the perspective of the regional scale, the diffusion of the QJC to the Guanzhong basin, northwest Sichuan Plateau, Inner Mongolia Plateau, and Hexi Corridor was far away from the cultural source; therefore, it was primarily relocation diffusion. The diffusion of culture in the regions close to the cultural hearth and southeast Gansu was mainly expansion diffusion. On the small regional scale, the culture in Gansu and Qinghai provinces not only has the pattern of expansion diffusion, but also relocation diffusion.

(2) The QJC forms a circular structure in the process of spatial diffusion, and different diffusion patterns form different cultural circles. Expansion diffusion is the main diffusion pattern of the core cultural region. The cultural sphere is the common result of expansion diffusion and relocation diffusion, and the pattern of relocation diffusion forms the cultural domain. The culture in the southeast of Gansu province is close to the cultural hearth and is mainly expansion diffusion, thus forming the cultural core region. Cultural diffusion in central Qinghai, Gansu, and southern Ningxia, which are far away from the cultural hearth, forms the sphere region. Sichuan, Henan, and Inner Mongolia, which are far from the cultural hearth, form the domain region of cultural diffusion.

(3) The spatial diffusion of the QJC is closely related to the climatic and social environments of the time. In the late Neolithic period, temperature and precipitation decreased and the climate tended to be dry and cool. During the same period, Sichuan, Henan, and Shaanxi had a warmer and more humid climate and sufficient water resources than Gansu, which was conducive to the survival of the QJC ancestors. Meanwhile, central and southern Inner Mongolia, southern Ningxia, and the Hexi Corridor could meet the needs of the livelihood patterns of semi-farming and semi-grazing. Cultural exchange across Eurasia, the introduction of copper smelting technology, and tribal wars may have also promoted the diffusion of QJC. In addition, the wars between tribes may have been an influencing factor.

This study illustrates the spatial diffusion pattern of the QJC at different regional scales and its influencing factors. The results of its spatial diffusion pattern reproduce the spatial diffusion process of the QJC. They also deepen and expand the research dynamics of prehistoric cultural space and enrich research on the QJC. The result of cultural diffusion is that culture moves from pluralism to integration, with advanced cultural factors from various places converging in the same region. Therefore, it is helpful to study the origins of Chinese civilisation. The role of climate in cultural diffusion may provide a historical reference for contemporary man to predict and address future climate change. However, the diffusion pattern of the prehistoric culture in China and globally, which still lacks time information, needs further examination.

Author Contributions: Conceptualization, Y.W. (Yuanyuan Wang); Proposed the idea and designed the study, N.W.; Implemented the method, X.Z. and X.L.; software, Y.W. (Yuanyuan Wang); validation, Y.W. (Yuanyuan Wang) and N.W.; formal analysis, Y.W. (Yuanyuan Wang); investigation, Y.W. (Yuanyuan Wang) and X.L.; resources, N.W.; data curation, N.W. and Y.W. (Yuanyuan Wang); writing—original draft preparation, Y.W. (Yuanyuan Wang), X.Z., J.L. and P.Y.; writing—review and editing, Y.W. (Yipeng Wang), N.W. and X.Z.; visualization, Y.W. (Yixin Wang); supervision, N.W.; project administration, N.W.; funding acquisition, N.W. All authors have read and agreed to the published version of the manuscript.

Funding: This research was funded by National Natural Science Foundation of China, grant number 41871021 and the Key Project of the National Natural Science Foundation of China, grant number 41530745.

Institutional Review Board Statement: Not applicable.

Informed Consent Statement: Not applicable.

Data Availability Statement: The data presented in this study are available on request from the corresponding author.

Conflicts of Interest: The authors declare no conflict of interest. The funders had no role in the design of the study; in the collection, analyses, or interpretation of data; in the writing of the manuscript, or in the decision to publish the results.

References

1. Urbanus, J. China's Hidden City Recent discoveries at an isolated northern settlement are challenging traditional narratives about the origins of Chinese civilization. *Archaeology* **2019**, *72*, 34–37.
2. Jing, Y.; Campbell, R. Recent archaeometry research on 'the origins of Chinese civilisation'. *Antiquity* **2009**, *83*, 96–109. [CrossRef]
3. Gao, S.; Yang, Y.; Xu, Y.; Zhang, Q.; Zhu, H.; Zhou, H. Tracing the genetic history of the Chinese people: Mitochondrial DNA analysis of a neolithic population from the Lajia Site. *Am. J. Phys. Anthropol.* **2007**, *133*, 1128–1136. [CrossRef] [PubMed]
4. Ma, Z.; Tang, S. *Qijia Culture and Chinese Civilization*, 1st ed.; Publisher Gansu Nationalities Publishing House: Lanzhou, China, 2015; pp. 360–361.
5. Xu, J.; Jia, Y.; Ma, C.; Zhu, C.; Wu, L.; Li, Y.; Wang, X. Geographic distribution of archaeological sites and their response to climate and environmental change between 10.0–2.8 ka BP in the Poyang Lake Basin, China. *J. Geogr. Sci.* **2016**, *26*, 603–618. [CrossRef]
6. Li, F.; Wu, L.; Zhu, C.; Zheng, C.; Sun, W.; Wang, X.; Shao, S.; Zhou, Y.; He, T.; Lie, S. Spatial-temporal distribution and geographic context of Neolithic cultural sites in the Hanjiang River Basin, Southern Shaanxi, China. *J. Archaeol. Sci.* **2013**, *40*, 3141–3152. [CrossRef]
7. Wagner, M.; Tarasov, P.; Hosner, D.; Fleck, A.; Ehrich, R.; Chen, X.; Leipe, C. Mapping of the spatial and temporal distribution of archaeological sites of northern China during the Neolithic and Bronze Age. *Quat. Int.* **2013**, *290*, 344–357. [CrossRef]
8. Li, Y.; Qiang, M.; Huang, X.; Zhao, Y.; Leppanen, J.; Weckstrom, J.; Valiranta, M. Lateglacial and Holocene climate change in the NE Tibetan Plateau: Reconciling divergent proxies of Asian summer monsoon variability. *Catena* **2021**, *199*, 7. [CrossRef]
9. Cai, Y.; Cheng, X.; Ma, L.; Mao, R.; Breitenbach, S.F.M.; Zhang, H.; Xue, G.; Cheng, H.; Edwards, R.L.; An, Z. Holocene variability of East Asian summer monsoon as viewed from the speleothem delta O-18 records in central China. *Earth Planet. Sci. Lett.* **2021**, *558*, 12. [CrossRef]
10. Feng, Z.; Sun, A.; Abdusalih, N.; Ran, M.; Kurban, A.; Lan, B.; Zhang, D.; Yang, Y. Vegetation changes and associated climatic changes in the southern Altai Mountains within China during the Holocene. *Holocene* **2017**, *27*, 683–693. [CrossRef]
11. Xie, D. *Prehistoric Archaeology in the Gansu and Qinghai Region*, 1st ed.; Cultural Relics Publishing House: Beijing, China, 2002; pp. 111–130.
12. Lu, H.; Yang, X.; Ye, M.; Liu, K.; Xia, Z.; Ren, X.; Cai, L.; Wu, N.; Liu, T. Millet noodles in Late Neolithic China—A remarkable find allows the reconstruction of the earliest recorded preparation of noodles. *Nature* **2005**, *437*, 967–968. [CrossRef]
13. Zhang, Y.; Huang, C.; Pang, J.; Guo, Y.; Zhou, Q. Identification of the prehistoric catastrophes at the Lajia Ruins using micromorphological analysis within the Guanting Basin, Minhe County, Qinghai Province. *Archaeol. Anthropol. Sci.* **2018**, *10*, 711–723. [CrossRef]
14. Han, J. The Development and Evolution of Qijia Culture: Cultural Interaction and Eurasian Background. *Cult. Relics* **2019**, *7*, 60–65.

15. Dong, G.; Jia, X.; Elston, R.; Chen, F.; Li, S.; Wang, L.; Cai, L.H.; An, C. Spatial and temporal variety of prehistoric human settlement and its influencing factors in the upper Yellow River valley, Qinghai Province, China. *J. Archaeol. Sci.* **2013**, *40*, 2538–2546. [CrossRef]
16. Gonzalez-Tennant, E. Recent Directions and Future Developments in Geographic Information Systems for Historical Archaeology. *Hist. Archaeol.* **2016**, *50*, 24–49. [CrossRef]
17. Gaffney, V.; Stančič, Z. GIS approaches to regional analysis: A case study of the island of Hvar. *Am. J Archaeol.* **1991**, *97*, 799–800.
18. Lu, G.; Batty, M.; Strobl, J.; Lin, H.; Zhu, A.; Chen, M. Reflections and speculations on the progress in Geographic Information Systems (GIS): A geographic perspective. *Int. J. Geogr. Inf. Sci.* **2019**, *33*, 346–367. [CrossRef]
19. Liu, J. The Research of Settlements Archaeology Supported by GIS. Ph.D. Thesis, China University of Geosciences, Beijing, China, May 2007.
20. Wu, L.; Zhu, C.; Zheng, C.; Ma, C.; Wang, X.; Li, F.; Li, B.; Li, K. Impact of Holocene climate change on the prehistoric cultures of Zhejiang region, East China. *J. Geogr. Sci.* **2014**, *24*, 669–688. [CrossRef]
21. Hosner, D.; Wagner, M.; Tarasov, P.; Chen, X.; Leipe, C. Spatiotemporal distribution patterns of archaeological sites in China during the Neolithic and Bronze Age: An overview. *Holocene* **2016**, *26*, 1576–1593. [CrossRef]
22. Jiang, S.; Zhang, H.; Wang, H.; Zhou, L.; Tang, G. Using Restaurant POI Data to Explore Regional Structure of Food Culture Based on Cuisine Preference. *ISPRS Int. J. Geo-Inf.* **2021**, *10*, 38. [CrossRef]
23. Liu, Y.; Goodchild, M.F.; Guo, Q.; Tian, Y.; Wu, L. Towards a General Field model and its order in GIS. *Int. J. Geogr. Inf. Sci.* **2008**, *22*, 623–643. [CrossRef]
24. Zhang, R.; Lukasczyk, J.; Wang, F.; Ebert, D.; Shakarian, P.; Mack, E.A.; Maciejewski, R. Exploring geographic hotspots using topological data analysis. *T Gis.* **2021**, *25*, 3188–3209. [CrossRef]
25. Zhang, H.; Zhou, X.; Tang, G.; Ji, G.; Zhang, X.; Xiong, L. Inference method for cultural diffusion patterns using a field model. *T Gis.* **2020**, *24*, 1578–1601. [CrossRef]
26. Zhang, H.; Zhou, X.; Tang, G.; Xiong, L.; Dong, K. Mining spatial patterns of food culture in China using restaurant POI data. *T Gis.* **2021**, *25*, 579–601. [CrossRef]
27. Womack, A.; Jaffe, Y.; Zhou, J.; Hung, L.; Wang, H.; Li, S.; Chen, P.; Flad, R. Mapping Qijiaping: New Work on the Type-Site of theQJC (2300–1500 BC) in Gansu Province, China. *J. Field Archaeol.* **2017**, *42*, 488–502. [CrossRef]
28. Qi, Y. Relics of Qijia Culture found in Baiyinhot, Inner Mongolia. *Archaeology* **1962**, *1*, 22.
29. Jiang, X. A comparative study of Dayangdui site in Xichang. *Sichuan Cult. Relics* **2007**, *5*, 26–35.
30. Lu, H.; Li, Y.; Zhang, J.; Yang, X.; Ye, M.; Li, Q.; Wang, C.; Wu, N. Component and simulation of the 4000-year-old noodles excavated from the archaeological site of Lajia in Qinghai, China. *Chin. Sci. Bull.* **2014**, *59*, 5136–5152. [CrossRef]
31. Cui, Y.; Dong, G.; Li, H.; An, T.; Liu, X.; Wang, J.; Wang, H.; Ren, X.; Li, X.; Chen, F. Early ceramic trade in Gansu and Qinghai regions, northwest China: A comparative elemental analysis on sherds of Majiayao culture, Yangshao culture and Qijia Culture. *J. Archaeol. Sci.-Rep.* **2015**, *3*, 65–72. [CrossRef]
32. Pang, X.; Wang, L. Analysis on the communication between Qijia Culture and Erlitou Culture. *Cult. Relics Cent. China* **2019**, *4*, 89–98.
33. Chen, W. Bulletin of archaeological survey in Kangba area in 2005. *Sichuan Cult. Relics* **2005**, *6*, 3–9.
34. An, C.; Feng, Z.; Tang, L. Environmental change and cultural response between 8000 and 4000 cal. yr BP in the western Loess Plateau, northwest China. *J. Quat. Sci.* **2004**, *19*, 529–535. [CrossRef]
35. Jia, X.; Sun, Y.; Wang, L.; Sun, W.F.; Zhao, Z.J.; Lee, H.F.; Huang, W.B.; Wu, S.Y.; Lu, H.Y. The transition of human subsistence strategies in relation to climate change during the Bronze Age in the West Liao River Basin, Northeast China. *Holocene* **2016**, *26*, 781–789. [CrossRef]
36. Dong, G.; Wang, L.; Cui, Y.; Elston, R.; Chen, F. The spatiotemporal pattern of the Majiayao cultural evolution and its relation to climate change and variety of subsistence strategy during late Neolithic period in Gansu and Qinghai Provinces, northwest China. *Quatern Int.* **2013**, *316*, 155–161. [CrossRef]
37. Bureau of National Cultural Relics. *Atlas of Chinese Cultural Relics—Fascicule of Gansu Province*; Sinomap Press: Beijing, China, 2011.
38. Bureau of National Cultural Relics. *Atlas of Chinese Cultural Relics—Fascicule of Qinghai Province*; Sinomap Press: Beijing, China, 1996.
39. Bureau of National Cultural Relics. *Atlas of Chinese Cultural Relics—Fascicule of Ningxia Province*; Cultural Relics Publishing House: Beijing, China, 2010.
40. Bureau of National Cultural Relics. *Atlas of Chinese Cultural Relics—Fascicule of InnerMongolia*; Xi'an Cartographic Publishing House: Xi'an, China, 2003.
41. Bureau of National Cultural Relics. *Atlas of Chinese Cultural Relics—Fascicule of Shaanxi Province*; Sinomap Press: Beijing, China, 1998.
42. Bureau of National Cultural Relics. *Atlas of Chinese Cultural Relics—Fascicule of Shanxi Province*; Sinomap Press: Beijing, China, 2006.
43. Bureau of National Cultural Relics. *Atlas of Chinese Cultural Relics—Fascicule of Sichuan Province*; Cultural Relics Publishing House: Beijing, China, 2009.
44. Bureau of National Cultural Relics. *Atlas of Chinese Cultural Relics—Fascicule of Henan Province*; Sinomap Press: Beijing, China, 1991.
45. Fort, J. Demic and cultural diffusion propagated the Neolithic transition across different regions of Europe. *J. R. Soc. Interface* **2015**, *12*, 165–166. [CrossRef]

46. Krishnan, S.; AlSudiary, M. Cultural Practices and Virtual Social Networks Diffusion: An International Analysis Using GLOBE Scores. *J. Glob. Inf. Technol. Man.* **2016**, *19*, 154–173. [CrossRef]
47. Valverde, S.; Sole, R. A Cultural Diffusion Model for the Rise and Fall of Programming Languages. *Hum. Biol.* **2015**, *87*, 224–234. [CrossRef]
48. Li, X.; Sun, N.; Dodson, J.; Zhou, X. Human activity and its impact on the landscape at the Xishanping site in the western Loess Plateau during 4800–4300 cal yr BP based on the fossil charcoal record. *J. Archaeol. Sci.* **2012**, *39*, 3141–3147. [CrossRef]
49. Li, X.; Zhou, X.; Zhou, J.; Dodson, J.; Zhang, H.; Shang, X. The earliest archaeobiological evidence of the broadening agriculture in China recorded at Xishanping site in Gansu Province. *Sci. China Ser. D-Earth Sci.* **2007**, *50*, 1707–1714. [CrossRef]
50. An, C.; Tang, L.; Barton, L.; Chen, F. Climate change and cultural response around 4000 cal yr BP in the western part of Chinese Loess Plateau. *Quat. Res.* **2005**, *63*, 347–352. [CrossRef]
51. Zhou, S.; Kong, X.; Zhu, H. *Cultural Geography*, 1st ed.; Higher Education Press: Beijing, China, 2004; pp. 175–205.
52. Yu, W.; Ai, T.; He, Y.; Shao, S. Spatial co-location pattern mining of facility points-of-interest improved by network neighborhood and distance decay effects. *Int. J. Geogr. Inf. Sci.* **2017**, *31*, 280–296. [CrossRef]
53. MacDonald, K.; Scherjon, F.; van Veen, E.; Vaesen, K.; Roebroeks, W. Middle Pleistocene fire use: The first signal of widespread cultural diffusion in human evolution. *Proc. Natl. Acad. Sci. USA* **2021**, *118*, 8. [CrossRef]
54. Allen, J.; Weinrich, M.; Hoppitt, W.; Rendell, L. Network-Based Diffusion Analysis Reveals Cultural Transmission of Lobtail Feeding in Humpback Whales. *Science* **2013**, *340*, 485–488. [CrossRef]
55. Cheng, C.; Cheung, S.-F.; Chio, J.H.-M.; Chan, M.-P.S. Cultural Meaning of Perceived Control: A Meta-Analysis of Locus of Control and Psychological Symptoms Across 18 Cultural Regions. *Psychol. Bull.* **2013**, *139*, 152–188. [CrossRef]
56. Fort, J. Synthesis between demic and cultural diffusion in the Neolithic transition in Europe. *Proc. Natl. Acad. Sci. USA* **2012**, *109*, 18669–18673. [CrossRef]
57. Xu, W.; Park, J.; Kim, J.; Park, H. Networked Cultural Diffusion and Creation on YouTube: An Analysis of YouTube Memes. *J. Broadcast Electron.* **2016**, *60*, 104–122. [CrossRef]
58. Fort, J. Vertical cultural transmission effects on demic front propagation: Theory and application to the Neolithic transition in Europe. *Phys. Rev. E.* **2011**, *83*, 56–124. [CrossRef]
59. Schmitt, D.; Diniz, G.; Alcalay, L.; Durkin, K.; Allensworth, M.; Echegaray, M.; Allik, J.; Eremsoy, E.; Ault, L.; Euler, H.; et al. Patterns and universals of adult romantic attachment across 62 cultural regions—Are models of self and of other pancultural constructs? *J. Cross Cult. Psychol.* **2004**, *35*, 367–402. [CrossRef]
60. Brady, M.; McGrath, V. Making Tuba in the Torres Strait Islands: The Cultural Diffusion and Geographic Mobility of an Alcoholic Drink. *J. Pac. Hist.* **2010**, *45*, 315–330. [CrossRef]
61. Koenig, F. Analyst Behavior: The Geography of Social Interaction. *J. Behav. Financ.* **2016**, *17*, 201–216. [CrossRef]
62. Kwon, R. Hegemonic Stability, World Cultural Diffusion, and Trade Globalization. *Sociol. Forum.* **2012**, *27*, 324–347. [CrossRef]
63. Wen, B.; Li, H.; Lu, D.; Song, X.; Zhang, F.; He, Y.; Li, F.; Gao, Y.; Mao, X.; Zhang, L.; et al. Genetic evidence supports demic diffusion of Han culture. *Nature* **2004**, *431*, 302–305. [CrossRef] [PubMed]
64. La Duc, E.; Chang, A. Analysis and Replication Studies of Prehistoric Chinese Ceramics from the Qijia Culture. *MRS Adv.* **2017**, *2*, 1849–1867. [CrossRef]
65. Jia, M.; Li, C.; Mao, X.; Zhang, D.; Wei, C.; Liu, H. Climate-human-environment interactions in the middle Yangtze Basin (central China) during the middle Holocene, based on pollen and geochemical records from the Sanfangwan Site. *Catena* **2021**, *204*, 9. [CrossRef]
66. Huang, X.; Xiang, L.; Lei, G.; Sun, M.; Qiu, M.; Storozum, M.; Huang, C.; Munkhbayar, C.; Demberel, O.; Zhang, J.; et al. Sedimentary Pediastrum record of middle-late Holocene temperature change and its impacts on early human culture in the desert-oasis area of northwestern China. *Quat. Sci. Rev.* **2021**, *265*, 11. [CrossRef]
67. Xu, B.; Gu, Z.; Pei, Y.; Lv, Y. Radiocarbon dating of the Pleistocene/Holocene climatic transition across the Chinese Loess Plateau. *Quat. Geochronol.* **2017**, *41*, 112–118. [CrossRef]
68. Zhao, H.; Huang, C.; Wang, H.; Liu, W.; Qiang, X.; Xu, X.; Zheng, Z.; Hu, Y.; Zhou, Q.; Zhang, Y.; et al. Mid-late Holocene temperature and precipitation variations in the Guanting Basin, upper reaches of the Yellow River. *Quat. Int.* **2018**, *490*, 74–81. [CrossRef]
69. Li, L.; Wu, L.; Zhu, C.; Li, F.; Ma, C. Relationship between archaeological sites distribution and environment from 1.15 Ma BP to 278 BC in Hubei Province. *J. Geogr. Sci.* **2011**, *21*, 909–925. [CrossRef]
70. Ren, X.; Xu, J.; Wang, H.; Storozum, M.; Lu, P.; Mo, D.; Li, T.; Xiong, J.; Kidder, T. Holocene fluctuations in vegetation and human population demonstrate social resilience in the prehistory of the Central Plains of China. *Environ. Res. Lett.* **2021**, *16*, 12. [CrossRef]
71. Xu, Q.; Zhang, S.; Gaillard, M.-j.; Li, M.; Cao, X.; Tian, F.; Li, F. Studies of modern pollen assemblages for pollen dispersal-deposition- preservation process understanding and for pollen-based reconstructions of past vegetation, climate, and human impact: A review based on case studies in China. *Quat. Sci. Rev.* **2016**, *149*, 151–166. [CrossRef]
72. Yang, L.; Shi, Z.; Zhang, S.; Lee, H. Climate Change, Geopolitics, and Human Settlements in the Hexi Corridor over the Last 5,000 Years. *Acta Geol. Sin.-Engl. Ed.* **2020**, *94*, 612–623. [CrossRef]
73. Marcott, S.; Shakun, J.; Clark, P.; Mix, A. A Reconstruction of Regional and Global Temperature for the Past 11,300 Years. *Science* **2013**, *339*, 1198–1201. [CrossRef]

74. Chen, F.; Xu, Q.; Chen, J.; Birks, H.J.B.; Liu, J.; Zhang, S.; Jin, L.; An, C.; Telford, R.; Cao, X.; et al. East Asian summer monsoon precipitation variability since the last deglaciation. *Sci. Rep.* **2015**, *5*, 11. [CrossRef]
75. Zhou, W.; Lu, X.; Wu, Z.K. The peat record and accelerator radiocarbon dating of climate change in the Zoige Plateau during the Holocene. *Chin. Sci. Bull.* **2001**, *46*, 1040–1044.
76. Cai, Y.; Tan, L.; Cheng, H.; An, Z.; Edwards, R.; Kelly, M.; Kong, X.; Wang, X. The variation of summer monsoon precipitation in central China since the last deglaciation. *Earth Planet Sci. Lett.* **2010**, *291*, 21–31. [CrossRef]
77. Maher, B. Holocene variability of the East Asian summer monsoon from Chinese cave records: A re-assessment. *Holocene* **2008**, *18*, 861–866. [CrossRef]
78. Zeng, L.; Yi, S.; Lu, H.; Chen, Y.; Lei, F.; Xu, Z.; Wang, X.; Zhang, W. Response of dune mobility and pedogenesis to fluctuations in monsoon precipitation and human activity in the Hulunbuir dune field, northeastern China, since the last deglaciation. *Glob. Planet Chang.* **2018**, *168*, 1–14. [CrossRef]
79. Peterse, F.; Prins, M.; Beets, C.; Troelstra, S.; Zheng, H.; Gu, Z.; Schouten, S.; Damste, J. Decoupled warming and monsoon precipitation in East Asia over the last deglaciation. *Earth Planet Sci. Lett.* **2011**, *301*, 256–264. [CrossRef]
80. Hu, C.; Henderson, G.; Huang, J.; Xie, S.; Sun, Y.; Johnson, K. Quantification of Holocene Asian monsoon rainfall from spatially separated cave records. *Earth Planet Sci. Lett.* **2008**, *266*, 221–232. [CrossRef]
81. Li, J.; Dodson, J.; Yan, H.; Cheng, B.; Zhang, X.; Xu, Q.; Ni, J.; Lu, F. Quantitative precipitation estimates for the northeastern Qinghai-Tibetan Plateau over the last 18,000 years. *J. Geophys. Res.-Atmos.* **2017**, *122*, 5132–5143. [CrossRef]
82. Guo, Y.; Mo, D.; Mao, L.; Wang, S.; Li, S. Settlement distribution and its relationship with environmental changes from the Neolithic to Shang-Zhou dynasties in northern Shandong, China. *J. Geogr. Sci.* **2013**, *23*, 679–694. [CrossRef]
83. Ma, M.; Dong, G.; Jia, X.; Wang, H.; Cui, Y.; Chen, F. Dietary shift after 3600 cal yr BP and its influencing factors in northwestern China: Evidence from stable isotopes. *Quat. Sci. Rev.* **2016**, *145*, 57–70. [CrossRef]
84. Jia, X.; Dong, G.; Li, H.; Brunson, K.; Chen, F.; Ma, M.; Wang, H.; An, C.; Zhang, K. The development of agriculture and its impact on cultural expansion during the late Neolithic in the Western Loess Plateau, China. *Holocene* **2013**, *23*, 85–92. [CrossRef]
85. Yang, Y.; Dong, G.; Zhang, S.; Cui, Y.; Li, H.; Chen, G.; Dodson, J.; Chen, F. Copper content in anthropogenic sediments as a tracer for detecting smelting activities and its impact on environment during prehistoric period in Hexi Corridor, Northwest China. *Holocene* **2017**, *27*, 282–291. [CrossRef]
86. Spengler, R.; Frachetti, M.; Doumani, P.; Rouse, L.; Cerasetti, B.; Bullion, E.; Mar'yashev, A. Early agriculture and crop transmission among Bronze Age mobile pastoralists of Central Eurasia. *Proc. R. Soc. B-Biol. Sci.* **2014**, *281*, 7. [CrossRef]
87. Doumani, P.; Frachetti, M.; Beardmore, R.; Schmaus, T.; Spengler, R.N., III; Mar'yashev, A.N. Burial ritual, agriculture, and craft production among Bronze Age pastoralists at Tasbas (Kazakhstan). *Archaeol. Res. Asia* **2015**, *1*, 17–32. [CrossRef]
88. Liu, X.; Jones, P.J.; Matuzeviciute, G.M.; Hunt, H.; Lister, D.; An, T.; Przelomska, N.; Kneale, C.; Zhao, Z.; Jones, M. From ecological opportunism to multi-cropping: Mapping food globalisation in prehistory. *Quat. Sci. Rev.* **2019**, *206*, 21–28. [CrossRef]
89. Flad, R.; Li, S.; Wu, X.; Zhao, Z. Early wheat in China: Results from new studies at Donghuishan in the Hexi Corridor. *Holocene* **2010**, *20*, 955–965. [CrossRef]
90. Brunson, K.; Ren, L.; Zhao, X.; Dong, X.; Wang, H.; Zhou, J.; Flad, R. Zooarchaeology, ancient mtDNA, and radiocarbon dating provide new evidence for the emergence of domestic cattle and caprines in the Tao River Valley of Gansu Province, northwest China. *J. Archaeol. Sci.-Rep.* **2020**, *31*, 14. [CrossRef]
91. Linduff, K.M.; Mei, J. Metallurgy in Ancient Eastern Asia: Retrospect and Prospects. *J. World Prehist.* **2009**, *22*, 265–281. [CrossRef]
92. Pernicka, E. Development of metallurgy in Eurasia. *Quat. Int.* **2020**, *560*, 38–44. [CrossRef]
93. Stevens, C.J.; Murphy, C.; Roberts, R.; Lucas, L.; Silva, F.; Fuller, D.Q. Between China and South Asia: A Middle Asian corridor of crop dispersal and agricultural innovation in the Bronze Age. *Holocene* **2016**, *26*, 1541–1555. [CrossRef] [PubMed]
94. Dong, G.; Yang, Y.; Liu, X.; Li, H.; Cui, Y.; Wang, H.; Chen, G.; Dodson, J.; Chen, F. Prehistoric trans-continental cultural exchange in the Hexi Corridor, northwest China. *Holocene* **2018**, *28*, 621–628. [CrossRef]
95. Jones, M.; Hunt, H.; Lightfoot, E.; Lister, D.; Liu, X.; Motuzaite-Matuzeviciute, G. Food globalization in prehistory. *World Archaeol.* **2011**, *43*, 665–675. [CrossRef]
96. Dittmar, J.; Yeh, H.; Berger, E.; Austen, J.; Zhan, X.; Hernandez, M.; Mao, R.; Hui, W.; Mitchell, P. Violent death and the formation of social structure: Skeletal evidence of violence from the Qijia Culture (2,300-1,400BCE), Gansu Provence, China. *Am. J. Phys. Anthropol.* **2018**, *165*, 69.
97. Clackson, J. The Horse, the Wheel, and Language. How Bronze-Age Riders from the Eurasian Steppes shaped the Modern World. *J. H Ellenic Study* **2010**, *130*, 286–287. [CrossRef]
98. Qian, Y.; Mao, R. Gansu Lintan Mogou Qijia cultural cemetery excavation and main harvest. *Collect. Stud. Archaeol.* **2002**, 638–656.
99. Ye, M. Excavation of minhe Lajia prehistoric site in Qinghai province. *Archaeology* **2002**, *7*, 3–5.
100. Zeng, Y. Social Forms in Shenna Relics. *J. Qinghai Minzu Univ.* **2013**, *3*, 35–37.
101. Huang, C.; Su, H. Climate change and Zhou relocations in early Chinese history. *J. Hist. Geogr.* **2009**, *35*, 297–310. [CrossRef]
102. Dong, G.; Li, R.; Lu, M.; Zhang, D.; James, N. Evolution of human-environmental interactions in China from the Late Paleolithic to the Bronze Age. *Prog. Phys. Geog.* **2020**, *44*, 233–250. [CrossRef]

remote sensing

MDPI

Article

Vegetation Browning Trends in Spring and Autumn over Xinjiang, China, during the Warming Hiatus

Moyan Li [1], Junqiang Yao [2], Jingyun Guan [1,3] and Jianghua Zheng [1,*]

[1] Key Laboratory of Oasis Ecology of Ministry of Education, Institute of Arid Ecology and Environment, College of Geographical Sciences, Xinjiang University, Urumqi 830046, China; 107556519105@stu.xju.edu.cn (M.L.); 107556519104@stu.xju.edu.cn (J.G.)

[2] Institute of Desert Meteorology, China Meteorological Administration, Urumqi 830002, China; yaojq@idm.cn

[3] College of Tourism, Xinjiang University of Finance and Economics, Urumqi 830012, China

* Correspondence: zheng.jianghua@xju.edu.cn; Tel.: +86-13579880590

Abstract: Satellite-derived vegetation records (GIMMS3g-NDVI) report that climate warming promotes vegetation greening trends; however, the climate impacts on vegetation growth during the global warming hiatus period (1998–2012) remain unclear. In this study, we focused on the vegetation change trend in Xinjiang in spring and autumn before and during the recent warming hiatus period, and their climate-driving mechanisms, which have not been examined in previous studies. Based on satellite records, our results indicated that the summer normalized difference vegetation index (NDVI) in Xinjiang experienced a greening trend, while a browning trend existed in spring and autumn during this period. The autumn NDVI browning trend in Xinjiang was larger than that in spring; however, the spring NDVI displayed a higher correlation with climatic factors than did the autumn NDVI. During the warming hiatus, spring climatic factors were the main controlling factors of spring NDVI, and spring vapor pressure deficit (VPD) had the highest positive correlation with spring NDVI, followed by spring temperature. The larger increase in air temperature in spring than in autumn resulted in increased VPD differences in spring and autumn. In autumn, summer climatic factors (e.g., VPD, WS, RH, and precipitation) were significantly correlated with the autumn NDVI during the warming hiatus. However, the autumn temperature was weakly correlated with the autumn NDVI. Our results have significant implications for understanding the response of vegetation growth to recent and future climatic conditions.

Keywords: vegetation browning; warming hiatus; vapor pressure deficit; spring; autumn; Xinjiang

Citation: Li, M.; Yao, J.; Guan, J.; Zheng, J. Vegetation Browning Trends in Spring and Autumn over Xinjiang, China, during the Warming Hiatus. *Remote Sens.* **2022**, *14*, 1298. https://doi.org/10.3390/rs14051298

Academic Editors: Zhongwen Hu, Yangyi Wu, Jie Zhang and Jingzhe Wang

Received: 1 January 2022
Accepted: 5 March 2022
Published: 7 March 2022

1. Introduction

The global air temperature dataset indicates that the warming rate did not increase significantly during the period of 1998–2012 [1–8], and this period has become known as the "warming hiatus" or "global slowdown" [2,9]. Many previous reports have indicated that the "warming hiatus" phenomenon has also been reported in China and the Xinjiang region [10–12]. A warming hiatus has profound effects on ecosystems, water resources, and sustainable development.

Many studies have found that air temperature is a major controlling factor for vegetation growth changes [13–16]. Generally, vegetation growth is consistent with changes in temperature [17]. Kaufmann et al. [18] reported that global warming was the main contributor to vegetation greening in the Northern Hemisphere using the normalized difference vegetation index (NDVI). Other studies have indicated that global warming has advanced spring vegetation green-up and delayed autumn vegetation senescence [19,20]. Wang et al. [21] found that the spring phenology advancing rate over the Northern Hemisphere from 2000 to 2012 was larger than that during 1982–1992, similar to the autumn phenology. The long-term monitoring of the global NDVI has shown significant greening since 1982 [22]. However, the NDVI trend has indicated a greening stall and subsequent vegetation browning since the late 1990s [22,23]. Approximately 59% of the global vegetated area experienced a significant browning trend from 1998 to 2015 [22]. In addition, greening to browning reversals were experienced in southeast and northwest High Mountain Asia from 1982 to 2015 due to the continuous warming shift [24].

Global dryland vegetation has undergone a significant greening trend since the 1980s [25–28]. Central Asia is one of the driest regions in the world, and the vegetation NDVI showed a greening trend during 1982–1998, followed by a browning trend after 1999 [29,30]. Vegetation NDVI trends in arid regions of China have decreased since 2000 [11,31]. In Xinjiang, a previous study reported that the increased warming and reduced precipitation exemplified by the vegetation NDVI trends have reversed since the late 1990s [29]. Another study found that increased drought severity led to a decreased NDVI [12]. In addition, Yao et al. [31] found that climate extremes were the major drivers of the decreased NDVI in Xinjiang. However, information regarding vegetation growth changes and their climate-driving mechanisms is still scarce, especially for spring and autumn during the warming hiatus.

The primary objective of this study is to quantify the trend in vegetation growth changes in spring and autumn in Xinjiang during the global warming hiatus, and to assess the effects of the warming hiatus on vegetation changes trends. In this study, we used satellite-derived vegetation records (GIMMS3g-NDVI) to investigate the latest vegetation growth changes in spring and autumn in Xinjiang and to reveal the effects of the warming hiatus on vegetation NDVI trends. We examined vegetation NDVI trends in spring and autumn from 1982 to 2015. We also compared vegetation NDVI trends before and during the warming period. The climatic controls on vegetation NDVI trends were investigated based on correlation and partial correlation analyses. We also discuss the climate-driving mechanisms during the warming hiatus.

2. Materials and Methods

2.1. Climate Data

Monthly meteorological datasets were provided by the China Meteorological Administration (CMA) (http://data.cma.cn/, accessed on 1 May 2020). The dataset contained monthly air temperature (Tair), maximum temperature (Tmax), minimum temperature (Tmin), precipitation (Prec), relative humidity (RH), wind speed (WS), and sunshine duration (SD). This dataset consisted of 89 observed sites over Xinjiang and was available for the period from to 1982–2015 (Figure 1). These datasets were strictly quality-controlled and homogeneity-tested with high accuracy [32]. However, the monthly RH variable was inhomogeneous from the early 2000s, mainly due to data collection changes from manual to automatic systems [33]. Thus, we selected the homogenized RH series from the ChinaRHv1.0 dataset (http://www.sciencedb.cn/dataSet/handle/804, accessed on 25 May 2021), which was developed by Li et al. [34]. These meteorological variables were interpolated with the spatial resolution of 10 km by the gradient inverse distance square method (GIDS) from the data measured at the meteorological stations, which consider the effects of topographical factors.

We used the self-calibrating Palmer drought severity index (sc_PDSI; [35,36]) to analyze the dry or wet conditions in Xinjiang from 1982–2015. The monthly time series of sc_PDSI was calculated using monthly temperature and precipitation data from the CRU TS4.04 version (https://doi.org/10.1038/s41597-020-0453-3, accessed on 20 March 2021), together with fixed parameters related to soil/surface characteristics at each location [36]. The sc_PDSI dataset (https://crudata.uea.ac.uk/cru/data/drought/, accessed on 20 March 2021) was obtained from the CRU, with a spatial resolution of 0.5° [37]. In addition, the temperature of the Northern Hemisphere (NH) was obtained from the HadCRUTS5 version for 1982–2015.

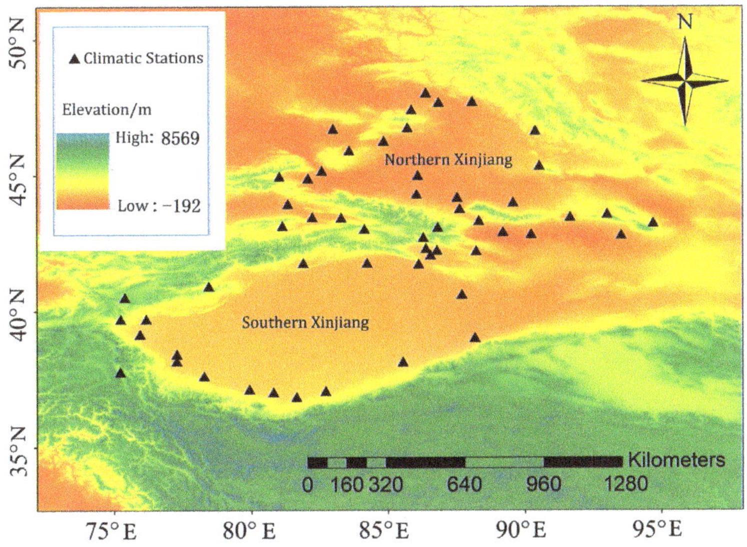

Figure 1. Meteorological stations in Xinjiang.

2.2. Satellite-Derived NDVI Data

The newest NASA GIMMS3g-NDVI dataset was used to indicate vegetation growth in Xinjiang during 1982–2015 [31,38]. The GIMMS NDVI3g dataset was obtained from the NASA Ames Ecological Forecasting Lab (https://ecocast.arc.nasa.gov/data/pub/gimms/, accessed on 1 November 2019) with a spatial resolution of 1/12° and a temporal resolution of 15-days [22]. The regional mean NDVI was calculated by averaging the annual growing season (GS) and seasonal NDVI, excluding unvegetated regions. Pixels with multi-year average (1982–2015) GS NDVI values greater than 0.1 were used for analysis, and GS is defined as April–October. The maximum value composite (MVC) method is utilized to construct the monthly and seasonal NDVI datasets [38].

2.3. Methods

2.3.1. Calculation of VPD

The monthly vapor pressure deficit (VPD) is calculated according to the Allen et al. [39] equation:

$$VPD = e_s - e_a \quad (1)$$

where e_s is the saturation vapor pressure (hPa) and e_a is the actual vapor pressure (hPa). The es value is calculated by:

$$e_s = 0.611\exp[(17.27 \times T_a)/(T_a + 237.3)] \quad (2)$$

where T_a is the surface air temperature (°C). The e_a value is calculated by:

$$e_a = RH \times e_s/100 \quad (3)$$

where RH is relative humidity (%).

2.3.2. Trend Analysis

In this study, a linear least squares regression analysis was conducted to calculate trends in NDVI and climatic factors. The slope of the regression line was defined as the trend in the NDVI and climatic factors. The trend significance was examined by the Mann–Kendall test [40,41].

For the vegetation NDVI and climatic factors (VPD, air temperature, maximum temperature, minimum temperature, precipitation, RH, WS, and SD), the seasonal and annual growing season (GS) values were calculated using the average of the monthly NDVI and climatic factor values. The annual GS was defined from April to October, and other seasons were defined as spring (March, April, May), summer (June, July, August), and autumn (September, October, and November).

2.3.3. Correlation and Partial Correlation Analysis

Pearson's correlation analysis was employed to examine the correlation coefficient between the four seasonal climatic factors and spring NDVI (or autumn NDVI), and a t-test was used to determine the significance of the correlation. Partial correlation analysis (PCA) was employed to determine the correlation between NDVI and climate factors at the annual GS and seasonal scales, excluding the impacts of other climatic factors. The PCA results describe the relationship between two variables independent of the impact of other variables [42,43].

3. Results

3.1. Temperature Change and Warming Hiatus Period

The global and regional surface air temperature did not increase significantly between 1998 and 2012, and this period can be regarded as the "warming hiatus" [1–8]. Figure 2 indicate the change in annual mean air temperature averaged across Xinjiang and the NH from 1982 to 2015, especially during the warming hiatus period (1998–2012). This indicates that the "warming hiatus" was also reporting the surface air temperature changes in Xinjiang during 1998–2012.

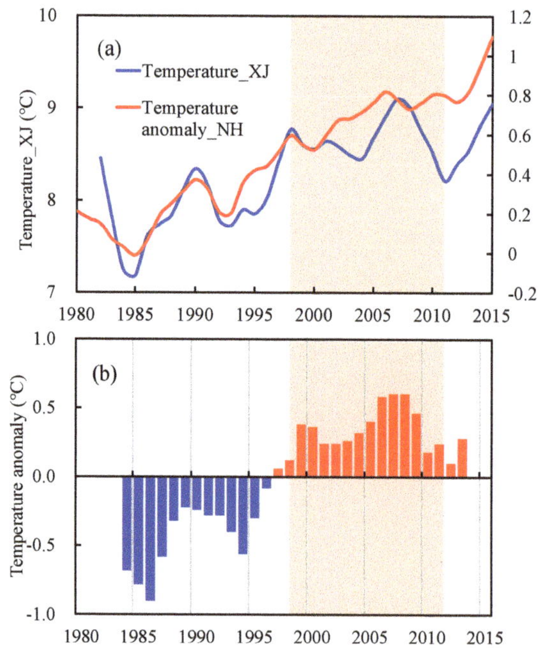

Figure 2. (**a**) Long-term changes in temperature _XJ and the Northern Hemisphere (NH) temperature anomaly from 1982 to 2015, showing the low-pass Loess filter design. (**b**) Five-year moving average of temperature anomaly in Xinjiang. The light gray background indicates the warming hiatus (1998–2012).

Trends in the regional mean time series of Tair, VPD, Tmax, Tmin, Prec, and RH had different characteristics during 1998–2012 (Figure 3). For Tair, the annual mean warming rate during 1998–2012 decreased by −0.14 °C/decade, and the decreases in the annual mean Tmax and Tmin, −0.13 °C/decade and −0.08 °C/decade, respectively, were much weaker and insignificant (Figure 3a,c,d). The annual total precipitation trends show an insignificant decline of −7.66 mm/decade between 1998–2012, which is contrasts markedly with the earlier period (1982–1997) (Figure 3e). The decrease in annual RH, −3.3%/decade, was significant during the warming hiatus (Figure 3f). For the VPD, the significant positive trends of 0.08 hPa/decade were impressive during 1998–2012 (Figure 3b), which is inconsistent with the reduction in other variables during the same time.

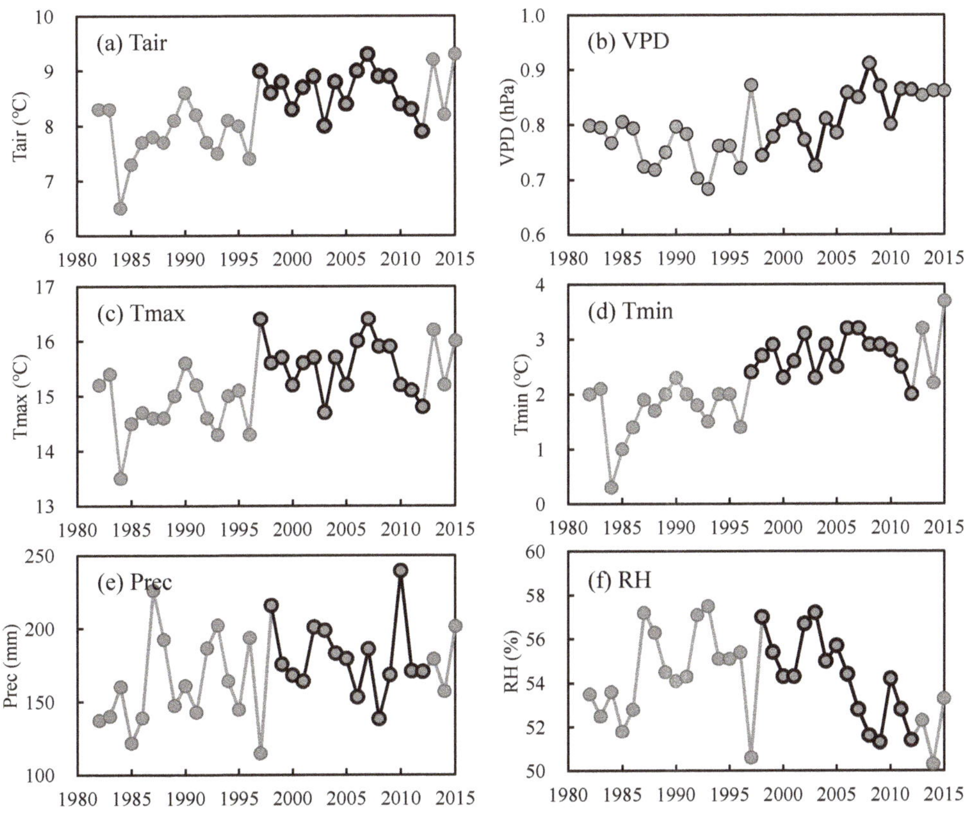

Figure 3. Regional mean time series of Tair, VPD, Tmax, Tmin, Prec, and RH in Xinjiang during the period 1998–2012 (1982–1997 and 2013–2015 series are shown by gray lines for comparison).

The seasonal SAT trends had different characteristics before and after 1998. The greatest increase in Tair occurred during the spring, 0.52 °C/decade, and the summer and autumn had slight warming trends, 0.22 °C/decade and 0.14 °C/decade, respectively; however, there was a decreasing trend, −1.36 °C/decade in winter. This contrasts markedly with 1982–1998, during which there was little change (0.17 °C/decade) in summer, while the other three seasons experienced significant warming trends, 0.76 °C/decade in spring, 0.93 °C/decade in autumn, and 0.98 °C/decade in winter (Table 1).

Table 1. Change trend for GIMMS3g-NDVI and temperature for the periods 1982–1998 and warming hiatus (1998–2012).

Data	Season	Warming Period (1982–1998)		Warming Hiatus (1998–2012)	
		Slope (/10a)	*p*-Value	Slope (/10a)	*p*-Value
GIMMS3g-NDVI	Annual	0.004	0.009	−0.002	0.151
	Spring	0.004	0.023	−0.002	0.218
	Summer	0.007	0.002	0.001	0.030
	Autumn	0.006	0.001	−0.003	0.072
Tair (°C)	Annual	0.80	0.002	−0.14	0.579
	Spring	0.76	0.047	0.52	0.450
	Summer	0.17	0.229	0.22	0.328
	Autumn	0.93	0.007	0.14	0.778

VPD is a direct indicator of atmospheric desiccation strength and is dependent on temperature and RH. The Xinjiang climate experienced remarkable changes during the warming hiatus. The annual mean temperature after 1998 increased by 9.2% compared with that during 1982–1998 (Figure 4). Seasonally, spring had the largest change at 12.4%, followed by autumn at 11.3%, and summer had the weakest change of 3.3% during 1998–2015 relative to 1982–1998 in Xinjiang (Figure 4). For VPD, the relative change was approximately 8.0% before and after 1998, and the seasonal VPD changes showed the greatest increase in spring (12.3%), followed by 6.9% in autumn, while summer showed the smallest change of 6.7% (Figure 4). Thus, on seasonal timescales, spring and autumn became warmer and drier during the warming hiatus period, and summer changed little.

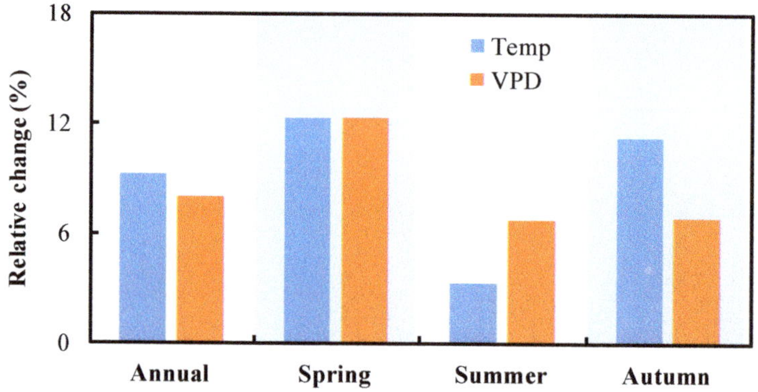

Figure 4. Change (%) in temperature during 1998–2015 relative to that in 1982–1998 in Xinjiang. The light gray background shows the spring and autumn.

3.2. *Vegetation Browning Trends Based on Remote Sensing Data*

The monthly vegetation NDVI for each grid cell in Xinjiang during 1982–2015 was estimated using the satellite-derived GIMMS3g dataset. To determine the effects of the warming hiatus (1998–2012) on vegetation change, we investigated NDVI trends in Xinjiang before and after 1998. The average vegetation NDVI and temperature trends in Xinjiang before 1998, after 1998, and during the warming hiatus periods, are shown in Figure 5 and Table 1.

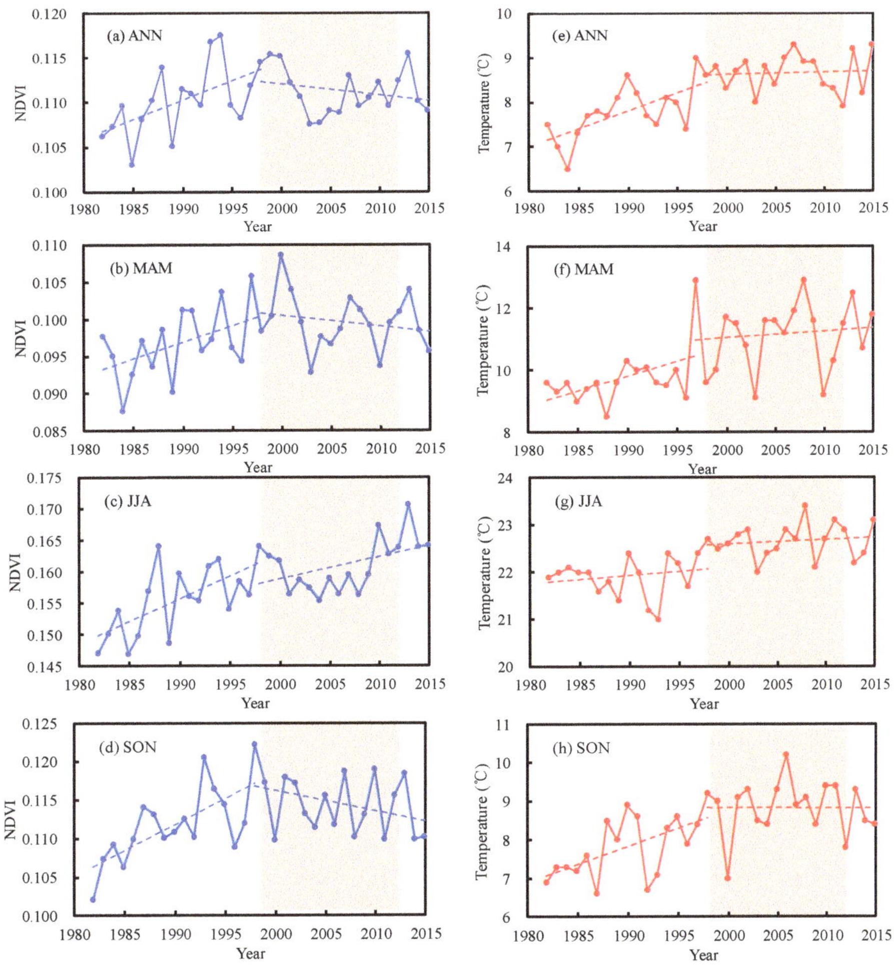

Figure 5. GIMMS3g-NDVI and temperature trends in Xinjiang from 1982 to 2015. A linear trend was determined for the periods before and after 1998, separately. The light gray background shows the warming hiatus (1998–2012).

The annual NDVI showed a significant greening trend of 0.004/decade ($p < 0.01$) before 1998 but had an insignificant browning trend during 1998–2015 (slope = −0.001/decade, $p > 0.1$), and an NDVI browning rate of −0.002/decade occurred during the warming hiatus period (Figure 5a). Seasonally, the greatest greening rate of 0.007/decade ($p < 0.01$) occurred in summer, followed by a considerable greening rate of 0.06/decade ($p < 0.01$) in autumn, and 0.04/decade ($p < 0.01$) in spring during 1982–1998 (Figure 5b–d, Table 1).

During the warming hiatus period, the reductions in the spring and autumn NDVI, −0.002/decade and −0.003/decade ($p > 0.05$), respectively, were much weaker and insignificant (Figure 5b,d, Table 1). However, the summer NDVI increased significantly, with a greening rate of 0.001/decade ($p < 0.05$) (Figure 4c, Table 1). We also examined the annual and seasonal warming rates (Figure 5). During 1998–2012, the average temperature warming rates reversed at the annual scale (from 0.80 °C/decade to −0.14 °C/decade) and decreased in spring and autumn (from 0.76 °C/decade to 0.52 °C/decade and from

0.93 °C/decade to 0.14 °C/decade, respectively) but continuously increased in summer (from 0.17 °C/decade to 0.22 °C/decade) (Figure 5, Table 1).

The spatial patterns of the vegetation NDVI were also compared before and during the warming hiatus (Figures 6 and 7). During 1982–1998, approximately 80% of the vegetation-covered area in Xinjiang exhibited a greening trend in spring (Figure 6c,d). In comparison, the greening trend of the spring NDVI reversed in southern Xinjiang, the Junggar Basin, and parts of the Tianshan Mountains, where a significant spring NDVI browning trend occurred during the warming hiatus (Figure 6e,f). In addition, the spring NDVI significantly increased between 1998 and 2012 in the mountainous areas, mainly in the Altay Mountains, northern Tianshan Mountains, and part of the Ili River valley, where there is abundant snowfall in winter.

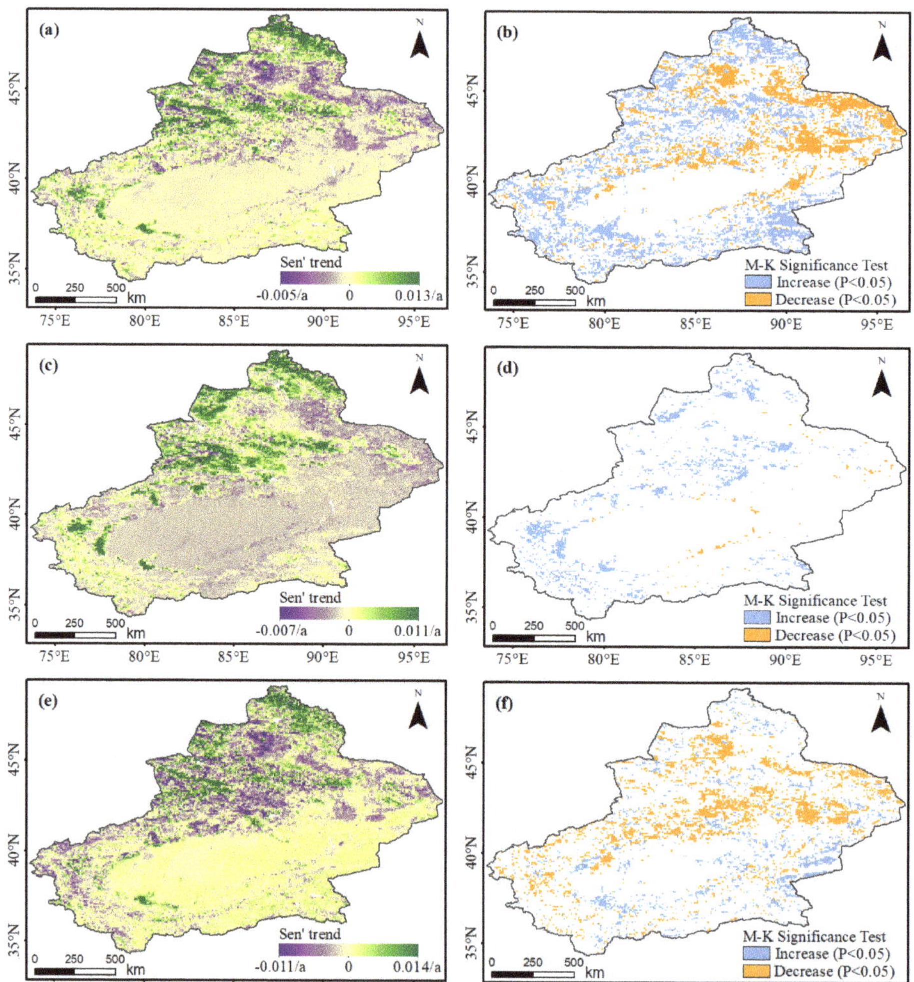

Figure 6. Vegetation NDVI trends (**left**) and M–K significance test (**right**) in spring over Xinjiang, China during 1982–2015 (**a**,**b**) and the two periods of 1982–1998 (**c**,**d**) and 1998–2012 (**e**,**f**).

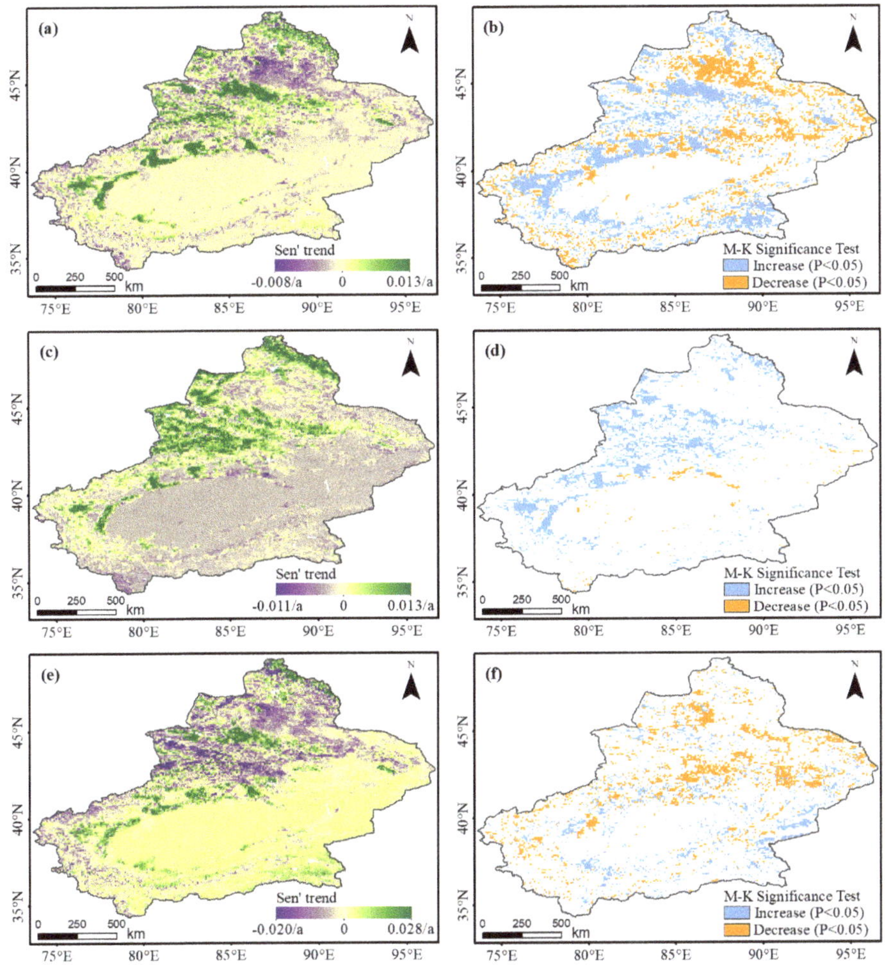

Figure 7. Vegetation NDVI trends (**left**) and M–K significance test (**right**) in autumn over Xinjiang, China during 1982–2015 (**a**,**b**) and the two periods of 1982–1998 (**c**,**d**) and 1998–2012 (**e**,**f**).

In comparison, the NDVI greening trends in autumn were larger than those in spring during 1982–1998, and almost all vegetation-covered areas in Xinjiang exhibited a greening trend in autumn (Figure 7c,d). However, the autumn NDVI showed a significant browning trend during the warming hiatus, and the NDVI browning trend in autumn was greater than that in spring. The browning NDVI area was mainly found in the Tianshan Mountains, northern Xinjiang, and the Kunlun Mountains, which are dominated by natural vegetation (Figure 7e,f). The most dominant vegetation cover type is grassland, which is mainly driven by climate factors. In addition, the autumn NDVI greening trend during 1998–2012 was generally found in the artificial vegetation area, including the northern slope of the Tianshan Mountains and the lower Tarim River Basin.

The lower Tarim River is an ecologically fragile region due to its sensitivities to climate variations and anthropogenic activities. Due to the intensified human influence, a serious decline in vegetation cover occurred in the lower Tarim River during the recent 30 years. The autumn NDVI also showed a significant browning trend in the lower Tarim River during 1982–1998 (Figure 7c,d). The Chinese government has implemented the ecological water diversion project (EWDP) since 2000 to restore the riparian vegetation in the lower Tarim River. To evaluate the ecological effects of EWDP, Figure 8 shows the time series of the vegetation area and total water conveyance volume over the lower Tarim River from 2000 to 2015 corresponding to the 15 ecological water deliveries. Until the end of 2015, the total water conveyance volume reached 51.11×10^8 m^3 (Figure 8). The vegetation processes significantly responded to the water conveyance, the NDVI of the lower Tarim River increased by 28.6%, and the vegetation area increased by 161.2% from 2000 to 2015 (Figure 8).

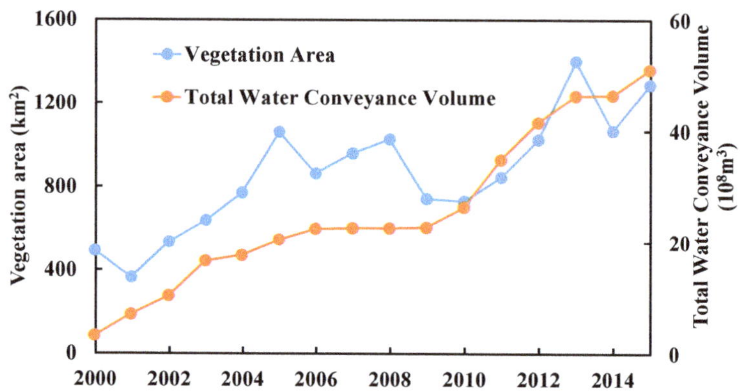

Figure 8. The time series of the vegetation area (km^2) and total water conveyance volume (10^8 m^3) over the lower Tarim River during 2000–2015.

3.3. Climatic Controls of Vegetation NDVI

To examine the effects of climatic factors on vegetation change, we calculated the correlation coefficients between seasonal climatic factors (Tair, Tmax, Tmin, Prec, RH, WS, and SD) and vegetation NDVI in Xinjiang for the 1982–2015, and warming hiatus periods. For the spring vegetation NDVI, we used spring and winter climatic factors, and spring, summer, and autumn climatic factors were used for the autumn vegetation NDVI. The influence of climatic factors on vegetation change varied greatly among the seasonal scales. Generally, the spring NDVI displayed a higher correlation with climatic factors than did the autumn NDVI (Figures 9 and 10).

For the spring vegetation NDVI during 1982–2015, the spring air temperature showed the highest correlation among the examined climatic factors, with correlation coefficients of 0.62 with Tair, 0.61 with Tmax, and 0.56 with Tmin, followed by the spring VPD (correlation coefficient of 0.54, $p < 0.01$) (Figure 9a). During the warming hiatus period, the spring VPD had the highest positive correlation with spring NDVI, followed by spring Tair and Tmax. Moreover, spring precipitation and RH were significantly negatively correlated with spring NDVI compared to any other climatic factor (Figure 9b). The spring browning NDVI area was mainly found in the alpine regions. Spring precipitation is mainly in the form of snowfall in alpine regions of Xinjiang, and snow cover is an essential component of alpine land cover [44]. Snow cover phenology has a strong control on alpine vegetation dynamics, for example, the length of snowing days and snowmelt dates have a significant impact on the NDVI variation in spring over large areas [45]. In summary, spring climatic factors (e.g., air temperature and VPD) were the main controlling factors of the spring NDVI.

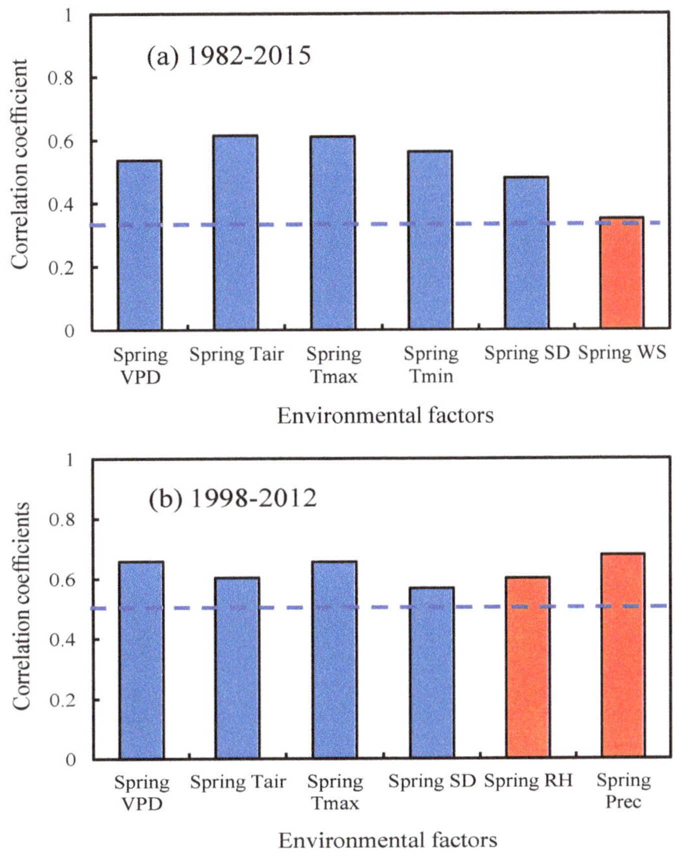

Figure 9. Correlation coefficients between spring NDVI and climatic factors. The 95% confidence limits for the correlation are ±0.35 and ±0.51 for 1982–2015 and 1998–2012, respectively. The blue background indicates a positive correlation, and red indicates a negative correlation.

Among the various factors, summer precipitation, and RH had the highest correlation with autumn vegetation NDVI during the 1982–2015 and warming hiatus periods, and antecedent and contemporaneous WS were significantly negatively correlated with autumn NDVI during 1982–2015 (Figure 10a,b). The correlation between temperature and NDVI in autumn was much weaker than that in spring. In addition, summer VPD was significantly negatively correlated with autumn NDVI for the warming hiatus period (Figure 10b). Thus, summer climatic factors (e.g., VPD, WS, RH, and precipitation) were significantly correlated with the autumn NDVI.

To illustrate the joint controls of the multiple climatic factors on NDVI change, we used PCA to determine the relationship between vegetation NDVI and climatic factors during the warming hiatus period. Spring temperature (Tair and Tmax) had the highest partial correlation with spring NDVI, while summer precipitation had the highest partial correlation with autumn NDVI. Climatic factors displayed a larger effect on autumn NDVI than spring NDVI. These results were generally like those based on univariate correlation analysis.

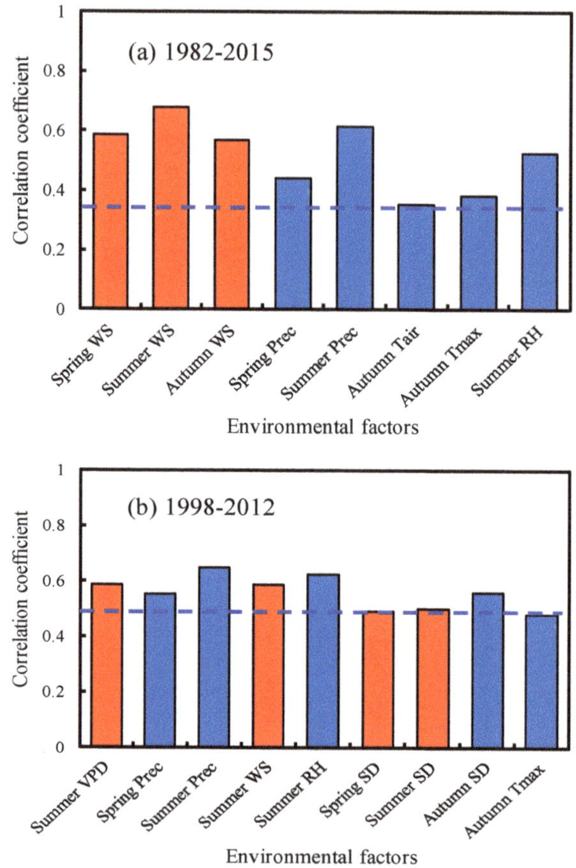

Figure 10. Correlation coefficients between autumn NDVI and climatic factors. The 95% confidence limits for the correlation are ±0.35 and ±0.51 for 1982–2015 and 1998–2012, respectively. The blue background indicates a positive correlation, and red indicates a negative correlation.

3.4. Impacts of Changes in Drought on Vegetation NDVI

To investigate the effects of dry and wet conditions on vegetation NDVI changes, we examined the correlation coefficients between vegetation detrended NDVI and the drought index (sc_PDSI). The vegetation NDVI semi-arid regions tend to respond to drought at longer time scales [46]. For this purpose, we used the sc_PDSI, which has been reasonably successful at quantifying long-term drought [47], as it uses surface air temperature and a physical water balance model, and can capture the basic effect of global warming on drought through changes in potential evapotranspiration [47]. The procedure followed to calculate the correlations between the standardized vegetation NDVI series in spring and autumn and the series of the sc_PDSI series, which are illustrated in Figures 11 and 12. For the spring NDVI, we used the spring and winter of the year before (pre-winter) sc_PDSI, and the summer and autumn sc_PDSI were used for autumn NDVI. The effects of dry and wet conditions on vegetation NDVI change varied greatly between the spring and autumn. Generally, the autumn NDVI showed a higher correlation with dry and wet conditions than the spring NDVI during 1982–2015 (Figures 11 and 12).

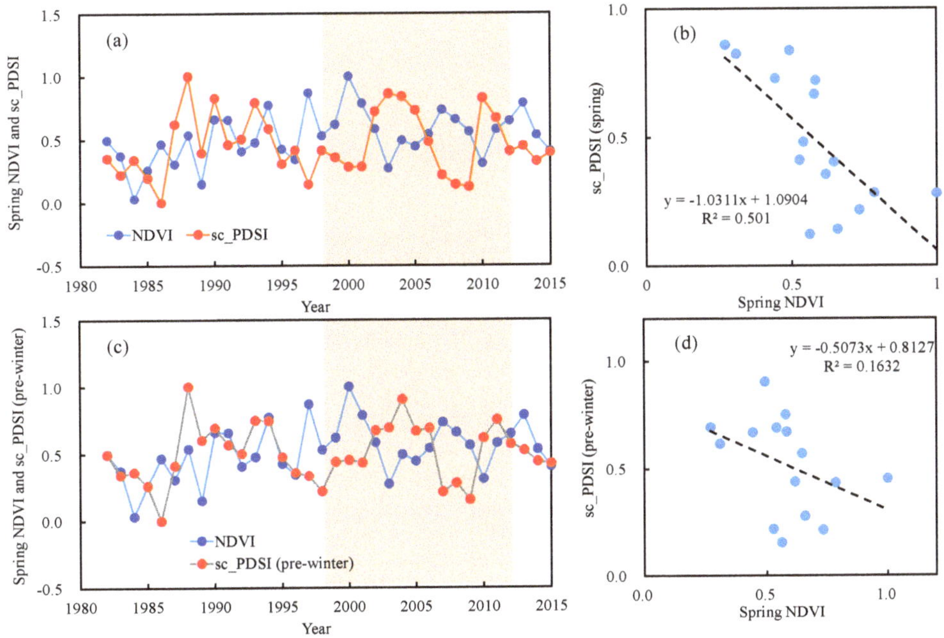

Figure 11. Relationships between spring standardized NDVI and standardized sc_PDSI in Xinjiang. The left graph is a comparison of the time series of spring NDVI and spring sc_PDSI (**a**), and winter of the year before (pre-winter) sc_PDSI (**c**) during 1982–2015. The right graph is the scatter plot between the spring NDVI and spring sc_PDSI (**b**) and between the spring NDVI and previous winter sc_PDSI (**d**) during 1998–2012. The gray shaded regions in (**a**,**c**) denote the warming hiatus (1998–2012).

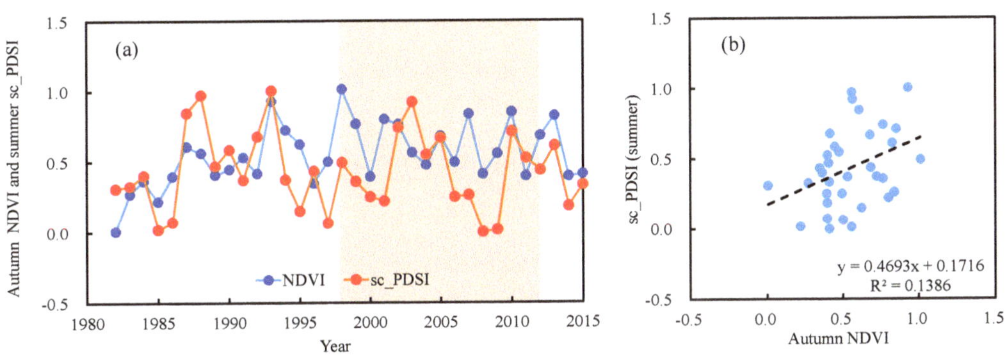

Figure 12. Relationships between autumn standardized NDVI and standardized sc_PDSI in Xinjiang. The left graph (**a**) is the time series of autumn NDVI and summer sc_PDSI and the right graph (**b**) is the scatter plot between the autumn NDVI and summer sc_PDSI during 1982–2015. The gray shaded regions in (**a**) denote the warming hiatus (1998–2012).

During 1982–2015, the spring NDVI had a slight negative correlation with spring sc_PDSI (CC = −0.16, $p > 0.05$) and an insignificant positive correlation with pre-winter sc_PDSI (CC = 0.14, $p > 0.05$) (Figure 11). However, the spring NDVI was significantly negatively correlated with spring sc_PDSI (CC = −0.71, $p < 0.01$) and slightly negatively correlated with pre-winter sc_PDSI (CC = −0.40, $p > 0.05$) during the warming hiatus (1998–2012) (Figure 11). In summary, dry and wet conditions in spring and pre-winter were not the key factors controlling spring vegetation changes. For the autumn NDVI during 1982–2015, the summer sc_PDSI showed a higher correlation (CC = 0.37, $p < 0.05$), followed by autumn sc_PDSI (CC = 0.33, $p > 0.05$) (Figure 12). However, the correlation between the autumn NDVI and summer (autumn) sc_PDSI during the warming hiatus was much weaker than that during 1982–2015. Thus, dry and wet conditions in summer were key factors controlling autumn vegetation changes.

4. Discussion

Based on satellite-derived vegetation records (GIMMS3g-NDVI), a significant greening trend was observed in our study during 1982–1998, and an insignificant browning trend during 1998–2015. Many studies based on satellite records indicated vegetation greening trends in most of the global vegetation areas before the late 1990s [22–24,48], and a recent study showed a pronounced vegetation browning trend over approximately 59% of the global vegetation surface from 1998 to 2015 [22]. Many other studies have reported vegetation greening towards browning in Xinjiang and High Mountain Asia from the late 1990s to the early 2010s [11,12,24,49,50].

Compared to previous studies, we focused on the vegetation change trend in spring and autumn during the recent warming hiatus (1998–2012), which has not been examined in previous studies. According to the definition of the recent warming hiatus, we divided the study period into two shorter periods: before 1998 and after 1998. A previous study using satellite records found that the advancing (or delaying) rate of spring (or autumn) phenology from 2002 to 2012 was faster than that in 1982–1992 [21,51]. Another study using long-term satellite-derived vegetation records and FLUXNET datasets in NH proposed that spring and autumn phenology showed no trends during the global warming hiatus [17]. Our analysis showed that the summer NDVI significantly increased; however, there was an insignificant reduction in the spring and autumn NDVI during the warming hiatus period. We also found that the autumn NDVI browning trend was greater than that in spring. The 56 site FLUXNET database indicated that more records showed significant phenology trends in autumn than in spring [17], and these studies were consistent with our results based on satellite records. We also examined the vegetation change trend in seasons that were consistent with those in seasonal temperature during 1998–2012, and the temperature in spring and autumn decreased but continuously increased in summer.

Our results using the GIMMS NDVI3g database indicated that the vegetation greening trend stalled or reversed during the warming hiatus period. This vegetation browning area is mainly dominated by natural vegetation. In addition, the GIMMS NDVI3g dataset contains uncertainties in data quality, which are caused by noise, sensor degradation, and calibration between different sensors [52].

Many studies have indicated that the spring temperature is the key controlling factor of vegetation change, and rising spring temperatures lead to advancing trends in spring phenology [17,53,54]. Previous studies have also indicated that winter temperature is negatively correlated with spring vegetation NDVI and phenology change [20,55].

Our results suggested that spring NDVI was more largely correlated with spring VPD than with spring temperature during the warming hiatus. Previous studies have shown that an increase in VPD can reduce global vegetation growth [22], and other studies have reported that an increase in VPD is a driver of drought-related forest mortality [56,57]. We also used PCA to confirm that the spring NDVI was mainly dominated by spring temperature, including mean and extreme variables. The VPD represents the difference between saturation vapor pressure (e_s) and the actual vapor pressure (e_a) at a given temperature [43]. Rising air temperatures may increase the e_s, and the e_a is constrained at the upper end by e_s so that as air temperature increases, so does the maximum amount of water vapor [58]. Thus, higher VPD is driven by rising air temperature following a nonlinear relationship [58]. Li et al. [43] also indicated that the VPD increase in Xinjiang was jointly determined by the increased saturation vapor pressure from rising temperatures and decreased actual vapor pressure. Thus, rising air temperatures resulted in increased VPD, which corresponded to the vegetation browning trend. In addition, our results indicated that spring precipitation was significantly negatively correlated with the spring NDVI. Li et al. [29] found that the reversed trends in NDVI after the late 1990s can be attributed to increased warming and reduced precipitation. Yao et al. [12] found that increased drought severity exacerbated the loss of soil moisture and led to a decreased NDVI. These results are consistent with those of our work during the warming hiatus period.

Compared to spring vegetation NDVI, climatic factors displayed a stronger control on autumn NDVI. Wang et al. [17] also proposed that autumn vegetation growth is determined by a combination of climatic factors and phenology statuses. Richardson et al. [59] reported the effects of temperature on autumn vegetation change and neglected other climatic factors [59]. In contrast, our results indicate that summer climatic factors (e.g., VPD, WS, RH, and precipitation) were significantly correlated with autumn NDVI during the warming hiatus. However, the autumn temperature was weakly correlated with the autumn NDVI. Wang et al. [17] found that temperature, VPD, shortwave radiation, precipitation, and soil moisture had strong effects on autumn vegetation among the 56 FLUXNET sites in the NH. These results emphasize the complexity of the impacts on autumn vegetation changes during the warming hiatus period.

Figure 13 shows the mechanisms of the impact of climatic variables on spring and autumn vegetation growth. Our results provide evidence for the different impacts of climatic variables on spring and autumn NDVI during the warming hiatus periods. The difference in spring and autumn temperature has played a key role in recent decades, and a larger increase in air temperature in spring than in autumn has resulted in increased VPD differences in spring and autumn [60,61]. The VPD and extreme drought stress can trigger vegetation growth retardation through hydraulic failure and carbon starvation [57]. In spring, an increased spring VPD enhances evaporative water loss, which triggers stomatal closure. Increased VPD and reduced stomatal conductance result in restricted plant carbon uptake and productivity and drive widespread vegetation browning [62]. Our study suggests the importance of summer climatic variables in controlling autumn vegetation growth in Xinjiang. The autumn NDVI had a higher correlation with dry and wet conditions. Furthermore, increased ET directly influences soil moisture and causes extreme drought. Soil moisture is a direct reservoir for vegetation, and its deficit plays a large role in autumn vegetation browning in Xinjiang. Extreme drought may decrease the amount of water that can be taken up by plant roots, increase catheter embolism, stagnate water transportation, and can drive vegetation browning.

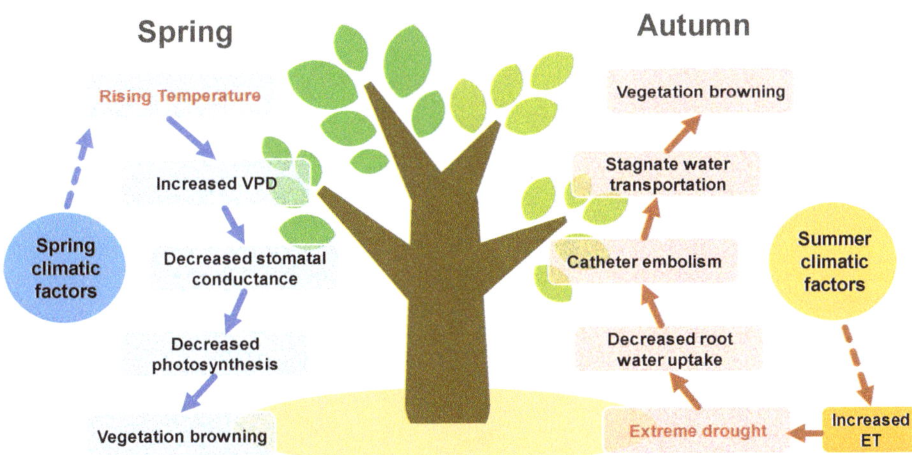

Figure 13. Schematic diagram of the impacts of climatic variables on spring and autumn vegetation growth.

The AVHRR-based GIMMS3g-NDVI is a fundamental dataset for the study of global vegetation growth and ecosystem processes. Guay et al. [63] compared the long-term NDVI trends in high northern latitudes (>50° N), between the GIMMS3g dataset and four commonly used NDVI datasets. and found that the marked differences were both in mean seasonal NDVI and long-term trends. The vegetation greening was more prominent in GIMMS 3g-NDVI than in the older dataset, especially in the tundra, needle-leaved evergreen and mixed-leaf forests [63–65], while the browning was across sizeable areas of boreal forest [66,67]. Consistent with the MODIS record, the spatial patterns of GIMMS 3g-NDVI trends showed roughly half greening and half browning across high northern latitudes [63]. In the High Mountain Asia region, the GIMMS 3g-NDVI showed a browning trend, whereas MODIS and SPOT exhibited a greening trend during 2001–2015 [68]. Thus, it was necessary to provide a comprehensive comparison with different NDVI datasets, as well as to validate these NDVI data against in situ measurements of ecosystem processes.

5. Conclusions

This study focused on the latest changes in spring and autumn vegetation growth in Xinjiang and their climate-driving mechanisms, using satellite-derived vegetation records (GIMMS3g-NDVI) as indicators. The results indicated that the vegetation NDVI showed a significant greening trend during 1982–1998 and an insignificant browning trend during 1998–2015. We focused on the vegetation change trends in spring and autumn during the recent warming hiatus (1998–2012), which had not been examined in previous studies.

During the recent warming hiatus, the summer NDVI increased significantly; however, there were insignificant reductions in the spring and autumn NDVI. The autumn NDVI browning trend in Xinjiang was greater than that in spring. The seasonal vegetation change trend was consistent with the seasonal temperature, and the temperature slowed down in spring and autumn but continuously increased in summer during the warming hiatus.

The correlation coefficients between climatic factors and vegetation NDVI showed that the climatic factors controlling vegetation change varied greatly at the seasonal scale. Spring NDVI displayed a stronger correlation with climatic factors than did the autumn NDVI. During the warming hiatus, the spring VPD had the highest positive correlation with the spring NDVI, followed by spring temperature. In addition, spring precipitation has a significantly negative correlation with the spring NDVI. Thus, spring climatic factors (e.g., air temperature and VPD) were the main controlling factors of the spring NDVI.

Compared to the spring vegetation NDVI, climatic factors displayed a stronger control over the autumn NDVI. Summer climatic factors (e.g., VPD, WS, RH, and precipitation) were significantly correlated with the autumn NDVI during the warming hiatus. However, the autumn temperature was weakly correlated with the autumn NDVI. These results emphasize the complexity of the impacts on autumn vegetation changes during the warming hiatus.

We also provided evidence for the different impacts of climatic variables on spring and autumn vegetation NDVI during the warming hiatus period. The larger increase in air temperature in spring than in autumn resulted in increased VPD differences in spring and autumn. Increased spring VPD can trigger spring vegetation browning through mechanisms of carbon starvation; however, summer drought may decrease the amount of water that can be taken up by plant roots and can drive autumn vegetation browning.

Author Contributions: Conceptualization, M.L., J.Y. and J.Z.; methodology, M.L.; formal analysis, M.L. and J.G.; writing—original draft preparation, M.L.; writing—review and editing, M.L., J.Y., J.G. and J.Z.; funding acquisition, J.Y. and J.Z. All authors have read and agreed to the published version of the manuscript.

Funding: This work was funded by the National Natural Science Foundation of China (U1903113), the Major Program of National Social Science Foundation of China (No. 17ZDA064), and the Xinjiang Tianshan Cedar Project (No. 2020XS04).

Institutional Review Board Statement: Not applicable.

Informed Consent Statement: Not applicable.

Data Availability Statement: The data presented in this study are available on request from the corresponding author or the reference website.

Acknowledgments: We thank "anonymous" reviewers for their original insights. We are also immensely grateful to the editor for their comments on the manuscript.

Conflicts of Interest: The authors declare no conflict of interest.

References

1. IPCC. Climate change 2013: The physical science basis. In *Contribution of Working Group I to the Fifth Assessment Report of the Intergovernmental Panel on Climate Change*; Stocker, T.F., Qin, D., Plattner, G.K., Tignor, M., Allen, S.K., Boschung, J., Nauels, A., Xia, Y., Bex, V., Midgley, P.M., Eds.; Cambridge University Press: Cambridge, UK; New York, NY, USA, 2013.
2. Fyfe, J.; Gillett, N.; Zwiers, F. Overestimated global warming over the past 20 years. *Nat. Clim. Chang.* **2013**, *3*, 767–769. [CrossRef]
3. Kosaka, Y.; Xie, S.P. Recent global-warming hiatus tied to equatorial Pacific surface cooling. *Nature* **2013**, *501*, 403–407. [CrossRef] [PubMed]
4. England, M.H.; McGregor, S.; Spence, P.; Meehl, G.A.; Timmermann, A.; Cai, W.; Gupta, A.S.; McPhaden, M.J.; Purich, A.; Santoso, A. Recent intensification of wind-driven circulation in the Pacific and the ongoing warming hiatus. *Nat. Clim. Chang.* **2014**, *4*, 222–227. [CrossRef]
5. Meehl, G.A.; Teng, H.; Arblaster, J.M. Climate model simulations of the observed early-2000s hiatus of global warming. *Nat. Clim. Chang.* **2014**, *4*, 898–902. [CrossRef]
6. Trenberth, K.E.; Fasullo, J.T.; Branstator, G.; Phillips, A.S. Seasonal aspects of the recent pause in surface warming. *Nat. Clim. Chang.* **2014**, *4*, 911–916. [CrossRef]
7. Dai, A.; Fyfe, J.C.; Xie, S.P.; Dai, X. Decadal modulation of global surface temperature by internal climate variability. *Nat. Clim. Chang.* **2015**, *5*, 555–559. [CrossRef]
8. Guan, X.; Huang, J.; Guo, R.; Lin, P. The role of dynamically induced variability in the recent warming trend slowdown over the Northern Hemisphere. *Sci. Rep.* **2015**, *5*, 12669. [CrossRef]
9. Guemas, V.; Doblas-Reyes, F.J.; Andreu-Burillo, I.; Asif, M. Retrospective prediction of the global warming slowdown in the past decade. *Nat. Clim. Chang.* **2013**, *3*, 649–653. [CrossRef]
10. Li, Q.; Yang, S.; Xu, W.; Wang, X.L.; Jones, P.; Parker, D.; Zhou, L.; Feng, Y.; Gao, Y. China experiencing the recent warming hiatus. *Geophys. Res. Lett.* **2015**, *42*, 889–898. [CrossRef]
11. Chen, Y.N.; Li, Z.; Fan, Y.; Wang, H.; Deng, H. Progress and prospects of climate change impacts on hydrology in the arid region of northwest China. *Environ. Res.* **2016**, *139*, 11–19. [CrossRef]
12. Yao, J.; Zhao, Y.; Chen, Y.; Yu, X.; Zhang, R. Multi-scale assessments of droughts: A case study in Xinjiang, China. *Sci. Total Environ.* **2018**, *630*, 444–452. [CrossRef]
13. Walther, G.R.; Post, E.; Convey, P.; Menzel, A.; Parmesan, C.; Beebee, T.J.; Fromentin, J.M.; Guldberg, O.H.; Bairlein, F. Ecological responses to recent climate change. *Nature* **2002**, *416*, 389–395. [CrossRef]
14. Osborne, C.; Chuine, I.; Viner, D.; Woodward, F. Olive phenology as a sensitive indicator of future climatic warming in the Mediterranean. *Plant Cell Environ.* **2000**, *23*, 701–710. [CrossRef]
15. Keenan, T.F.; Gray, J.; Friedl, M.A.; Toomey, M.; Bohrer, G.; Hollinger, D.Y.; Munger, J.W.; O'Keefe, J.; Schmid, H.P.; Wing, I.S.; et al. Net carbon uptake has increased through warming-induced changes in temperate forest phenology. *Nat. Clim. Chang.* **2014**, *4*, 598–604. [CrossRef]
16. Fu, Y.H.; Zhao, H.; Piao, S.; Peaucelle, M.; Peng, S.; Zhou, G.; Ciais, P.; Huang, M.; Menzel, A.; Peñuelas, J.; et al. Declining global warming effects on the phenology of spring leaf unfolding. *Nature* **2015**, *526*, 104–107. [CrossRef]
17. Wang, X.; Xiao, J.; Li, X.; Cheng, G.; Ma, M.; Zhu, G.; Altaf Arain, M.; Andrew Black, T.; Rachhpal, S.J. No trends in spring and autumn phenology during the global warming hiatus. *Nat. Commun.* **2019**, *10*, 2389. [CrossRef]
18. Kaufmann, R.K.; Zhou, L.; Tucker, C.J.; Slayback, D.; Shabanov, N.V.; Myneni, R.B. Reply to Comment on 'Variations in northern vegetation activity inferred from satellite data of vegetation index during 1981–1999 by JR Ahlbeck. *J. Geophys. Res.* **2002**, *107*, 10-1029. [CrossRef]
19. Menzel, A.; Sparks, T.H.; Estrella, N.; Koch, E.; Aasa, A.; Ahas, R.; Alm-Kübler, K.; Bissolli, P.; Braslavská, O.G.; Briede, A.; et al. European phenological response to climate change matches the warming pattern. *Glob. Chang. Biol.* **2006**, *12*, 1969–1976. [CrossRef]
20. Körner, C.; Basler, D. Phenology under global warming. *Science* **2010**, *327*, 1461–1462. [CrossRef]
21. Wang, S.; Yang, B.; Yang, Q.; Lu, L.; Wang, X.; Peng, Y. Temporal trends and spatial variability of vegetation phenology over the Northern Hemisphere during 1982–2012. *PLoS ONE* **2016**, *11*, e0157134. [CrossRef]
22. Yuan, W.; Zheng, Y.; Piao, S.; Ciais, P.; Lombardozzi, D.; Wang, Y.; Ryu, Y.; Chen, G.; Dong, W.; Hu, Z.; et al. Increased atmospheric vapor pressure deficit reduces global vegetation growth. *Sci. Adv.* **2019**, *5*, eaax1396. [CrossRef] [PubMed]
23. Piao, S.; Yin, G.; Tan, J.; Cheng, L.; Huang, M.; Li, Y.; Liu, R.; Mao, J.; Myneni, R.B.; Peng, S.; et al. Detection and attribution of vegetation greening trend in China over the last 30 years. *Glob. Chang. Biol.* **2015**, *21*, 1601–1609. [CrossRef] [PubMed]

24. Liu, Y.; Li, Z.; Chen, Y.N. Continuous warming shift greening towards browning in the southeast and northwest high mountain Asia. *Sci. Rep.* **2021**, *11*, 17920. [CrossRef] [PubMed]
25. Donohue, R.J.; Roderick, M.L.; McVicar, T.R.; Farquhar, G.D. Impact of CO_2 fertilization on maximum foliage cover across the globe's warm, arid environments. *Geophys. Res. Lett.* **2013**, *40*, 3031–3035. [CrossRef]
26. Fensholt, R.; Langanke, T.; Rasmussen, K.; Reenberg, A.; Prince, S.D.; Tucker, C.; Scholes, R.J.; Le, Q.B.; Bondeau, A.; Eastman, R.; et al. Greenness in semi- arid areas across the globe 1981–2007—An Earth Observing Satellite based analysis of trends and drivers. *Remote Sens. Environ.* **2012**, *121*, 144–158. [CrossRef]
27. Andela, N.; Liu, Y.Y.; van Dijk, A.I.J.M.; de Jeu, R.A.M.; McVicar, T.R. Global changes in dryland vegetation dynamics (1988–2008) assessed by satellite remote sensing: Comparing a new passive microwave vegetation density record with reflective greenness data. *Biogeosciences* **2013**, *10*, 6657–6676. [CrossRef]
28. Piao, S.; Wang, X.; Park, T.; Chen, C.; Lian, X.U.; He, Y.; Bjerke, J.W.; Chen, A.; Ciais, P.; Tømmervik, H.; et al. Characteristics, drivers and feedbacks of global greening. *Nat. Rev. Earth Environ.* **2020**, *1*, 14–27. [CrossRef]
29. Li, Z.; Chen, Y.; Li, W.; Deng, H.; Fang, G. Potential impacts of climate change on vegetation dynamics in Central Asia. *J. Geophys. Res. Atmos.* **2015**, *120*, 12345–112356. [CrossRef]
30. Yao, J.; Chen, Y.; Chen, J.; Zhao, Y.; Tuoliewubieke, D.; Li, J.; Yang, L.M.; Mao, W. Intensification of extreme precipitation in arid Central Asia. *J. Hydrol.* **2021**, *598*, 125760. [CrossRef]
31. Yao, J.Q.; Chen, Y.N.; Zhao, Y.; Mao, W.; Xu, X.; Liu, Y.; Yang, Q. Response of vegetation NDVI to climatic extremes in the arid region of Central Asia: A case study in Xinjiang, China. *Theor. Appl. Climatol.* **2018**, *131*, 1503–1515. [CrossRef]
32. Yao, J.; Chen, Y.; Zhao, Y.; Guan, X.; Mao, W.; Yang, L. Climatic and associated atmospheric water cycle changes over the Xinjiang, China. *J. Hydrol.* **2020**, *585*, 124823. [CrossRef]
33. Zhu, Y.N.; Cao, L.J.; Tang, G.L.; Zhou, Z.J. Homogenization of Surface Relative Humidity over China. *Adv. Clim. Chang. Res.* **2015**, *11*, 379–386. [CrossRef]
34. Li, Z.; Yan, Z.W.; Zhu, Y.N.; Freychet, N.; Tett, S. Homogenized daily relative humidity series in China during 1960–2017. *Adv. Atmos. Sci.* **2020**, *37*, 318–327. [CrossRef]
35. Barichivich, J.; Osborn, T.J.; Harris, I.; van der Schrier, G.; Jones, P.D. Drought [in "State of the Climate in 2018"]. *Bull. Am. Meteorol. Soc.* **2019**, *100*, S1–S306. [CrossRef]
36. Van der Schrier, G.; Barichivich, J.; Briffa, K.R.; Jones, P.D. A scPDSI-based global data set of dry and wet spells for 1901–2009. *J. Geophys. Res. Atmos.* **2013**, *118*, 4025–4048. [CrossRef]
37. Barichivich, J.; Osborn, T.J.; Harris, I.; van der Schrier, G.; Jones, P.D. Drought [in "State of the Climate in 2019"]. *Bull. Am. Meteorol. Soc.* **2020**, *101*, S1–S429. [CrossRef]
38. Pinzon, J.E.; Tucker, C.J. A non-stationary 1981–2012 AVHRR NDVI 3g time series. *Remote Sens.* **2014**, *6*, 6929–6960. [CrossRef]
39. Allen, R.G.; Pereira, L.S.; Raes, D.; Smith, M. *Crop Evapotranspiration—Guidelines for Computing Crop Water Requirements-FAO Irrigation and Drainage Paper 56. 300*; FAO: Rome, Italy, 1998; p. 6541.
40. Mann, H.B. Non-Parametric Test Against Trend. *Econometrica* **1945**, *13*, 245–259. [CrossRef]
41. Kendall, M.G. *Rank Correlation Methods*, 4th ed.; Charles Grifin: London, UK, 1975.
42. Yadolah, D. *The Concise Encyclopedia of Statistics*; Springer: New York, NY, USA, 2008.
43. Li, M.; Yao, J.Q.; Guan, J.Y.; Zheng, J.H. Observed changes in vapor pressure deficit suggest a systematic drying of the atmosphere in Xinjiang of China. *Atmos. Res.* **2020**, *248*, 105199. [CrossRef]
44. Yao, J.; Chen, Y.; Guan, X.; Zhao, Y.; Chen, J.; Mao, W. Recent climate and hydrological changes in a mountain—Basin system in Xinjiang, China. *Earth-Sci. Rev.* **2022**, *226*, 103957. [CrossRef]
45. Wang, X.; Wu, C.; Peng, D.; Gonsamo, A.; Liu, Z. Snow cover phenology affects alpine vegetation growth dynamics on the Tibetan Plateau: Satellite observed evidence, impacts of different biomes, and climate drivers. *Agric. For. Meteorol.* **2018**, *256*, 61–74. [CrossRef]
46. Vicente-Serrano, S.M.; Gouveia, C.; Camarero, J.J.; Beguería, S.; Trigo, R.; López-Moreno, J.; Azorín-Molina, C.; Pasho, E.; Lorenzo-Lacruz, J.; Revuelto, J.; et al. Response of vegetation to drought time-scales across global land biomes. *Proc. Natl. Acad. Sci. USA* **2013**, *110*, 52–57. [CrossRef]
47. Dai, A. Characteristics and trends in various forms of the Palmer Drought Severity Index (PDSI) during 1900–2008. *J. Geophys. Res.* **2011**, *116*, D12115. [CrossRef]
48. Li, Y.; Chen, Y.; Sun, F.; Li, Z. Recent vegetation browning and its drivers on Tianshan Mountain, Central Asia. *Ecol. Indic.* **2021**, *129*, 107912. [CrossRef]
49. Chen, Y.; Zhang, X.; Fang, G.; Li, Z.; Wang, F.; Qin, J.; Sun, F. Potential risks and challenges of climate change in the arid region of northwestern China. *Reg. Sustain.* **2021**, *1*, 20–30. [CrossRef]
50. Fang, S.; Yan, J.; Che, M.; Zhu, Y.; Liu, Z.; Pei, H.; Lin, X. Climate change and the ecological responses in Xinjiang, China: Model simulations and data analyses. *Quat. Int.* **2013**, *311*, 108–116. [CrossRef]
51. Zeng, H.; Jia, G.; Epstein, H. Recent changes in phenology over the northern high latitudes detected from multi-satellite data. *Environ. Res. Lett.* **2011**, *6*, 045508. [CrossRef]
52. Jiang, C.; Ryu, Y.; Fang, H.; Myneni, R.; Claverie, M.; Zhu, Z. Inconsistencies of interannual variability and trends in long-term satellite leaf area index products. *Glob. Chang. Biol.* **2017**, *23*, 4133–4146. [CrossRef] [PubMed]

53. Millerrushing, A.J.; Primack, R.B. Global warming and flowering times in Thoreau's Concord: A community perspective. *Ecology* **2008**, *89*, 332–341. [CrossRef] [PubMed]
54. Vitasse, Y.; Delzon, S.; Dufrêne, E.; Pontailler, J.Y.; Louvet, J.M.; Kremer, A.; Michalet, R. Leaf phenology sensitivity to temperature in European trees: Do within-species populations exhibit similar responses? *Agric. For. Meteorol.* **2009**, *149*, 735–744. [CrossRef]
55. Richardson, A.D.; Bailey, A.; Denny, E.; Martin, C.; O'Keefe, J. Phenology of a northern hardwood forest canopy. *Glob. Chang. Biol.* **2006**, *12*, 1174–1188. [CrossRef]
56. McDowell, N.; Pockman, W.T.; Allen, C.D.; Breshears, D.D.; Cobb, N.; Kolb, T.; Plaut, J.; Sperry, J.; West, A.; Williams, D.G.; et al. Mechanisms of plant survival and mortality during drought: Why do some plants survive while others succumb to drought? *New Phytol.* **2008**, *178*, 719–739. [CrossRef]
57. Allen, C.D.; Macalady, A.K.; Chenchouni, H.; Bachelet, D.; McDowell, N.; Vennetier, M.; Kitzberger, T.; Rigling, A.; Breshears, D.D.; Hogg, E.H.T.; et al. A global overview of drought and heat-induced tree mortality reveals emerging climate change risks for forests. *For. Ecol. Manag.* **2010**, *259*, 660–684. [CrossRef]
58. Grossiord, C.; Buckley, T.N.; Cernusak, L.A.; Novick, K.A.; Mcdowell, N.G. Plant responses to rising vapor pressure deficit. *New Phytol.* **2020**, *226*, 1550–1556. [CrossRef]
59. Richardson, A.D.; Keenan, T.F.; Migliavacca, M.; Ryu, Y.; Sonnentag, O.; Toomey, M. Climate change, phenology, and phenological control of vegetation feedbacks to the climate system. *Agric. For. Meteorol.* **2013**, *169*, 156–173. [CrossRef]
60. Xia, J.; Chen, J.; Piao, S.; Ciais, P.; Luo, Y.; Wan, S. Terrestrial carbon cycle affected by non-uniform climate warming. *Nat. Geosci.* **2014**, *7*, 173–180. [CrossRef]
61. Wang, Y.; Xu, W.; Yuan, W.; Chen, X.; Zhang, B.; Fan, L.; He, B.; Hu, Z.; Liu, S.; Liu, W.; et al. Higher plant photosynthetic capability in autumn responding to low atmospheric vapor pressure deficit. *Innov.* **2021**, *2*, 100163. [CrossRef]
62. McDowell, N.G.; Allen, C.D. Darcy's law predicts widespread forest mortality under climate warming. *Nat. Clim. Chang.* **2015**, *5*, 669–672. [CrossRef]
63. Guay, C.K.; Beck, P.S.; Berner, L.; Goetz, C.J.; Baccini, A.; Buermann, W. Vegetation productivity patterns at high northern latitudes: A multi-sensor satellite data assessment. *Glob. Chang. Biol.* **2014**, *20*, 3147–3158. [CrossRef]
64. Forbes, B.C.; Fauria, M.M.; Zetterberg, P. Russian Arctic warming and 'greening' are closely tracked by tundra shrub willows. *Glob. Chang. Biol* **2010**, *16*, 1542–1554. [CrossRef]
65. Verbyla, D. The greening and browning of Alaska based on 1982–2003 satellite data. *Glob. Ecol. Biogeogr.* **2008**, *17*, 547–555. [CrossRef]
66. Beck, P.S.A.; Goetz, S.J. Satellite observations of high northern latitude vegetation productivity changes between 1982 and 2008: Ecological variability and regional differences. *Environ. Res. Lett.* **2011**, *7*, 029501. [CrossRef]
67. Bunn, A.G.; Goetz, S.J. Trends in satellite-observed circumpolar photosynthetic activity from 1982 to 2003: The influence of seasonality, cover type, and vegetation density. *Earth Interact.* **2006**, *10*, 1–19. [CrossRef]
68. Liu, Y.; Li, Z.; Chen, Y.; Li, Y.; Li, H.; Xia, Q.Q.; Kayumba, P.M. Evaluation of consistency among three NDVI products applied to High Mountain Asia in 2000–2015. *Remote Sens. Environ.* **2022**, *269*, 112821. [CrossRef]

remote sensing

Article

Spatial-Temporal Variation in Paddy Evapotranspiration in Subtropical Climate Regions Based on the SEBAL Model: A Case Study of the Ganfu Plain Irrigation System, Southern China

Guangfei Wei [1], Jingjing Cao [2], Hua Xie [1], Hengwang Xie [3], Yang Yang [4], Conglin Wu [4], Yuanlai Cui [1] and Yufeng Luo [1,*]

1 State Key Laboratory of Water Resources and Hydropower Engineering Science, Wuhan University, Wuhan 430072, China; guangfei.wei@whu.edu.cn (G.W.); xiehua@whu.edu.cn (H.X.); ylcui@whu.edu.cn (Y.C.)
2 Yellow River Engineering Consulting Co., Ltd., Zhengzhou 450003, China; caojingjing@whu.edu.cn
3 Jiangxi Center Station of Irrigation Experiment, Nanchang 330201, China; xhw2208@163.com
4 Changjiang Institute of Survey, Planning, Design and Research, Wuhan 430010, China; yangyang5@cjwsjy.com.cn (Y.Y.); wuconglin@cjwsjy.com.cn (C.W.)
* Correspondence: yfluo@whu.edu.cn

Abstract: The surface energy balance algorithm for land (SEBAL) is a commonly used method for estimating evapotranspiration (ET) at a regional scale; however, the cloudy and rainy characteristics of subtropical monsoon regions pose a greater challenge for estimating paddy field ET based on remote sensing technology. To this end, a typical subtropical climate region in southern China (Ganfu Plain irrigation system) was selected as the study area. Subsequently, we evaluated the applicability of the SEBAL model for estimating the ET of paddy fields at the daily scale; derived the interannual variation (2000–2017) characteristics of early, middle, and late rice ET; and finally analyzed the spatial distribution patterns of rice in different hydrological years. The results demonstrated that: (1) the SEBAL model estimated ET accurately on a daily scale, with R^2, NSE, and RMSE values of 0.85, 0.81, and 0.84 mm/day, respectively; (2) the ET of paddy fields in the irrigated area was higher in July and August and the interannual trend of ET of early rice was not obvious, with a declining trend observed in middle rice and late rice from 2000 to 2009, which was followed by an increasing trend from 2009 to 2017; and (3) variations in the spatial distribution of ET were significant for early and late rice at different precipitation levels and less obvious for middle rice in wet years but significant in dry years. Overall, this study verified the applicability of the SEBAL model for estimating ET in paddy fields in subtropical regions and provided a basis and reference for the rational allocation of water resources at a regional scale.

Keywords: evapotranspiration; paddy rice; SEBAL; interannual variation; spatial distribution; Ganfu Plain irrigation system

Citation: Wei, G.; Cao, J.; Xie, H.; Xie, H.; Yang, Y.; Wu, C.; Cui, Y.; Luo, Y. Spatial-Temporal Variation in Paddy Evapotranspiration in Subtropical Climate Regions Based on the SEBAL Model: A Case Study of the Ganfu Plain Irrigation System, Southern China. *Remote Sens.* **2022**, *14*, 1201. https://doi.org/10.3390/rs14051201

Academic Editors: Jingzhe Wang, Zhongwen Hu, Yangyi Wu and Jie Zhang

Received: 26 January 2022
Accepted: 25 February 2022
Published: 28 February 2022

1. Introduction

Paddy rice is a dominant staple food that feeds over 50% of the population worldwide. However, paddy rice needs a large amount of water to maintain high production. Statistics have shown that paddy field irrigation accounts for 80% of agricultural water consumption in southern China [1]. Evapotranspiration (ET) represents the sum of vegetation transpiration and soil evaporation, which is the major component of farmland water consumption, and it includes the amount of evaporation on the ground surface and plant surface and the amount of transpiration that occurs at the plant surface and inside the plant. ET not only plays a huge role in hydrological cycles and energy balance systems, but also represents a crucial link between ecological and hydrological processes; furthermore, it is a significant parameter for calculating the regional water balance and a convenient means to retrieve water amounts applied for irrigation over anthropized regions [2–4]. Water consumption dominated by ET is the final form of water resource consumption, which often approximates the regional water resource consumption [5]. With economic and social development and population growth, the problem of water resource shortages is becoming increasingly prominent. The unbalanced spatial and temporal distribution of water resources is one of the challenges of water resource management. Therefore, understanding the spatial and temporal variability in ET is of great significance for rationally allocating regional water resources and scheduling irrigation [6].

With the continuous development of remote sensing (RS) technology in recent decades [7–9], satellite sensors with multiple spatial, temporal, and spectral resolutions (e.g., Land Remote Sensing Satellite (Landsat), Moderate Resolution Imaging Spectroradiometer (MODIS), Advanced Very-High-Resolution Radiometer (AVHRR), and Sentinel) have become ideal tools for estimating ET at a regional scale [10–13]. Various models have been reported for estimating ET with RS datasets [14–16]. According to different theoretical bases, RS-based ET models can be roughly classified as: (i) empirical statistical methods, which have the advantages of requiring less data and being easy to use; however, they have significant regional limitations because their statistical parameters need to be recalibrated when used in different regions [17,18]; (ii) feature space methods, which refers to the method of obtaining feature points and interpolating the ET of the feature points to derive the ET of each image element based on each surface parameter obtained from remote sensing images—such as surface albedo, vegetation index, surface temperature, etc.—presenting a certain shape on the scatter plot, and using that shape to obtain the ET of the feature points [18,19]; (iii) residual surface energy balance methods, which calculate the net radiation, soil heat flux, and sensible heat flux of the surface energy balance equation and finally calculate the latent heat flux as the residual term, which can be divided into single-source models and dual-source models according to whether soil evaporation and vegetation transpiration are distinguished. Studies have shown that vegetation cover can affect the water cycle and accurate estimation of ET [20,21]. Since the dual-source model considers the parameters of vegetation and soil separately, it can describe the surface fluxes of nonuniform subsurfaces in greater detail; however, it is difficult to obtain the corresponding parameters for the model [22]. The uncertainty and errors of the dual-source model will increase, and the estimation results may not be more accurate than those of the single-source model [23]. In contrast, the single-source model simplifies the treatment of the subsurface, which makes the model parameters easier to obtain and more flexible to apply.

As a typical single-source model, the surface energy balance algorithm for land (SEBAL) based on the surface energy balance has been applied to estimate ET in different regions with multiple data sources [24–30], since it does not excessively rely on ground-assisted data. The accuracy of ET estimated by the SEBAL model can reach 85% for the daily ET, 95% for the whole-season ET at a field scale, and 96% for the annual ET in large watersheds [31]. The SEBAL-based ET model has been gradually applied to water use and water management in irrigation districts in recent years. The SEBAL model has been used to estimate ET across various land uses in Nansi Lake Wetland of China using Landsat-7 ETM+ images, and the results showed that the SEBAL model can calculate the spatial distribution of ET despite little ground-based weather data over large areas [32]. Gao et al. (2020) [33] estimated the actual ET of the Loess Plateau in China from 1990 to 2015 using the MODIS and AVHRR datasets based on the SEBAL model, and the R^2 values were above 0.87 in 161 subbasins, which indicated the simulation results are reliable and effective; meanwhile, they obtained an interesting finding that the SEBAL model overestimates regional ET, the main reason mainly being due to the overestimation of regional latent heat or underestimation of sensible heat. Ding (2020) [34] compared the performance of the SEBAL model and Surface Energy Balance System model (SEBS, a dual-source model) in estimating regional ET, and results showed that the SEBAL model is suitable for estimating

ET in plain areas, while the SEBS model is suitable in composite terrain. The above studies show that the SEBAL model has been used for estimating ET at field scale to regional scale with encouraging results, but the application of the SEBAL model to rice-dominated irrigation areas has not been sufficiently investigated.

The rice-planting areas are mainly distributed in the monsoon regions of East Asia, Southeast Asia, and South Asia, as well as the tropical rainforests of Southeast Asia. The SEBAL model has been applied to several regions in China: the northeast Sanjiang River Basin [35,36], the North China Plain [37], the Loess Plateau [33,38], the agro-pastoral ecotone of Northwestern China [39], the Taklimakan Desert [40], and the Yellow River Delta [21]. Additionally, the SEBAL model has been widely applied over a multitude of different climatic conditions, including subtropical areas [41–43], but the cloudy and rainy weather in the subtropical climate zone—which is represented by the southern region of China, where mixed double-season early and late rice and single-season middle rice are cultivated—presents a greater challenge for accurate estimations of ET using the SEBAL model.

Therefore, the Ganfu Plain irrigation system (GFPIS) located in southern China, which is a subtropical monsoon climate region, was selected as the study area. The objectives of this study are to: (1) assess the applicability of the SEBAL model for estimating ET in rice-dominated irrigation areas; (2) analyze the daily variation in paddy rice ET over the growing season; and (3) analyze the interannual variation (2000–2017) characteristics of early, middle, and late rice ET and the spatial distribution pattern in different hydrological years. Hence, this study estimated ET of the rice-dominated irrigation areas in southern China using Landsat and the SEBAL model, and derived the spatial-temporal variation patterns of ET in the region, which will provide a reference for regional farmland water consumption and a decision-making basis for the rational and efficient allocation of regional water resources.

2. Material and Methods

2.1. Study Area

The GFPIS (8°10′–28°48′ N, 115°48′–116°17′ E) is the largest diversion irrigated area south of the Yangtze River in China (Figure 1) and one of the main grain-producing areas in Jiangxi Province, with an effective irrigated area of approximately 67,000 Hm^2. The main crop planted in this region is rice, which can be divided into double-season early and late rice and single-season middle rice. The growing season is shown in Figure 2 [44]. Among the rice crops, double-season early-late rice may only be planted with early-late rice, which is rotated with cash crops, such as vegetables, melons, and fruits.

The distribution of water conservancy projects in the GFPIS is shown in Figure 1d, with a total of 10 major trunk canals. With the expansion and economic development of Nanchang city in recent years, the water consumption of urban residents and industry in irrigated areas has been growing and the demand for water resources from various water-using units has been increasing; therefore, the contradiction between water supply and consumption in irrigated areas will be more prominent.

The GFPIS is located in a humid subtropical climate zone, with an average annual temperature of 17–18 °C, an average annual wind speed of 3.4 m/s, an average annual sunshine of 1603.4 h, a frost-free period of 260–280 days, an average annual precipitation of approximately 1662.5 mm, and an average annual land evaporation of approximately 800 mm. The irrigated area has four distinct seasons, with low temperatures and rain in spring, greater precipitation and flooding in late spring and early summer, high temperatures and little rain in summer, mild temperatures but little rain in autumn, and cold temperatures in winter.

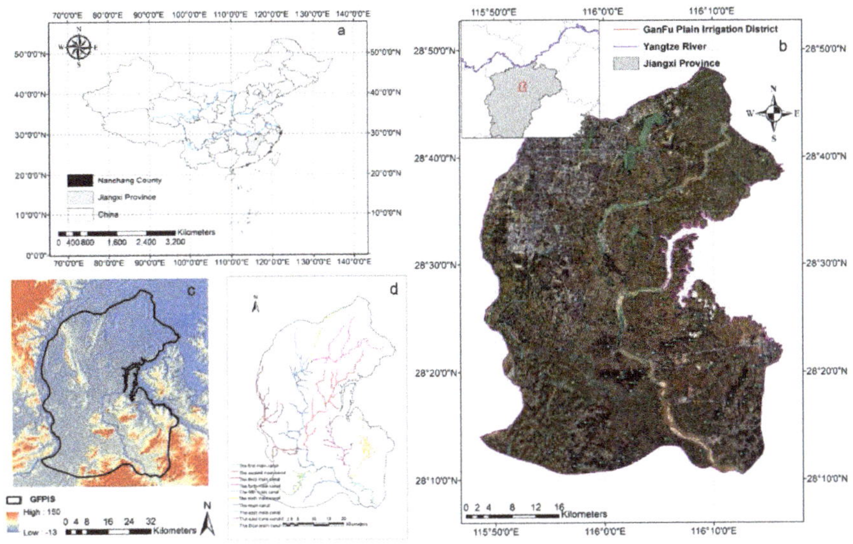

Figure 1. Overview of study area. (**a**) Geographical location of Jiangxi Province; (**b**) Location of the study area; (**c**) DEM of the study area; and (**d**) Distribution of the water conservancy project in the GFPIS.

Month	March		April			May			June			July			August			September			October	
Ten-day	M	L	E	M	L	E	M	L	E	M	L	E	M	L	E	M	L	E	M	L	E	M
Early rice		1	2	2	3	4	4	5	5	6	6/7	8	9									
Middle rice									1	2	2	3	4	4	4	5	5	6	6/7	7/8	8	9
Late rice											1	2	2	3	4	4	5	5	6	6/7	8	9

Figure 2. Common growth stages of early, late, and middle rice in the GFPIS: 1–sowing; 2–seeding; 3–transplanting; 4–tillering; 5–booting; 6–heading; 7–milk; 8–mature; and 9–harvest.

2.2. Data Used

The input data (Table 1) in the SEBAL model include satellite imagery, meteorological data, and digital elevation data (DEM).

Table 1. Information of datasets used.

Datasets	Source	Time Period
Satellite imagery	Landsat, U.S. Geological Survey	From 2000 to 2017
Meteorological data	China Meteorological Data Sharing Service Network (http://data.cma.cn)	From 2000 to 2017
DEM	SRTM data, NASA, and the National Geospatial-Intelligence Agency (https://earthexplorer.usgs.gov/)	/

The Landsat images have the same spatial resolution and continuous temporal coverage. All of the Landsat images are level 1 T data, with terrain corrections referencing the World Reference System-2 and intercalibrated across the various Landsat sensors; thus, these images are suitable for time series processing analysis. According to the rice growing season in the GFPIS, satellite imagery acquisition occurred between 20 April and 31 October from 2000 to 2017. All cloud-free Landsat 5 TM, 7 ETM+, and 8 OLI images (Path 121 and Row 40) from the U.S. Geological Survey (USGS) Earth Explorer data portal (http://earthexplorer.usgs.gov/, accessed on 10 January 2022) were downloaded. A total of 63 images were used, and Figure 3 shows the temporal distribution of the available Landsat images.

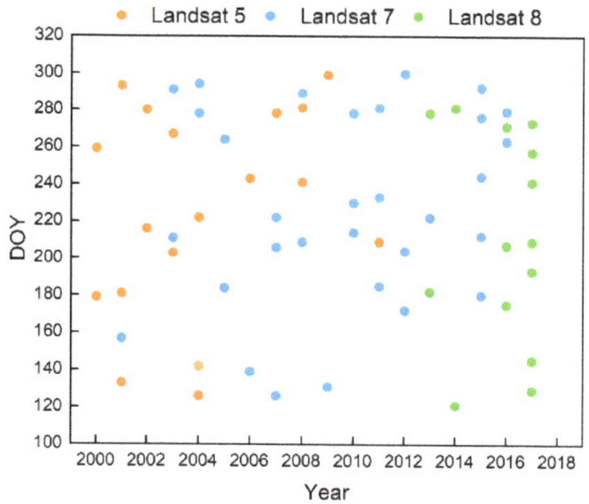

Figure 3. Temporal distribution of the available Landsat images for the rice growing season from 2000 to 2017.

The meteorological data in this study are from the China Meteorological Data Sharing Service Network (http://data.cma.cn, accessed on 10 January 2022), and the Daily Climatic Dataset (V3.0) of the China International Ground Exchange station was selected as the basic meteorological data. Meteorological data were collected from Nanchang Station from 2000 to 2017, and they included the maximum temperature, minimum temperature, average temperature, average wind speed, relative humidity, and sunshine duration.

DEM data are derived from SRTM data, which are jointly measured by NASA and the National Geospatial-Intelligence Agency (https://earthexplorer.usgs.gov/, accessed on 10 January 2022). The spatial resolutions of SRTM data are 30 and 90 m. SRTM data from radar images have the highest resolution and represent the most widely used terrain data thus far; furthermore, global SRTM data have a unified coordinate system. In this study, DEM data with 30 m resolution were mosaicked together with two sheets and then cropped to a rectangle containing the range size using the GFPIS vector file, as shown in Figure 1c.

2.3. Selection of Typical Hydrological Years

Precipitation and irrigation are the main sources of water in paddy fields, and ET is the main consumer of water in paddy fields; thus, it is also the basis of irrigation management. Therefore, it is of great guiding significance for water resource management to investigate the distribution characteristics of ET under different precipitation conditions in irrigated areas.

First, the total precipitation during the growing seasons of early, middle, and late rice from 2000 to 2017 was statistically analyzed, and the results are shown in Figure 4. Overall, the total precipitation during the growing season of early rice was the highest and ranged from 351.8 mm to 1025.9 mm, with the minimum value occurring in 2009 and the maximum value in 2010. The second highest was the total precipitation during the growing season of middle rice, with a minimum value of 216.2 mm (2016) and a maximum value of 563.8 mm (2014). The total precipitation of late rice was the lowest and ranged from 58.9 mm to 401.2 mm. These results indicate that precipitation is abundant in the early rice growing season and less abundant in the late rice growing season, and the total precipitation in the middle rice growing season is between that of early rice and late rice.

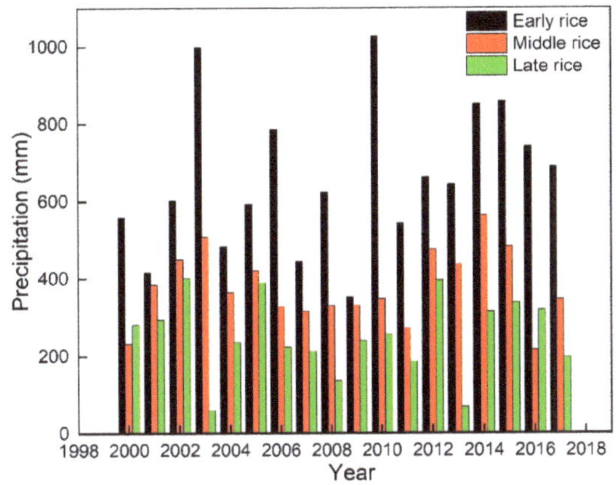

Figure 4. Interannual variation in precipitation during the early, middle, and late rice growing seasons from 2000 to 2017.

In this study, the precipitation data for the GFPIS from 1952 to 2017 were sorted by frequency, the Pearson type III curve was used to fit the line, and the precipitation values corresponding to each frequency case were obtained by fitting the line. Generally, the precipitation values corresponding to P = 25%, 50%, and 75% can be taken as the design values for wet, normal, and dry years, respectively, and then the real years equal to or close to the design values can be selected as the corresponding wet, normal, and dry representative years in the existing actual measurement series. The wet years selected in this paper were 2002 and 2014; the normal years were 2004, 2008, and 2010; and the dry years were 2007, 2008, and 2016.

Consistent with the previous conclusion, the total precipitation during the growing season of rice in different hydrological years was ranked as early rice, middle rice, and late rice. The precipitation during the growing season of early rice was higher than 400 mm in different hydrological years; the precipitation during the growing season of middle rice was as high as 564 mm in wet years, 365 mm in normal years, and only 216 mm in dry years; and the precipitation during the growing season of late rice was 401 mm in wet years, 255 mm in normal years, and as low as 136 mm in dry years.

2.4. SEBAL Algorithm Description

Since the SEBAL model was proposed by Bastiaanssen, it has been successfully applied to a variety of ecosystems through continuous improvement and development [45]. The SEBAL model calculates the latent heat flux by estimating the net radiation, soil heat flux, and sensible heat flux to obtain the instantaneous ET. The theoretical basis of the SEBAL model is the surface energy balance formula, and the calculation formula is

$$\lambda \mathrm{ET} = \mathrm{R_n} - \mathrm{G} - \mathrm{H} \tag{1}$$

where λET is the latent heat flux (W/m^2), R_n is the net radiation flux (W/m^2), G is the soil heat flux (W/m^2), and H is the sensible heat flux (W/m^2).

R_n is the main energy source in the transport and exchange of energy, momentum, and water on Earth's surface, and it is calculated as follows:

$$\mathrm{R_n} = (1 - \alpha_0)\mathrm{R_s} + \mathrm{R_L}\downarrow - \mathrm{R_L}\uparrow - (1 - \varepsilon_0)\mathrm{R_L}\downarrow \tag{2}$$

$$R_{L\downarrow} = \varepsilon_a \times \sigma \times {T_a}^4 \tag{3}$$

$$R_{L\uparrow} = \varepsilon_0 \times \sigma \times {T_s}^4 \tag{4}$$

where R_s is the incoming shortwave radiation (W/m^2), $R_L \downarrow$ is the incoming longwave radiation, $R_L \uparrow$ is the outgoing longwave radiation (W/m^2), $(1-\varepsilon_0)R_L \downarrow$ is the incident longwave radiation emitted through the surface (W/m^2), α_0 is the surface albedo (dimensionless), ε_a is the atmospheric radiance, ε_0 is the thermal surface emissivity (dimensionless), T_a is the temperature of the air near the surface, and T_s is the surface temperature (K).

$$\varepsilon_0 = 1.009 + 0.047ln(NDVI),\ NDVI \geq 0.002 \tag{5}$$

$$\varepsilon_0 = 1, \cdot NDVI < 0.002, \tag{6}$$

$$T_s = \mathrm{K}_2 / \ln\left(\frac{\varepsilon_{NB} * K_1}{R_c} + 1\right) \tag{7}$$

where K_1 and K_2 are the constant values (K), R_c is the corrected thermal radiance of the surface, and ε_{NB} is the narrow band surface emissivity.

$$NDVI = \frac{\rho_{NIR} - \rho_{RED}}{\rho_{NIR} + \rho_{RED}} \tag{8}$$

where ρ_{NIR} and ρ_{RED} are DN values of the near-infrared band and red band, respectively.

It should be pointed out that the radiance changes at different times in the year and between years as a result of different solar incidences over the surface, especially in rough terrain. The ratio properties of NDVI enable the great mass of noise due to changing sun angles, clouds, topography, or shadow to be cancelled, making NDVI less susceptible to illumination conditions [46].

$$\alpha_0 = \sum \gamma_i \alpha_i \tag{9}$$

where α_i and γ_i are the surface reflectance and the weight coefficient of band i, respectively.

G is the soil heat flux, and its equation is

$$G/R_n = T_s * (0.0038 + 0.0074 * \alpha) * \left(1 - 0.98 * NDVI^4\right) \tag{10}$$

where T_s represents the surface temperature (C); α represents albedo; and $NDVI$ is the normalized difference vegetation index (dimensionless).

$$H = \frac{\rho_{\mathrm{air}} * C_p * dT}{\gamma_{ah}} \tag{11}$$

where ρ_{air} is the air density (kg/m^3), C_p is the air specific heat (J/kg*K), γ_{ah} is the aerodynamic resistance against heat transfer (s/m), and dT is the temperature difference ($\mathrm{T}_1 - \mathrm{T}_2$) at heights of Z_1 and Z_2 (°C).

$$\gamma_{\mathrm{ah}} = \frac{ln(\frac{z_2}{z_1})}{k * U_*} \tag{12}$$

where k is the von Karman constant (0.41) and U_* is the fraction velocity at heights of 2 m and 0.1 m.

$$U_* = \frac{k * U_{200}}{ln(\frac{z_{200}}{z_{0m}})} \tag{13}$$

where z_{0m} is the surface momentum roughness height (m). The equation for z_{0m} is

$$Z_{0m} = \exp(\frac{a * NDVI}{\alpha} + b) \tag{14}$$

where a and b are empirical coefficients.

$$dT = aT_s + b \tag{15}$$

where a and b are linear parameters, and dT is the temperature difference ($\mathrm{T}_1 - \mathrm{T}_2$) at heights of Z_1 and Z_2 (°C), with Z_1 = 0.1 m, Z_2 = 2 m usually used in the SEBAL model. The purpose of defining the zero-plane displacement is to reduce the obstruction of wind by ground vegetation.

In Equation (1), ET represents instantaneous evapotranspiration. The evaporative fraction (Λ) was taken to extrapolate the instantaneous ET to the daily ET in the SEBAL model, and the equations are

$$\Lambda = \frac{ET}{R_n - G} \quad (16)$$

$$ET_{24} = \frac{\Lambda \cdot (R_{24} - G_{24})}{\lambda} \quad (17)$$

where λ is the latent heat of vaporization (MJ kg^{-1}); ET, Rn, and G are the instantaneous latent heat flux, net radiation and ground heat flux in each pixel, respectively; R_{24} is the daily net radiation (W m^{-2}); and G_{24} is the daily soil heat flux (W m^{-2}).

2.5. Calculation of ET in Paddy Fields

The paddy rice fields of the GFPIS were extracted in our previous study [44], and we took the average ET value calculated based on the extracted rice image elements as the daily average paddy rice ET value calculated in Section 3.2.

For the days without accessible RS images, the daily ET can be estimated by linearly interpolating over periods between two consecutive images [47,48]. The specific steps are as follows: (1) Calculate the daily crop reference ET (ET_0) using the Penman–Monteith method recommended by the FAO [49]; (2) Calculate the ratio of ET_{24} to ET_0 in two available RS images; and (3) Linearly interpolate the two-spoke image to determine the ratio on the missing dates. The ET_{24} of the missing date can be obtained by multiplying the ratio by the ET_0 on the missing date [50]. The Penman model was modified by the FAO in 1998, and the modified Penman model has a solid theoretical basis and high universality; thus, it is currently the most accurate method of calculating ET_0. The calculation formula of the Penman model is as follows:

$$ET_{0,\ PM} = \frac{0.408\Delta(R_n - G) + \gamma \frac{900}{T+273} u_2 (e_s - e_a)}{\Delta + \gamma(1 + 0.34u_2)} \quad (18)$$

$$ET_{season} = \sum_{i=start}^{i=end} ET_{daily\_i} = \sum_{i=start}^{i=end} ETF_i \times ET_{0\_PM\_i} \quad (19)$$

2.6. Evaluation of Model Accuracy

In this study, the daily average ET observed based on the eddy covariance of the Jiangxi Province Irrigation Experiment Station (116°00′03″ E, 28°26′27″ N) in the GFPIS from 2016–2017 was compared with the results calculated by the SEBAL model. To evaluate the prediction accuracy of the model for ET, the coefficient of determination (R^2), root-mean-square error (RMSE), and Nash–Sutcliffe efficiency coefficient (NSE) were chosen to analyze the ET derived by the model, and the calculation process is shown in Equations (20)–(22). A reliable model usually has high R^2 and NSE values and low RMSE values [51,52].

$$R^2 = 1 - \frac{\sum_{i=1}^{n} (ET_{EC\_i} - ET_{sat\_i})^2}{\sum_{i=1}^{n} (ET_{EC\_i})^2} \quad (20)$$

$$RMSE = \sqrt{\frac{\sum_{i=1}^{n} (ET_{EC\_i} - ET_{sat\_i})^2}{n}} \quad (21)$$

$$NSE = 1 - \left[\frac{\sum_{i=1}^{n} (ET_{EC\_i} - ET_{sat\_i})^2}{\sum_{i=1}^{n} \left(ET_{EC\_i} - ET_{\overline{EC\_i}}\right)^2} \right] \quad (22)$$

where ET_{EC_i} is the ET observed by vorticity on day i, ET_{sat_i} is the ET estimated by remote sensing on day i, $ET_{\overline{EC\_i}}$ is the mean ET value observed by vorticity, and n is the total number of observed samples.

In this study, ArcGIS (ESRI, West Redlands, CA, USA) and ENVI (Environment for Visualizing Images, Exelis Visual Information Solutions, Boulder, CO, USA) were used to process the RS images and PyCharm (JetBrains, Prague, Czech Republic) was used to construct the SEBAL model using the Python programming language.

3. Results and Discussion

3.1. Estimation Accuracy of the SEBAL Model for Daily ET in the GFPIS

In this paper, Landsat images on typical days with cloudless skies or less cloud cover were selected to analyze the estimated surface ET results in the growing season of 2017, which were 25 May, 12 July, 14 September, and 30 September 2017. Due to cloud cover and other influences, the processing results of some pixels were still null, as shown in Figure 5.

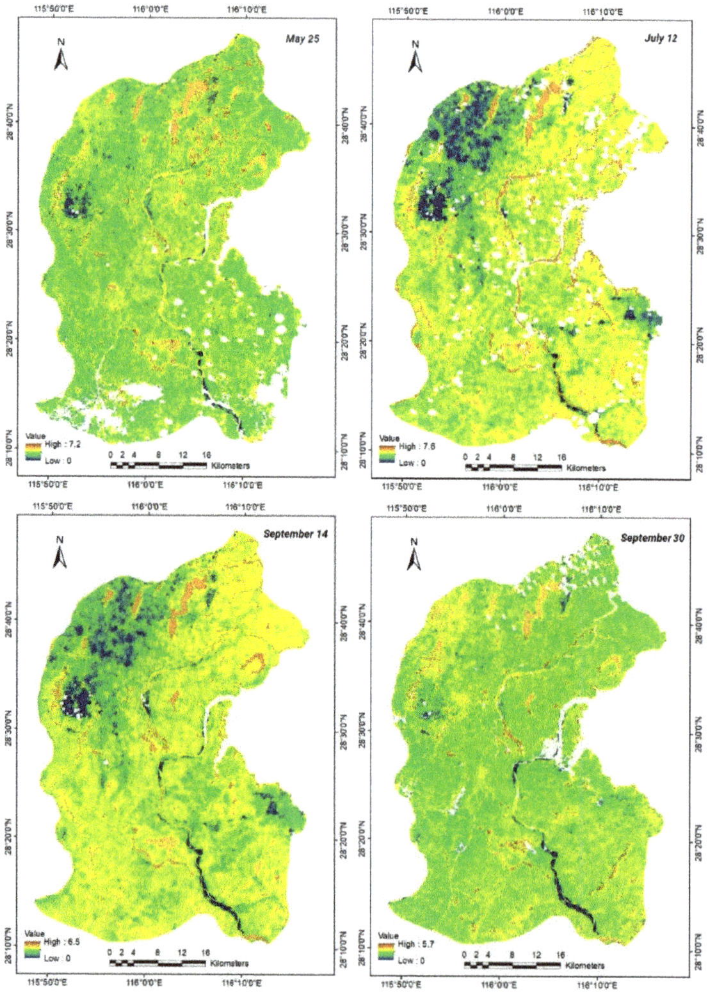

Figure 5. Daily ET estimated in the GFPIS in 2017 by the SEBAL model.

In terms of time series, the maximum ET value in the GFPIS was 7.6 mm/day on July 12, followed by those on 25 May and 14 September, and the minimum daily ET was 5.6 mm/day on 30 September. The ET of the farmland area on 12 July and 14 September was higher, and the ET of the farmland area on 30 September was lower than that on 14 September. In terms of the four seasons, the maximum and minimum ET values were observed in summer and winter, respectively; moderate values were observed in spring and autumn; and higher values were observed in spring than in autumn.

The ET was estimated by the SEBAL model in 2016 and 2017 and the ETs observed based on the eddy covariance at the experimental station are shown in Figure 6. The RMSE of the comparison between the model estimation results and the observation results was 0.84 mm/d, the R^2 was 0.85, and the NSE was 0.81. Figure 6 shows that the estimated ET results in 2016 and 2017 were consistent with the observed eddy covariance values. In general, the ET estimated by the SEBAL model was of high accuracy, and the adaptability and stability of the SEBAL model in the irrigated area of GFPIS were satisfactory.

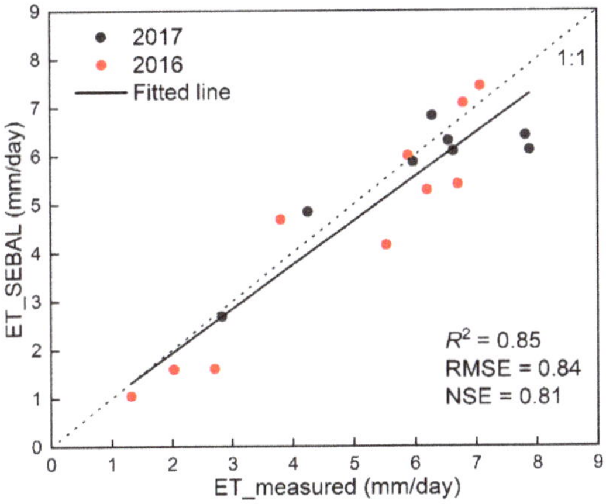

Figure 6. Validation of the estimated ET.

The results of this study indicate that the SEBAL model has reasonable accuracy when predicting the ET of subtropical regions. In Tan et al.'s previous study in 2021 [42], the SEBAL model was applied to estimate the ET characteristics in a subtropical region (Huaihe River Basin, China) and the SEBAL model performance was evaluated by fitting the regression with the daily reference ET calculated by multiple theoretical methods. The results showed that the bias between the ET estimated by the SEBAL model and daily reference ET was less than 1.5%. The results of this study have a slightly higher error than that in the previous study. The bias of the SEBAL model's estimation of ET is mainly due to the following factors: (1) The SEBAL model relies on instantaneous surface temperature, and the selection of hot and cold pixels in the imagery is subjective, which has a great impact on the accuracy of ET estimation; (2) The SEBAL model has been proven to be applicable for estimating ET in flat surfaces with relative accuracy, while it tends to have a high bias when applied to mountainous areas [53], which may lead to some increase in bias due to the presence of some hilly areas in our study area; and (3) Studies have shown that even a thin layer of clouds can lead to large errors when calculating sensible heat fluxes [54].

Landsat images with a resolution of 30 m can reflect spatial differences in the ET distribution. Moreover, ET is greatly affected by different ground object types. The greatest evapotranspiration occurred from water bodies—mainly lakes and rivers—whose ET was approximately 5 mm/day. The daily ET of the urban construction area and beach sand was the lowest at only 0–3 mm/day. The ET of forestland and farmland varied with different growth stages. Thus, the estimated spatial distribution of evapotranspiration in the irrigated area is consistent with the actual evapotranspiration.

3.2. Variation in the Daily Rice ET during the Growing Season

The daily variation characteristics of ET from paddy fields in the GFPIS were analyzed based on the ET of double-season rice in 2016 and 2017, and the results are shown in Table 2.

Table 2. SEBAL-estimated daily average evapotranspiration from early and late rice in 2016 and 2017.

Date	ETa (mm)	Date	ETa (mm)
7 June 2016	4.69	9 May 2017	5.87
23 June 2016	7.43	25 May 2017	6.31
9 July 2016	6.00	12 July 2017	6.42
25 July 2016	7.08	28 July 2017	6.82
19 September 2016	5.29	29 August 2017	6.12
27 September 2016	5.41	14 September 2017	6.09
5 October 2016	4.15	30 September 2017	4.85
16 December 2016	1.05	1 November 2017	2.69

In June 2016, as the rice grew and changed from the booting stage to the heading stage. The average ET value increased from 4.69 mm/day on June 7 to 7.43 mm/day on 23 June. On 9 July, as the double-season early rice was in the mature stage, the ET was weak and had a value of 6 mm/day. On 25 July, as the double-season late rice was about to enter the tiller stage, the ET increased to 7.08 mm/day. In September, the ET of double-season late rice was approximately 5.3 mm/day; and in October, it decreased to 4.15 mm/d as double-season late rice entered the mature stage.

In May 2017, as rice growth and transpiration increased, the daily ET value of rice also increased gradually from 5.87 mm/day on 9 May to 6.31 mm/day on 25 May. In July and August, the peak of the double-season early and late rice growing season occurred, the temperature in the GFPIS was high, and the daily ET of rice fields also maintained a high value, which was higher than 6 mm/day. On 14 September, when the late rice was in the heading stage, transpiration remained at 6.09 mm/day. However, the transpiration intensity gradually decreased on 30 September when the rice was in the milking stage and water was lacking in the paddy field due to evaporation by the sun; therefore, the transpiration value decreased significantly to 4.85 mm/day. In November, the transpiration value reached a minimum of 2.69 mm/day when there was no vegetation in the paddy field.

The results showed that the seasonal ET exhibited significant changes and was closely related to the climate. Overall, excluding November and December after rice harvest, the regional daily average double-season early and late rice ET values in 2016 and 2017 showed a pattern higher than 6 mm/day in July and August and lower than 5 mm/day in late September and October, while the daily average paddy ET values varied in the range of 5 to 6 mm/day for the rest of the year.

3.3. Interannual Variation in ET and Precipitation in the Rice Growing Season

Due to cloudy and rainy weather during the growing season of early rice, limited remote sensing image data are available, and no data are available to calculate the ET of the growing season of early rice in 2003, 2008, 2010, and 2015. Due to the lack of data, we cannot obtain the distribution map of early, middle, and late rice in each year. According to our extraction results for paddy rice in the GFPIS in a previous study [37], the rice planting structure changed, with 2010 as the boundary; according to the available quantity and image quality of satellite data sources during the rice growing period, the middle and double-season rice maps of 2003 were used to calculate the ET of paddy fields from 2000 to 2009 and the middle and double-season rice maps of 2017 were used to calculate the ET of paddy rice from 2010 to 2017. The division of the rice growing season was determined by the observation data from the Jiangxi Irrigation Experiment Station, where the early rice growing season was from 22 April to 10 July, the middle rice growing season was from 26 June to 7 October, and the late rice growing season was from 19 July to 24 October.

Figure 7 shows the interannual variation in ET and precipitation in the rice growing season by the SEBAL model combined with the paddy rice distribution map. The interannual variation in evapotranspiration in the GFPIS from 2000 to 2017 shows that the ET of middle rice is generally the largest, with a multiyear average value of 434 mm; followed by the ET of the late rice growing season, with a multiyear average value of 380 mm; and the ET of early rice is the smallest, with a multiyear average value of 287 mm. The ET in the early rice growing season ranged from 263 mm to 349 mm, with maximum and minimum values occurring in 2013 and 2014, respectively. The average ET value during the growing season of middle rice ranged from 378 mm to 520 mm, with maximum and minimum values occurring in 2016 and 2009, respectively. The mean ET value during the growing season of late rice ranged from 324 mm to 457 mm, with maximum and minimum values occurring in 2016 and 2012, respectively.

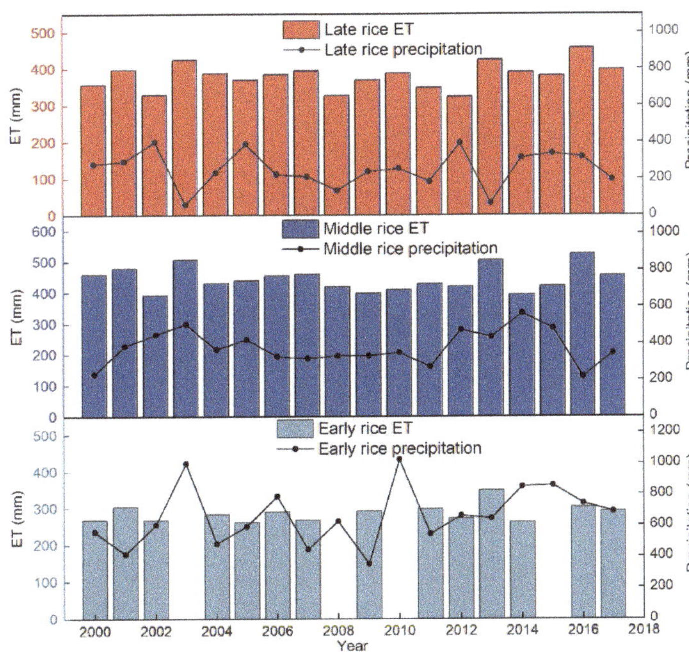

Figure 7. Interannual variation in ET and precipitation in the rice growing season in the GFPIS.

The interannual variation trend of ET during the rice growing season of early rice was not obvious, but the variation trend of ET of middle rice and late rice was consistent. The interannual change in ET of middle rice and late rice was not significant before 2003, and ET of middle rice and late rice showed a downward trend from 2003 to 2009, while ET of middle rice and late rice showed an upward trend from 2009 to 2017.

As can be seen from Figure 7, the interannual variation trend of precipitation is opposite to the interannual variation trend of ET overall in rice growing season, showing a negative correlation which is especially obvious in the early rice growth season. This may be that the study area is located in the subtropical monsoon region, and the east Asian summer circulation is enhanced during the early rice growth period, resulting in increased precipitation. ET can be regarded as one of the most important factors for indicating climate change at the catchment scale [55], and our research supports this view. Meanwhile, the interannual variation of ET is closely related with many ecological factors. This study's focus on the analysis of the time characteristics of ET and cannot directly explain the relationship between the interannual variation in ET and precipitation. If ET values are combined with meteorological factors, topographical factors, and vegetation changes, we may arrive at a more accurate and comprehensive understanding of the relationship between the interannual variation in ET and precipitation in the study area.

3.4. Spatial Distribution of ET during Rice Growth in Different Hydrological Years

Table 3 shows the precipitation and ET values during the rice growing season in different hydrological years, and the selection of a typical hydrological year is explained in Section 2.3. Almost no difference was observed between the ET of early rice in wet years and dry years, which might be due to the abundant precipitation of different hydrological years in the growing season of early rice. The ET of middle rice in different hydrological years showed significant differences, with a value of 393 mm in wet years, 423 mm in normal years, and 520 mm in dry years. The ET of late rice in wet years and normal years showed little difference, and the values in dry years were 330 mm and 367 mm. Overall, ET showed limited differences in the early and late rice in different hydrological years, but great differences in middle rice.

Table 3. ET of early, middle, and late rice during typical years in the GFPIS.

Rice Types	Wet Years			Normal Years			Dry Years		
	Year	Precipitation	ET	Year	Precipitation	ET	Year	Precipitation	ET
Early rice	2014	850	263	2008	624	*	2007	443	267
Middle rice	2014	564	393	2004	365	423	2016	216	525
Late rice	2002	401	330	2010	255	330	2008	136	367

* ET in 2008 could not be estimated because no images were available during the growing season of early rice.

Figures 8 and 9 show the ET frequency distribution and spatial distribution of rice in different hydrological years. In the wet year (2014), the early rice ET in the GFPIS was mainly within the range of 200–300 mm, with an average of 263 mm. The ET of the first, third, and fourth main canals of the irrigated area was between 263 mm and 283 mm. The early rice ET in the second main canal and the middle of the main trunk canal was lower than the average early rice ET in the whole irrigated area, and the early rice ET in the east trunk canal and the main canal was generally approximately 243–263 mm. In the dry year (2007), the ET of early rice was mainly in the range of 225–300 mm, with an average of 267 mm. Among them, the early rice ET in the first, second, and third main canal irrigated areas was higher than the average early rice ET value in the whole irrigated area, and the early rice ET in the fourth main canal and the end of the third main canal irrigated area was lower than the average value. These results indicate that the mean values of early rice ET do not differ significantly in different hydrological years, although significant differences occur in the spatial distribution of early rice ET.

The middle rice ET values in the wet year (2014) mainly ranged between 375 mm and 425 mm and the mean middle rice ET value in the whole irrigated area was 392 mm, with little spatial difference. In the normal year (2004), the middle rice ET mainly ranged from 360 mm to 450 mm and the average middle rice ET value in the whole irrigated area was 423 mm. In terms of spatial distribution, the middle rice ET in the east main trunk and the end of the fourth main canal was higher than 450 mm. In the dry year (2016), the middle rice ET value mainly ranged from 487–570 mm and the mean ET value was 525 mm in the whole irrigated area. There were differences in the spatial distribution of ET, with generally higher values at the end of the fourth main canal and the second main canal compared with the mean ET value in the whole irrigated area while the ET values in the middle section of the west main canal and the second main canal were lower than 525 mm.

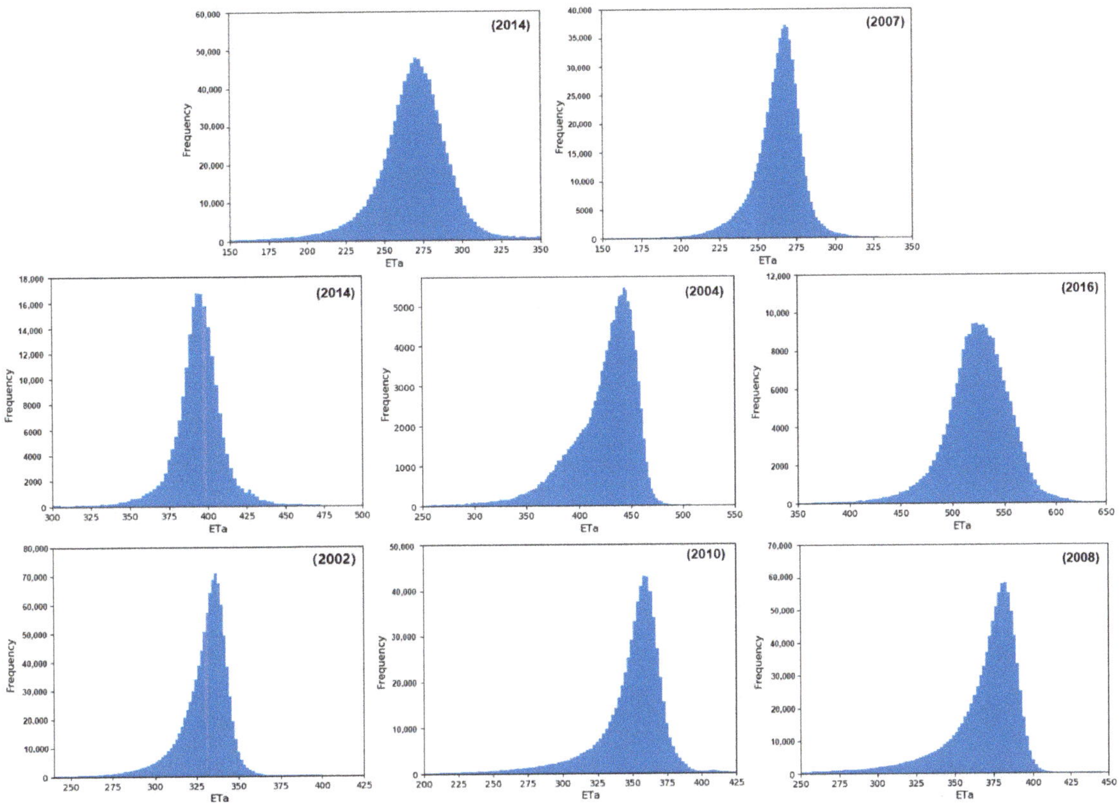

Figure 8. Frequency distribution of rice ET in different hydrological years.

The late rice ET values in the wet year (2002) in the GFPIS mainly ranged from 300 mm to 350 mm, with an average value of 328 mm. In the western GFPIS, the late rice ET was generally higher than that of early rice and the ET was evenly distributed. The eastern GFPIS is a hilly area, and the late rice ET was generally lower than the mean value. In the normal year (2010), the late rice ET values in the GFPIS areas mainly ranged from 300 mm to 380 mm, with an average value of 347 mm. The late rice ET at the end of the second main canal and the fourth main canal was higher than that in the whole irrigated area (347 mm). The late rice ET in the third main canal and the first main canal irrigated area was higher than that in the whole irrigated area. In the dry year (2008), the late rice ET in the whole irrigated area mainly ranged from 325 mm to 400 mm, with an average value of 366 mm. The spatial distribution of the late rice ET was similar to that of the wet year, and the late rice ET in the western irrigated area was higher than the average value, while the ET in the eastern irrigated area was lower than the average value overall. Compared with the wet year, the ET in the whole irrigated area increased in dry years and mainly ranged from 40 mm to 60 mm.

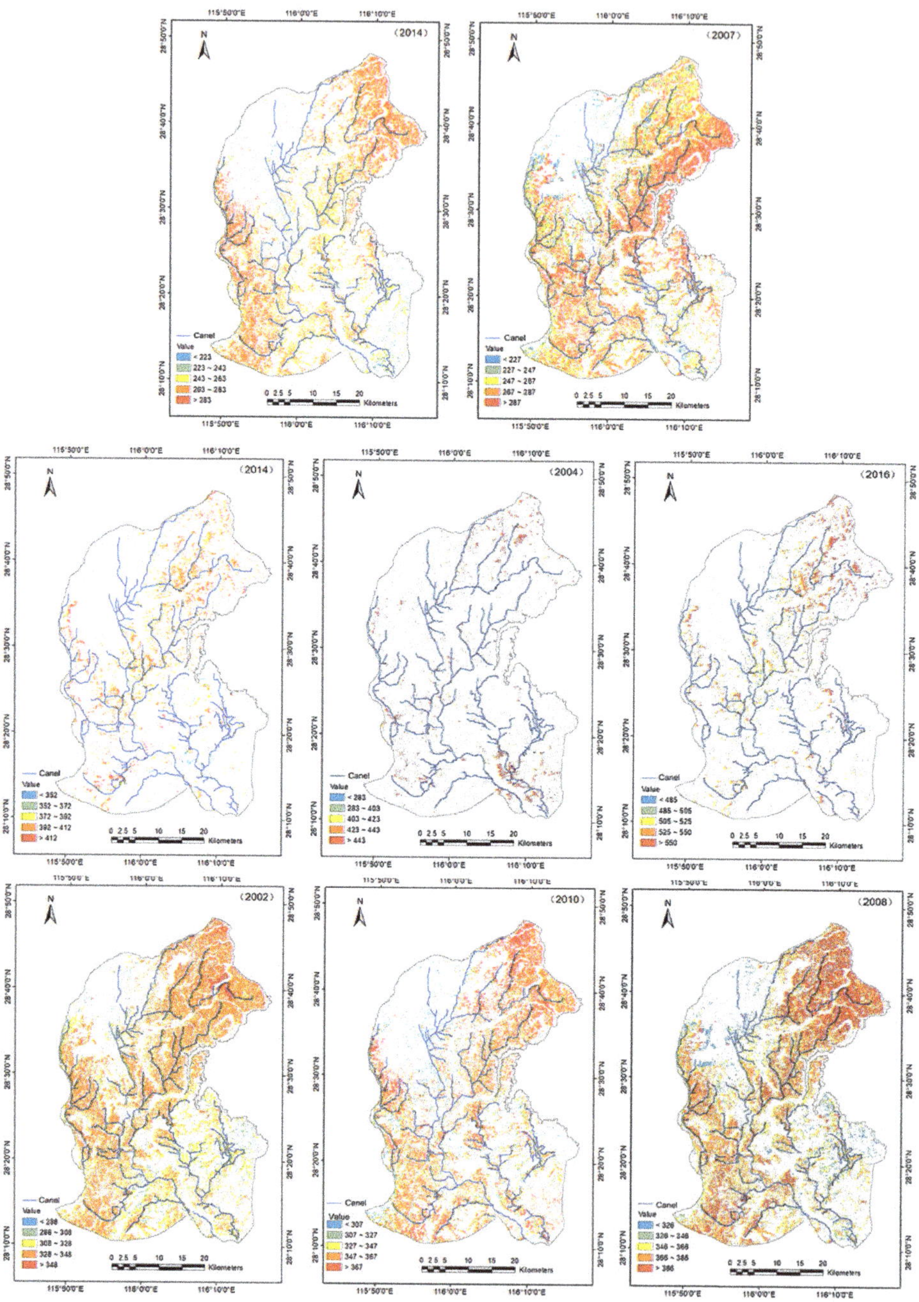

Figure 9. Spatial distribution of rice ET in different hydrological years.

In general, the late rice ET in the western irrigated areas was higher than the mean late rice value in irrigated areas in different hydrological years. The late rice ET in the irrigated areas of the first and second main canals and the third canal was slightly higher than the mean late rice ET in the whole irrigated area in the three hydrological years. In the normal and dry years, the late rice ET at the end of the fourth main canal and the second main canal was much higher than the mean late rice ET value in the whole irrigated area. This finding is related to water resource dispatching measures in the GFPIS. In the case of drought, the first, second, and third main canals on the east main canal and west main canal should be adopted to temporarily control water inflow so that the water output can guarantee water for production and living in the middle and lower reaches of the irrigated area. The late rice ET of the eastern irrigated area and the hilly area near the west main water inlet in the three hydrological years was generally lower than the mean late rice ET value of the GFPIS [56]. Agricultural water consumption accounts for approximately 8% of the actual water consumption in the eastern irrigated area; therefore, the irrigation conditions in the irrigated area are not as good as those in the western irrigated area.

4. Conclusions

In this paper, the SEBAL model was applied to the subtropical climate region of southern China (GFPIS) to assess its rationality and applicability for daily scale estimates of paddy ET and to analyze the spatial and temporal distribution characteristics of early, middle, and late rice evapotranspiration in the GFPIS in different hydrologic years. The main conclusions are as follows:

(1) On the daily scale, the ET estimated by the SEBAL model is consistent with the eddy covariance, with R^2, NSE, and RMSE values of 0.85, 0.81, and 0.84 mm/day, respectively.
(2) Based on the analysis of the temporal characteristics of paddy ET, the daily average evapotranspiration value in irrigated areas was higher in July and August but lower in other months. The interannual variation trend of ET during the growing season of early rice was not obvious from 2000 to 2017, while that during the growing season of middle rice and late rice generally showed a downward trend from 2000 to 2009 and an upward trend after 2009.
(3) In terms of spatial distribution, significant differences were observed between early rice and late rice in different hydrological years. The spatial difference of middle rice was not significant in the wet year but was significant in the dry year.

Due to the large amount of water evaporation in paddy fields, the single-source SEBAL model needs further research to estimate ET in rice flooding periods, which will improve the accuracy of the SEBAL model in estimating ET in rice growing areas.

Author Contributions: Conceptualization, Y.L., J.C. and G.W.; Methodology, Y.L., J.C. and G.W.; Software, J.C. and G.W.; Validation, H.X (Hua Xie). and Y.Y.; Formal analysis, J.C. and G.W.; Investigation, J.C. and H.X. (Hengwang Xie); Resources, C.W. and Y.C.; Data curation, Y.Y. and C.W.; Writing—original draft preparation, J.C. and G.W.; Writing—review and editing, Y.L., G.W. and Y.C.; Visualization, G.W.; Supervision, Y.L.; Project administration, Y.L.; Funding acquisition, Y.L. All authors have read and agreed to the published version of the manuscript.

Funding: This work was financially supported by the NSFC-MWR-CTGC Joint Yangtze River Water Science Research Project (No. U2040213) and the National Natural Science Foundation of China (nos. 51979201 and 51779187).

Institutional Review Board Statement: Not applicable.

Informed Consent Statement: Not applicable.

Data Availability Statement: Not applicable.

Conflicts of Interest: The authors declare no conflict of interest.

References

1. Mao, Z. Water Saving Irrigation for Rice and Its Effect on Environment. *Strateg. Study Chin. Acad. Eng.* **2002**, *4*, 8–16.
2. Shao, G.; Han, W.; Zhang, H.; Liu, S.; Wang, Y.; Zhang, L.; Cui, X. Mapping maize crop coefficient Kc using random forest algorithm based on leaf area index and UAV-based multispectral vegetation indices. *Agric. Water Manag.* **2021**, *252*, 106906. [CrossRef]

3. Dari, J.; Brocca, L.; Quintana-Seguí, P.; Escorihuela, M.J.; Stefan, V.; Morbidelli, R. Exploiting High-Resolution Remote Sensing Soil Moisture to Estimate Irrigation Water Amounts over a Mediterranean Region. *Remote Sens.* **2020**, *12*, 2593. [CrossRef]
4. Zappa, L.; Schlaffer, S.; Bauer-Marschallinger, B.; Nendel, C.; Zimmerman, B.; Dorigo, W. Detection and Quantification of Irrigation Water Amounts at 500 m Using Sentinel-1 Surface Soil Moisture. *Remote Sens.* **2021**, *13*, 1727. [CrossRef]
5. Yi, Z. Research and Application of Collaborating Acquisition of Evapotranspiration and Surface Soil Moisture over Irrigated Area. Ph.D. Thesis, China Institute of Water Resources and Hydropower Research, Beijing, China, 2019.
6. Cao, M.; Wang, W.; Xing, W.; Wei, J.; Chen, X.; Li, J.; Shao, Q. Multiple sources of uncertainties in satellite retrieval of terrestrial actual evapotranspiration. *J. Hydrol.* **2021**, *601*, 126642. [CrossRef]
7. Wang, J.; Ding, J.; Yu, D.; Ma, X.; Zhang, Z.; Ge, X.; Teng, D.; Li, X.; Liang, J.; Lizaga, I.; et al. Capability of Sentinel-2 MSI data for monitoring and mapping of soil salinity in dry and wet seasons in the Ebinur Lake region, Xinjiang, China. *Geoderma* **2019**, *353*, 172–187. [CrossRef]
8. Chen, Z.; Wang, W.; Woods, R.A.; Shao, Q. Hydrological effects of change in vegetation components across global catchments. *J. Hydrol.* **2021**, *595*, 125775. [CrossRef]
9. Wang, J.; Ding, J.; Yu, D.; Teng, D.; He, B.; Chen, X.; Ge, X.; Zhang, Z.; Wang, Y.; Yang, X.; et al. Machine learning-based detection of soil salinity in an arid desert region, Northwest China: A comparison between Landsat-8 OLI and Sentinel-2 MSI. *Sci. Total Environ.* **2020**, *707*, 136092. [CrossRef] [PubMed]
10. Wang, X.-G.; Wang, W.; Huang, D.; Yong, B.; Chen, X. Modifying SEBAL Model Based on the Trapezoidal Relationship between Land Surface Temperature and Vegetation Index for Actual Evapotranspiration Estimation. *Remote Sens.* **2014**, *6*, 5909–5937. [CrossRef]
11. Urso, G.D.; Bolognesi, S.F.; Kustas, W.P.; Knipper, K.R.; Anderson, M.C.; Alsina, M.M.; Hain, C.R.; Alfieri, J.G.; Prueger, J.H.; Gao, F.; et al. Determining Evapotranspiration by Using Combination Equation Models with Sentinel-2 Data and Comparison with Thermal-Based Energy Balance in a California Irrigated Vineyard. *Remote Sens.* **2021**, *13*, 3720.
12. Sobrino, J.A.; Souza Da Rocha, N.; Skoković, D.; Suélen Käfer, P.; López-Urrea, R.; Jiménez-Muñoz, J.C.; Alves Rolim, S.B. Evapotranspiration Estimation with the S-SEBI Method from Landsat 8 Data against Lysimeter Measurements at the Barrax Site, Spain. *Remote Sens.* **2021**, *13*, 3686. [CrossRef]
13. Xue, J.; Anderson, M.C.; Gao, F.; Hain, C.; Yang, Y.; Knipper, K.R.; Kustas, W.P.; Yang, Y. Mapping Daily Evapotranspiration at Field Scale Using the Harmonized Landsat and Sentinel-2 Dataset, with Sharpened VIIRS as a Sentinel-2 Thermal Proxy. *Remote Sens.* **2021**, *13*, 3420. [CrossRef]
14. Han, X.; Wei, Z.; Zhang, B.; Han, C.; Song, J. Effects of Crop Planting Structure Adjustment on Water Use Efficiency in the Irrigation Area of Hei River Basin. *Water* **2018**, *10*, 1305. [CrossRef]
15. Tang, J.; Han, W.; Zhang, L. UAV Multispectral Imagery Combined with the FAO-56 Dual Approach for Maize Evapotranspiration Mapping in the North China Plain. *Remote Sens.* **2019**, *11*, 2519. [CrossRef]
16. Srivastava, A.; Sahoo, B.; Raghuwanshi, N.; Singh, R. Evaluation of Variable-Infiltration Capacity Model and MODIS-Terra Satellite-Derived Grid-Scale Evapotranspiration Estimates in a River Basin with Tropical Monsoon-Type Climatology. *J. Irrig. Drain. Eng.* **2017**, *143*, 04017028. [CrossRef]
17. Jackson, R.D.; Reginato, R.J.; Idso, S.B. Wheat canopy temperature: A practical tool for evaluating water requirements. *Water Resour. Res.* **1977**, *13*, 651–656. [CrossRef]
18. Jiang, L.; Islam, S. A methodology for estimation of surface evapotranspiration over large areas using remote sensing observations. *Geophys. Res. Lett.* **1999**, *26*, 2773–2776. [CrossRef]
19. Moran, M.S.; Rahman, A.F.; Washburne, J.C.; Goodrich, D.C.; Weltz, M.A.; Kustas, W.P. Combining the Penman-Monteith equation with measurements of surface temperature and reflectance to estimate evaporation rates of semiarid grassland. *Agric. Forest Meteorol.* **1996**, *80*, 87–109. [CrossRef]
20. Srivastava, A.; Kumari, N.; Maza, M. Hydrological Response to Agricultural Land Use Heterogeneity Using Variable Infiltration Capacity Model. *Water Resour. Manag.* **2020**, *34*, 3779–3794. [CrossRef]
21. Ning, J.; Gao, Z.; Xu, F. Effects of land cover change on evapotranspiration in the Yellow River Delta analyzed with the SEBAL model. *J. Appl. Remote Sens.* **2017**, *11*, 16009. [CrossRef]
22. Bai, L.; Cai, J.; Liu, Y.; Chen, H.; Zhang, B.; Huang, L. Responses of field evapotranspiration to the changes of cropping pattern and groundwater depth in large irrigation district of Yellow River basin. *Agric. Water Manag.* **2017**, *188*, 1–11. [CrossRef]
23. Bai, L. Inversion and Application of Agriculture Evapotranspiration and Soil Moisture in Irrigation District Based on Multi-source Remote Sensing Data. Ph.D. Thesis, China Institute of Water Resources and Hydropower Research, Beijing, China, 2017.
24. Lima, C.E.S.D.; Costa, V.S.D.O.; Galvíncio, J.D.; Silva, R.M.D.; Santos, C.A.G. Assessment of automated evapotranspiration estimates obtained using the GP-SEBAL algorithm for dry forest vegetation (Caatinga) and agricultural areas in the Brazilian semiarid region. *Agric. Water Manag.* **2021**, *250*, 106863. [CrossRef]
25. Senkondo, W.; Munishi, S.E.; Tumbo, M.; Nobert, J.; Lyon, S.W. Comparing Remotely-Sensed Surface Energy Balance Evapotranspiration Estimates in Heterogeneous and Data-Limited Regions: A Case Study of Tanzania's Kilombero Valley. *Remote Sens.* **2019**, *11*, 1289. [CrossRef]
26. Kamali, M.I.; Nazari, R. Determination of maize water requirement using remote sensing data and SEBAL algorithm. *Agric. Water Manag.* **2018**, *209*, 197–205. [CrossRef]

27. Rahimzadegan, M.; Janani, A. Estimating evapotranspiration of pistachio crop based on SEBAL algorithm using Landsat 8 satellite imagery. *Agric. Water Manag.* **2019**, *217*, 383–390. [CrossRef]
28. García-Gutiérrez, V.; Stöckle, C.; Gil, P.M.; Meza, F.J. Evaluation of Penman-Monteith Model Based on Sentinel-2 Data for the Estimation of Actual Evapotranspiration in Vineyards. *Remote Sens.* **2021**, *13*, 478. [CrossRef]
29. Gobbo, S.; Presti, S.L.; Martello, M.; Panunzi, L.; Berti, A.; Morari, F. Integrating SEBAL with in-Field Crop Water Status Measurement for Precision Irrigation Applications—A Case Study. *Remote Sens.* **2019**, *11*, 2069. [CrossRef]
30. Grosso, C.; Manoli, G.; Martello, M.; Chemin, Y.; Pons, D.; Teatini, P.; Piccoli, I.; Morari, F. Mapping Maize Evapotranspiration at Field Scale Using SEBAL: A Comparison with the FAO Method and Soil-Plant Model Simulations. *Remote Sens.* **2018**, *10*, 1452. [CrossRef]
31. Bastiaanssen, W.G.M.; Noordman, E.J.M.; Pelgrum, H.; Davids, G.; Thoreson, B.P.; Allen, R.G. SEBAL Model with Remotely Sensed Data to Improve Water-Resources Management under Actual Field Conditions. *J. Irrig. Drain. Eng.* **2005**, *131*, 85–93. [CrossRef]
32. Sun, Z.; Wei, B.; Su, W.; Shen, W.; Wang, C.; You, D.; Liu, Z. Evapotranspiration estimation based on the SEBAL model in the Nansi Lake Wetland of China. *Math. Comput. Model.* **2011**, *54*, 1086–1092. [CrossRef]
33. Gao, X.; Sun, M.; Luan, Q.; Zhao, X.; Wang, J.; He, G.; Zhao, Y. The spatial and temporal evolution of the actual evapotranspiration based on the remote sensing method in the Loess Plateau. *Sci. Total Environ.* **2020**, *708*, 135111. [CrossRef] [PubMed]
34. Ding, C. Research and Implementation of Remote Sensing Ground Evapotranspiration Estimation Algorithm Based on SEBAL Model. Master's Thesis, Xidian University, Xi'An, China, 2020.
35. Du, J.; Zhang, B.; Song, K.; Wang, Z.; Zeng, L. Study on Daily evapotranspiration estimation of Sanjiang Plain based on MODIS product and SEBAL Model. *Chin. J. Agrometeorol.* **2010**, *31*, 104–110.
36. Du, J.; Song, K.; Wang, Z.; Zhang, B.; Liu, D. Evapotranspiration estimation based on MODIS products and surface energy balance algorithms for land (SEBAL) model in Sanjiang Plain, Northeast China. *Chin. Geogr. Sci.* **2013**, *23*, 73–91. [CrossRef]
37. Li, H.; Zheng, L.; Lei, Y.; Li, C.; Liu, Z.; Zhang, S. Estimation of water consumption and crop water productivity of winter wheat in North China Plain using remote sensing technology. *Agric. Water Manag.* **2008**, *95*, 1271–1278. [CrossRef]
38. Kimura, R.; Bai, L.; Fan, J.; Takayama, N.; Hinokidani, O. Evapo-transpiration estimation over the river basin of the Loess Plateau of China based on remote sensing. *J. Arid Environ.* **2007**, *68*, 53–65. [CrossRef]
39. Li, X.; Xu, X.; Wang, X.; Xu, S.; Tian, W.; Tian, J.; He, C. Assessing the Effects of Spatial Scales on Regional Evapotranspiration Estimation by the SEBAL Model and Multiple Satellite Datasets: A Case Study in the Agro-Pastoral Ecotone, Northwestern China. *Remote Sens.* **2021**, *13*, 1524. [CrossRef]
40. Wu, Z.; Wang, M.; Zhao, X. Retrieval of land surface heat fluxes based on SEBAL model in hyper-arid region. *Sci. Surv. Mapp.* **2015**, *40*, 26–29.
41. Kayser, R.H.; Ruhoff, A.; Laipelt, L.; Kich, E.D.M.; Roberti, D.R.; Souza, V.D.A.; Rubert, G.C.D.; Collischonn, W.; Neale, C.M.U. Assessing geeSEBAL automated calibration and meteorological reanalysis uncertainties to estimate evapotranspiration in subtropical humid climates. *Agric. Forest Meteorol.* **2022**, *314*, 108775. [CrossRef]
42. Tan, L.; Zheng, K.; Zhao, Q.; Wu, Y. Evapotranspiration Estimation Using Remote Sensing Technology Based on a SEBAL Model in the Upper Reaches of the Huaihe River Basin. *Atmosphere* **2021**, *12*, 1599. [CrossRef]
43. Laipelt, L.; Bloedow Kayser, R.H.; Santos Fleischmann, A.; Ruhoff, A.; Bastiaanssen, W.; Erickson, T.A.; Melton, F. Long-term monitoring of evapotranspiration using the SEBAL algorithm and Google Earth Engine cloud computing. *ISPRS J. Photogramm.* **2021**, *178*, 81–96. [CrossRef]
44. Cao, J.; Cai, X.; Tan, J.; Cui, Y.; Xie, H.; Liu, F.; Yang, L.; Luo, Y. Mapping paddy rice using Landsat time series data in the Ganfu Plain irrigation system, Southern China, from 1988–2017. *Int. J. Remote Sens.* **2021**, *42*, 1556–1576. [CrossRef]
45. Bastiaanssen, W.G.M.; Menenti, M.; Feddes, R.A.; Holtslag, A.A.M.; Hutjes, R.W.A.; Kabat, P.; Bass, B.; Field, C.; Running, S.W.; Shuttleworth, W.J. A remote sensing Surface Energy Balance Algorithm for Land (SEBAL); 1, Formulation. *J. Hydrol.* **1998**, *212–213*, 198–212. [CrossRef]
46. Martín-Ortega, P.; García-Montero, L.G.; Sibelet, N. Temporal Patterns in Illumination Conditions and Its Effect on Vegetation Indices Using Landsat on Google Earth Engine. *Remote Sens.* **2020**, *12*, 211. [CrossRef]
47. Teixeira, A.H.D.C.; Bastiaanssen, W.G.M.; Ahmad, M.D.; Bos, M.G. Reviewing SEBAL input parameters for assessing evapotranspiration and water productivity for the Low-Middle São Francisco River basin, Brazil. *Agric. Forest Meteorol.* **2009**, *149*, 462–476. [CrossRef]
48. Bastiaanssen, W.G.M.; Ahmad, M.; Chemin, Y. Satellite surveillance of evaporative depletion across the Indus Basin. *Water Resour. Res.* **2002**, *38*, 1–9. [CrossRef]
49. Allen, R.G.; Pereira, L.S.; Raes, D.; Smith, M. *Crop Evapotranspiration: Guidelines for Computing Crop Requirements*; FAO Irrigation and Drainage Paper 56; Food and Agricultural Organization of the U.N.: Rome, Italy, 1998.
50. Yang, Y.; Shang, S.; Jiang, L. Remote sensing temporal and spatial patterns of evapotranspiration and the responses to water management in a large irrigation district of North China. *Agric. Forest Meteorol.* **2012**, *164*, 112–122. [CrossRef]
51. Ge, X.; Ding, J.; Jin, X.; Wang, J.; Chen, X.; Li, X.; Liu, J.; Xie, B. Estimating Agricultural Soil Moisture Content through UAV-Based Hyperspectral Images in the Arid Region. *Remote Sens.* **2021**, *13*, 1562. [CrossRef]

52. Wang, J.; Shi, T.; Yu, D.; Teng, D.; Ge, X.; Zhang, Z.; Yang, X.; Wang, H.; Wu, G. Ensemble machine-learning-based framework for estimating total nitrogen concentration in water using drone-borne hyperspectral imagery of emergent plants: A case study in an arid oasis, NW China. *Environ. Pollut.* **2020**, *266*, 115412. [CrossRef] [PubMed]
53. Zhong, Y.; Zhong, M.; Mao, Y.; Ji, B. Evaluation of Evapotranspiration for Exorheic Catchments of China during the GRACE Era: From a Water Balance Perspective. *Remote Sens.* **2020**, *12*, 511. [CrossRef]
54. Liu, C.; Zhang, D.; Liu, X.; Zhao, C. Spatial and temporal change in the potential evapotranspiration sensitivity to meteorological factors in China (1960–2007). *J. Geogr. Sci.* **2012**, *22*, 3–14. [CrossRef]
55. Zhang, H.; Wang, L. Analysis of the variation in potential evapotranspiration and surface wet conditions in the Hancang River Basin, China. *Sci. Rep.* **2021**, *11*, 1–10. [CrossRef]
56. He, Z. The Study on Water Allocation and Related Engineering Measure Research of Ganfu Plain Irrigation Area. Master's Thesis, NanChang Univeresity, Nanchang, China, 2012.

remote sensing

Article

Effects of Mulching on Maize Yield and Evapotranspiration in the Heihe River Basin, Northwest China

Qianxi Shen [1,2,3], Jun Niu [1,2,3,*], Bellie Sivakumar [4] and Na Lu [1,2,3]

1 Center for Agricultural Water Research in China, China Agricultural University, Beijing 100083, China; qianxishen@cau.edu.cn (Q.S.); luna@cau.edu.cn (N.L.)
2 National Field Scientific Observation and Research Station on Efficient Water Use of Oasis Agriculture in Wuwei of Gansu Province, Wuwei 733000, China
3 Key Laboratory of Agricultural Water Saving of the Ministry of Water Resources, Beijing 100083, China
4 Department of Civil Engineering, Indian Institute of Technology Bombay, Powai, Mumbai 400076, India; b.sivakumar@iitb.ac.in
* Correspondence: niuj@cau.edu.cn; Tel.: +86-10-62737911

Abstract: Plastic film mulching is an effective way to manage agricultural fields in water shortage areas. Through increasing the soil surface temperature at the early stage of crop growth and reducing the soil evaporation during the whole growth period, plastic film mulching can realize the effect of water saving and yield increase. This study examined the effects of plastic film mulching on crop yield and evapotranspiration (*ET*) in the Heihe River basin in Northwest China. By using remote sensing data, the gridding G-AquaCrop model was built to simulate the maize yield and *ET* in the basin under conditions of film mulching and no-film mulching. Through an analysis of changes in maize yield and *ET* before and after film mulching, suitable areas for film mulching in the whole basin were identified. Through comparative analysis, it was found that after plastic film mulching, maize yield in 12–41% of the Heihe River basin increased to a certain extent, reaching 8%. Furthermore, film mulching decreased *ET* by 5–30% in 34–41% of areas planted with maize. Based on these results, suggestions were made on suitable areas for expansion of maize cultivation to balance the benefits of water saving and production increase with environmental pollution. Furthermore, the way of assessing the suitable mulching area is obtained by examining the meteorological condition directly. The results of this study are of great significance for rational allocation of agricultural production resources and efficient utilization of agricultural water resources.

Keywords: maize yield; evapotranspiration; plastic film mulching; G-AquaCrop model; Heihe River basin

Citation: Shen, Q.; Niu, J.; Sivakumar, B.; Lu, N. Effects of Mulching on Maize Yield and Evapotranspiration in the Heihe River Basin, Northwest China. *Remote Sens.* **2022**, *14*, 700. https://doi.org/10.3390/rs14030700

Academic Editor: Pradeep Wagle

Received: 21 December 2021
Accepted: 31 January 2022
Published: 2 February 2022

1. Introduction

Plastic film mulching is a field management technique that results from advances in the field of material science. In addition to all the benefits of traditional biological mulching (i.e., to increase temperature, conserve soil moisture, and reduce evaporation) [1,2], plastic film mulching has several advantages, such as low cost, easy production, and good performance [3]. As a result, plastic film mulching has been widely used in agriculture in recent years [4–7]. It is likely that the use of plastic film mulching will continue to increase in the future, especially considering the increasing food demands due to population growth and the anticipated impacts of climate change.

In China, increasing population and climate change have already raised concerns about food security, and the situation is anticipated to only worsen [8–10]. Projections of more frequent and greater magnitude droughts [11–14] emphasize the need for water saving in the agricultural sector, especially since the sector is the largest consumer of water. In this regard, plastic film mulching offers an important means, as it has the benefits of water saving, especially with low cost and greater efficiency in production. Needless to say, the benefits of plastic film mulching are largest when it is used in regions where water resources are already scarce and agriculture is a key sector, such as the Heihe River basin in Northwest China.

The Heihe River basin, located in the arid Hexi Corridor of Northwest China, is an important commodity grain production base in China [15,16]. Due to its geography and climatic conditions, this area has been facing a severe shortage of water resources for a long time [17]. Theoretically speaking, plastic film mulching has a great potential for application in this area. However, due to differences in climate and soil types over the region, not all areas in the basin can produce satisfactory results with film mulching [18]. Another issue is that the plastic film cannot be degraded in a short time after being discarded [19], which not only results in environmental pollution but also reduces crop yield to varying degrees. Jiang et al. (2017) used the dye tracer method to track soil water movement, and the results highlighted that the presence of residual plastic film fragments significantly influenced soil physical properties, altered soil water distribution, and decreased the matching degree of the flow distribution region and the maize root (densely rooted) zone [20]. Gao et al. (2019) used meta-analysis to show that the crop yield decreased significantly with increased time and residual plastic film exceeded 240 kg/ha [21]. Thus, the use of plastic film mulching without considering its applicability may aggregate negative effects. Therefore, in actual production, it is necessary to allocate agricultural production according to local conditions in order to realize efficient utilization of resources.

Producing more food with less water has been the goal of agricultural development in arid areas, especially where mulch is widely used. Crop water productivity (*WP*) refers to the ratio of crop yield to crop evapotranspiration (*ET*) during yield acquisition [22–24]. By definition, a larger *WP* is the goal pursued by researchers and farmers. For this reason, *WP* can be used as a key index to explore the potential and suitability of plastic film mulching in the Heihe River basin.

The AquaCrop model published by the Food and Agriculture Organization (FAO) is a typical water-driven crop growth simulation model [25–27]. It accurately simulates not only the crop yield but also the *ET* in the process of crop growth. Further, it divides the *ET* into soil evaporation (*E*) and crop transpiration (*Tr*). At the same time, compared with other crop growth simulation models (e.g., EPIC, DSSAT, WOFOST, APSIM, and WAVES) [28–32], this model requires fewer input parameters with stable performance and can simulate different field management practices, such as different irrigation methods, film mulching methods, and fertilization methods [25,26]. It is indeed a good choice to use the AquaCrop model to explore the effects of different field management practices on *WP* in arid areas. Ran et al. (2017) showed that the simulation performance of the AquaCrop model was better than that of the SIMDualKc model under deficit irrigation conditions through a 5-year field experiment data verification [33]. Compared with SIMDualKc model [34], the AquaCrop model not only considered the effect of soil water content, but also took into account the physiological damage (for example, a decrease in canopy coverage) caused by long-term water shortage when simulating *Tr*., Shen et al. (2019) used the AquaCrop model to simulate the changes in *ET* of seed maize with and without film mulching in the Shiyang River basin in Northwest China and verified the results with field data [35]. The results not only showed that the model had good simulation performance under both conditions but also indicated that plastic film mulching significantly reduced soil evaporation, thereby reducing water consumption during the growth period of the crop. The current AquaCrop model does not take into account the effects of plant diseases and pests on crop growth and yield. In this case, the simulation results are often the potential yield (defined by solar radiation and temperature) or attainable yield (limited by water and nutrient availability) rather than the actual yield (reduced by diseases, pests, and environmental stressors); this seems to be a common flaw in most crop growth models [36]. Tigkas et al. (2015) and Li et al. (2019) showed that when the independent variables were meteorological conditions or management measures, the crop model was an ideal tool. In this way, other influencing factors can be set to have a neutral role in crop growth without affecting the interpretation of the results [37,38].

Despite these encouraging results, it is important to note that most simulation models such as AquaCrop are applicable only at the field scale. This presents certain limitations when applied to the Heihe River basin. This is because, due to the wide distribution of maize planting areas in the whole Heihe River basin, the spatial heterogeneities meteorological, soil and management variables in the region are large and are difficult to calculate and measure [39]. Therefore, although the AquaCrop model has been found to perform well for field-scale studies, it cannot be directly applied to regional studies due to limitations and difficulties in parameter estimation.

Advances in remote sensing technology have provided effective support for exploring the potential of plastic film mulching in the Heihe River basin using the AquaCrop model [40,41]. For example, many studies have proposed different important indexes in agricultural production using remote sensing data. Jin et al. (2020) and Toshihiro et al. (2006) estimated crop physiological indicators such as canopy coverage (*CC*) and growth period by using remote sensing data such as the enhanced vegetation index (*EVI*) and the three-band water index (*TBWI*) [42,43]. Lorite et al. (2013) and Han et al. (2019, 2020) used remote sensing technology to obtain *CC*, sowing date, aboveground biomass and other crop physiological indicators preliminarily. Then, the above indicators were introduced into the crop model and distributed simulation was carried out [44–46]. These results indicate that the coupling of remote sensing data and crop models can extend the application scale of existing crop models.

According to the above review, plastic film mulching has great application potential in arid areas of northwest China. However, few studies attempt to clarify the application effect and applicability of plastic film mulching in the whole Heihe River basin; this provided the motivation for the present study. In this study, the spatial distribution of water-saving and yield-increasing effects of plastic film mulching in the Heihe River basin was investigated and suggestions for suitable expansion areas were put forward. First, the maize sowing date and *CC* without statistical data were retrieved from *EVI* data and normalized difference vegetation index (*NDVI*) data, respectively. This lays a foundation for subsequent simulation and analysis. Due to the limited availability of land use data, only data for the period of 2011–2014 were considered. Further, combined with other products for the grids (i.e., meteorological data, land use data, soil variables), the gridding G-AquaCrop model was constructed. The G-AquaCrop model outputted maize yield and *ET* during the whole growth period under different scenarios (mulching/no-mulching). On the basis of reliable output results, the application effect and applicability of plastic film mulching of maize field in the Heihe River basin were quantified. At the same time, a new method to quickly obtain the applicability of plastic film mulching was proposed.

2. Materials and Methods

2.1. Study Area

This study focused on the maize cultivation area of the Heihe River basin (Figure 1) in the arid inland region of Northwest China. The basin is located in the middle of the Hexi Corridor, at 97°42′~102°04′ E, 37°41′~42°42′ N and covering an area of 142,900 square kilometers [47]. The maize cultivation areas in this study are mainly distributed in nine counties: Shandan, Minle, Ganzhou, Linze, Gaotai, Sunan, Suzhou, Jiayuguan, and Jinta in the middle and lower reaches of the Heihe River basin, as shown in Figure 1. In this area, the climate is mainly temperate continental arid climate; the annual average temperature is 4–8 °C; the annual average precipitation decreases from more than 250 mm in the east to less than 50 mm in the west; the annual average evaporation increases from the east to the west, from less than 2000 mm to more than 4000 mm; and the annual average sunshine duration reaches 3000–4000 h [48–50]. It is a relatively suitable area for agricultural development in the whole Heihe River basin. The main crops in the region are maize, spring wheat, spring barley, and spring rapeseed. Among these, maize occupies the largest area [51].

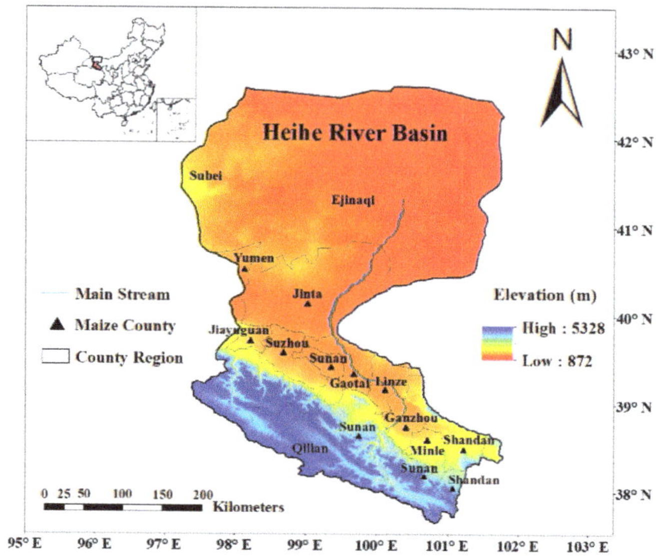

Figure 1. Location map of the Heihe River basin and the nine counties considered in this study.

2.2. Acquisition of Spatial Heterogeneity Variables

2.2.1. Land Use Data

The land use data used in this study were from the Landsat remote sensing data obtained by the Heihe Plan Science Data Center (http://heihedata.org/ (accessed on 10 December 2019)), with a spatial resolution (*SR*) of 30 m. In order to matching other available forcing data, the 500 m *SR* is applied. The 500 m *SR* can also reflect spatial heterogeneity for the relatively flat agriculture land. The area cultivated under maize was re-samples to 500 m resolution using NEAREST by ArcGIS. This data was used as the basic data for this study and also as a reference for other data processing.

2.2.2. Meteorological Data

The ERA (*ECMWF* re-analysis-interim) data of precipitation (P, mm), maximum air temperature (T_{max}, °C), minimum air temperature (T_{min}, °C), wind speed at 10 m height (u_2, m/s), solar net radiation (R_n, MJ m^{-2} d^{-1}), surface pressure (SP, kPa), and dew point temperature (T_{dew}, °C) was obtained from the European Centre for Medium-Range Weather Forecast (*ECMWF*, https://www.ecmwf.int/ (accessed on 11 August 2021)), with a spatial resolution of 0.125° (13.89 km) and a temporal resolution of one day. In this study, the spatial resolution was re-sampled to 500 m using NEAREST with the land use data as the reference.

The reference crop evapotranspiration (ET_0, mm/d) was calculated by the Penman–Monteith method [52]:

$$ET_0 = \frac{0.408 \cdot \Delta(R_n - G) + \gamma \cdot \frac{900}{T+273} \cdot u_2 \cdot (e_s - e_a)}{\Delta + \gamma \cdot (1 + 0.34 \cdot u_2)} \quad (1)$$

where R_n is the average net radiation of the crop surface; G is the soil heat flux, which is approximately equal to 0 in a day, γ is the hygrometer constant, T is the average temperature; u_2 is the average wind speed at a height of 2 m, e_s is the saturated water pressure at average temperature, e_a is the actual water pressure; and Δ is the slope of the water pressure curve.

2.2.3. Soil Variables

The main soil types in the Heihe River basin include clay, sand, and silt. The distribution of soil types was obtained from the Digital soil mapping dataset of soil texture in the Heihe River basin (2012–2014) (spatial resolution was 100 m) [53] downloaded from the National Tibetan Plateau Data Center (https://data.tpdc.ac.cn/zh-hans/ (accessed on 25 August 2020)). The distribution of the soil hydraulic properties (including soil saturated water content, field water capacity, and saturated water conductivity) were also derived from the Digital soil mapping dataset of hydrological features in the Heihe River basin (2012) (spatial resolution is 90 m) [54] downloaded from the same website. In this study, the spatial resolution was uniformly aggregated to 500 m by NEAREST.

2.2.4. Sowing Date Retrieval

The sowing date is the starting point of time for crop growth and yield simulation by the AquaCrop model. Accurate acquisition of the spatial distribution data of sowing date in grids is of great importance for maize yield and *ET*. With the help of *MODIS-NDVI*, Han et al. (2019) obtained the sowing date of maize in the middle reaches of the irrigation area in the Heihe River through the threshold method [45]. However, many studies have shown that *EVI* has been improved to overcome the problems of atmospheric noise, soil background, and saturation on the basis of *NDVI*, and more and more studies choose to use *EVI* for phenological monitoring [55–59]. Therefore, in this study, MYDEV1D China 250 M *EVI* daily product of Geospatial Data Cloud (http://www.gscloud.cn/#page1/3 (accessed on 18 September 2020)) was used for sowing date retrieval. The sowing date was calculated using [60]:

$$\begin{cases} EVI_i > \beta \\ EVI_{i+5} - EVI_i > 0 \\ EVI_{i+10} - EVI_{i+5} > 0 \\ sowing\ date = i - 10 \end{cases} \tag{2}$$

where EVI_i is the *EVI* value of day *i*, and β is the empirical threshold, which is closely related to crop variety and planting area. Due to the lack of empirical data, this study used the empirical sowing date data of the Yingke Irrigation District in the Heihe River basin, i.e., 22 April [61], to carry out the backward deduction so that the calculated sowing date generally falls on 22 April. Here, β is 0.22.

2.2.5. Canopy Coverage Retrieval

Canopy coverage (*CC*) is an important variable connecting the maize plant body and the AquaCrop model, which reflects crop growth [62]. In this study, MYDND1M China 500 M *NDVI* monthly composite product of the Geospatial Data Cloud (http://www.gscloud.cn/#page1/3 (accessed on 23 September 2020)) was adopted. The spatial resolution of data was re-sampled to 500 m using NEAREST by ArcGIS. The CC was calculated using:

$$WDRVI = \frac{[(\alpha+1)NDVI + (\alpha-1)]}{[(\alpha-1)NDVI + (\alpha+1)]} \tag{3}$$

$$LAI = LAI_{max}\frac{WDRVI - WDRVI_{min}}{WDRVI_{max} - WDRVI_{min}} \tag{4}$$

$$CC = 1.005 \times [1 - exp(-0.6LAI)]^{1.2} \tag{5}$$

where *WDRVI* is the wide dynamic threshold vegetation index [63], LAI_{max} is the maximum leaf area index of maize, which was set as 5 [64]; and α is the empirical coefficient, which was set as 0.2.

2.3. *G-AquaCrop Model Parameterization & Yield and ET Simulation*

2.3.1. G-AquaCrop Model

Since the AquaCrop model came into being, it has been widely used and recognized, especially in arid areas [25,26,33,35].

In the simulation of the AquaCrop model, the key process of conversion from CC to yield (Y) may be summarized as [26]:

$$CC^* = 1.72CC - CC^2 + 0.30CC^3 \tag{6}$$

$$T_r = K_S CC^* Kc_{Tr,x} ET_0 \tag{7}$$

$$B = Ks_b WP^* \sum \frac{T_r}{ET_0} \tag{8}$$

$$Y = f_{HI} HI_0 B \tag{9}$$

where CC^* is the adjusted canopy coverage, $Kc_{Tr,x}$ is the largest standard crop transpiration coefficient, K_s is the soil water stress coefficient, T_r is the plant transpiration, WP^* is the normalized water productivity, Ks_b is the air temperature stress coefficient, HI_0 is the reference harvest index, and f_{HI} is the water stress adjustment factor. In this study, water stress was not considered, and so both K_s and f_{HI} were 1. According to Equations (5)–(8), accurate simulation of CC is a necessary prerequisite for the accurate simulation of maize Y and ET (For more detailed model mechanism process, please refer to Raes et al. (2009)).

However, since the Heihe River basin covers a large area and the meteorological, soil, vegetation, and other variables have significant spatial heterogeneity, it is impossible to use the point-scale AquaCrop model for simulation. Therefore, this study combined the advantages of AquaCrop plug-in v6.0, ArcMap 10.0, and Matlab R2018a software to build a G-AquaCrop model. The G-AquaCrop model uses ArcGIS 10.0 software to grid all spatial heterogeneity variables obtained by remote sensing. Each grid in the Heihe River basin was taken as an independent simulation unit and was simulated by the AquaCrop model. Then, the ArcGIS 10.0 software was used to present the simulation results of each grid in the basin map. Due to the large number of grids in the Heihe River basin, this study prepared each input file required for the model operation in advance and used Matlab R2018a to call AquaCrop plug-in v6.0 to simulate. Calendar Day was used for simulation in this study.

2.3.2. Setting Simulation Scenarios for the G-AquaCrop Model

In order to compare the effects of plastic film mulching on Y and ET in the Heihe River basin, two simulation scenarios were set up in this study: (1) plastic film mulching; and (2) no plastic film mulching. In order to control the variables and highlight the influence of plastic film mulching, the irrigation schedule of the two scenarios was unified. Due to the lack of actual irrigation statistics, the irrigation schedule was set based on the irrigation experience of Yingke irrigation area. Yingke irrigation area is located in the middle reaches of the Heihe River basin, which could be regarded as a good representative for the region. The irrigation amount in Yingke irrigation area was 507–819 mm, and its average value (663 mm) was taken as the total amount of irrigation here. The irrigation events were four times, with border irrigation. The water stress was excluded from all scenarios in this study. It is assumed that there is no fertility stress, i.e., fertility is not limited. At the same time, due to the widespread application of pesticides in agricultural production for the study period, plant diseases and insect pests were not included in the two scenarios. In addition, according to the FAO recommended value, under the film mulching condition, the mulching proportion parameter was set to 70% and the mulching effect parameter was set to 100%. Under the condition of no film mulching, the film mulching proportion and the film mulching effect parameters were set as 0 (Table 1). Due to the lack of spatial distribution data of plastic film mulching in the Heihe River basin, this study considered the non-plastic film mulching scenario as the current basic scenario, and the simulation results of Y under the non-plastic film mulching condition were compared with the data in the 2012–2015

Statistical Yearbook of the Gansu Provincial Bureau of Statistics (http://tjj.gansu.gov.cn/ (accessed on 13 December 2020)). Table 1 shows the management parameter settings in different scenarios.

Table 1. Parameters for characterizing on-farm water management interventions in AquaCrop.

	Technology Name	Water Application Efficiency (%)	Surface Soil Wetting Area (%)
Irrigation technology	Border √ Sprinkler Drip	60 75 90	80 100 30
	Mulching Method	**Material Factor (f_m)**	**Mulching Cover Area (%)**
Soil management	No mulching √ Organic mulching Plastic mulching √	0 0.5 1	0 100 70
	Soil Fertility limiting	**Soil Fertility Stress (%)**	
Soil fertility	No limiting √ 20–99%	0 74–2	
	Date	**Water Irrigation Quantity (mm)**	
Irrigation schedule	5/31 6/30 7/31 8/31	166 166 166 165	

√: Management measures used in this study.

2.3.3. Tuning and Simulation of G-AquaCrop Model

Increasing the accuracy of the simulation parameter in each grid can improve the simulation of *Y* in the G-AquaCrop model and the accuracy of *ET* simulation. In addition to the spatial heterogeneity variables that can be obtained from remote sensing data, there are some mechanical parameters involved in the G-AquaCrop model. According to past studies [35], the nine parameters in Table 2 are the most sensitive parameters about maize *Y* and *ET* in the model, which would be adjusted in this study.

Table 2. Optimization parameters and their default values and calibrated values.

Parameters	Description	Default Value	Calibrated Value
$Kc_{Tr,x}$	Crop coefficient when canopy is complete but prior to senescence	1.05	1.00–1.25
cc0	Soil surface covered by an individual seedling at 90% emergence (cm^2/plant)	6.5	5.00–8.00
Numpph	Number of plants per hectare	50,000–100,000	50,000–100,000
CGC	Canopy growth coefficient (fraction per day)	No reference is given on the calendar mode	0.03–0.07
CCx	Maximum canopy cover (%)	65–99	60–100
CDC	Canopy decline coefficient (fraction per day)	No reference is given on the calendar mode	0.03–0.07
mat	Time from sowing to maturity, i.e., length of crop cycle (day)	No reference is given on the calendar mode	153–183
wp	Water productivity normalized for ETo and CO_2 (gram/m^2)	33.7	20.0–40.0
HI_0	Reference harvest index (%)	48–52	30–55

As mentioned above, accurate simulation of CC is the premise of accurate simulation of maize Y and ET. Therefore, monthly CC values of remote sensing retrieval were taken as the verification data of each grid's parameter in the basic scenario (that is, without film mulching). With the help of Matlab R2018a software, a particle swarm optimization (*PSO*) algorithm was used to conduct gridding optimization and parameter tuning [65,66]. In order to highlight the effect of film mulching and simplify the simulation process, the same set of parameters was used for the corresponding grid of film mulching and no-film mulching scenarios, and only the proportion parameter and effect parameter of film mulching were changed.

The objective of parameter optimization is to minimize the difference between the CC values obtained by remote sensing and which output by the model on the corresponding date. The objective function is:

$$\min f_{CC} = \sqrt{\frac{1}{n}\sum(CC_{RS} - CC_S)^2} \quad (10)$$

where CC_{RS} refers to the CC value of remote sensing retrieval, CC_S refers to the CC value of the corresponding date output by AquaCrop model, and n is the number of CC_{RS} during the growth period of maize in each grid.

Parameters were optimized via the server. The CPU was Intel (R) Xeon (R) Silver 4214 @2.20 GHz 2.19 GHz, and the operating memory was 128 GB. In the *PSO* algorithm, the number of particles and iterations were both set to 30. The elapsed time of optimization was 210 s for each grid. Due to the large number of grids in the research area, taking the year 2014 as an example, the number of grids was 9901. Therefore, all grids of each year were first grouped according to the sowing date to improve optimization efficiency. Some grids were selected for parameter optimization in each group, which (about 60) with the nearest grid number shared a set of parameter optimization results. The spatial distribution of parameter optimization objective function is shown in Figure 2, where, $\min f_{CC} < 0.2$ means the optimization result is Very good, $0.2 \leq \min f_{CC} < 0.6$ means the optimization result is Good, and $\min f_{CC} > 0.6$ means the optimization result is Normal. It can be seen from Figure 2 that satisfactory results were obtained during parameter optimization for 4 years, providing a solid foundation for the accurate simulation of Y and ET. When $\min f_{CC}$ is reached, the input parameters of this model are adopted as the optimal parameters, as shown in Table 2.

After the parameters were adjusted, each grid-specific parameter was brought into the G-AquaCrop model to carry out the simulation of Y and ET. Through extensive literature reading, we obtained some ground experimental observations from published papers. The data were collected from three ground stations in the Heihe River basin, including maize yield and leaf area index (*LAI*). Using Equation (5), *LAI* was converted into CC. The data details are shown in Table 3. Then, we fitted the simulation results of the corresponding grid with the observation data to evaluate the simulation effect of G-AquaCrop model. Due to the limited amount of ground observation data, this study also averaged the Y per unit area of grid in the same county without film mulching and compared it with the Y per unit area of the county obtained through the Statistical Yearbook. Here, we used the following statistical evaluation indicators: the root mean squared error (*RMSE*), normalized root mean squared error (*NRMSE*), Willmott's index of agreement (*d*), and mean bias error (*MBE*).

$$RMSE = \left[\frac{\sum_{i=1}^{n}(M_i - S_i)^2}{n}\right]^{0.5} \quad (11)$$

$$NRMSE = \frac{100}{\overline{M}} \left[\frac{\sum_{i=1}^{n} (M_i - S_i)^2}{n} \right]^{0.5} \tag{12}$$

$$d = 1 - \frac{\sum_{i=1}^{n} (M_i - S_i)^2}{\sum_{i=1}^{n} \left(|S_i - \overline{M}| + |M_i - \overline{M}| \right)^2} \tag{13}$$

$$MBE = \frac{1}{n} \sum_{i=1}^{n} (S_i - M_i) \tag{14}$$

where M_i is field observation data or yearbook data, and S_i is the simulated data. When M_i represents field observation data, n represents the quantity of observation data; when M_i represents yearbook data, n is the number of counties. For the *RMSE*, when the *RMSE* value is close to 0, the simulation results are in good agreement with the measured results. While for the d, the closer it is to 1, the better the simulation result is. The *NRMSE* can indicate the degree of overall consistency, and the *MBE* can represent positive and negative attributes of model bias.

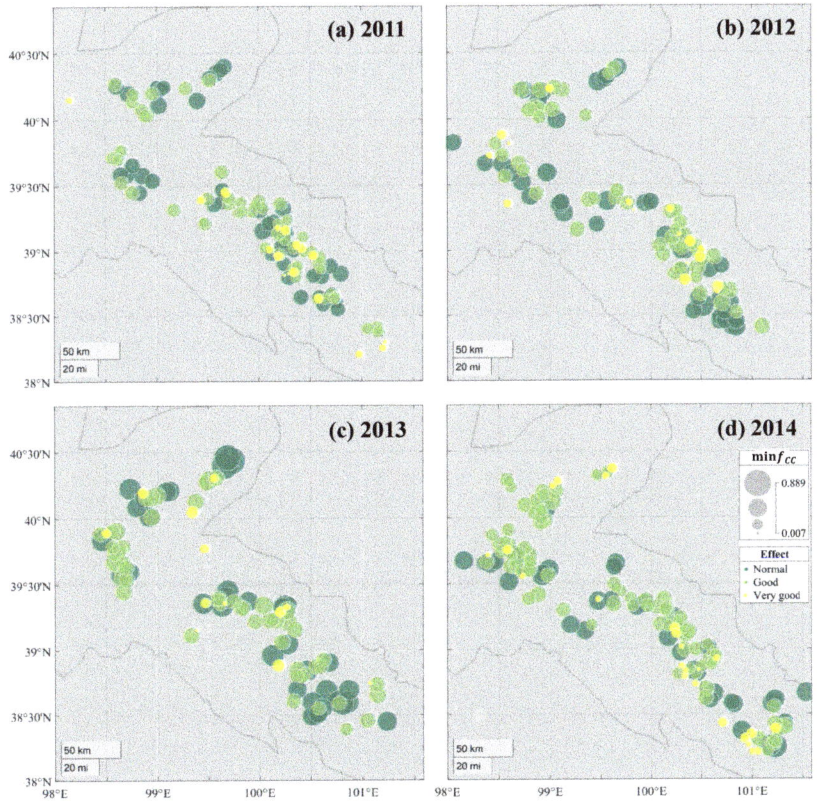

Figure 2. Spatial distribution of parameter optimization objective function (minf_{CC}) in the Heihe River basin for four different years: (**a**) 2011, (**b**) 2012, (**c**) 2013, and (**d**) 2014.

Table 3. Field observation data details.

Field Stations	Coordinate	Crop	Parameter	Year	n	Data Sources
1	(100°07′ E, 39°21′ N)	Spring Maize	Yield	2011	1	Zhang (2012) [67]
2	(100°20′ E, 38°51′ N)	Spring Maize	Yield Canopy Coverage	2012, 2013 2013	2 7	Jiang (2017) [68]
3	(100°34′ E, 38°57′ N)	Spring Maize	Yield Canopy Coverage	2012, 2013 2013	2 3	Jiang (2017) [68]

2.4. Relationship between Mulching Areas and Meteorological Variables

The suitable areas for mulching were given by comparing the model output results under different conditions. Based on this result, the study attempts to explore the relationship between the mulching areas and their meteorological variables (accumulated temperature and precipitation). Here, the relationship between mulching areas and meteorological variables was expressed by a ratio, which of the number of grids recommended for mulching to the total number of grids under each meteorological condition:

$$r_i = \frac{N_{mulch}}{N_{mulch} + N_{no-mulch}} \quad (15)$$

where i refers to the specific total accumulated temperature value or total precipitation value in each grid during the whole crop growth period. When i stands for accumulated temperature, 1200 d°C $< i <$ 3000 d°C; when i represents precipitation, 0 mm $< i <$ 1000 mm. N_{mulch} is the number of grids mulching under condition of i meteorological value; $N_{no-mulch}$ is the number of grids without mulching under i condition; r_i is the ratio of the recommended grid number for film mulching to the total grid number under the same accumulated temperature value or precipitation value.

3. Results

3.1. Spatial Distribution of Sowing Date

Through the inversion method of the sowing date introduced above, the spatial distribution map of the maize sowing dates from 2011 to 2014 in the Heihe River basin was obtained, as shown in Figure 3. As can be seen from Figure 3, the sowing date of maize in the Heihe River basin was approximately from the 91st day to the 142nd day of a year, with a span of 52 days. In most areas, the sowing dates are concentrated on the 111th–125th day, around 22 April. From the spatial perspective, this part of the region is densely distributed in the Ganzhou and Linze counties. The maize areas in these two counties are irrigated areas, where unified field management measures are implemented. The sowing date of maize in the irrigated areas is about 22 April, which is consistent with the results obtained in this study. However, in areas outside the irrigated areas, such as Jinta and Suzhou, there are certain differences between the sowing dates and the irrigated areas due to differences in geography, climate, and crop varieties. It is obvious that the sowing date in the northern part of the Ganzhou region is generally early, which may be related to the maize varieties planted in this region. Ganzhou is the largest maize seed production base at the county level in China. However, some studies have shown that in actual production, the seed maize should be separated from the sowing date of field maize with a flexible time of 30–40 days [69] to better carry out seed production. This is why the sowing date in the northern region of Ganzhou is regularly advanced.

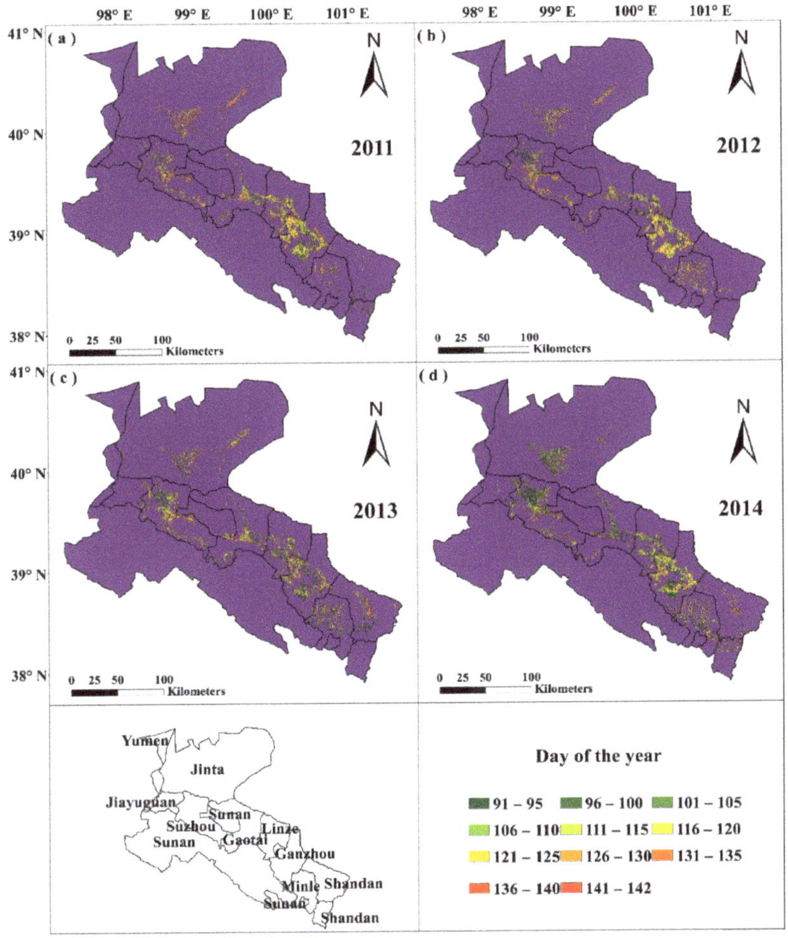

Figure 3. Sowing date of maize in the Heihe River basin for four different years: (**a**) 2011, (**b**) 2012, (**c**) 2013, and (**d**) 2014.

3.2. Spatial Distribution of Y under Different Scenarios

Figure 4 shows the spatial distribution of maize *Y* per unit area in the Heihe River basin from 2011 to 2014 under the two scenarios. Figure 4a,c,e,g represents the no-film mulching scenario for the four respective years (top to bottom), and Figure 4b,d,f,h represents the film mulching scenario. In general, the maximum *Y* per unit area of maize in the Heihe River basin was 17.6–20.7 t/ha, while the average *Y* was only 8.6–10.0 t/ha. These values indicated that there was a large difference in the spatial distribution of maize *Y* per unit area. At the same time, after film mulching, the maximum *Y* did not increase significantly, but in a local range, film mulching did improve the *Y* of maize to a certain extent. From one point of view, plastic film mulching is already in use in some areas. If plastic film mulching was covered in a wide range, there would be two consequences: first, plastic film mulching could not reach the desired effect; second, it would put great pressure on the ecological environment.

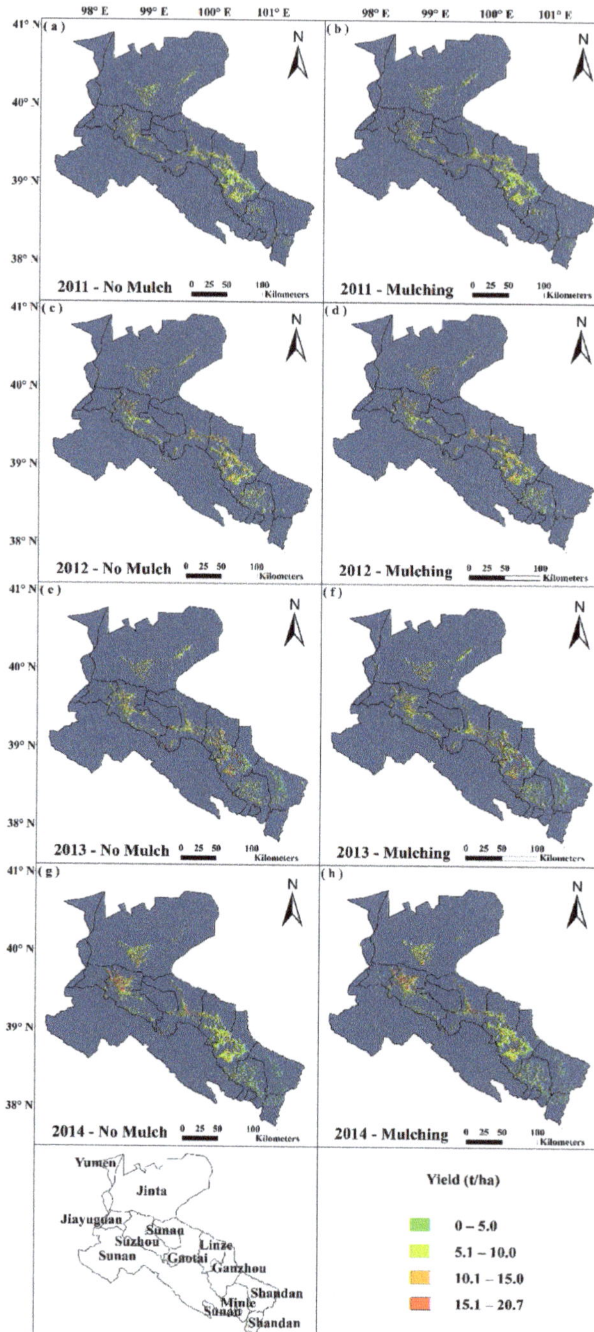

Figure 4. Spatial distribution of simulated maize yield in the Heihe River basin from 2011 (top) to 2014 (bottom): (**a**,**c**,**e**,**g**) are the *Y* values without mulching; and (**b**,**d**,**f**,**h**) are the *Y* values with mulching.

Figure 5 shows the comparison between the observed and simulated *Y* and *CC* of maize at the three field stations in the Heihe River basin. Through consulting the published literature, we obtained some data of maize planting experiment in Heihe River basin. After excluding the data with inconsistent planting years and unclear planting locations, data of the three field stations in Table 3 were finally selected for comparison. The *d* and R^2 values for maize *Y* and *CC* were larger than 0.90, indicating good agreement for both *Y* and *CC*. The simulated *Y* slightly underestimated the observed values, with *MBE* of −0.51 t/ha. While the simulated *CC* slightly overestimated the observed values, with *MBE* of 1.6%. Considering all the statistical indicators, the G-AquaCrop model has good simulation performance. The results of Zhang et al. (2021) also included 5 *Y* values [70]. All the statistical indicators about *Y* in our study are superior to the results of Zhang et al.

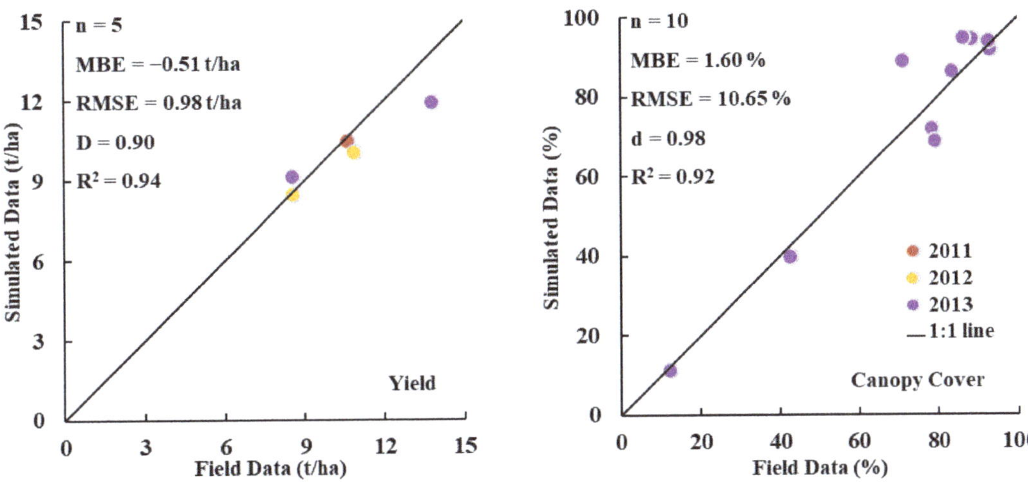

Figure 5. Comparison between the observed and simulated (**left**) *Y* and (**right**) *CC* of maize at the three field stations in the Heihe River basin.

Table 4 shows the fitting effect of maize *Y* per unit area at the county level in the Heihe River basin from 2011 to 2014. The results showed that, under the condition of no-film mulching, the *RMSE* and *NRMSE* of spatial average *Y* reached 1.29 t/ha and 14.37 (dimensionless), respectively. The absolute values of *MBE* were not larger than 0.77 t/ha. However, the absolute value of maize yield *MBE* was 1.19 t/ha in the study of Zhang et al. (2021) [70]. For 9901 grids in the Heihe River basin, the fitting accuracy of our study is acceptable. After film mulching, the *Y* in each county was generally improved, to a certain extent. However, the simulated *Y* in Shandan in 2013 and 2014 was abnormally low. According to the meteorological data of these two years, spring drought of different degrees occurred in Shandan from January to April in 2013 and 2014. The government and agricultural management departments issued early warning information in advance and suggested farmers to take countermeasures. Therefore, the Statistical Yearbook data on *Y* did not show much decline. However, in the G-AquaCrop model simulation, empirical irrigation data was still used, resulting in abnormally low values. Similarly, spatial averaging weakened the local effect of film mulching, as will be described in detail in Section 3.3.

Table 4. Maize yield simulation in counties in the Heihe River basin for four different years (2011–2014). (Y_Y represents the yearbook yield per unit area, Y_N represents the yield per unit area of the model without film mulching, and Y_M represents the yield per unit area of the model without film mulching in t/ha.).

Counties /Indicators	2011			2012			2013			2014		
	Y_Y	Y_N	Y_M	Y_Y	Y_N	Y_M	Y_Y	Y_N	Y_M	Y_Y	Y_N	Y_M
Shandan	9.75	7.27	7.27	9.76	8.15	8.15	9.15	5.46	5.46	9.57	3.50	3.51
Minle	8.22	7.90	7.90	8.08	8.27	8.53	8.25	8.29	8.29	8.59	6.81	6.81
Ganzhou	8.34	8.60	8.59	8.08	9.87	9.87	8.08	9.88	9.88	8.65	8.07	8.07
Linze	8.22	9.23	9.44	6.95	8.53	10.96	7.03	10.58	10.58	7.40	9.01	9.01
Gaotai	8.42	9.86	10.14	7.29	11.36	11.35	7.28	9.86	9.86	7.77	10.31	10.32
Sunan	7.77	8.69	8.76	7.55	9.93	9.93	6.32	9.31	9.31	6.88	8.95	8.97
Suzhou	8.59	9.60	10.02	9.45	10.08	10.08	9.29	9.79	9.79	9.25	10.59	10.61
Jiayuguan	10.57	10.57	10.58	7.90	9.99	9.99	11.48	9.73	9.73	11.86	11.62	11.64
Jinta	10.52	8.79	8.80	11.91	10.47	10.48	11.35	9.34	9.36	10.96	10.01	10.02
Yumen	9.94	8.33	8.33	10.64	10.47	10.47	9.37	8.04	8.07	8.90	7.54	7.55
RMSE (t/ha)	1.29		-	2.03		-	2.33		-	2.44		-
NRMSE	14.37		-	23.17		-	26.57		-	27.13		-
MBE (t/ha)	−0.15		-	0.77		-	0.27		-	−0.37		-

In general, the distribution of maize in the irrigated areas of Ganzhou and Linze was relatively concentrated and the *Y* value was relatively stable, which is closely related to the mature planting system and field management measures in the irrigated areas. Han et al. (2019) obtained the spatial distribution of maize *Y* in the Yingke Irrigated Area in 2012 and 2015 [45]. In terms of quantity, the maize *Y* in the Yingke Irrigated Area ranged from 7.8 to 13.7 t/ha, which was consistent with the results obtained in this study. At the same time, Han et al. (2019) also found that the reason why the *Y* of 2015 was higher than that of 2012 might be that the sowing date of 2015 was about 10 days earlier than that of 2012. Compared with Figures 3 and 4, in this study, the sowing date of northern Suzhou was earlier and its *Y* was higher, which was consistent with the results of Han et al. (2019).

3.3. Change in Y after Film Mulching

The Figure 6 shows the spatial distribution of *Y* change ratio after plastic film mulching in the Heihe River basin from 2011 to 2014. It can be clearly seen from the spatial three-dimensional diagrams that the *Y* change of most areas in the basin is −2–0%, and the negative value mainly comes from the random error of the model. In this study, the maize *Y* in the irrigated area after film mulching did not increase significantly, while the *Y* in Yumen, Jiayuguan, Jinta, Suzhou, Sunan, Shandan, and Minle counties significantly increased after film mulching, with the maximum increase of nearly 8%. There are two main reasons. First, the *Y* of maize in the irrigated area hardly increased after plastic film mulching, which may be due to the improvement of irrigation system, water and fertilizer management, and field management measures in the irrigated area at the present stage. Plastic film mulching may have been used in a large range, and the *Y* potential of maize has been fully tapped. However, since the spatial distribution of the film-mulched areas in the Heihe River basin could not be obtained at the present stage, this study assumed that the whole Heihe River basin was not film-mulched during parameter adjustment, and the CC data of remote sensing retrieval used in parameter adjustment was the actual CC. It is possible that the CC used for the parameter adjustment may be relatively larger in the irrigated area. Second, Jiayuguan, Jinta, and Suzhou are located in the middle and lower reaches of the Heihe River basin, whose latitudes are higher than that of the other counties, so the corresponding temperature is relatively low and the annual average precipitation is also less. Film mulching can increase soil surface temperature and reduce soil evaporation, thus promoting various physiological processes of maize plants. Therefore, after film mulching, the maize *Y* of Jiayuguan, Jinta, and Suzhou improved to a certain extent.

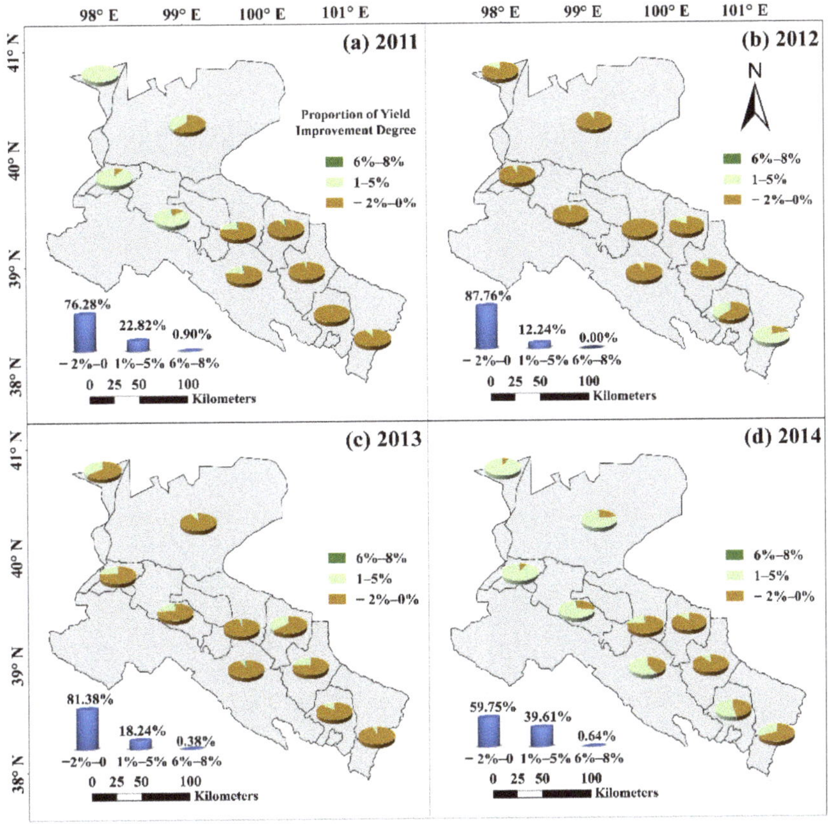

Figure 6. Spatial distribution of maize *Y* improvement degree after mulching and the proportion of each improvement degree in the Heihe River basin for four different years: (**a**) 2011, (**b**) 2012, (**c**) 2013, and (**d**) 2014.

3.4. ET Changes after Film Mulching

Figure 7 shows the spatial distribution of changes in *ET* after plastic film mulching in the Heihe River basin from 2011 to 2014. As can be seen from these diagrams, after plastic film mulching, *ET* in the whole maize planting area in the Heihe River basin decreased, with the maximum reduction reaching 30%. Further, *ET* in most areas decreased between 0.9 and 10.0%, including the irrigated areas with relatively perfect field management measures. The areas with large *ET* reduction were concentrated in Shandan and Minle, especially in Shandan, where the *ET* reduction reached 20–30% in 2014. This was mainly due to a large degree of spring drought in Shandan from January to April 2014, which reduced the soil evaporation by plastic film mulching.

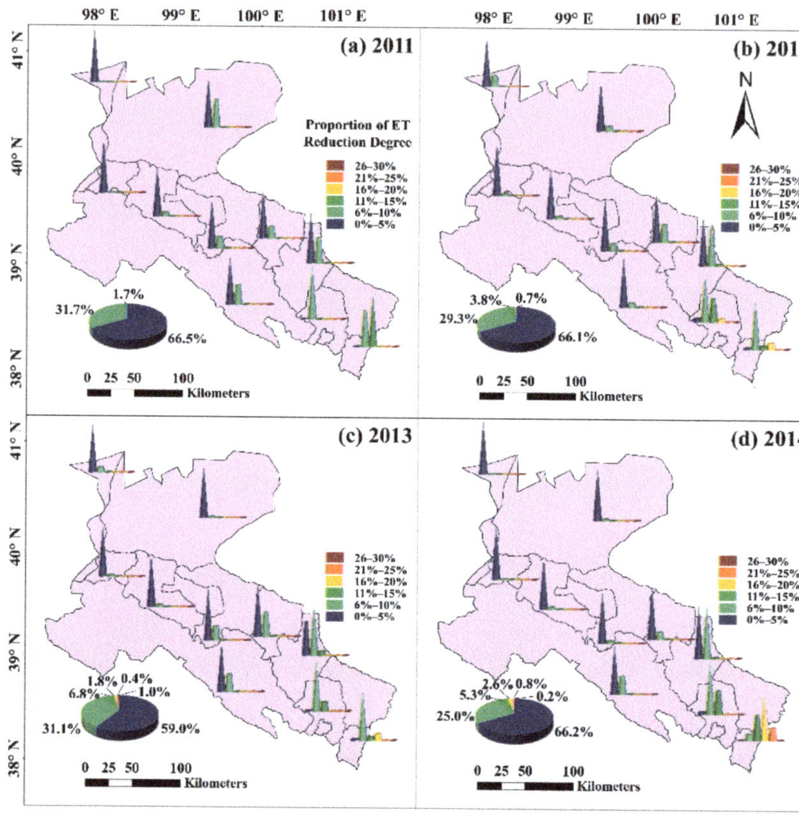

Figure 7. Spatial distribution of maize *ET* reduction degree after mulching and the proportion of each reduction degree in the Heihe River basin for four different years: (**a**) 2011, (**b**) 2012, (**c**) 2013, and (**d**) 2014.

3.5. Suggestions for Film Mulching According to Changes in Yield

On the basis of the spatial distribution in Figure 6, a statistical chart of *Y* change range after plastic film mulching in the Heihe River basin from 2011 to 2014 is shown in the bar chart of each plot in Figure 6.

As can be seen, in four years, the areas where *Y* barely changed after plastic film mulching accounted for 59.8–87.8% of the maize cultivation area in the entire Heihe River basin. Among the areas where *Y* increased, the areas where *Y* increased by 0.1–5.0% accounted for the largest proportion, and the areas where *Y* increased by 5.1–8.0% accounted for less than 0.9% of the maize planting area. As mentioned above, the areas with no significant change in maize *Y* after plastic film mulching was mainly the irrigated areas in the middle reaches of the river basin, where various field management measures are relatively mature and crop *Y* potential is largely developed. On this basis, the areas with improved *Y* in the simulation should be encouraged to expand the area mulched with film in the actual production at the present stage to increase the maize yield. Figure 8 presents the proposed map of the mulched area according to different targets, including changes of maize *Y* (bottom layer in each plot), *ET* (middle layer), and *Y* and *ET* (top layer). As can be seen from the bottom layer in each plots, it is recommended to continue to expand the area of plastic film mulching in the southern areas of Jinta, Suzhou, Minle, and Shandan to obtain a higher yield. Irrigated areas continue to maintain the current status of field management.

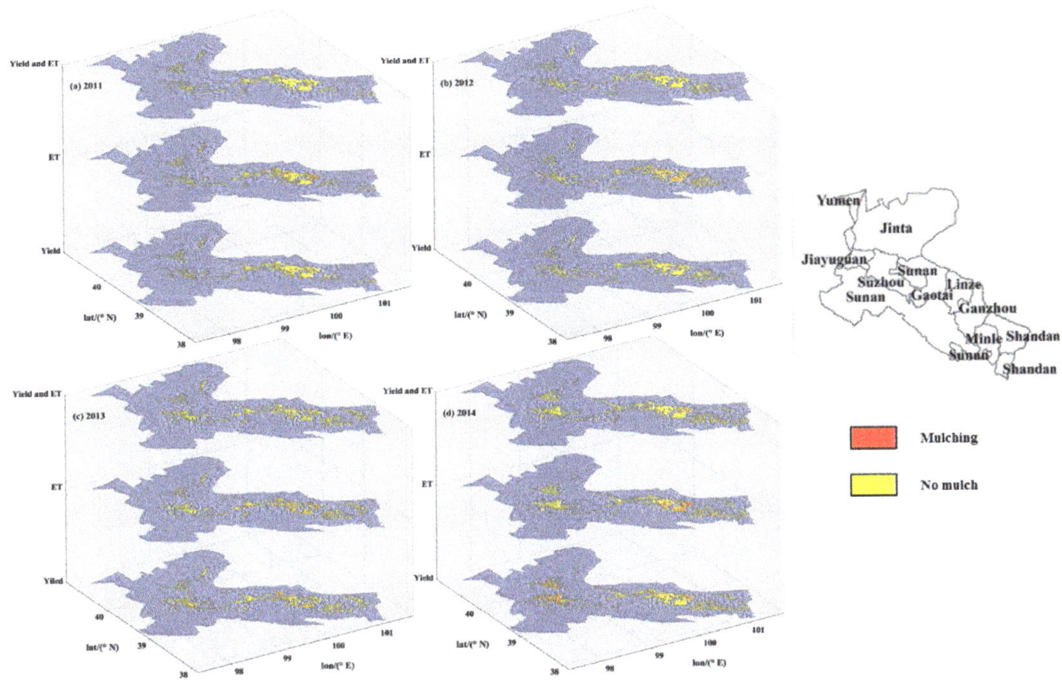

Figure 8. The recommendations for mulching according the change of maize Y (bottom map in each figure); ET (middle map in each figure); and maize Y and ET (top map in each figure) after mulching for four different years: (**a**) 2011, (**b**) 2012, (**c**) 2013, and (**d**) 2014. The red area is recommended for mulching, and the yellow area is not recommended for mulching.

3.6. Suggestions Based on ET Changes

On the basis of the spatial distribution in Figure 7, we summarized the statistical chart of changes in ET after plastic film mulching in the Heihe River basin from 2011 to 2014, as shown in the pie chart of Figure 7. From 2011 to 2014, after plastic film mulching, ET in the whole Heihe River basin decreased by varying degrees, while 59.8–66.6% of the region only decreased by 5.0% (the pie chart of Figure 7). Considering that the plastic film cannot be degraded in a short time after use, leading to white pollution to the ecological environment of the area, the water-saving benefits brought by the plastic film mulching will be greatly reduced. Therefore, it is not necessary to expand film mulching in areas where ET reduction is less than 5%. Increasing the plastic film mulching area in the areas where ET reduction exceeds 5% can reduce the water consumption of crops to a certain extent, thus promoting the sustainable development of water-saving agriculture. With the above results, a proposed film mulching map for the Heihe River basin was drawn based on the distribution of ET reduction during 2011–2014, as shown in the middle layer of each plot in Figure 8. The results indicate that it is more suitable to increase the area of film mulching in Ganzhou, Minle, Shandan, and southeast of Linze in the Heihe River basin, while it is not necessary to increase it in the other areas.

3.7. Suggestions Based on Y and ET Changes

The ability of plastic mulch to reduce soil evaporation and increase yield provides a feasible solution to improve crop water productivity. Therefore, combined with the lower and middle layers of each plot in Figure 8, it is recommended to apply film mulching in areas with increased Y and reduction of more than 5% in ET, mainly distributed in some areas of Ganzhou, Minle, Shandan, and other counties. Mulching film in this area can improve *WP* and achieve water saving and yield increase (the upper layer of each plot in Figure 8).

3.8. Relationship between Mulching Areas and Meteorological Variables

Relationship between mulching areas and meteorological variables were studied during the period 2011–2014 (Figure 9). As can be seen from Figure 9a, the proportion of film mulching is the highest when the accumulated temperature is between 2100 and 2400 d°C. When the accumulated temperature exceeds 2700 d°C and below 1800 d°C, there is almost no film mulching. The reason is obvious: due to full irrigation, a large accumulated temperature can support the growth of maize independently, while too small accumulated temperature cannot meet the physiological requirements of maize germination. Even with measures such as film mulching, maize cannot grow normally. Figure 9b shows the relationship between film mulching areas and precipitation. As the study area is located in an arid area, the highlights in the figure are concentrated in the part with low precipitation. It can be seen that the proportion of film mulching is highest when the precipitation is between 300 and 400 mm. Figure 9c shows the comprehensive influence of accumulated temperature and precipitation. With the increase of accumulated temperature and the decrease of precipitation, the proportion of film mulching increases. Meteorological conditions suitable for mulching are concentrated in the lower right corner of Figure 9c. By referring to this figure, you can clearly know the accumulated temperature and precipitation most suitable for plastic film mulching. Conversely, when you want to know whether the climatic conditions of an area are suitable for mulching, you can also easily obtain the results by referring to it. Therefore, Figure 9c provides a new method for judging whether mulching is suitable for each maize cultivation area in the Heihe River basin, which greatly improves the efficiency of determining.

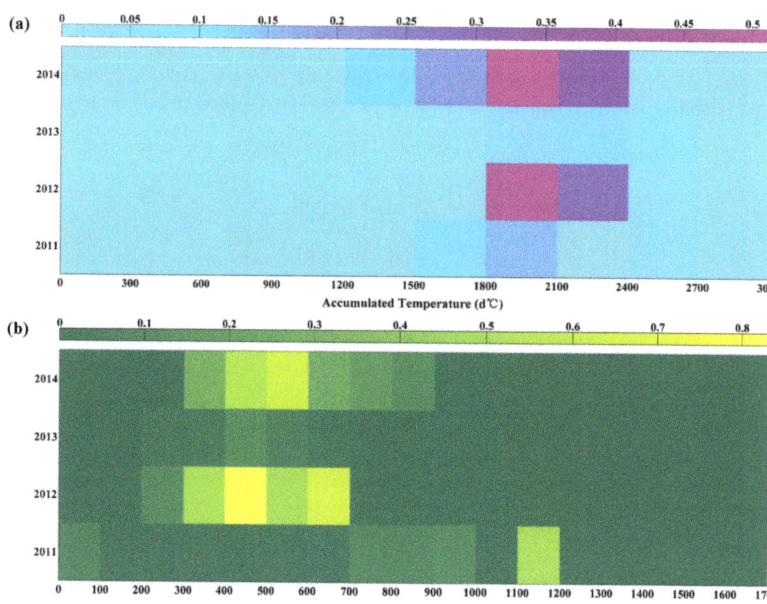

Figure 9. *Cont.*

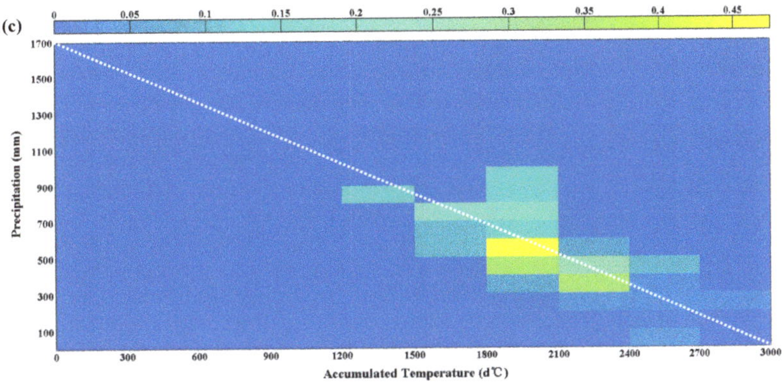

Figure 9. Proportion of mulching under different meteorological conditions in the Heihe River basin for four years: (**a**) accumulated temperature, (**b**) precipitation, (**c**) accumulated temperature and precipitation.

4. Discussion and Conclusions

To study the effects of plastic film mulching on maize yield and *ET* in arid areas, we chose the Heihe River basin as a typical representative to explore the impacts at the regional scale. The gridding large-scale G-AquaCrop model was established fully considering spatial heterogeneity by using remote sensing data. The parameters of the G-AquaCrop model were optimized by the particle swarm optimization method. The yield and canopy coverage of maize were verified by field data, and the R^2 was larger than 0.92.

There is a large amount of plastic-mulched farmland in China [71]. However, the records of the current situation of plastic film mulch used on all statistical yearbooks only included the mulching weight used. The spatial distribution of plastic film mulch was not recorded. This would undoubtedly cause some obstacles to exploring the application effect and applicability of plastic film mulching in the region. Some scholars used remote sensing technology to identify the situation of plastic film mulch [72–74]. They often obtained original images from UVA or high-resolution satellites. For a large area, a large amount of input is required. At the same time, the crop canopy coverage was extensive when it reached the middle and late growth period. The crop canopy would cover the ground. This could cause the identification results to be not reasonable.

The skillful strategy in our modeling approach realizes relatively reliable simulation in evaluating the effects of film mulching in the studied region. We set up two scenarios: no-mulching and plastic film mulching, and the comparison of the difference between the two scenarios is focused. There were two possible situations if there was little difference between the two scenarios. In the first situation, the area was already mulched with plastic film. The second situation was that plastic film mulch was not applicable in the area. In either situation, the area did not need additional mulching. Too much plastic film mulching not only alters soil moisture distribution, but also reduces crop yields over time. As mentioned earlier, the G-AquaCrop model does not take into account crop disease and insect infestation. Meanwhile, as pesticides have been widely used in the Heihe River basin, the effects of disease and pests are not considered here. From the perspective of the model mechanism, the limitation caused by field data loss was avoided by comparing the two scenarios.

Compared with no mulching, plastic film mulching could retain soil moisture and significantly reduce soil evaporation. The biggest drop is 30%. Plastic film mulching could improve maize yield, but the effect of it on maize yield varied spatially. Considering the environmental pressure caused by plastic film mulching, the suggestions for mulching expansion were given from different angles. In general, expanding the plastic film mulching area in the upper and lower reaches of the Heihe River basin could produce greater positive benefits. The middle reaches of the Heihe River basin where the irrigation area is located have been fully explored. This research provides a quantitative and spatial assessment of plastic film mulching application, which helps better to understand management practices.

There are also some limitations of the current study work. Although the results of this study by comparing the differences between the two scenarios are reliable, they can be further improved. With the emergence of additional higher-resolution remote sensing products, we will continue to carry out remote sensing retrieval of actual film mulching conditions and bring data into our research to obtain more accurate and reliable simulation results. The study period is relatively short due to the available land use data on spatial crop varieties, which is important for the regional simulation study. The used data for the period of 2011–2014 was well evaluated by using Google Earth high-resolution images and field survey data, and the overall accuracy reached 92.19%. Meanwhile, the 4 years for the period of 2011–2014 vary in terms of the precipitation variability. For example, the year 2013 and 2014 is the spring drought year, and the year 2012 is with abundant precipitation. The regional precipitation variability has been reflected to a certain degree. Long-term data support would be expected for the next-step investigation.

In terms of research results, the G-AquaCrop model has the original advantages of the AquaCrop model, while overcoming the limitation that the AquaCrop model can only be used on field points. Due to the water-driven attribute of the G-AquaCrop model, its outputs were sensitive to water. Ganzhou, Linze, Gaotai, Sunan, and Suzhou counties are located in the middle reaches of Heihe River basin, which are closed to the Yingke irrigation area. The irrigation schedule in this study was set according to the experience of Yingke irrigation area. Therefore, the yield simulation results in these counties were relatively good. As can be seen from Table 4, the G-AquaCrop model only slightly overestimated the maize yield of these counties. This was also a reflection of the full irrigation (no water stress) setup in this study. Jiayuguan, Jinta, and Yumen counties are located in the lower reaches of the Heihe River basin and have less precipitation. The maize yields of the three counties in Table 4 were underestimated. However, it also reflected that the G-AquaCrop model could well capture water changes, which was suitable for the northwest arid region with water shortage.

Moreover, based on mathematical statistics, the accumulated temperature and precipitation conditions with the maximum probability of plastic film mulching were obtained. The result provided a new method for quickly judging whether a certain area was suitable for plastic film mulching.

Author Contributions: Conceptualization, J.N.; methodology, J.N. and Q.S.; software, Q.S.; validation, N.L.; formal analysis, Q.S.; investigation, Q.S.; resources, J.N.; data curation, Q.S. and N.L.; writing—original draft preparation, Q.S.; writing—review and editing, J.N., N.L. and B.S.; visualization, Q.S.; supervision, J.N.; funding acquisition, J.N. All authors have read and agreed to the published version of the manuscript.

Funding: This work was financially supported by the National Natural Science Foundation of China (51621061, 51679233), the Program of Introducing Talents of Discipline to Universities (B14002) provided by Ministry of Education of the People's Republic of China.

Institutional Review Board Statement: Not applicable.

Informed Consent Statement: Not applicable.

Data Availability Statement: The data used in this research will be available (by the corresponding author), upon reasonable request.

Acknowledgments: Thanks for land use data provided by the Heihe Plan Science Data Center; meteorological data provided by the European Centre for Medium-Range Weather Forecast; soil types and hydraulic properties data provided by the National Tibetan Plateau Data Center; MODIS-EVI/NDVI data provided by Geospatial Data Cloud. The authors thank the reviewers for their constructive comments and useful suggestions on earlier versions of this manuscript.

Conflicts of Interest: The authors declare no conflict of interest.

References

1. Liu, Y.; Li, Y.F.; Li, J.S. Effects of nitrogen management on the growth and yield of mulched and drip-irrigated maize in Northeast Black Soil Regions. *J. Hydraul. Eng.* **2014**, *45*, 529–536. (In Chinese)
2. Li, S.X.; Wang, Z.H.; Li, S.Q.; Gao, Y.J.; Tian, X.H. Effect of plastic sheet mulch, wheat straw mulch, and maize growth on water loss by evaporation in dryland areas of China. *Agric. Water Manag.* **2013**, *116*, 39–49. [CrossRef]
3. Qin, W.; Hu, C.; Oenema, O. Soil mulching significantly enhances yields and water and nitrogen use efficiencies of maize and wheat: A meta-analysis. *Sci. Rep.* **2015**, *5*, 16210. [CrossRef] [PubMed]
4. Lal, R. Sustainable intensification of China's agroecosystems by conservation agriculture. *Int. Soil Water Conserv. Res.* **2018**, *6*, 1–12. [CrossRef]
5. Li, Q.; Li, H.; Zhang, L.; Zhang, S.; Chen, Y. Mulching improves yield and water-use efficiency of potato cropping in China: A meta-analysis. *Field Crop. Res.* **2018**, *221*, 50–60. [CrossRef]
6. Gao, H.; Yan, C.; Liu, Q.; Li, Z.; Yang, X.; Qi, R. Exploring optimal soil mulching to enhance yield and water use efficiency in maize cropping in China: A meta-analysis. *Agric. Water Manag.* **2019**, *225*, 105741. [CrossRef]
7. Huang, Y.; Liu, Q.; Jia, W.; Yan, C.; Wang, J. Agricultural plastic mulching as a source of microplastics in the terrestrial environment. *Environ. Pollut.* **2020**, *260*, 114096. [CrossRef]
8. Kang, S.; Hao, X.; Du, T.; Tong, L.; Su, X.; Lu, H.; Li, X.; Huo, Z.; Li, S.; Ding, R. Improving agricultural water productivity to ensure food security in China under changing environment: From research to practice. *Agric. Water Manag.* **2017**, *179*, 5–17. [CrossRef]
9. Jiang, X. Study on China's Food Security and the "Three Rural Issues". *J. Shanxi Agric. Sci.* **2014**, *42*, 771–785. (In Chinese)
10. Wang, J.; Li, Y.; Huang, J.; Yan, T.; Sun, T. Growing water scarcity, food security and government responses in China. *Glob. Food Secur.* **2017**, *14*, 9–17. [CrossRef]
11. He, J.; Yang, X.; Li, Z.; Zhang, X.; Tang, Q. Spatiotemporal Variations of Meteorological Droughts in China During 1961–2014: An Investigation Based on Multi-Threshold Identification. *Int. J. Disaster Risk Sci.* **2016**, *7*, 63–76. [CrossRef]
12. Ayantobo, O.O.; Li, Y.; Song, S.; Yao, N. Spatial comparability of drought characteristics and related return periods in mainland China over 1961–2013. *J. Hydrol.* **2017**, *550*, 549–567. [CrossRef]
13. Fu, J.; Niu, J.; Kang, S.; Adeloye, A.; Du, T. Crop production in the Hexi Corridor challenged by future climate change. *J. Hydrol.* **2019**, *579*, 124197. [CrossRef]
14. Niu, J.; Kang, S.; Zhang, X.; Fu, J. Vulnerability analysis based on drought and vegetation dynamics. *Ecol. Indic.* **2019**, *105*, 329–336. [CrossRef]
15. Zhang, W.; Nie, Z.; Liu, S.; Wang, J. Threshold of influence of water resources in the Heihe River valley, northwestern Gansu, China on the ecological environment variation of the lower reaches. *Geol. Bull. China* **2006**, *25*, 244–250. (In Chinese)
16. Li, X.; Lu, L.; Cheng, G.; Xiao, H. Quantifying landscape structure of the Heihe River Basin, north-west China using FRAGSTATS. *J. Arid Environ.* **2001**, *48*, 521–535. [CrossRef]
17. Zhou, J.; Hu, B.X.; Cheng, G.; Wang, G.; Li, X. Development of a three-dimensional watershed modelling system for water cycle in the middle part of the Heihe rivershed, in the west of China. *Hydrol. Process.* **2011**, *25*, 1964–1978. [CrossRef]
18. Yaghi, T.; Arslan, A.; Naoum, F. Cucumber (*Cucumis sativus*, L.) water use efficiency (WUE) under plastic mulch and drip irrigation. *Agric. Water Manag.* **2013**, *128*, 149–157. [CrossRef]
19. Liu, E.K.; He, W.Q.; Yan, C.R. 'White revolution' to 'white pollution'—agricultural plastic film mulch in China. *Environ. Res. Lett.* **2014**, *9*, 091001. [CrossRef]
20. Jiang, X.J.; Liu, W.; Wang, E.; Zhou, T.; Xin, P. Residual plastic mulch fragments effects on soil physical properties and water flow behavior in the Minqin Oasis, northwestern China. *Soil Tillage Res.* **2017**, *166*, 100–107. [CrossRef]
21. Gao, H.; Yan, C.; Liu, Q.; Ding, W.; Chen, B.; Li, Z. Effects of plastic mulching and plastic residue on agricultural production: A meta-analysis. *Sci. Total Environ.* **2019**, *651*, 484–492. [CrossRef] [PubMed]
22. Rockström, J. Biomass production in dry tropical zones: How to increase water productivity. In *Land and Water Integration and River Basin Management*; FAO: Rome, Italy, 1995; pp. 31–48.
23. Kijne, J.W.; Barker, R.; Molden, D. *Water Productivity in Agriculture: Limits and Opportunities for Improvement*; CAB International: Wallingford, UK, 2003.
24. Comas, L.H.; Trout, T.J.; DeJonge, K.C.; Zhang, H.; Gleason, S.M. Water productivity under strategic growth stage-based deficit irrigation in maize. *Agric. Water Manag.* **2019**, *212*, 433–440. [CrossRef]
25. Steduto, P.; Hsiao, T.C.; Raes, D.; Fereres, E. AquaCrop—The FAO crop model to simulate yield response to water: I. Concepts and underlying principles. *Agron. J.* **2009**, *101*, 426–437. [CrossRef]
26. Raes, D.; Steduto, P.; Hsiao, T.C.; Fereres, E. AquaCrop—The FAO crop model to simulate yield response to water: II. Main algorithms and software description. *Agron. J.* **2009**, *101*, 438–447. [CrossRef]
27. Hsiao, T.C.; Heng, L.; Steduto, P.; Rojas-Lara, B.; Raes, D.; Fereres, E. AquaCrop—The FAO crop model to simulate yield response to water: III. Parameterization and testing for maize. *Agron. J.* **2009**, *101*, 448–459. [CrossRef]
28. van Diepen, C.A.; Wolf, J.; van Keulen, H.; Rappoldt, C. WOFOST: A simulation model of crop production. *Soil Use Manag.* **1989**, *5*, 16–24. [CrossRef]
29. Sharpley, A.N.; Williams, J.R. *EPIC-Erosion/Productivity Impact Calculator: 1. Model Documentation*; U.S. Department of Agriculture, Agricultural Research Service: Springfield, VA, USA, 1990; pp. 206–207.

30. Jones, J.W.; Hoogenboom, G.; Porter, C.H.; Boote, K.J.; Batchelor, W.D.; Hunt, L.A.; Wilkens, P.W.; Singh, U.; Gijsman, A.J.; Ritchie, J.T. The DSSAT cropping system model. *Eur. J. Agron.* **2003**, *18*, 235–265. [CrossRef]
31. Keating, B.A.; Carberry, P.S.; Hammer, G.L.; Probert, M.E.; Robertson, M.J.; Holzworth, D.; Huth, N.I.; Hargreaves, J.N.G.; Meinke, H.; Hochman, Z.; et al. An overview of APSIM, a model designed for farming systems simulation. *Eur. J. Agron.* **2003**, *18*, 267–288. [CrossRef]
32. Kang, S.; Zhang, L.; Liang, Y.; Dawes, W. Simulation of winter wheat yield and water use efficiency in the Loess Plateau of China using WAVES. *Agric. Syst.* **2003**, *78*, 355–367. [CrossRef]
33. Ran, H.; Kang, S.; Li, F.; Tong, L.; Ding, R.; Du, T.; Li, S.; Zhang, X. Performance of AquaCrop and SIMDualKc models in evapotranspiration partitioning on full and deficit irrigated maize for seed production under plastic film-mulch in an arid region of China. *Agric. Syst.* **2017**, *151*, 20–32. [CrossRef]
34. Rosa, R.D.; Paredes, P.; Rodrigues, G.C.; Alves, I.; Fernando, R.M.; Pereira, L.S.; Allen, R.G. Implementing the dual crop coefficient approach in interactive software. 1. Background and computational strategy. *Agric. Water Manag.* **2012**, *103*, 8–24. [CrossRef]
35. Shen, Q.; Ding, R.; Du, T.; Tong, L.; Li, S. Water Use Effectiveness Is Enhanced Using Film Mulch Through Increasing Transpiration and Decreasing Evapotranspiration. *Water* **2019**, *11*, 1153. [CrossRef]
36. Donatelli, M.; Magarey, R.D.; Bregaglio, S.; Willocquet, L.; Whish, J.P.M.; Savary, S. Modelling the impacts of pests and diseases on agricultural systems. *Agric. Syst.* **2017**, *155*, 213–224. [CrossRef] [PubMed]
37. Tigkas, D.; Tsakiris, G. Early Estimation of Drought Impacts on Rainfed Wheat Yield in Mediterranean Climate. *Environ. Process.* **2015**, *2*, 97–114. [CrossRef]
38. Li, F.; Yu, N.; Zhao, Y. Irrigation Scheduling Optimization for Cotton Based on the AquaCrop Model. *Water Resour. Manag.* **2019**, *33*, 39–55. [CrossRef]
39. Xing, Y.; Liu, D.; Wang, P. Advances of the coupling application of remote sensing information and crop growth model. *Adv. Earth Sci.* **2009**, *24*, 444–451. (In Chinese)
40. Meng, J.-H.; Wu, B.-F.; Du, X.; Zhang, F.-F.; Zhang, M.; Dong, T.-F. A review and outlook of applying remote sensing to precision agriculture. *Remote Sens. Land Resour.* **2011**, *90*, 1–7. (In Chinese)
41. Li, C.; Wang, J.; Wang, X.; Liu, F.; Li, R. Methods for integration of remote sensing data and crop model and their prospects in agricultural application. *Trans. Chin. Soc. Agric. Eng.* **2008**, *24*, 295–301. (In Chinese)
42. Jin, X.; Li, Z.; Feng, H.; Ren, Z.; Li, S. Estimation of maize yield by assimilating biomass and canopy cover derived from hyperspectral data into the AquaCrop model. *Agric. Water Manag.* **2020**, *227*, 105846. [CrossRef]
43. Sakamoto, T.; Nguyen, N.V.; Ohno, H.; Ishitsuka, N.; Yokozawa, M. Spatio–temporal distribution of rice phenology and cropping systems in the Mekong Delta with special reference to the seasonal water flow of the Mekong and Bassac rivers. *Remote Sens. Environ.* **2006**, *100*, 1–16. [CrossRef]
44. Lorite, I.J.; García-Vila, M.; Santos, C.; Ruiz-Ramos, M.; Fereres, E. AquaData and AquaGIS: Two computer utilities for temporal and spatial simulations of water-limited yield with AquaCrop. *Comput. Electron. Agric.* **2013**, *96*, 227–237. [CrossRef]
45. Han, C.; Zhang, B.; Chen, H.; Wei, Z.; Liu, Y. Spatially distributed crop model based on remote sensing. *Agric. Water Manag.* **2019**, *218*, 165–173. [CrossRef]
46. Han, C.; Zhang, B.; Chen, H.; Liu, Y.; Wei, Z. Novel approach of upscaling the FAO AquaCrop model into regional scale by using distributed crop parameters derived from remote sensing data. *Agric. Water Manag.* **2020**, *240*, 106288. [CrossRef]
47. Feng, Q.; Liu, W.; Su, Y.; Zhang, Y.; Si, J. Distribution and evolution of water chemistry in Heihe River basin. *Environ. Earth Sci.* **2004**, *45*, 947–956. [CrossRef]
48. Wang, J.; Meng, J.J. Characteristics and Tendencies of Annual Runoff Variations in the Heihe River Basin during the Past 60 years. *Sci. Geogr. Sin.* **2008**, *28*, 83–88. (In Chinese)
49. Chen, X.-L. The Ecological Water Dispatching in the Heihe River from 2000 to 2004. *J. Arid. Land Res. Environ.* **2006**, *20*, 104–108. (In Chinese)
50. Lai, Z.-Q.; Li, S.; Li, C.-G.; Nan, Z.-T.; Yu, W.-J. Improvement and Applications of SWAT Model in the Upper-middle Heihe River Basin. *J. Nat. Res.* **2013**, *28*, 1404–1413. (In Chinese)
51. Niu, J.; Liu, Q.; Kang, S.; Zhang, X. The response of crop water productivity to climatic variation in the upper-middle reaches of the Heihe River basin, Northwest China. *J. Hydrol.* **2018**, *563*, 909–926. [CrossRef]
52. Allen, R.G.; Pereira, L.S.; Raes, D.; Smith, M. *Crop Evapotranspiration–Guidelines for Computing Crop Water Requirements FAO Irrigation and Drainage Paper 56*; FAO: Rome, Italy, 1998; p. 300.
53. Zhang, G.; Song, X. *Digital Soil Mapping Dataset of Soil Texture in the Heihe River Basin (2012–2014)*; National Tibetan Plateau Data Center: Beijing, China, 2017.
54. Zhang, G.; Song, X. *Digital Soil Mapping Dataset of Hydrological Parameters in the Heihe River Basin (2012)*; National Tibetan Plateau Data Center: Beijing, China, 2017.
55. Pan, Y.; Li, L.; Zhang, J.; Liang, S.; Zhu, X.; Sulla-Menashe, D. Winter wheat area estimation from MODIS-EVI time series data using the Crop Proportion Phenology Index. *Remote Sens. Environ.* **2012**, *119*, 232–242. [CrossRef]
56. Pan, Z.; Huang, J.; Zhou, Q.; Wang, L.; Cheng, Y.; Zhang, H.; Blackburn, G.A.; Yan, J.; Liu, J. Mapping crop phenology using NDVI time-series derived from HJ-1 A/B data. *Int. J. Appl. Earth Obs. Geoinf.* **2015**, *34*, 188–197. [CrossRef]

57. Manfron, G.; Delmotte, S.; Busetto, L.; Hossard, L.; Ranghetti, L.; Brivio, P.A.; Boschetti, M. Estimating inter-annual variability in winter wheat sowing dates from satellite time series in Camargue, France. *Int. J. Appl. Earth Obs. Geoinf.* **2017**, *57*, 190–201. [CrossRef]
58. Dubovyk, O.; Landmann, T.; Erasmus, B.F.; Tewes, A.; Schellberg, J. Monitoring vegetation dynamics with medium resolution MODIS-EVI time series at sub-regional scale in southern Africa. *Int. J. Appl. Earth Obs. Geoinf.* **2015**, *38*, 175–183. [CrossRef]
59. Son, N.-T.; Chen, C.-F.; Chen, C.-R.; Duc, H.-N.; Chang, L.-Y. A Phenology-Based Classification of Time-Series MODIS Data for Rice Crop Monitoring in Mekong Delta, Vietnam. *Remote Sens.* **2013**, *6*, 135–156. [CrossRef]
60. Vyas, S.; Nigam, R.; Patel, N.K.; Panigrahy, S. Extracting Regional Pattern of Wheat Sowing Dates Using Multispectral and High Temporal Observations from Indian Geostationary Satellite. *J. Indian Soc. Remote Sens.* **2013**, *41*, 855–864. [CrossRef]
61. Kang, S.Z. *Multi-Process Coupling and Efficient Water Use Regulation of Agricultural Water Conversion in Oasis in Arid of Northwest China*; Science Press: Beijing, China, 2019; pp. 36–37.
62. Maas, S.J. Use of remotely-sensed information in agricultural crop growth models. *Ecol. Model.* **1988**, *41*, 247–268. [CrossRef]
63. Gitelson, A.A.; Wardlow, B.; Keydan, G.P.; Leavitt, B. An evaluation of MODIS 250-m data for green LAI estimation in crops. *Geophys. Res. Lett.* **2007**, *34*, 20403. [CrossRef]
64. Lei, H.; Yang, D.; Cai, J.; Wang, F. Long-term variability of the carbon balance in a large irrigated area along the lower Yellow River from 1984 to 2006. *Sci. China Earth Sci.* **2013**, *56*, 671–683. [CrossRef]
65. Kennedy, J.; Eberhart, R. Particle swarm optimization. In Proceedings of the International Conference on Neural Networks (ICNN 95), Perth, WA, Australia, 27 November–1 December 1995; pp. 1942–1948.
66. Jakubik, J.; Binding, A.; Feuerriegel, S. Directed particle swarm optimization with Gaussian-process-based function forecasting. *Eur. J. Oper. Res.* **2021**, *295*, 157–169. [CrossRef]
67. Zhang, Z. Effects of Regulated Deficit Irrigation on Seed Maize. Master's Thesis, Gansu Agricultural University, Lanzhou, China, 2012. (In Chinese)
68. Jiang, Y. Simulation Analysis and Optimal Regulation for Agro-Hydrological Processes and Water Use Efficiency on Multiple Scales of the Middle Heihe River Basin. Ph.D. Thesis, China Agricultural University, Beijing, China, 2017. (In Chinese)
69. Wu, T. Several key links of maize seed production. *Nongcun Shiyong Keji Xinxi* **2011**, *5*, 16. (In Chinese)
70. Zhang, W.; Tian, Y.; Sun, Z.; Zheng, C. How does plastic film mulching affect crop water productivity in an arid river basin? *Agric. Water Manag.* **2021**, *258*, 107218. [CrossRef]
71. Chen, Z.; Wang, L.; Liu, J. Selecting appropriate spatial scale for mapping plastic-mulched farmland with satellite remote sensing imagery. *Remote Sens.* **2017**, *9*, 265.
72. Wang, H.H. Study on the Polarized Reflectance Characteristics of Agricultural Thin Membrane. Master's Thesis, Northeast Normal University, Changchun, China, 2007. (In Chinese)
73. Lu, L.; Di, L.; Ye, Y. A Decision-Tree Classifier for Extracting Transparent Plastic-Mulched Landcover from Landsat-5 TM Images. *IEEE J. Sel. Top. Appl. Earth Obs. Remote Sens.* **2014**, *7*, 4548–4558. [CrossRef]
74. Lu, L.; Hang, D.; Di, L. Threshold model for detecting transparent plastic-mulched landcover using moderate-resolution imaging spectroradiometer time series data: A case study in southern Xinjiang, China. *J. Appl. Remote Sens.* **2015**, *9*, 097094. [CrossRef]

remote sensing

MDPI

Article

Multiple Effects of Topographic Factors on Spatio-Temporal Variations of Vegetation Patterns in the Three Parallel Rivers Region, Southeast Qinghai-Tibet Plateau

Chunya Wang [1,2,3], Jinniu Wang [1,2,*], Niyati Naudiyal [1], Ning Wu [1], Xia Cui [4], Yanqiang Wei [5] and Qingtao Chen [3]

1 Chengdu Institute of Biology, Chinese Academy of Sciences, Chengdu 610041, China; wangcy9610@gmail.com (C.W.); naudiyal.niyati@gmail.com (N.N.); wuning@cib.ac.cn (N.W.)

2 Mangkang Ecological Monitoring Station, Tibet Ecological Security Barrier Ecological Monitoring Network, Qamdo 854000, China

3 Earth Sciences College, Chengdu University of Technology, Chengdu 610059, China; cqt@cdut.edu.cn

4 School of Resources and Environmental Sciences, Lanzhou University, Lanzhou 730000, China; xiacui@lzu.edu.cn

5 Northwest Institute of Eco-Environment and Resources, Chinese Academy Sciences, Lanzhou 730000, China; weiyq@lzb.ac.cn

* Correspondence: wangjn@cib.ac.cn

Citation: Wang, C.; Wang, J.; Naudiyal, N.; Wu, N.; Cui, X.; Wei, Y.; Chen, Q. Multiple Effects of Topographic Factors on Spatio-Temporal Variations of Vegetation Patterns in the Three Parallel Rivers Region, Southeast Qinghai-Tibet Plateau. *Remote Sens.* **2022**, *14*, 151. https://doi.org/10.3390/rs14010151

Academic Editors: Jingzhe Wang, Zhongwen Hu, Yangyi Wu and Jie Zhang

Received: 7 December 2021
Accepted: 26 December 2021
Published: 30 December 2021

Abstract: Topographic factors are critical for influencing vegetation distribution patterns, and studying the interactions between them can enhance our understanding of future vegetation dynamics. We used the Moderate-resolution Imaging Spectroradiometer Normalized Differential Vegetation Index (MODIS NDVI) image dataset (2000–2019), combined with the Digital Elevation Model (DEM), and vegetation type data for trend analysis, and explored NDVI variation and its relationship with topographic factors through an integrated geographically-weighted model in the Three Parallel Rivers Region (TPRR) of southeastern Qinghai-Tibet Plateau (QTP) in the past 20 years. Our results indicated that there was no significant increase of NDVI in the entire basin between 2000–2019, except for the Lancang River basin. In the year 2004, abrupt changes in NDVI were observed across the entire basin and each sub-basin. During 2000–2019, the mean NDVI value of the whole basin increased initially and then decreased with the increasing elevation. However, it changed marginally with variations in slope and aspect. We observed a distinct spatial heterogeneity in vegetation patterns with elevation, with higher NDVI in the southern regions NDVI than those in the north as a whole. Most of the vegetation cover was concentrated in the slope range of 8~35°, with no significant difference in distribution except flat land. Furthermore, from 2000 to 2019, the vegetation cover in the TPRR showed an improving trend with the changes of various topographic factors, with the largest improvement area (36.10%) in the slightly improved category. The improved region was mainly distributed in the source area of the Jinsha River basin and the southern part of the whole basin. Geographically weighted regression (GWR) analysis showed that elevation was negatively correlated with NDVI trends in most areas, especially in the middle reaches of Nujiang River basin and Jinsha River basin, where the influence of slope and aspect on NDVI change was considerably much smaller than elevation. Our results confirmed the importance of topographic factors on vegetation growth processes and have implications for understanding the sustainable development of mountain ecosystems.

Keywords: MODIS NDVI; vegetation cover; trend analysis; Sen + Mann-Kendall; topography; GWR; Qinghai-Tibet Plateau

1. Introduction

Being an important component of terrestrial ecosystems, vegetation plays a crucial role in maintaining ecosystem stability, functioning, and services [1,2]. The dynamics of vegetation patterns is a complex and prolonged process, driven by multiple biotic, abiotic, and anthropogenic factors such as climate change, land use change, and ecological engineering measures amongst others [3]. Being one of the most fragile terrestrial ecosystems, mountainous ecosystems are rather sensitive to global climatic change and human activities. A large vertical gradient and unique topographical conditions make mountainous regions the most abundant land unit on the earth and a key area for global biodiversity conservation [4,5]. Vegetation cover change of mountainous ecosystems have a much more significant and sensitive feedback on climate and other factors than plain areas. Therefore, the analysis of the process and driving factors of mountain vegetation dynamics has always been a key concern amongst global climatic change and mountain research community [6,7].

The topographic variability in a mountain ecosystem forms a micro-habitat with diverse microclimates influencing the variations in local and regional vegetation patterns thereby supporting high levels of biodiversity [5]. In recent years, international research on mountain vegetation has broadened from earlier studies that focused mostly on temperate zones of Europe (e.g., the Alps and the Scandes) [8] to other mountainous regions in the world including Alaska and the State of California, and the Rocky Mountains in North America [9,10], tropical mountainous regions of Africa (Kenya, Mount Kilimanjaro, Mount Wilhelm in New Guinea, etc.) [11,12], the subtropical Andean regions [13,14], and the southern and eastern Himalayas [15]. Meanwhile, the research field has also expanded from the physiological ecology of alpine plants to issues of elevational zonality and the impact of global climate change on alpine vegetation patterns [16,17]. In particular, studies of alpine vegetation in the eastern Himalayas and across the QTP have made great progress in recent years through the efforts of scholars in China, which not only studied the vegetation pattern in the Three River Headwater Region, Hengduan Mountains, but also included high latitudes, such as the Tianshan Mountains, Changbai Mountains, and Qinling Mountains, and mainly deliberated the response mechanisms for conservation of mountain vegetation in the face of global climate change and human activities, in particular those major ecological projects [18–21].

Several environmental factors have been widely identified as drivers for vegetation cover changes. Natural (e.g., climatic factors, topography and etc.) and social factors (e.g., socioeconomics, ecological engineering policies and settlements) can have a strong influence on vegetation growth processes, the extent of which varies from one place to another [5,22]. Most studies believe that precipitation and temperature are the main climatic factors affecting regional vegetation change [23], in addition to topography [24]. In mountainous areas with complex topography, topographic features (elevation, slope, aspect, etc.) have noticeable effects on vegetation patterns by themselves but also by controlling other environmental factors, such as solar radiation, wind, precipitation, snow cover, and edaphology distribution, a combination of these environmental factors jointly determines the spatial heterogeneity of vegetation pattern [17,19]. Besides, human activities (such as roads construction, settlements, and hydraulic engineering) also have an impact on mountain vegetation cover [22], which may have a negative impact on vegetation in low elevation regions, while less interference on the vegetation in the middle elevation [7,24]. At present, the analysis of drivers affecting vegetation patterns in the Tibetan Plateau region is mostly concentrated in the analysis of climatic factors, and relatively little research has been conducted on topographic control mechanisms [25–27]. In the context of global climate change, topography is the only relatively constant environmental factor, and deeper insights on how it controls the vegetation change pattern is particularly necessary to understand the vegetation dynamics and strengthen the response analysis of vegetation growth and distribution to topographic factors.

Given their inaccessibility in most mountainous regions, both remote sensing (RS) and geographic information system (GIS) technology stand out jointly as powerful tools for monitoring vegetation cover changes in mountainous areas by providing continuous, spatially detailed satellite data on mountainous vegetation cover [28]. With the development of RS technology, the types of remote sensing sensors have become more and more diversified, and the large-scale long time series vegetation dynamics has gradually developed into a hot spot for global change research, among which MODIS vegetation cover product data is regarded as one of the most effective data products for vegetation productivity analysis [29]. NDVI has been proved to be a comprehensive index to describe the ecological functional characteristics of vegetation growth, net primary productivity and phenology at regional, continental, and global scales [22,30]. Thus far, it has been widely used to detect the response of vegetation dynamics to climate change, human activities and other driving factors at multiple spatio-temporal scales [31,32]. However, many drivers are known to exhibit spatial non-stationarity with distinct spatio-temporal characteristics [33], while geographical weighted regression (GWR) model has improved the traditional model by combining spatial correlation with linear regression to better reflect the spatial non-stationarity of the relationship between NDVI and drivers by calculating the local parameters of the regression model [34].

The TPRR, located in the heart of the Hengduan Mountains in the southeastern Tibetan Plateau, has been paid widespread attention for being a world-renowned scenic spot and World Natural Heritage Site. The region is characterized by paralleling alternately high mountains and canyons, and the influence of topographic elevation differences has resulted in significant vertical temperature variation and contributed to the rich biodiversity of the region. Mountain ecosystems are signal "amplifiers" of biological responses to global change [4,35], and exploring the unique geographical environment and vertical zone spectrum of mountain bioclimates favors our understanding of regional biodiversity, ecosystem functions and services, vegetation restoration and reconstruction, as well as other issues. Existing studies from the TPRR mainly focus on geology, dynamic changes of alpine vegetation, land use/cover changes, and biodiversity conservation [36–39], and tourism resources [40–42]. Although a few scholars have studied vegetation changes and its driving factors analysis in the whole basin of the TPRR and various sub-basins (e.g., Lancang river and Jinsha river basins and etc.) [43], both of the vegetation dynamics in the Jinsha River Basin and the Lancang River Basin are jointly affected by temperature and precipitation, and the influence of precipitation is more significant in the Jinsha River Basin, while in the Lancang River Basin is more sensitive to temperature [44–46]. Furthermore, anthropological factors such as the establishment of large dams and reservoirs, hydropower stations, and the implementation of ecological projects have great and obvious specific effects on the vegetation pattern of the basin [2]. Only a few previous studies highlight the relationship between vegetation change and meteorological factors in the source areas of the Three Parallel Rivers which confirmed a significant correlation of vegetation cover with air temperature in this area [43–46]. The TPRR is one of the important ecological barriers and water resources conservation areas in China with unique arid valleys and distinct topography differences. Therefore, it is of vital importance to understand the influence of topographic factors on vegetation cover in this typical geographical unit.

Based on the MODIS NDVI image data from 2000 to 2019, combined with DEM and vegetation type data for trend method and geographically weighted model, our study discusses the spatio-temporal dynamic changes of vegetation and its response to topographic factors (i.e., elevation, slope, aspect, and other factors) in the TPRR. In terms of the spatial heterogeneity, the difference of vegetation restoration effect and its causes, and multiple drivers were analyzed by using the geographically weighted model. The specific research questions addressed in this study include: (1) What are the dynamic characteristics of NDVI at spatial and temporal scale? (2) How do different topographic factors effect on vegetation pattern? (3) What is the overall performance of all topographic factors by generalized analysis in terms of Geographically Weighted Regression modelling?

2. Materials and Methods

2.1. Study Area

The TPRR (24°00′~36°00′ N, 90°20′~102°20′ E) is located in the Hengduan Mountains in southwest China, spanning four provinces/Autonomous Region, Qinghai, Sichuan, Yunnan, and Tibet, bordering Qinghai-Tibet provincial boundary in the north, Sichuan-

Yunnan provincial boundary in the east, and China-Myanmar national boundary in the west, with a total watershed area of about 50.07 million km^2 (Figure 1). In this paper, the division of the study area was obtained based on the DEM data and watershed data provided by the Resource and Environment Science and Data Center of the Chinese Academy of Sciences (available at https://www.resdc.cn accessed on 12 August 2020). Coupled with the existing reference, a series of spatial data analyses were conducted using DEM data, including fill, flow direction, and flow accumulation, which ultimately determined the watershed extent of the TPRR [47–49]. Three major Asian rivers, the Yangtze (Jinsha River), the Mekong River (Lancang River) and the Salween River (Nujiang River), run parallel through the longitudinal valleys from north to south, with a vertical elevation difference of nearly 6000 m [47]. The unique environmental gradients make the region rich in biodiversity and one of the most unique areas of species diversity in China [50]. The climate type in the basin is diverse, including cold, temperate, subtropical, and tropical zones from north to south, with temperature and precipitation decreasing from south to north with increasing elevation. The variable climate types and geographical environment make the vegetation distribution in the basin have obvious latitudinal and vertical zonality characteristics (Figure 1).

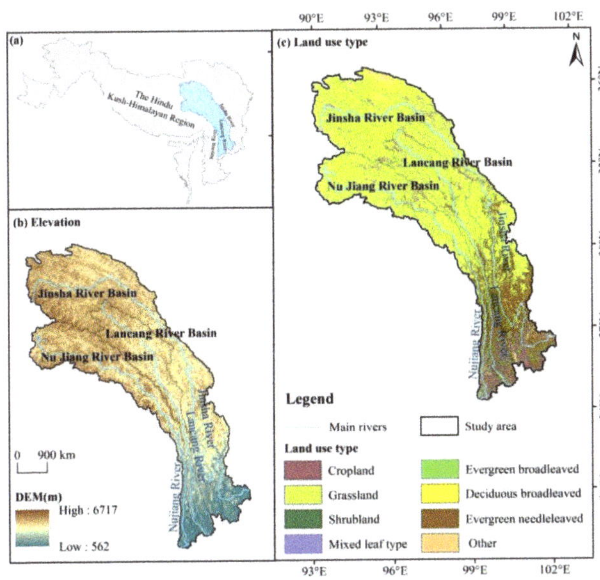

Figure 1. The location elevation and vegetation type maps of the study area. (**a**) the location of TPRR at the eastern part of Hindu-Kush-Himalayan Region; (**b**) Elevation gradients across the TPRR; (**c**) the vegetation types of TPRR.

2.2. *Data Sources and Preprocessing*

2.2.1. Remote Sensing Data

NDVI data were provided by the National Aeronautics and Space Administration (NASA) MODIS Terra (MOD13Q1) 16-day vegetation index product (2000–2019) at a spatial resolution of 250 m (available at https://ladsweb.Modaps.eosdis.nasa.gov/search/order/, accessed on 12 August 2020). MODIS Reprojection Tool (MRT) was used to pre-process the original data (e.g., format and projection conversion, clipping, etc.). Besides, the maximum NDVI value was calculated using the maximum value composition method (MVC), and we obtained the optimal vegetation cover for the study area and further reduced the influence of cloud cover and atmospheric scattering. Finally, regions with NDVI value greater than 0.05 were identified as vegetated areas to avoid the interference of underlying surface information on NDVI in areas with low vegetation cover.

2.2.2. DEM Data

DEM data at a spatial resolution of 90 m × 90 m in the TPRR were collected from the SRTM dataset (https://srtm.csi.cgiar.org/srtmdata/, accessed on 12 August 2020). The primary topographic features based on DEM data, such as elevation, slope, and aspect layers were computed using the "Slope/Aspect Tool" in the ArcGIS 10.8 Spatial Analyst module. According to the Chinese classification standards for potential hazard of soil erosion and some other relevant references [51–53], they were further divided into several categories (Table S1), and three maps of elevation, slope and aspect were generated (Figure 2). In addition, spatial analysis tools were also used to overlay both the topographic factors layers and NDVI data, to obtain the spatio-temporal distribution of NDVI for each topographic factor over the past 20 years.

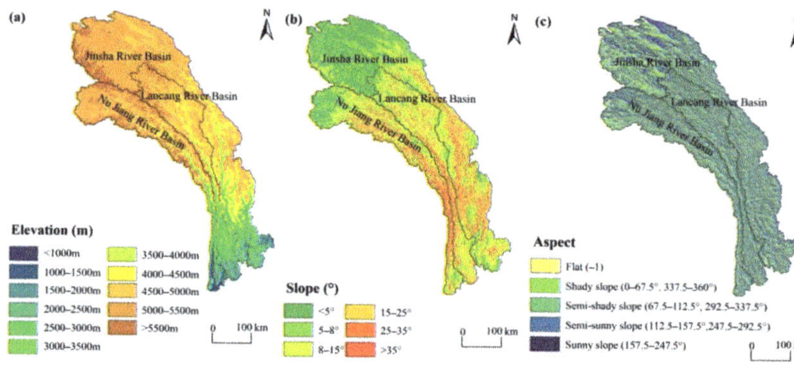

Figure 2. The hierarchical spatial distribution of topographic factors in the study area. (**a**) elevation gradients, (**b**) slope distribution, and (**c**) aspect classification across the TPRR.

2.2.3. Vegetation Type Data

ESA CCI-LC land cover types were provided by the European space agency (http://maps.elie.ucl.ac.be/CCI/viewer/, accessed on 12 August 2020) with a 300 m spatial resolution. With more than 70% overall accuracy, the global ESA CCI-LC data has been widely used in land use/land cover dynamic studies, around the world monitoring [54–56]. The vegetation types of the region comprise of cropland, grassland, shrub grass, evergreen broad-leaved forest, deciduous broad-leaved forest, evergreen coniferous forest, and others (mainly water, bare land, snow, and ice, etc.), a total of seven categories [57,58]. Referring to relevant literature [59], the areas with unchanged vegetation types from 2000 to 2015 were extracted here to represent the vegetation cover status of the study area in recent years, so to minimize the influence of land use change on vegetation type distribution.

In this paper, we resampled all the remote sensing data with 300 m resolution to facilitate correlation analysis based on pixel scale.

2.3. Methods

2.3.1. Trend Analysis and Significance Test

Firstly, we obtained the spatial distribution of the regional annual mean NDVI using MVC method for trend analysis and significance test. Then, we classified NDVI into six classes according to previous literature and actual vegetation conditions in the region [51], i.e., non-vegetation cover (NDVI < 0.05), low vegetation cover (0.05–0.30), medium and low vegetation cover (0.30–0.45), medium and high vegetation cover (0.45–0.60), medium and high vegetation cover (0.60–0.75), and high vegetation cover (NDVI > 0.75).

Theil-Sen + Mann-Kendall method has been widely used in meteorology and hydrology to get the variation characteristics of NDVI analysis [30,60], which has advantages to avoid outlier data disturbance or measurement error. The formula is:

$$S_{NDVI} = median\frac{x_j - x_i}{j - i} \qquad (j > i > 1) \tag{1}$$

where, x_i and x_j denote the *i-th* and *j-th* year time series data. S_{NDVI} >0 indicates that the vegetation shows an upward trend and vice versa indicates a downward trend. Meanwhile, Mann-Kendall method can test whether the trend is significant and find the abrupt point of NDVI changes, which formula is as follows:

$$Z = \begin{cases} \frac{S-1}{\sqrt{Var(S)}} & (S > 0) \\ 0 & (S = 0) \\ \frac{S+1}{\sqrt{Var(S)}} & (S < 0) \end{cases} \tag{2}$$

Among them,

$$S = \sum_{i=1}^{n=1} \sum_{j=i+1}^{n} sign(x_j - x_i), \quad sign = \begin{cases} 1 & (\theta > 0) \\ 0 & (\theta = 0) \\ -1 & (\theta < 0) \end{cases} \tag{3}$$

where, Z is a standard normal distribution, and S is the test statistic. x_i and x_j are time series data; n is the sample number; When $n \geq 8$, S is approximately normal distribution, and the variance calculation formula is:

$$Var(S) = \frac{n(n-1)(2n+5)}{18} \tag{4}$$

Z is a standard normal distribution, when $|Z| > Z_{1-\alpha}/2$, it showed significant change trend. Where, $Z_{1-\alpha/2}$ is the corresponding value of the distribution table of the standard normal distribution function at the confidence level α. The trends could be estimated by combining Sen's slope and the Mann–Kendall test in light of the NDVI series. Given confidence level $\alpha = 0.01$ and $\alpha = 0.05$, whether the trend changed significantly depended on $|Z|$, which can be divided into six classes: at 0.01 confidence level, (1) very significant degradation ($S_{NDVI} < 0$, $|Z| > 2.58$); (2) very significant improvement ($S_{NDVI} \geq 0$, $|Z| > 2.58$); at 0.05 confidence level, (3) significant degradation ($S_{NDVI} < 0$, $1.96 < |Z| < 2.58$); (4) significant improvement ($S_{NDVI} \geq 0$, $1.96 < |Z| < 2.58$); (5) non-significant degradation ($S_{NDVI} < 0$, $|Z| \leq 1.96$); (6) non-significant improvement ($S_{NDVI} \geq 0$, $|Z| \leq 1.96$).

2.3.2. Geographically Weighted Regression Model

GWR model can analyze the spatial nonstationary characteristics of the data and explore the spatial heterogeneity between vegetation changes and its driving factors [6,61,62]. The kernel and bandwidth of the GWR model, which are key parameters affecting the accuracy of the model, are now generally determined by Gaussian kernel function, and the optimal bandwidth is generally determined based on the estimated distance of the spatial autocorrelation of the dependent variable [63]. In addition, variance Inflation Factor (VIF) was generally used to eliminate highly correlated variables from the model.

Ordinary Least Squares regression (OLS) model is used to check for the presence of multicollinearity between variables to improve the accuracy of GWR model [34,64]. Compared with OLS model, GWR model is an extension of the traditional global logistic regression model, which includes spatial factors and geographic location information. The weighted least square method is used to estimate the parameters of each sample point, so that each of them has a corresponding estimation coefficient. In GWR4.0 software, the GWR model was constructed with the same variables as the OLS model [65,66], and the optimal bandwidth of 102 was determined according to the minimum *AIC* value. Besides, we divided the region into a 10 km × 10 km grid and extracted NDVI trend slope (S_{NDVI}), elevation, slope, and aspect values for each grid-cell [66,67]. Assuming that the dependent variable (Slope of NDVI variation trend from 2000 to 2019 is represented by S_{NDVI}) is y, and the elevation, slope and aspect value are x_1, x_2, and x_3 respectively, the GWR model can be expressed as:

$$y = b_0(u_i, v_i) + \sum_{i=1}^{k} b_i(u_i, v_i)x_{ij} + \varepsilon_i \quad (5)$$

where (u_i, v_i) is the geographic coordinate center of a certain region, b_0 is a constant, b_1 is the regression coefficient of independent variable, and x_{ij} is the independent variable. Figure 3 summarizes the analysis procedures.

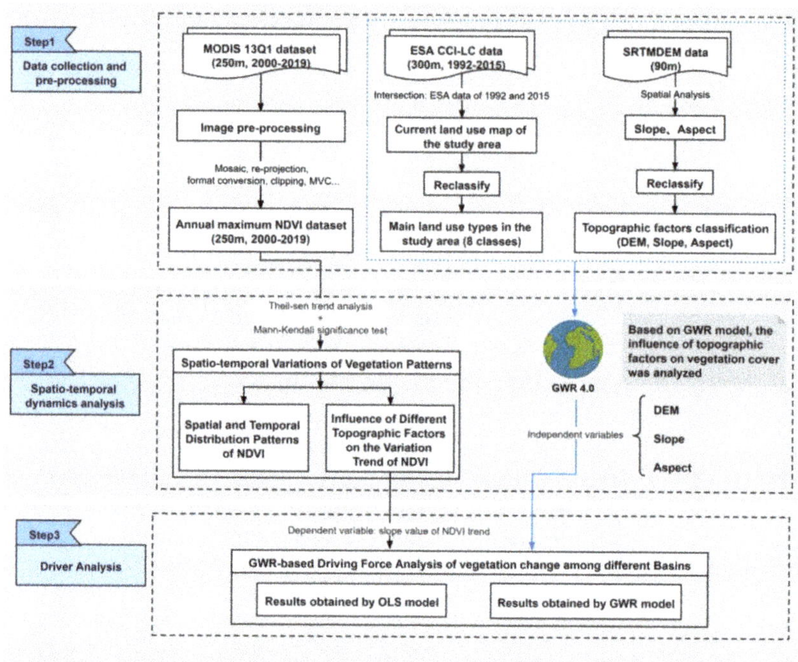

Figure 3. Conceptual framework of study design.

3. Results

3.1. Spatial and Temporal Distribution Patterns of NDVI

Inter-annual variation analysis results revealed that regional NDVI value fluctuated between 0.56 and 0.61 in the past 20 years, with an overall growth rate of 0.007/10a ($p > 0.05$), and only the Lancang River basin showed a significant upward trend ($p < 0.05$) (Figure 4). Additionally, the results of Mann-Kendall test showed an abrupt change of NDVI in the whole basin occurred around the year 2002, 2004, and 2008, respectively. Combined with the abrupt changes of NDVI in each sub-basin, these changes occurred almost simultaneously around 2004.

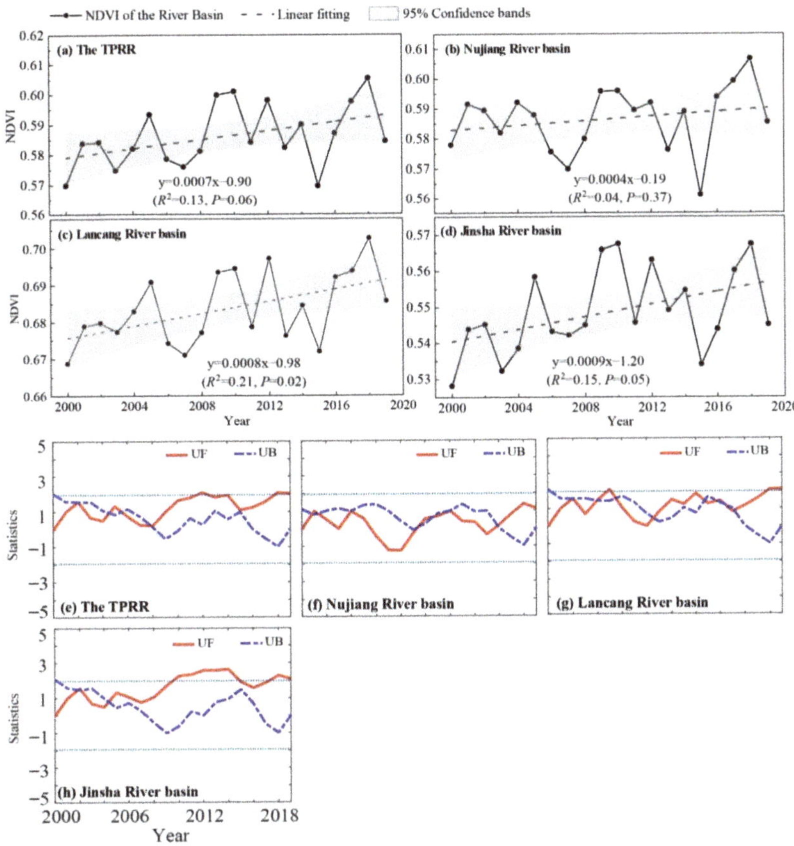

Figure 4. Inter-annual variation and Mann-Kendall test of NDVI in the basin from 2000 to 2019. (**a**–**d**) represent inter-annual NDVI variations in the TPRR and its sub-basins (Nujiang, Lancang and Jinsha river basins, respectively), while (**e**–**h**) represent the M-K results of them.

The mean NDVI in the basin tended to increase and then decrease with increasing elevation, and its value reached the maximum at 3000 m (about 0.80) and the minimum at 5500 m (about 0.05) (Figure S1a). Vegetation in the area below 2500 m was dominated by high vegetation cover, mainly in the southern part of the region, and the vegetation types were mainly cropland and evergreen coniferous forest (Table S2). Besides, the spatial distribution of mean NDVI in different elevation zones showed that regional vegetation was affected by elevation showing significant north-south differences (Figure S2a). The proportion of area occupied by each class indicated that the vegetation cover of all elevation zones is dominated by high vegetation cover, especially in the southern regions of the TPRR and in its core area (Figure 5). However, as the elevation continued to rise, NDVI started to decline from 4500 m, and the vegetation cover gradually transitions from low vegetation cover to unvegetated snow-capped mountain tops.

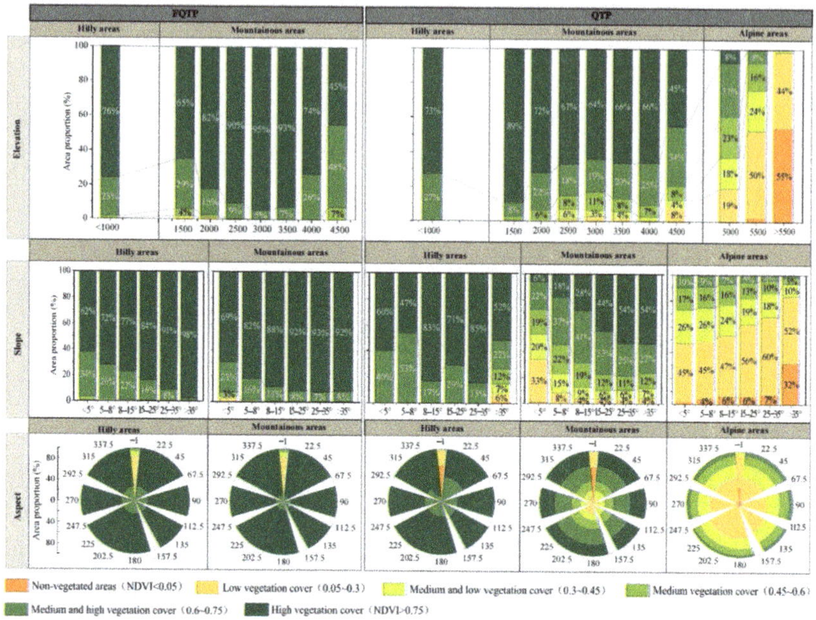

Figure 5. Area statistics of multi-year mean NDVI for topography factors during 2000–2019.

With the increase of slope, NDVI showed an increasing trend with gradually slower growth rate (Figure S1b). The mean NDVI with slope <5° was the lowest (0.40) and the value was between 0.05 and 0.3, which was mainly low vegetation cover type dominated by grassland and distributed in the river source area of the basin (Figure 5, Figure S2b, Table S2). The NDVI for regions within the slope range of 5–8° ranges from 0.6–0.75, which was mainly middle and high vegetation types. NDVI increased with the increasing of slope grades and each slope is gradually dominated by high vegetation cover types, whose distribution gradually shifted to the dry-hot valley and eventually reached its maximum value (about 0.80) at slope >35°.

Furthermore, Figure S1c showed the NDVI changes with aspect from shady slope to sunny slope that the mean NDVI values were uniformly distributed in all aspects (approximately 0.60), and the smallest (about 0.01) existed on flat land. The proportion of high vegetation cover in each aspect was the largest (Figure 5), which distribution pattern was relatively consistent (Figure S2c). The brief descriptions were the proportion of high vegetation cover in semi-shady slope was the largest (35.25%), followed by semi-sunny slope, while the proportion of sunny slope was the smallest (27.83%) (Table S2).

3.2. Influence of Different Topographic Factors on the Variation Trend of NDVI

There was a significant spatial variation in NDVI of the TPRR from 2000 to 2019, showing that the vegetation cover in the basin had slightly improved (Figure 6). However, 31.74% of the region showed no changes and 20.28% areas indicated slight degradation. Overall, the areas with improvement were mainly concentrated in the source area of Jinsha river basin in the north and the southern part of the TPRR. Meanwhile, degradation trends were primarily observed in north of the Jinsha river basin and the valley of the TPRR as well as other areas with significant response to vegetation growth changes.

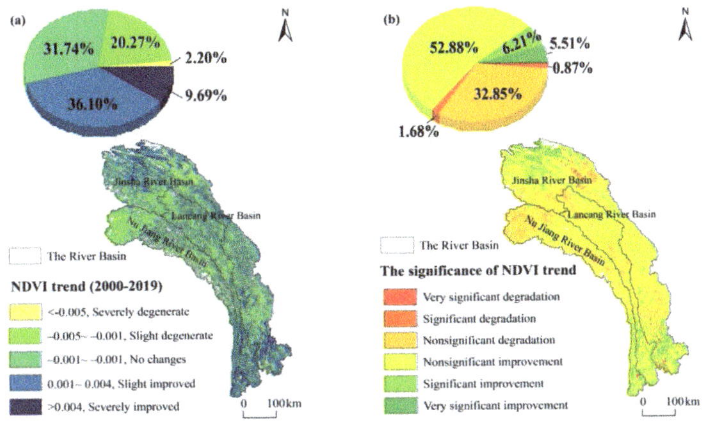

Figure 6. Spatial distribution of (**a**) NDVI trends and (**b**) significance during 2000–2019.

Most of NDVI in the TPRR showed increasing trends under different topographic factors (Figure S3). Among them, NDVI value at all elevation zones showed increasing trends except for regions above 5500 m (Figure S3a), and we also observed that both slope and aspect did not play any significant role in determining NDVI trend distributional patterns (Figure S3b,c).

More specifically, except for the three elevation zones, 3500–4000 m, 4000–4500 m, and above 5500 m, with the largest proportion of no changes at 34.36%, 35.04%, and 43.81%, respectively, and the majority of elevation zones (mainly 1000–3500 m) showed a slight improvement trend (Figure 7a, Table S3).

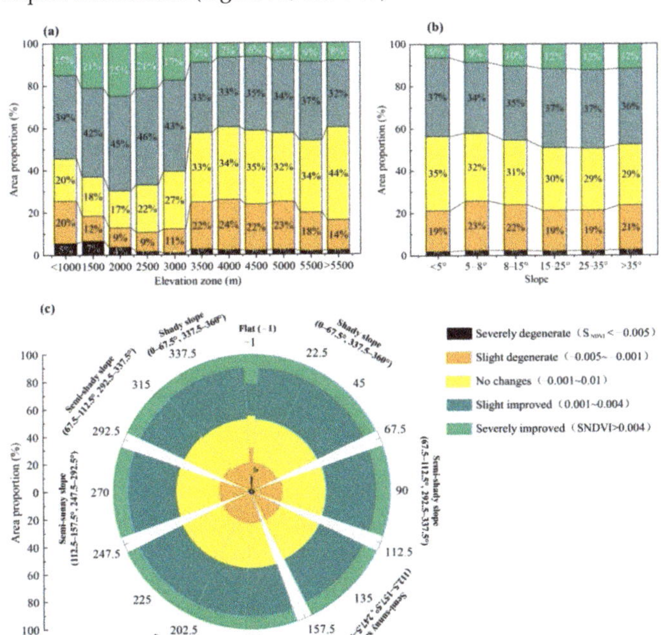

Figure 7. NDVI trends area statistics for different topography factors from 2000 to 2019. (**a**–**c**) represent elevation, slope and aspect, respectively.

For different slopes, approximately 35% of the area under all slope grades showed a slight improvement (Figure 7b, Table S3). The proportion of vegetation improvement area increased initially and eventually decreased with the steeper slope. Among them, the slope in the northern part of the basin and the source area of TPRR was relatively small (<5°), while the vegetation improvement trend was obvious, with grassland as mainly vegetation type (Figure S4a). The slope in the valley was mainly between 8–35°, which dominated by grassland and evergreen coniferous forest. The slope more than 35° in a few river valleys, especially in the Yunnan section of the TPRR, was mainly evergreen coniferous forest. Besides, the proportion of improved areas (37.12%) was the largest in regions with 15–25° slope. However, we observed that the proportion of area with no changes decreased gradually with the increasing slope (Figure S4b).

According to different aspects, although there was little difference in the proportion of areas between vegetation improvement, stability, and degradation, on an average we could witness slight improvement (Figure 7c, Table S3). Given that aspect had little influence on the vegetation variation trend in the basin, particularly in flat land, the area proportion of vegetation improvement had an inconspicuous increasing trend from shady slope to sunny slope (Figure S4c).

3.3. GWR-Based Driving Force Analysis of Vegetation Change among Different Basins

Taking S_{NDVI} as the dependent variable, and topographic factors (elevation, slope, and aspect) as independent variables, the traditional multiple linear regression model based on OLS could not well explain the spatial heterogeneity of vegetation changes and topographic factors (Table S4; Figure S5; OLS model, $R^2 = 0.03$, $F = 60.76$), given variance inflation factors (VIF) of each variable were all less than 10.

However, GWR model provided a better overall fitting than OLS (Table 1, $R^2 = 0.33$), which is more proper than OLS in explaining vegetation changes. Moreover, the $AICc$ difference between GWR and OLS was much larger than 3 ($\Delta AICC = 1458.77$), indicating that the fit degree of GWR model is significantly higher than that of OLS, but the Residuals SS of the GWR model also decreases, which indicates that the fitting effect of the GWR model is greatly improved compared with the OLS model (Residuals $F = 4.\ 31$) (Table S5).

Table 1. Parameter estimation and test results of the GWR model.

Variable	Mean	Min	Max	Lwr Quartile	Median	Upr Quartile	SD
DEM	−1.13	−20.66	10. 52	−1.97	−0.16	0.00	2.98
Slope	21.67	−580.16	824. 38	−0.00	1.15	30.08	81.37
Aspect	0.04	−11.05	10. 54	−0.55	0.00	0.53	2.35

The elevation in the headwaters of the TPRR (in particular, Nujiang river and Jinsha river basins), the middle reaches of the Lancang river, and a few areas in the south of the TPRR (mainly in the south of the Jinsha river) has a significant positive correlation with S_{NDVI}, while the remaining vast majority of areas showed a negative correlation, e.g., the middle reaches of Nujiang river and Jinsha river basin (Figure 8a).

Slope also has a certain regularity on vegetation change since the slope steepness affects the soil and water conservation so that will directly affect the plant growth and development on the earth surface. Slope has location-specific effects positively or negatively on vegetation change in the study area, given whose influence intensity is much smaller than that of elevation. Among them, positive correlation trend mainly existed in the headwaters of the Jinsha river basin and a few areas in the middle reaches, and the lower reaches of the Jinsha river and Nujiang river basins. However, there is a significant negative correlation trend between slope and vegetation variation trend in the junction of the Jinsha river and Nujiang river basins and a small part of the northwest of the source area in Jinsha river basin (Figure 8b).

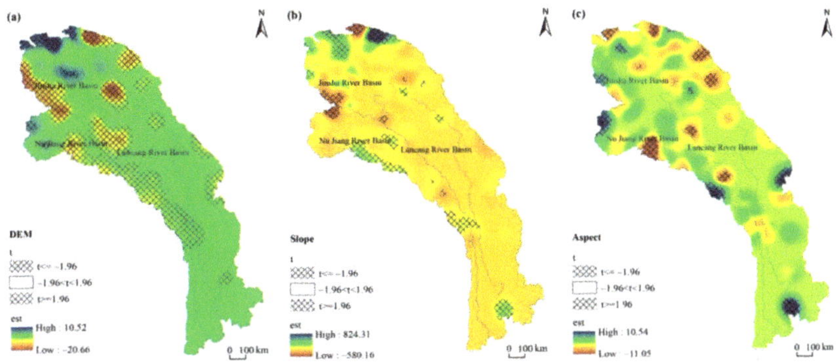

Figure 8. Spatial distribution maps of regression coefficients and significances for all independent variables. (**a**–**c**) are the spatial distribution maps of the local regression coefficients and the significances for all independent variables, e.g., (**a**) elevation, (**b**) slope and (**c**) aspect respectively.

In terms of aspect, some significant positive correlations exist in a few source areas of the Nujiang river basin, a small part of the TPRR, as well as the junction of the south lower reaches between Jinsha river basin and Nujiang river basin given a significant negative correlation presented in some part of the Nujiang river basin and the source areas of the Jinsha river basin (Figure 8c).

4. Discussion

4.1. Temporal and Spatial Distribution of Vegetation Patterns

Our results show that from 2000 to 2019, the vegetation in the TPRR shows an improving trend in a majority of the areas, while a few areas remain degraded, especially in the source areas of the entire Three Parallel Rivers basin where degradation is more obvious. Compared with the degraded areas, the proportion of areas with improved vegetation is much larger (45.79%). Our findings are consistent with those of other scholars in the QTP or TPRR, who also emphasized that vegetation dynamics in most areas of the QTP have an upward trend, while some areas such as the Nagqu, a few areas in Qinghai Province and several areas on the southeastern edge of the QTP including the Hengduan Mountains are at degradation trends. Natural environmental changes and anthropogenic disturbances jointly increased vegetation degradation [18,24]. Previous studies of vegetation pattern changes in the TPRR and showed that the significant reduction of vegetation cover was mainly in the source area of the Nujiang river and Lancang river as well as some areas in the central part of the study area, which corroborates with our study's results [20,43]. Different vegetation types of mountainous vegetation are also significantly influenced by spatial scales [47,68,69]. With higher elevations in its northern region and relatively lower elevation in the southern regions the geographical structure in TPRR supports better vegetative growth in the southern region, with cropland and evergreen coniferous forest being main vegetation types. Most of the vegetation is distributed within the elevation range of 1500 to 5500 m, with a myriad of landscape and vegetation types overlaid and intertwined with one another [23]. Along an increasing elevational gradient, the vegetation types transition from subalpine forests, low scrub, alpine grassland to unvegetated areas. This variation in vegetation types is driven by topographic factors that alter regional hydrothermal conditions which in turn indirectly affect vegetation distribution [20,46]. Peng et al. [24] showed the zonal characteristics of vegetation dynamics of different vegetation types, specifically the severe degradation of forests in the southeastern Tibetan Plateau, the increase of alpine grassland vegetation cover in some humid areas of southern Tibet, and the rising trend of cropland under the influence of agricultural activities in the Ali region and the plateau hinterland, all of which undoubtedly confirmed the complex vegetation changes in TPRR.

4.2. Spatial Heterogeneity of Topographic Factors on Vegetation

The results of GWR model strengthen the intuitive perception of the spatial heterogeneity of different topographic factors on NDVI variation in the TPRR, with results indicating that elevation itself has the greatest influence on local vegetation patterns, followed by slope and aspect, which is consistent with previous studies [24,26]. Altitude mostly influences vegetation patterns by affecting mountain climate conditions [70,71]. Our study area is in the QTP region with large elevation differences, and the distribution of heat conditions is successively controlled by the law of decreasing elevation and solar radiation differences, while precipitation distribution mainly affected by the trend of water vapor channel orientation, which is mainly related to the distribution of mountains. Therefore, analyzing the influence of topographic factors on vegetation change can better understanding mechanism of the climate-vegetation interaction, which is often ignored in current research [72]. Notably, the relative importance of elevation is generally reflected in its effect on the gradient of species' ecological niches, which determines the distributed of vegetation types in the region and the vegetation growth by representing laterally such environmental factors as slope or the amount of soil moisture [73,74]. Besides, slope can be one predictor variable to investigate plant diversity or capture potential changes in elevation in a region [73–76]. Slope can affect species richness by influencing soil moisture, soil erosion rates, and litter accumulation. However, related studies have shown that alpine treeline elevation changes in the Three Parallel Rivers region of southwest China are less affected by slope, further confirming the validity of our findings that slope has little effect on NDVI trends [39]. Moreover, Yirdaw et al. (2015) suggested that aspect affected woody plant species diversity, especially for woody plants at higher elevations, and that sunny slopes tended to receive higher amounts of solar radiation, which could accelerate greater air and soil temperatures [27]. Nevertheless, our results indicated the effect of aspect on NDVI trend was not significant, and the highest vegetation cover (35.25%) was found on the semi-shady slopes. Few scholars have explored the effect of aspect on vegetation changes and even the distribution of some major species in our study area, which may require further research.

4.3. Analysis of the Other Drivers Influencing Changes in Vegetation Dynamics

Beyond the influence of topographic factors on the vegetation pattern mentioned in this study, climatic conditions are the most well discussed factors for vegetation change in the region. Studies have shown that temperature and precipitation indirectly affect the productivity of terrestrial ecosystems by altering nutrient availability [17,77], and several scholars have reported the influence of temperature and precipitation these two climatic factors on the photosynthesis and respiration of vegetation changes by influencing soil moisture and microbial activities in the QTP. Obviously, their short-term effect on vegetation degradation is not significant [26,78,79]. However, the direct influence of topography on climatic conditions in mountainous areas was mentioned in the previous subsection, and our study confirms the strong influence of elevation on vegetation distribution, but the relationship between topographic and climatic factors and the joint influence of both on vegetation distribution is lacking in this study area and needs further analysis.

In addition, the impact of human activities on the vegetation pattern should not be underestimated. Hengduan Mountain area in the southeast QTP acts as a critical ecological barrier due to its unique ecological location and unique characteristics. In this region, several key national ecological projects have been implemented in the past to protect its natural forests, returning cropland to forest, biodiversity conservation, and ecological restoration, which have shown positive outcomes, and can also be the main reason for the improvement in NDVI observed in this study. Lu et al.compared the effects of three large-scale ecological programs (National Nature Reserves, Three North Shelter Forest Program, and the Natural Forest Protection Program) on Chinese vegetation change and revealed that the effectiveness of ecological restoration projects can vary geographically even under the same incentive policy context [80]. Moreover, several national and provincial protected areas, including the Three Parallel Rivers National Park, have been established in the TPRR. These protected areas can play a positive role in maintaining biodiversity, enhancing the regulation function of ecosystem, and building a scientific and reasonable ecological space. Grazing intensity is also one of the main driving factors affecting vegetation cover in grasslands in the QTP. Grasslands in QTP region have been considered as natural pastures for grazing by herders for many years [81,82]. Studies have shown that long-term grazing directly affects the vegetation cover and productivity of grasslands, but with the implementation of a series of ecological measures related to grassland restoration, e.g., grazing prohibition, and pastureland rehabilitation, the degradation of grasslands has been mitigated, which may also be one of the main reasons why the vegetation trends in our study area tends to improve [83,84].

5. Conclusions

In this paper, we explored the influence of topographic factors on the vegetation change patterns in the TPRR of southeastern QTP over the past 20 years. We analyzed the spatial and temporal distribution of vegetation patterns and the magnitude of the driving forces of each topographic factor with the GWR model, and the main conclusions are as follows:

During 2000–2019, the NDVI in the basin showed an overall increasing trend, with the most significant upward trend in the Lancang River basin, and the year of sudden change in NDVI in each basin was around 2004. Besides, the spato-temporal distribution of regional multi-year NDVI was influenced by elevation with significant north-south differences, while slope and aspect had little influence. Moreover, NDVI trends in the basin showed an overall improvement trend, mainly with slight improvement. The improvement areas are generally distributed in the southern section and the source area of the Jinsha River basin, and there is a degradation trend in the river source area of the TPRR. Specifically, the vegetation in all elevation zones was slightly improved, and the area of vegetation improvement increased and then decreased with increasing slope, while aspect has the least effect. The GWR model showed that elevation and NDVI were negatively correlated in most areas, and the effect of slope and aspect on vegetation changes was smaller, which were significantly correlated in the middle reaches of the Nujiang River basin and some parts of the Jinsha River basin. The interactions between topographic and climatic factors and their joint effects on vegetation distribution is lacking and needs further study in TPRR. Furthermore, the impact of human activities on the vegetation pattern should also be highlighted and integrated with climate change and other drivers for comprehensive analysis.

Supplementary Materials: The following are available online at https://www.mdpi.com/article/10.3390/rs14010151/s1, Table S1: Classification of topographic factors in the study area, Figure S1: Variation characteristics of mean NDVI with different topographic factors, Figure S2: Spatial distribution of mean NDVI with different topographic factors during 2000–2019, Table S2: Annual mean NDVI area statistics for different topographic factors during 2000–2019, Figure S3: NDVI trends during 2000–2019 with different topographic factors in the basin (a, b, c show elevation, slope and aspect respectively), Figure S4: Spatial distribution of NDVI trends for different topographic factors from 2000 to 2019, Table S3: Area statistics of NDVI trends for different topographic factors from 2000 to 2019, Table S4: Parameter estimation and test results of the OLS model, Figure S5: Estimation result of regression coefficients obtained through OLS modeling. (Notes: for the entire regression equation, the adjustment R2 is 0.03, F value is 60.76, and $P < 0.01$; the two dashed lines denote that the t-value is equal to -1.96 and 1. 96 respectively; at 0.05 significance level, $t < -1.96$ means a significantly negative correlation, while $t > 1.96$ represents a positive correlation.), Table S5: Comparison of model performance between the GWR and OLS models.

Author Contributions: Conceptualization, J.W., N.W. and C.W.; methodology, C.W., X.C. and Y.W.; software, C.W.; validation, J.W. and C.W.; formal analysis, C.W.; investigation, C.W.; resources, C.W.; data curation, C.W.; writing—original draft preparation, C.W.; writing—review and editing, C.W., J.W., N.N., N.W., X.C., Y.W. and Q.C.; visualization, C.W.; supervision, J.W.; project administration, J.W.; funding acquisition, J.W. All authors have read and agreed to the published version of the manuscript.

Funding: This research was funded by the National Science Foundation of China (No. 31971436 and 41661144045); State Key Laboratory of Cryospheric Science, Northwest Institute of Eco-Environment and Resources, Chinese Academy Sciences (SKLCSOP-2018-07) and China Biodiversity Observation Networks (Sino BON)".

Data Availability Statement: The data presented in this study are available on request from the corresponding author.

Acknowledgments: The authors would like to thank the researchers who have provided the open-source algorithms, which have been extremely helpful to the research in this paper.

Conflicts of Interest: The authors declare no conflict of interest.

References

1. Wen, Z.F.; Wu, S.J.; Chen, J.L.; Lu, M.Q. NDVI indicated long-term interannual changes in vegetation activities and their responses to climatic and anthropogenic factors in the Three Gorges Reservoir Region, China. *Sci. Total Environ.* **2017**, *574*, 947–959. [CrossRef]
2. Zhang, P.P.; Cai, Y.P.; Yang, W.; Yi, Y.J.; Yang, Z.F.; Fu, Q. Contributions of climatic and anthropogenic drivers to vegetation dynamics indicated by NDVI in a large dam-reservoir-river system. *J. Clean Prod.* **2020**, *256*, 120477. [CrossRef]
3. Qu, S.; Wang, L.C.; Lin, A.W.; Zhu, H.J.; Yuan, M.X. What drives the vegetation restoration in Yangtze River basin, China: Climate change or anthropogenic factors? *Ecol. Indic.* **2018**, *90*, 438–450. [CrossRef]
4. Julio Camarero, J.; Manzanedo, R.D.; Sanchez-Salguero, R.; Navarro-Cerrillo, R.M. Growth response to climate and drought change along an aridity gradient in the southernmost Pinus nigra relict forests. *Ann. For. Sci.* **2013**, *70*, 769–780. [CrossRef]
5. Palombo, C.; Marchetti, M.; Tognetti, R. Mountain vegetation at risk: Current perspectives and research reeds. *Plant Biosyst.* **2014**, *148*, 35–41. [CrossRef]
6. Lu, L.; Xu, Y.; Huang, A.; Liu, C.; Marcos-Martinez, R.; Huang, L. Influences of Topographic Factors on Outcomes of Forest Programs and Policies in a Mountain Region of China: A Case Study. *Mt. Res. Dev.* **2020**, *40*, R48–R60. [CrossRef]
7. Xu, M.; Ma, L.; Jia, Y.; Liu, M. Integrating the effects of latitude and altitude on the spatial differentiation of plant community diversity in a mountainous ecosystem in China. *PLoS ONE* **2017**, *12*, e0176866. [CrossRef]
8. Grytnes, J.-A.; Kapfer, J.; Jurasinski, G.; Birks, H.H.; Henriksen, H.; Klanderud, K.; Odland, A.; Ohlson, M.; Wipf, S.; Birks, H.J.B. Identifying the driving factors behind observed elevational range shifts on European mountains. *Glob. Ecol. Biogeogr.* **2014**, *23*, 876–884. [CrossRef]
9. Elliott, G.P. Extrinsic regime shifts drive abrupt changes in regeneration dynamics at upper treeline in the Rocky Mountains, USA. *Ecology* **2012**, *93*, 1614–1625. [CrossRef]
10. Moyes, A.B.; Germino, M.J.; Kueppers, L.M. Moisture rivals temperature in limiting photosynthesis by trees establishing beyond their cold-edge range limit under ambient and warmed conditions. *New Phytol.* **2015**, *207*, 1005–1014. [CrossRef]
11. Carbutt, C.; Edwards, T.J. Reconciling ecological and phytogeographical spatial boundaries to clarify the limits of the montane and alpine regions of sub-Sahelian Africa. *S. Afr. J. Bot.* **2015**, *98*, 64–75. [CrossRef]
12. Hemp, A. Climate change-driven forest fires marginalize the impact of ice cap wasting on Kilimanjaro. *Glob. Change Biol.* **2005**, *11*, 1013–1023. [CrossRef]
13. Hertel, D.; Wesche, K. Tropical moist Polylepis stands at the treeline in East Bolivia: The effect of elevation on stand microclimate, above- and below-ground structure, and regeneration. *Trees Struct. Funct.* **2008**, *22*, 303–315. [CrossRef]
14. Monteiro, J.A.F.; Hiltbrunner, E.; Koerner, C. Functional morphology and microclimate of Festuca orthophylla, the dominant tall tussock grass in the Andean Altiplano. *Flora* **2011**, *206*, 387–396. [CrossRef]
15. Klimes, L.; Dolezal, J. An experimental assessment of the upper elevational limit of flowering plants in the western Himalayas. *Ecography* **2010**, *33*, 590–596. [CrossRef]
16. Koerner, C. Global Statistics of "Mountain" and "Alpine" Research. *Mt. Res. Dev.* **2009**, *29*, 97–102. [CrossRef]
17. Körner, C. *Alpine Plant Life: Functional Plant Ecology of High Mountain Ecosystems*, 3rd ed.; Springer Nature Switzerland AG: Cham, Switzerland, 2021; pp. 53–85. [CrossRef]
18. Yu, B.H.; Lv, C.H.; Lv, T.T.; Yang, A.Q.; Liu, C. Regional differentiation of vegetation change in the Qinghai-Tibet Plateau. *Prog. Geog.* **2009**, *28*, 391–397.
19. Guo, D.; Zhang, H.Y.; Hou, G.L.; Zhao, J.J.; Liu, D.Y.; Guo, X.Y. Topographic controls on alpine treeline patterns on Changbai Mountain, China. *J. Mt. Sci. Engl.* **2014**, *11*, 429–441. [CrossRef]
20. Liu, X.F.; Zhang, J.S.; Zhu, X.F.; Pan, Y.Z.; Liu, Y.X.; Zhang, D.H.; Lin, Z.H. Spatiotemporal changes in vegetation coverage and its driving factors in the Three-River Headwaters Region during 2000–2011. *J. Geogr. Sci.* **2014**, *24*, 288–302. [CrossRef]
21. Wang, B.; Xu, G.; Li, P.; Li, Z.; Zhang, Y.; Cheng, Y.; Jia, L.; Zhang, J. Vegetation dynamics and their relationships with climatic factors in the Qinling Mountains of China. *Ecol. Indic.* **2020**, *108*, 105719. [CrossRef]

22. Zoungrana, B.J.B.; Conrad, C.; Thiel, M.; Amekudzi, L.K.; Da, E.D. MODIS NDVI trends and fractional land cover change for improved assessments of vegetation degradation in Burkina Faso, West Africa. *J. Arid. Environ.* **2018**, *153*, 66–75. [CrossRef]
23. Peng, J.; Li, Y.; Tian, L.; Liu, Y.X.; Wang, Y.L. Vegetation Dynamics and Associated Driving Forces in Eastern China during 1999–2008. *Remote Sens.* **2015**, *7*, 13641–13663. [CrossRef]
24. Xu, M.; Li, X.; Liu, M.; Shi, Y.; Zhou, H.; Zhang, B.; Yan, J. Spatial variation patterns of plant herbaceous community response to warming along latitudinal and altitudinal gradients in mountainous forests of the Loess Plateau, China. *Environ. Exp. Bot.* **2020**, *172*, 103983. [CrossRef]
25. Peng, J.; Liu, Z.; Liu, Y.; Wu, J.; Han, Y. Trend analysis of vegetation dynamics in Qinghai–Tibet Plateau using Hurst Exponent. *Ecol. Indic.* **2012**, *14*, 28–39. [CrossRef]
26. Wang, H.; Qi, Y.; Huang, C.; Li, X.; Deng, X.; Zhang, J. Analysis of vegetation changes and dominant factors on the Qinghai-Tibet Plateau, China. *Sci. Cold Arid. Reg.* **2019**, *11*, 150–158. [CrossRef]
27. Yirdaw, E.; Starr, M.; Negash, M.; Yimer, F. Influence of topographic aspect on floristic diversity, structure and treeline of afromontane cloud forests in the Bale Mountains, Ethiopia. *J. For. Res.* **2015**, *26*, 919–931. [CrossRef]
28. Mohapatra, J.; Singh, C.P.; Tripathi, O.P.; Pandya, H.A. Remote sensing of alpine treeline ecotone dynamics and phenology in Arunachal Pradesh Himalaya. *Int. J. Remote Sens.* **2019**, *40*, 7986–8009. [CrossRef]
29. Jacquin, A.; Sheeren, D.; Lacombe, J.-P. Vegetation cover degradation assessment in Madagascar savanna based on trend analysis of MODIS NDVI time series. *Int. J. Appl. Earth Obs. Geoinf.* **2010**, *12*, S3–S10. [CrossRef]
30. Gao, Q.Z.; Li, Y.; Wan, Y.F.; Zhang, W.N.; Borjigdai, A. Challenges in disentangling the influence of climatic and socio-economic factors on alpine grassland ecosystems in the source area of Asian major rivers. *Quatern. Int.* **2013**, *304*, 126–132. [CrossRef]
31. Sun, Y.L.; Yang, Y.L.; Zhang, Y.; Wang, Z.L. Assessing vegetation dynamics and their relationships with climatic variability in northern China. *Phys. Chem. Earth* **2015**, *87–88*, 79–86. [CrossRef]
32. Zewdie, W.; Csaplovics, E.; Inostroza, L. Monitoring ecosystem dynamics in northwestern Ethiopia using NDVI and climate variables to assess long term trends in dryland vegetation variability. *Appl. Geogr* **2017**, *79*, 167–178. [CrossRef]
33. Ren, Y.; Lu, Y.; Fu, B.; Comber, A.; Li, T.; Hu, J. Driving Factors of Land Change in China's Loess Plateau: Quantification Using Geographically Weighted Regression and Management Implications. *Remote Sens.* **2020**, *12*, 453. [CrossRef]
34. Brunsdon, C.; Fotheringham, A.S.; Charlton, M.E. Geographically Weighted Regression: A Method for Exploring Spatial Nonstationarity. *Geogr. Anal.* **1996**, *28*, 281–298. [CrossRef]
35. Dullinger, S.; Dirnbock, T.; Grabherr, G. Modelling climate change-driven treeline shifts: Relative effects of temperature increase, dispersal and invasibility. *J. Ecol.* **2004**, *92*, 241–252. [CrossRef]
36. Diao, Y.X.; Wang, J.J.; Yang, F.L.; Wu, W.; Zhou, J.; Wu, R.D. Identifying optimized on-the-ground priority areas for species conservation in a global biodiversity hotspot. *J. Environ. Manag.* **2021**, *290*, 112630. [CrossRef] [PubMed]
37. Liang, J.; Liu, Y.; Ying, L.; Li, P.; Xu, Y.; Shen, Z. Road impacts on spatial patterns of land use and landscape fragmentation in three parallel rivers region, Yunnan Province, China. *Chin. Geogr. Sci.* **2014**, *24*, 15–27. [CrossRef]
38. Ou, X.; Replumaz, A.; van der Beek, P. Contrasting exhumation histories and relief development within the Three Rivers Region (south-east Tibet). *Solid Earth* **2021**, *12*, 563–580. [CrossRef]
39. Wang, W.; Körner, C.; Zhang, Z.; Wu, R.; Geng, Y.; Shi, W.; Ou, X. No slope exposure effect on alpine treeline position in the Three Parallel Rivers Region, SW China. *Alp. Bot.* **2013**, *123*, 87–95. [CrossRef]
40. Li, H.; Zhao, K.; Zhu, X.; Yang, S. Protecting and using of ecotourism resource of moon mountain scenic spot in Three Parallel River Area. *Ecol. Econ.* **2008**, *5*, 75–80.
41. Guo, Y.; Yang, F.; Wang, J.; Wu, R. Assessment of the tourism and recreation cultural ecosystem services in Three Parallel Rivers Region. *Acta Ecol.* **2020**, *40*, 4351–4361.
42. Guo, Y.; Wu, R.; Yang, F.; Wang, J. The spatial patterns of scenic spots in Three Parallel Rivers Region. *J. Yunnan Univ.* **2020**, *42*, 992–1003.
43. Pan, X.; Wu, X.; Shen, Y.; Liu, F.; Zhang, C. Responses of Vegetation Coverage Changes to Climate Factors in the Source Regions of Three Parallel Rivers. *Mt. Res.* **2015**, *33*, 218–226. [CrossRef]
44. Deng, C.; Zhang, W. Spatiotemporal distribution and the characteristics of the air temperature of a river source region of the Qinghai-Tibet Plateau. *Environ. Monit. Assess* **2018**, *190*, 368. [CrossRef] [PubMed]
45. Huang, C.; Li, Y.F.; Liu, G.H.; Zhang, H.L.; Liu, Q.S. Recent climate variability and its impact on precipitation, temperature, and vegetation dynamics in the Lancang River headwater area of China. *Int. J. Remote Sens.* **2014**, *35*, 2822–2834. [CrossRef]
46. Li, J.P.; Dong, S.K.; Peng, M.C.; Li, X.Y.; Liu, S.L. Vegetation distribution pattern in the dam areas along middle-low reach of Lancang-Mekong River in Yunnan Province, China. *Front. Earth Sci.* **2012**, *6*, 283–290. [CrossRef]
47. Cheng, Z.J.; Weng, C.Y.; Guo, J.Q.; Dai, L.; Zhou, Z.Z. Vegetation responses to late Quaternary climate change in a biodiversity hotspot, the Three Parallel Rivers region in southwestern China. *Palaeogeogr. Palaeocl.* **2018**, *491*, 10–20. [CrossRef]
48. Yao, X.; Deng, J.; Liu, X.; Zhou, Z.; Yao, J.; Dai, F.; Ren, K.; Li, L. Primary Recognition of Active Landslides and Development Rule Analysis for Pan Three-river-parallel Territory of Tibet Plateau. *Adv. Eng. Sci.* **2020**, *52*, 16–37. [CrossRef]
49. Gao, Y.; Zhao, S.; Deng, J. Developing Law of Damming Landslide and Challenges for Disaster Prevention and Mitigation in the Three-river-parallel Territory in the Tibetan Plateau. *Adv. Eng. Sci.* **2020**, *52*, 50–61. [CrossRef]
50. Myers, N.; Mittermeier, R.A.; Mittermeier, C.G.; da Fonseca, G.A.B.; Kent, J. Biodiversity hotspots for conservation priorities. *Nature* **2000**, *403*, 853–858. [CrossRef]

51. Guo, T.; Duan, Q.; Hua, S.; Shi, D.; Xu, C.; Tong, W.; Ning, D.; Lu, S.; Qin, B.; Yu, J.; et al. *Ministry of Water Resources the People's Republic of China. Standards for Classification and Gradation of Soil Erosion*; China Water & Power Press: Beijing, China, 2008; pp. 8–9. (In Chinese)
52. He, W.; Ye, C.; Sun, J.; Xiong, J.; Wang, J.; Zhou, T. Dynamics and Drivers of the Alpine Timberline on Gongga Mountain of Tibetan Plateau-Adopted from the Otsu Method on Google Earth Engine. *Remote Sens.* **2020**, *12*, 2651. [CrossRef]
53. Jiang, L.G.; Liu, X.N.; Feng, Z.M. Remote sensing identification and spatial pattern analysis of the alpine timberline in the three parallel rivers region. *Resour. Sci.* **2014**, *36*, 259–266.
54. Yang, Y.K.; Xiao, P.F.; Feng, X.Z.; Li, H.X. Accuracy assessment of seven global land cover datasets over China. *ISPRS J. Photogramm.* **2017**, *125*, 156–173. [CrossRef]
55. Kuppel, S.; Fan, Y.; Jobbagy, E.G. Seasonal hydrologic buffer on continents: Patterns, drivers and ecological benefits. *Adv. Water Resour.* **2017**, *102*, 178–187. [CrossRef]
56. Gang, C.C.; Zhao, W.; Zhao, T.; Zhang, Y.; Gao, X.R.; Wen, Z.M. The impacts of land conversion and management measures on the grassland net primary productivity over the Loess Plateau, Northern China. *Sci. Total Environ.* **2018**, *645*, 827–836. [CrossRef]
57. Bontemps, S.; Boettcher, M.; Brockmann, C.; Kirches, G.; Lamarche, C.; Radoux, J.; Santoro, M.; Vanbogaert, E.; Wegmüller, U.; Herold, M.; et al. Multi-year global land cover mapping at 300 m and characterization for climate modelling: Achievements of the Land Cover component of the ESA Climate Change Initiative. *ISPRS Int. Arch. Photogramm. Remote Sens. Spat. Inf. Sci.* **2015**, *XL-7/W3*, 323–328. [CrossRef]
58. Gregory, D.; Josh, H.; Alessandro, C. A dataset mapping the potential biophysical effects of vegetation cover change. *Sci. Data* **2018**, *5*, 180014. [CrossRef]
59. Liu, J.; Wen, Z.; Gang, C. Normalized difference vegetation index of different vegetation cover types and its responses to climate change in the Loess Plateau. *Acta Ecol. Sin.* **2020**, *40*, 678–691. [CrossRef]
60. Li, H.; Xie, M.; Wang, H.; Li, S.; Xu, M. Spatial Heterogeneity of Vegetation Response to Mining Activities in Resource Regions of Northwestern China. *Remote Sens.* **2020**, *12*, 3247. [CrossRef]
61. Zhi, Y.; Shan, L.; Ke, L.; Yang, R. Analysis of Land Surface Temperature Driving Factors and Spatial Heterogeneity Research Based on Geographically Weighted Regression Model. *Complexity* **2020**, *2020*, 2862917. [CrossRef]
62. Comber, A.J.; Brunsdon, C.; Radburn, R. A spatial analysis of variations in health access: Linking geography, socio-economic status and access perceptions. *Int. J. Health Geogr.* **2011**, *10*, 44. [CrossRef] [PubMed]
63. Yang, C.; Li, R.; Sha, Z. Exploring the Dynamics of Urban Greenness Space and Their Driving Factors Using Geographically Weighted Regression: A Case Study in Wuhan Metropolis, China. *Land* **2020**, *9*, 500. [CrossRef]
64. Zhao, R.; Yao, M.; Yang, L.; Qi, H.; Meng, X.; Zhou, F. Using geographically weighted regression to predict the spatial distribution of frozen ground temperature: A case in the Qinghai-Tibet Plateau. *Environ. Res. Lett.* **2021**, *16*, 024003. [CrossRef]
65. Nakaya, T. GWR4.09 User Manual. WWW Document. Available online: https://raw.githubusercontent.com/gwrtools/gwr4/master/GWR4manual_409.pdf (accessed on 16 May 2021).
66. Zhang, D.; Jia, Q.; Wang, P.; Zhang, J.; Hou, X.; Li, X.; Li, W. Analysis of spatial variability in factors contributing to vegetation restoration in Yan'an, China. *Ecol. Indic.* **2020**, *113*, 106278. [CrossRef]
67. Xue, R.; Yu, X.; Li, D.; Ye, X. Using geographically weighted regression to explore the effects of environmental heterogeneity on the space use by giant pandas in Qinling Mountains. *Acta Ecol. Sin.* **2020**, *40*, 2647–2654. [CrossRef]
68. Levin, S.A. The Problem of Pattern and Scale in Ecology: The Robert H. MacArthur Award Lecture. *Ecology* **1992**, *73*, 1943–1967. [CrossRef]
69. Zhang, W.J.; Liu, L.; Song, K.C.; Li, X.D.; Wang, Y.D.; Tang, Y.; Jiang, H.R. Remote sensing the orographic effects of dry-hot valley on vegetation distribution in the southeast Tibetan Plateau. *Int. J. Remote Sens.* **2019**, *40*, 8589–8607. [CrossRef]
70. Grytnes, J.A. Species-richness patterns of vascular plants along seven altitudinal transects in Norway. *Ecography* **2003**, *26*, 291–300. [CrossRef]
71. Lenoir, J.; Graae, B.J.; Aarrestad, P.A.; Alsos, I.G.; Armbruster, W.S.; Austrheim, G.; Bergendorff, C.; Birks, H.J.B.; Brathen, K.A.; Brunet, J.; et al. Local temperatures inferred from plant communities suggest strong spatial buffering of climate warming across Northern Europe. *Glob. Chang. Biol.* **2013**, *19*, 1470–1481. [CrossRef]
72. White, A.B.; Kumar, P.; Tcheng, D. A data mining approach for understanding topographic control on climate-induced inter-annual vegetation variability over the United States. *Remote Sens. Environ.* **2005**, *98*, 1–20. [CrossRef]
73. Jia, L.; Li, Z.; Xu, G.; Ren, Z.; Li, P.; Cheng, Y.; Zhang, Y.; Wang, B.; Zhang, J.; Yu, S. Dynamic change of vegetation and its response to climate and topographic factors in the Xijiang River basin, China. *Environ. Sci. Pollut. Res.* **2020**, *27*, 11637–11648. [CrossRef]
74. Moeslund, J.E.; Arge, L.; Bocher, P.K.; Dalgaard, T.; Odgaard, M.V.; Nygaard, B.; Svenning, J.C. Topographically controlled soil moisture is the primary driver of local vegetation patterns across a lowland region. *Ecosphere* **2013**, *4*, 91. [CrossRef]
75. Jones, M.M.; Szyska, B.; Kessler, M. Microhabitat partitioning promotes plant diversity in a tropical montane forest. *Glob. Ecol. Biogeogr.* **2011**, *20*, 558–569. [CrossRef]
76. Zhang, C.S.; Li, X.Y.; Chen, L.; Xie, G.D.; Liu, C.L.; Pei, S. Effects of Topographical and Edaphic Factors on Tree Community Structure and Diversity of Subtropical Mountain Forests in the Lower Lancang River Basin. *Forests* **2016**, *7*, 222. [CrossRef]
77. Jobbágy, E.G.; Sala, O.E.; Paruelo, J.M. Patterns and Controls of Primary Production in the Patagonian Steppe: A Remote Sensing Approach. *Ecology* **2002**, *83*, 307–319. [CrossRef]

78. Mou, X.; Yu, Y.; Li, X.; Degen, A. Presence frequency of plant species can predict spatial patterns of the species in small patches on the Qinghai-Tibetan Plateau. *Glob. Ecol. Conserv.* **2020**, *21*, e00888. [CrossRef]
79. Sun, J.; Cheng, G.W.; Li, W.P. Meta-analysis of relationships between environmental factors and aboveground biomass in the alpine grassland on the Tibetan Plateau. *Biogeosciences* **2013**, *10*, 1707–1715. [CrossRef]
80. Lu, Y.; Zhang, L.; Feng, X.; Zeng, Y.; Fu, B.; Yao, X.; Li, J.; Wu, B. Recent ecological transitions in China: Greening, browning, and influential factors. *Sci. Rep.* **2015**, *5*, 8732. [CrossRef] [PubMed]
81. Cao, J.; Adamowski, J.F.; Deo, R.C.; Xu, X.; Gong, Y.; Feng, Q. Grassland Degradation on the Qinghai-Tibetan Plateau: Reevaluation of Causative Factors. *Rangel. Ecol. Manag.* **2019**, *72*, 988–995. [CrossRef]
82. Liu, Y.; Liu, S.; Sun, Y.; Li, M.; An, Y.; Shi, F. Spatial differentiation of the NPP and NDVI and its influencing factors vary with grassland type on the Qinghai-Tibet Plateau. *Environ. Monit. Assess.* **2021**, *193*, 48. [CrossRef]
83. Zhang, W.; Ganjurjav, H.; Liang, Y.; Gao, Q.; Wan, Y.; Li, Y.; Baima, Y.; Xirao, Z. Effect of a grazing ban on restoring the degraded alpine meadows of Northern Tibet, China. *Rangel. J.* **2015**, *37*, 89–95. [CrossRef]
84. Fayiah, M.; Dong, S.; Li, Y.; Xu, Y.; Gao, X.; Li, S.; Shen, H.; Xiao, J.; Yang, Y.; Wessell, K. The relationships between plant diversity, plant cover, plant biomass and soil fertility vary with grassland type on Qinghai-Tibetan Plateau. *Agric. Ecosyst. Environ.* **2019**, *286*, 106659. [CrossRef]

remote sensing

MDPI

Article

Analysis on the Spatio-Temporal Changes of LST and Its Influencing Factors Based on VIC Model in the Arid Region from 1960 to 2017: An Example of the Ebinur Lake Watershed, Xinjiang, China

Nigenare Amantai [1,2] and Jianli Ding [1,2,*]

1 College of Resources & Environmental Science, Xinjiang University, Urumqi 830046, China; ngnr113@stu.xju.edu.cn
2 Key Laboratory of Oasis Ecology, Xinjiang University, Urumqi 830046, China
* Correspondence: watarid@xju.edu.cn

Abstract: LST (Land surface temperature) is an important indicator for monitoring dynamic changes in the earth's resources and environment. However, the complexity of obtaining long-term, continuous LST data hinders the development of research on LST responses to meteorological factors or LUCC in areas where data is lacking. The objective of this research was to use the VIC-3L (Variable Infiltration Capacity) based on multi-source remote sensing data to simulate and explore spatio-temporal changes in the LST, to analyze the relationship between the LST and meteorological elements by using cross-wavelet transform (XWT) and wavelet coherence (WTC), the relationship between the LST and LUCC by using three-phase remote sensing images of LUCC. The following results were obtained. The annual average LST of the study area is increasing at a rate of 0.027 °C per year. The annual average LST level is relatively high in the central and eastern regions. The average temperature has an important influence on LST, which is mainly reflected in the period scale of 1~4a in 1963–1972, 1980–1996, and 2004–2010. The sharp decline in open shrubs may have exacerbated the increase in LST in the study area. This study provides a scientific reference for studying LST in arid areas.

Keywords: land surface temperature; VIC model; wavelet analysis; multiscale analysis; LUCC

Citation: Amantai, N.; Ding, J. Analysis on the Spatio-Temporal Changes of LST and Its Influencing Factors Based on VIC Model in the Arid Region from 1960 to 2017: An Example of the Ebinur Lake Watershed, Xinjiang, China. *Remote Sens.* **2021**, *13*, 4867. https://doi.org/10.3390/rs13234867

Academic Editors: Zhongwen Hu, Jingzhe Wang, Yangyi Wu and Jie Zhang

Received: 18 October 2021
Accepted: 24 November 2021
Published: 30 November 2021

1. Introduction

Land surface temperature (LST) is a typical indicator of the energy flow in the land surface-atmosphere-biosphere interactions. The UN Intergovernmental Panel on Climate Change (IPCC) Fifth Assessment Report pointed out that the global average surface temperature increased by 0.89 °C (0.69~1.08 °C) from 1901 to 2012 [1]. The change of LST is an intuitive regional climate response to global climate change. On the one hand, the LST variations will directly affect the soil moisture content and the soil evaporation, thereby indirectly affecting regional vegetation growth, agricultural production, and the water cycle [2,3]. On the other hand, the LST is an important element for predicting drought, climate change [4–6]. Therefore, it is of greater research significance in agriculture, hydrology, ecology, environment, climate, and biogeochemistry [4,7–11].

Research on changes in LST is more important in arid regions. One of the most prominent features of arid regions is the low atmospheric precipitation [12], which directly affects the soil moisture content and indirectly affects the vegetation conditions in these areas [13,14]. An increase in the LST can lead to reduce soil moisture and increase dust emission [15]. LST can affect wind erosion through changes in soil moisture content, as temperature increases, the evapotranspiration from the land surface increases, and the soil moisture content decreases [16]. As a result, the adhesion between the soil particles is reduced, and the soil particles are more easily separated from the soil, thereby increasing the concentration of the particles in the atmosphere [17]. Moreover, LST has shown a positive relationship with changes in the frequency of wind erosion events in Asia [18]. Therefore, studying the temporal and spatial changes of the LST and its driving factors is helpful to better understand the evolution of drought and water resources and provide theoretical support for sustainable development.

Under the global warming background, experts and scholars in different fields are paying more and more attention to changes in LST. The evolution of the global LST trend in the past century is diagnosed using the spatial–temporally multidimensional ensemble empirical mode decomposition method, the fastest warming in recent decades (>0.4 K per decade) occurred in northern mid-latitudes [19]. It also showed a particularly enhanced trend of LST in semi-arid regions [20]. However, the existing studies on LST dynamics mainly focused on the depiction and quantitative temporal comparison of LST spatial distribution or LST time series analysis over time [21]. The relationship between LST and other climate elements is less involved. Moreover, there are still some difficulties in the long-term large-scale continuous simulation of LST in data-scarce areas.

Exploring the mechanism of interaction between LUCC and local climate has become a popular research topic [22–24]. In the past few years, many studies based on the impact of LUCC on LST have focused on the urban heat island (UHI) [25–27]. The increase of the impermeable layer increases the absorption of solar radiation to a certain extent, thereby increasing the LST [28]. However, each land cover type possesses unique qualities in terms of energy radiation and absorption, in addition, LUCC also alters greenhouse gas emissions, such as CO_2, in the atmosphere [29]. Although the urbanization process in some areas is not fast, the effect of LUCC on LST is worth studying, especially for arid areas.

The Land surface model (LSM) is a key tool to predict the terrestrial hydrosphere and its atmospheric coupling and to understand this in-depth. Distributed LSM predicts hydrological states and fluxes, such as LST, at each grid cell [30]. The LSM based on multi-source remote sensing data can carry out large-scale simulations of LST to compensate for the discontinuity in time and space caused by only using remote sensing images and station data. The VIC (Variable Infiltration Capacity) model is one of the core application models of the GLDAS and NLDAS [31]. Therefore, it is effective to use VIC to carry out large-scale and long-term continuous LST simulations in scarce data areas.

The main objectives of this study are (1) to simulate daily LST from 1960 to 2017 by using multi-source data and to verify the accuracy of the model from both space and time; (2) to understand temporal and spatial evolution characteristics of LSTs and explore the multiscale, abrupt and periodic characteristics of LSTs in the time series; and (3) to analyze the impact of LUCC on LSTs. The results of this study can provide a reference for hydrological simulation research in arid areas and drought monitoring and management for meteorological delays and other aspects.

2. Materials and Methods

2.1. Study Area

The Ebinur Lake Watershed (43°38′~45°52′N and 79°53′~85°02′E) is located in the arid area of Northwest China (Figure 1). The study area has a typical continental arid climate, characterized by dramatic annual temperature change, scarce precipitation, and strong potential evaporation [32,33], and the lake basin is surrounded by mountains on the south, west, and north sides, which block outside airflow [34]. Climate change easily affects the basin's water resources and ecosystems. The Ebinur Lake watershed is an important ecological barrier for environmental changes in the Junggar Basin in Xinjiang Uygur Autonomous Region (XUAR), which is a typical inland lake basin in an arid region [35]. Some studies found that the LUCC of the watershed has changed greatly [36]. However, the influence of LUCC on LST in the Ebinur Lake Watershed is not yet clear and needs to be further investigated.

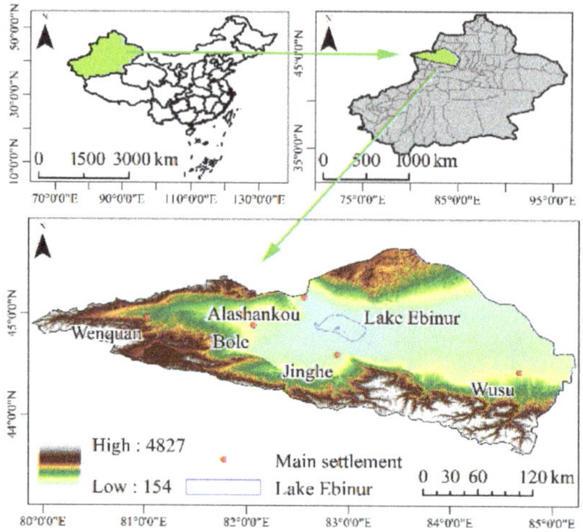

Figure 1. Location of the study area.

2.2. Data Sources

The data used in this study can be described as follows: (1) The elevation data are based on the 90-m digital elevation model (DEM) of Shuttle Radar Topographic Mission (SRTM) data and were obtained from the Geospatial Data Cloud Platform of the Computer Network Information Center of the Chinese Academy of Sciences (http://www.gscloud.cn/ accessed on 27 November 2021). (2) The global land cover data, with a resolution of 1 km, were obtained from the University of Maryland (http://app.earth-observer.org/data/basemaps/images/global/LandCover_512/LandCoverUMD_512/LandCoverUMD_512.html/ accessed on 27 November 2021), and the Terra and Aqua combined Moderate Resolution Imaging Spectroradiometer (MODIS) Land Cover Type (MCD12Q1) Version 6 data from the USGS EROS Center (http://doi.org/10.5067/MODIS/MCD12Q1.006/ accessed on 27 November 2021), has a resolution of 500 m and uses two images (2001 and 2017). (3) The soil data were taken from the 0.5-degree soil data available for the VIC model released by the University of Washington (https://vic.readthedocs.io/en/master/Datasets/Datasets/ accessed on 27 November 2021) and based on the World Soil Database (HWSD) 1:1 million Chinese soil dataset (v1.1). (4) The meteorological data were obtained from the daily dataset of surface climate data in China (V3.0), the time resolution is daily and the time scale is from 1 January 1960 to 31 December 2017. The data were obtained from the China Meteorological Data Network (http://data.cma.cn/ accessed on 27 November 2021). (5) The verification data of LST were obtained from a combined Terra and Aqua MODIS LST and meteorological station data product for China from 2003 to 2017 [37] with a resolution of 5.6 km. The model resolution is 8 km*8 km, and the basin is divided into 767 grids (Figure 2).

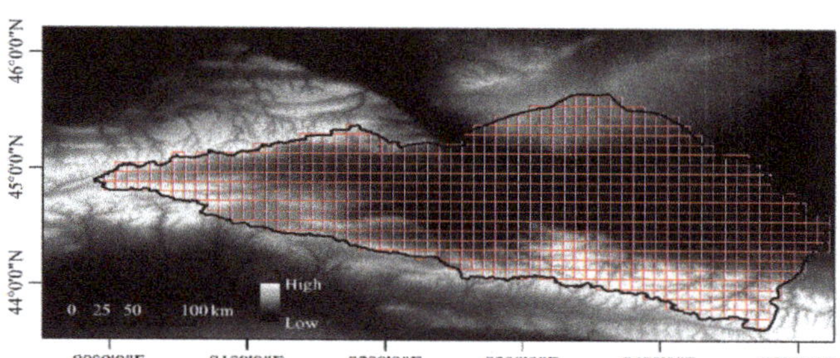

Figure 2. Schematic diagram of the model grid.

2.3. VIC Model

The VIC hydrological model is a large-scale hydrological model with a variable infiltration capacity. The VIC model is an open-source large-scale distributed conceptual hydrological model based on one kind of water and heat balance and physical dynamic mechanisms (VIC's homepage: http://vic.readthedocs.org/en/master/ accessed on 27 November 2021). Specific theoretical references include [38,39].

The energy balance calculation of the VIC-3L model mainly consists of four parts: LST, sensible heat flux, latent heat flux, and geothermal change. The other three parts are determined by LST, and the connecting factor of water balance and energy balance is latent heat flux. In an ideal situation, the energy balance calculation formula of each land cover type in the grid unit is as follows:

$$R_n = H + \rho_w L_e E + G \tag{1}$$

where, R_n is net radiation (W/m^2), H is sensible heat flux (W/m^2), $\rho_w L_e E$ is latent heat flux (W/m^2), and G is surface heat flux (W/m^2). The net radiation is calculated by the following formula:

$$R_n = (1-\alpha)R_s + \varepsilon\left(R_L - \sigma T_s^4\right) \tag{2}$$

where, α is the surface reflectance (%) of a certain land cover type, R_S is the downward short-wave radiation (µm), ε is the specific emissivity(%), R_L is the downward long-wave radiation (µm), and σ is the Stefan-Boltzmann constant (σ = 5.67 $\times$ 10^{-8} W/(m^2·°C^4)). The calculation formula of sensible heat is as follows:

$$H = \frac{\rho_a c_P}{r_h}(T_s - T_a) \tag{3}$$

where, Ts is the LST (°C), Ta is the atmospheric temperature (°C), and r_h is the aerodynamic impedance (s/m). The surface heat flux is estimated by the heat of two layers of soil. For the upper layer (the first layer + the second layer), the total depth is assumed to be D_1 (m), then its calculation formula is:

$$G = \frac{\kappa}{D_1}(Ts - T_1) \tag{4}$$

where, κ is the soil thermal conductivity(W/m·°C), T_1 is the temperature of the upper soil (°C). Assuming that the depth of the underlying soil is D_2 (m) and the temperature at the bottom of the soil is constant, then:

$$\frac{C_s\left(T_1^+ - T_1^-\right)}{2\Delta t} = \frac{G}{D_2} - \frac{k(T_1 - T_2)}{D_2^2} \tag{5}$$

where, C_S is the coefficient of soil heat conductivity (J/m^3·°C), T_1^+ and T_1^- is the soil temperature (°C) at the beginning and the end of the period in the upper soil respectively, and T_2 is the constant temperature (°C) of the lower soil. Currently, it is believed that κ and C_S do not change with the change of soil moisture content. According to Equations (4) and (5), the calculation formula of surface heat flux can be deduced as follows:

$$G_{(n)} = \frac{\frac{\kappa}{D_2}(T_s - T_2) + \frac{C_s D_2}{2\cdot\Delta t}\left(T_s - T_1^-\right)}{1 + \frac{D_1}{D_2} + \frac{C_s D_1 D_2}{2\Delta t k}} \tag{6}$$

Combined with the above formula, the LST, sensible heat flux and surface heat flux can be calculated through iteration.

2.4. Wavelet Analysis

2.4.1. Morlet Continuous Wavelet

The long-term LST changes have obvious periodic characteristics, and there is less analysis of the multi-scale time characteristics between LST and meteorological elements.

The wavelet analysis method has a powerful multi-scale resolution function, which can detect the response period of meteorological elements and LST, and can also show multiple change cycles hidden in the time series, reflecting the changing trend of the system within the same time range. Therefore, the wavelet analysis is more advantageous when studying the multi-scale characteristics of LST and meteorological elements. In this paper, we used the toolbox provided by Grinsted et al. [40], you can download it from this address: (http://www.glaciology.net/wavelet-coherence/ accessed on 27 November 2021).

The Morlet continuous wavelet function is a commonly used complex wavelet function. It has the advantages of rich phase information, good time aggregation, and high-frequency resolution. Therefore, it is widely used to study the correlation between two non-stationary sequences [41,42]. Therefore, this study selected the Morlet wavelet to analyze the degree of correlation between the LST and each meteorological element. Morlet wavelet is a complex wavelet, and its expressions in the time domain and frequency domain are:

$$\Psi(t) = \pi^{-1/4} e^{i\omega_0 t} e^{-t^2/2} \tag{7}$$

$$\hat{\psi}(\omega) = \pi^{-1/4}\sqrt{2}e - \frac{1}{2}(\omega - \omega_0) \tag{8}$$

In the formula, ω_0 is a constant, and i is an imaginary number. Morlet wavelet is a single-frequency negative sine function under the Gaussian envelope. When using wavelets for feature extraction purposes the Morlet wavelet (with $\omega_0 = 6$) is a good choice, since it provides a good balance between time and frequency localization.

2.4.2. Cross-Wavelet Transform (XWT)

Similar to the Fourier transform, according to the description of the transformed sequence energy in Perseval's theorem, that is, after the signal is transformed from the time domain to the frequency domain, the total energy remains unchanged, then the local time sequence $x(t)$ The expression of Wavelet Power Spectrum (Wavelet Power Spectrum, WPS) is:

$$(WPS)_x(a,b) = |W_x(a-b)|^2 \tag{9}$$

The above-mentioned continuous wavelet transform is only a one-dimensional sequence of the wavelet transform, and this research mainly involves the correlation of twotime series. In the early research, it is necessary to perform a one-dimensional wavelet transform on each sequence separately, while Hudgins et al. [43], the XWT can not only directly analyze the correlation between two time series at different frequencies, but also reflect their phase information and local characteristics in the time domain and frequency domain. For any two series of time $x(t)$ and $y(t)$, the expression of the XWT is:

$$W_t^{XY}(\omega) = W_t^X(\omega)\overline{W_t^Y(\omega)} \tag{10}$$

In the formula, the wavelet transforms of the two time series $x(t)$ and $y(t)$ are $W_t^X(\omega)$ and $W_t^Y(\omega)$ respectively. The corresponding wavelet cross power (Cross Wavelet Power, XWP) is $(XWP)xy = W_t^{XY}(\omega)$ [44].

2.4.3. Wavelet Coherence (WTC)

WTC is to identify the possible relationship between two time series by finding the common points of two time series in different time scales and frequency ranges. It is a kind of enhanced time-frequency correlation. Recognition tool that can express the local coherence of the XWT in the time-frequency domain [45]. Even if they correspond to the low-energy region of the XWT, their correlation in the WTC spectrum may be significant. The expression of complex wavelet coherency (Complex Wavelet Coherency) ρ_{xy} is:

$$\rho_{xy} = \frac{S(W_t^{XY}(\omega))}{\left[s(|W_t^X(\omega)|)^2 s\left(|W_t^Y(\omega)|^2\right)\right]^{1/2}} \tag{11}$$

In the formula, S is a smooth spectrum operation in the time domain and the frequency domain. WTC is the absolute value of complex WTC. The mathematical expression is:

$$R_{xy} = \frac{\left|S\left(W_t^{XY}(\omega)\right)\right|}{\left[s\left(\left|W_t^X(\omega)\right|\right)^2 s\left(\left|W_t^Y(\omega)\right|^2\right)\right]^{1/2}} \tag{12}$$

3. Results

3.1. Accuracy Verification

The selection of the best model parameters is the main factor that affects the simulation accuracy of the VIC model. There are seven main parameters that need to be input to the VIC model: the storage volume curve index B; the maximum base flow velocity D_m (m^3/s); the nonlinear base flow velocity D_s (m^3/s); the soil moisture content W_s (%); and the three layers of soil thickness d_1, d_2 and d_3 (m) during the nonlinear base flow. The selection of model parameters was based on traditional parameter calibration methods [46]. The above parameters were adjusted according to the empirical interval of the hydrological parameters to make the peak discharge value match the measured data. The above steps were repeated several times until a satisfactory result was obtained. The parameters are calibrated by repeatedly calculating the NSE and R^2 of the simulated and the measured values of runoff. The NSE and R^2 of the monthly runoff of Wenquan hydrological station during the verification period (2002–2017) are 0.44 and 0.47, respectively. The Final selected parameters are described as follows (Table 1):

Table 1. Seven parameters of the VIC model.

B	D_m	Ds	Ws	d_1	d_2	d_3
0.25	3.5	0.05	0.1	0.1	0.1	1.5

3.1.1. Accuracy Verification in Terms of Time

By correctly inputting various parameters and running the model, a dataset of daily values of the LST from 1960 to 2017 can be obtained. To ensure the accuracy of the simulation, the average monthly and yearly LSTs of 6 meteorological stations (Jinghe Station, Tuoli Station, Kelamayi Station, Wusu Station, and Alashankou Station) from 1960 to 2017 were calculated as the actual measured values for this time. The results were verified, the determination coefficient R^2, the Root Mean Squared Error RMSE, the Mean Absolute Error MAE, the Nash-Sutcliffe efficiency coefficient NSE were selected to evaluate the simulation accuracy of the model. The R^2, RMSE, MAE, NSE values of annual simulated LSTs are 0.51 2.21, 1.85, and 0.17, respectively (Figure 3b). The R^2, RMSE, MAE, NSE values of monthly simulated LSTs are 0.98, 3.93, 3.31, and 0.93, respectively (Figure 3d). It can be seen from the respective boxplots (Figure 3e,f) that the difference between the simulated and measured monthly data is not very large. The medians are 11.11 °C and 11.65 °C respectively. The measured values are slightly higher. The simulated and measured annual values have a large gap. They are 8.56 °C and 10.07 °C, the difference is 1.5 °C, our simulated values are underestimated.

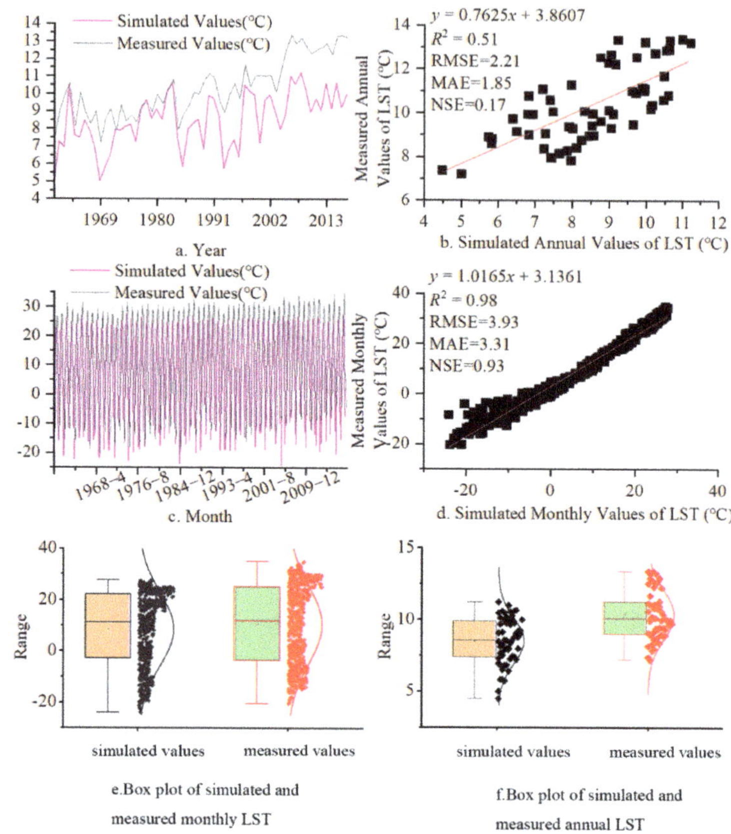

Figure 3. (**a**). Comparison of LST changes between measured and simulated annual values. (**b**). Verification accuracy of measured and simulated annual value LST. (**c**). Comparison of LST changes between measured and simulated monthly values. (**d**). Verification accuracy of measured and simulated monthly value LST. (**e**,**f**). Boxplot of monthly and annual simulated and measured value, respectively.

3.1.2. Accuracy Verification in Terms of Space

In addition to verifying the LST over time, its spatial accuracy should be verified as well. Therefore, according to the remote sensing LST inversion dataset in China, the simulated LST based on the VIC model and its spatial distribution were compared and studied. However, through verification, it was found that the accuracy of the LST simulated by the VIC model for cold months is inaccurate, especially in January and December, where the LST is even greater than 0 in high-altitude mountainous areas in winter. According to Koch's research [30], who found that the warmth of the VIC model in cold months is obvious, VIC has a distinct seasonality in its bias with a warm bias in winter (~3 °C) as opposed to a slight cool bias in summer (~−1 °C), which is related to an underestimation of the actual evapotranspiration and altitude. Our study area is located in an arid area, and the actual evapotranspiration in winter is small, which does not affect the accuracy of the LST simulated by the VIC model for areas other than high-altitude mountainous areas. To ensure the reasonableness of the data, the simulated LST values of the high-altitude mountainous area in November, January and December were deleted and converted to a no-data area. Taking 2017 as an example, the spatial resolutions were 0.05° and 1/12° respectively, as shown in Figure 4. It can be seen that the LST simulated by the VIC model has a high spatial consistency with the measured LST. The LST in the higher altitude area is significantly lower than that in other areas. As the time changes, the LST inside and outside the basin both "increase first and then decrease", and the simulated LST is slightly lower than the measured data.

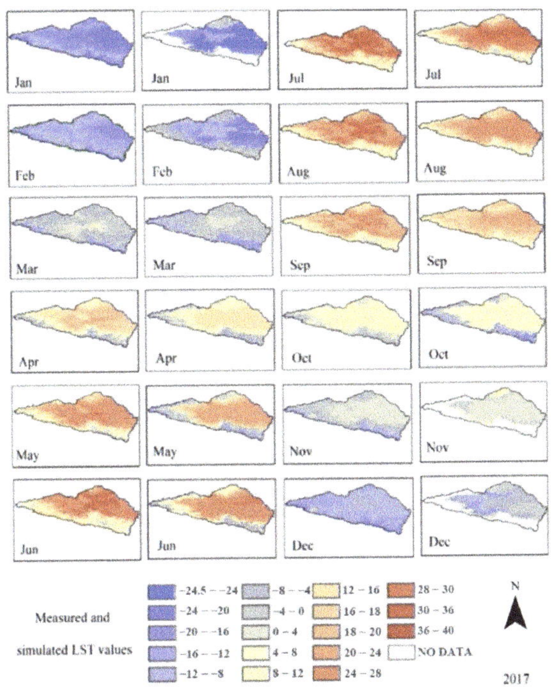

Figure 4. Spatial comparison map of simulated and measured LSTs in the Ebinur Lake Watershed.

Of course, the accuracy needs to be improved, but for the data-scarce area, the simulation results can provide continuous and long-term serial data support, which has certain advantages.

3.2. Temporal and Spatial Variation Characteristics of the LST

The overall trend of the daily average LST in the study area was a unimodal distribution. The daily average LST (Figure 5, left) began to increase at the end of February, reached the maximum in mid-July, began to decline at the end of August, and reached the minimum at the end of January. The peak in mid-July (July 13) was 25.92 °C, and the LST dropped rapidly to the lowest value in late January (January 22) at −18.99 °C. The daily average LST of the entire study area was 6.95 °C.

The monthly average LST in the study area also showed a single-peak trend (Figure 5, middle). The peak monthly average LST appeared in July. The order of the July peaks in each decadal period was 2010–2017 > 80s > 90s > 2000–2009 > 70s > 60s, and the peak values were 26.06 °C, 25.60 °C, 25.58 °C, 25.41 °C, 25.07 °C and 24.89 °C, respectively.

A histogram was used to represent the distribution of the annual average LST from 1960 to 2017 (Figure 5, right). The figure shows that the average annual LST is on the rise, with a rising rate of approximately 0.0279 °C/a and an increase of approximately 1.62 °C in the past 58 years. The annual average LST is 6.40 °C. The annual average LST reached a trough in 1960, with a value of 4.54 °C; the figure clearly shows that the annual average LST after 2000 was significantly higher than that before 2000, which may be caused by increasing human activities, the industrial and agricultural activities in the Ebinur Lake Watershed.

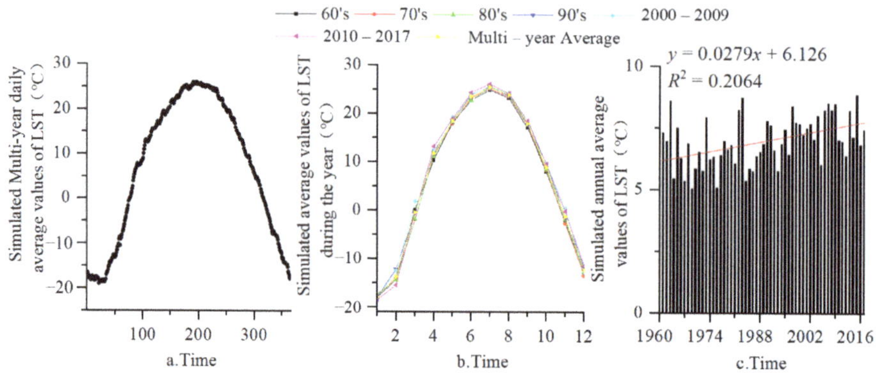

Figure 5. Temporal changes of daily (**a**), monthly (**b**) and annual average (**c**) LST in the Study area.

Based on the values of the daily LST simulated by the VIC model, the spatial distribution map of the inter-decadal and multi-year average LSTs from 1960 to 2017 was calculated. It can be seen that there are no obvious differences in the LST in the basin, but there are varying degrees of variation between different decadal periods.

The LST changes within the basin are not obvious, and the degree of change is different (Figure 6). The main manifestation is the gradual decrease of the high LST areas in the east and the gradual increase of the central LST. The LST range changed from −11.48–9.04 °C in the 1960s to −11.14–9.56 °C in the 1980s, and then to −3.5–10.51 °C in 2010–2017, with the average values being 3.28 °C, 5.24 °C, and 6.79 °C, respectively. These results show that the LST in this area has an increasing trend. Although the range of high-value areas in the east has decreased, gradually shrinking to the low vegetation coverage area near Lake Ebinur, the LST value has increased, and the multi-year average LST in other regions has increased compared with the values in the 1960s. Therefore, the LST of the basin is increasing.

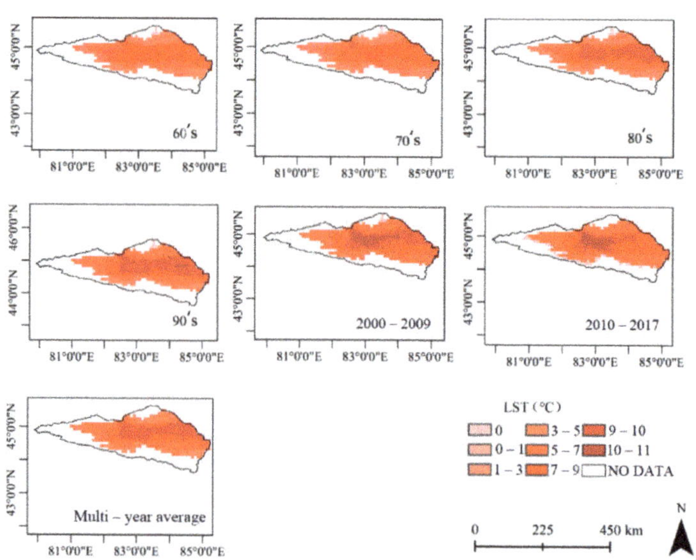

Figure 6. Interdecadal changes and annual mean LST spatial distribution in the Ebinur Lake Watershed.

3.3. Delayed Correlation Analysis of LST and Meteorological Elements

The XWT and WTC of the dual time series were applied to the coefficients after continuous wavelet transform, and the XWT spectrum between the LST of the Ebinur Lake Watershed and other meteorological elements was obtained (Figure 7) and crossed with the WTC spectrum (Figure 8). The XWT and WTC can well reflect the "correlation" between changes in two different time series. The XWT generally reflects the strength of the "shared period" between sequences. The WTC generally reflects the consistency of the periodic "change trend" between sequences but does not directly reflect the intensity relationship of the change period. As shown in Figures 7 and 8, the meteorological elements and LST of the basin are bounded by 1989. The two figures together show the following:

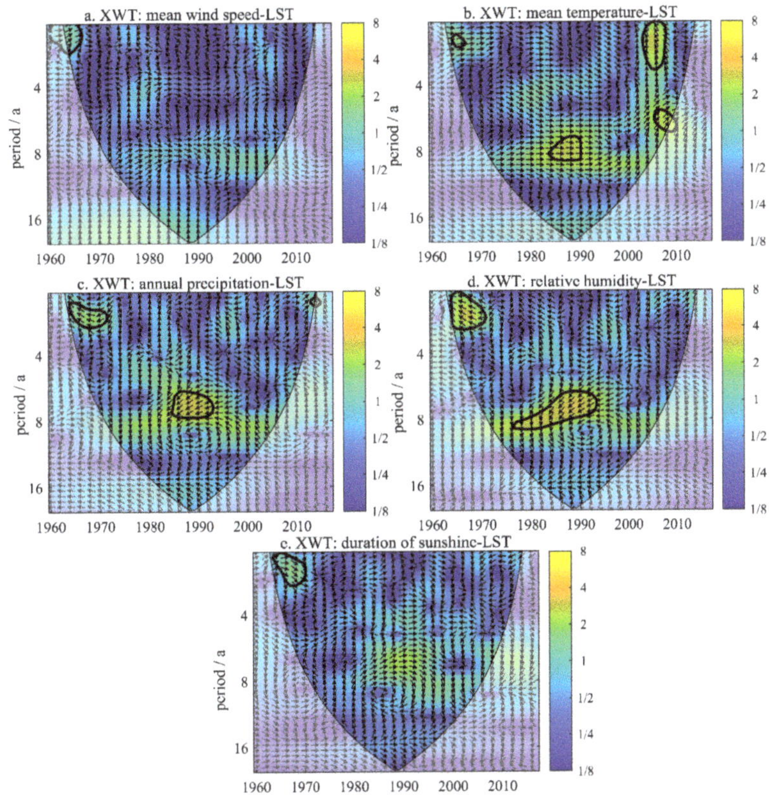

Figure 7. XWT analysis of mean wind speed and LST (**a**), XWT analysis of mean temperature and LST (**b**), XWT analysis of annual precipitation and LST (**c**), XWT analysis of relative humidity and LST (**d**), XWT analysis of duration of sunshine and LST (**e**). (Note: The horizontal axis of the wavelet period diagram corresponds to the time of the original time series, the vertical axis represents the period of change, and the color represents the intensity of the change period. In this figure, yellow represents the high intensity of the change cycle. The thin solid line is the Cone of Influence (COI), and only the energy spectrum within this solid line needs to be considered. In order to eliminate the interference of the boundary effect, the thick solid line represents the key value that passes the significance test with 95% confidence; the arrow indicates the relative phase difference: → indicates that the two changes in phase are consistent, ← indicates that the two changes in phase are opposite, ↑ indicates that the phase change of the meteorological element is 90° ahead of the LST phase, and ↓ indicates that the phase change of the meteorological element is 90° behind the LST phase [47,48]). (The same below).

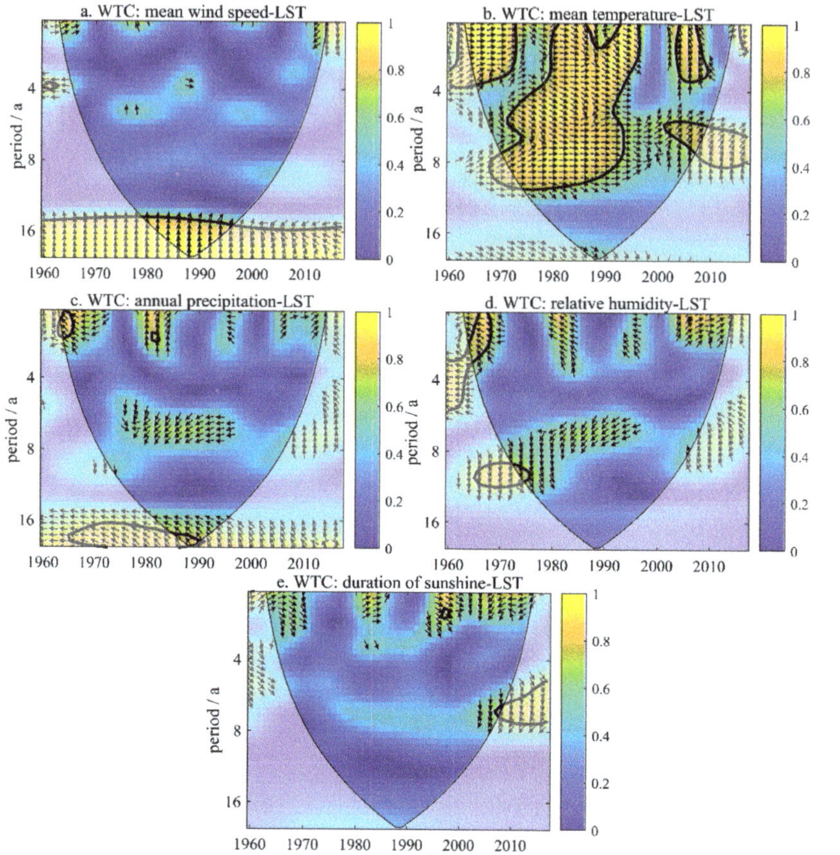

Figure 8. WTC analysis of mean wind speed and LST (**a**), WTC analysis of mean temperature and LST (**b**), WTC analysis of annual precipitation and LST (**c**), WTC analysis of relative humidity and LST (**d**), WTC analysis of duration of sunshine and LST (**e**). (Note: The right axis generally reflects the consistency of periodic "change trends" between sequences, similar to the correlation coefficient. The larger the coefficient, the higher the correlation and the more consistent the change.).

In the power spectrum of XWT, there was a significant area with a period of 1~2a in the average wind speed and LST from 1962 to 1966, the correlation is approximately positive; there are 4 significant areas between the average temperature and the LST, which are approximately 1.5a (1964–1967), 1~3a (2003–2008), 5.5–7a (2005–2009) and 7~9a (1984–1990). The two showed a positive correlation between the average temperature and the LST; the annual precipitation and the LST had significant areas of 0.5~2a and 6~8a in 1964–1972 and 1984–1994, respectively. Both have an approximately negative correlation; relative humidity and annual precipitation are the same, and there are significant areas of 1~2a and 6~9a in 1964–1971 and 1978–1994, respectively, and the two also show a similar negative correlation. Sunshine hours and LST only existed in a significant area of 1~2a from 1964 to 1970, and the two are in an approximately positive correlation.

In the coherence spectrum of WTC, the average wind speed and surface temperature have a significant area greater than 15 years from 1980 to 1996, and the average wind speed in this area is ahead of the surface temperature, that is, 1/4a ahead of time; the average temperature and the surface temperature in the area with the 95% confidence interval accounted for more than 60%, indicating that the average temperature is the main controlling factor of the surface temperature, and the two existed in 1963–1972, 1980–1996 and 2004–2010 on the time scale of 1~4a. For significant periods, most areas of the two are approximately positively correlated. Among them, there may be abrupt changes around 1980 and 1996 on the time scale of 1~2a; that is, there may be a "high-low" transition of surface temperature, and there are still two significant cycles of ~10a (1970–1994) and 5~7.5a (2000–2009); the annual precipitation and surface temperature only have a small cycle of 1~1.5a (1964–1966), and the two are negatively correlated; the other periodic feature is not obvious; the relative humidity and surface temperature have a significant area of 1~2.5a from 1964 to 1968 and a negative correlation, and the other two periodic features are not obvious; the sunshine hours and surface temperature have only one area where the periodic characteristics are not obvious.

3.4. The Impact of LUCC on LST

The statistics were classified according to the three-phase LUCC in this study. The specific changes in land cover types in the 1990s, 2001, and 2017 are shown in Figures 9 and 10, and the range distribution of the LST in the corresponding time period is shown in Figure 11.

The order of land cover types in the Ebinur Lake Watershed in the 1990s (Figure 9a) was open shrubland > grassland > bare ground > woodland grassland > crop land > water > woodland > evergreen needleleaf forest > mixed forest > urban and built-up > deciduous broadleaf forest. Among them (Figure 10), the largest area of open shrubland was 18,587.6 km^2, accounting for 35.1% of the total area, followed by grassland, with an area of 18,476.3 km^2, accounting for 34.9% of the total area. Bare ground accounted for 13.9% of the total area, and cropland accounted for 3.8% of the total area. Deciduous broadleaf forest occupied the smallest area, accounting for only 0.14% of the total area.

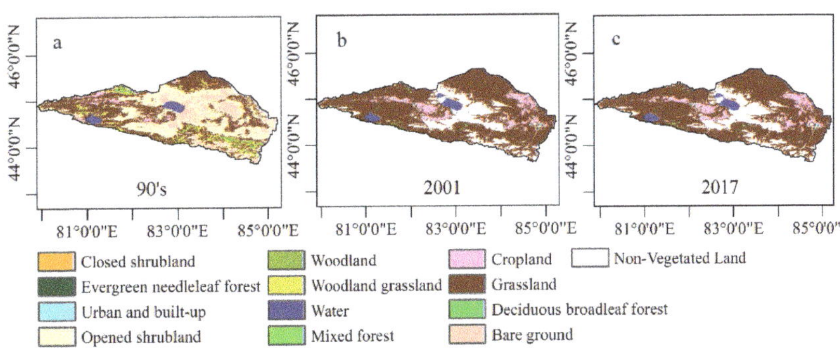

Figure 9. Distribution of LUCC in the study area in the 90's (**a**), in 2001 (**b**), in 2017 (**c**).

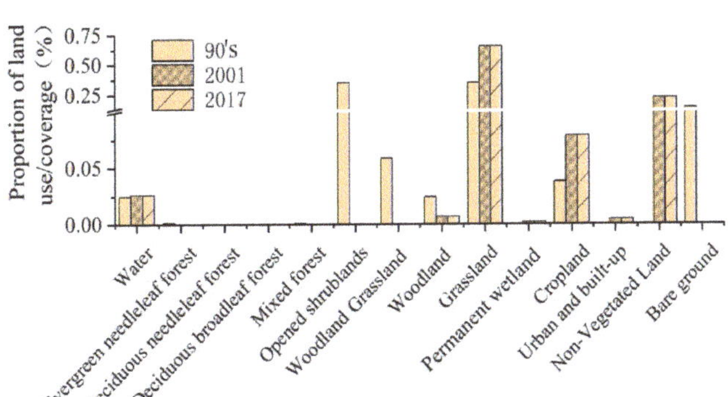

Figure 10. Changes in LUCC in the Ebinur Lake Watershed.

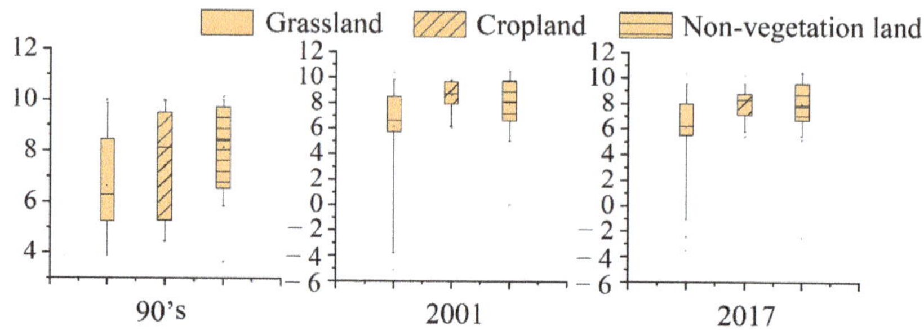

Figure 11. Box plots of LST.

The order of land cover types in the Ebinur Lake Watershed in 2001 (Figure 9b) was grassland > non-vegetation land > cropland > water > woodland > urban and built-up > permanent wetland > evergreen needleleaf forest > deciduous broadleaf forest > mixed forest > woodland grassland > open shrubland > deciduous needleleaf forest. Among them (Figure 10), the area of grassland has increased, reaching the largest area of 34,714.9 km^2 and accounting for 65.6% of the total area, followed by non-vegetation land with an area of 11,904.9 km^2, accounting for 22.5% of the total area. The two accounted for 88.1% of the total area. The area of cropland decreased to 4184.4 km^2, accounting for only 7.9% of the total area. The area of open shrubland dropped sharply, most of them were transformed into grassland or non-vegetation land.

The order of land cover types in the Ebinur Lake Watershed in 2017 (Figure 9c) was grassland > non-vegetation land > cropland > water > woodland > urban and built-up > permanent wetland > evergreen needleleaf forest > deciduous broadleaf forest > mixed forest > woodland grassland > open shrubland > deciduous needleleaf forest. This is consistent with the ranking in 2001, and the area has not changed much. Grassland and non-vegetation land are also the main land cover types in the study area (Figure 10).

In the 90s, the LST of grassland ranged from 5.2 to 8.4 °C, with a median of 6.3 °C; the LST of cropland ranged from 5.3 to 9.5 °C, with a median of 8.1 °C; the LST of non-vegetation land was in the range of 6.5–9.7 °C, with a median of 8.4 °C.

In 2001, the LST of grassland ranged from 5.7 to 8.5 °C, with a median of 6.6 °C; the LST of cropland ranged from 7.9 to 9.7 °C, with a median of 8.7 °C; the LST of non-vegetation land was in the range of 6.6–9.7 °C, with a median of 8.1 °C.

In 2017, the LST of grassland ranged from 5.5 to 8.0 °C, with a median of 6.2 °C; the LST of cropland ranged from 7.1 to 8.7 °C, with a median of 8.3 °C; the LST of non-vegetation land was in the range of 6.7–9.6 °C, with a median of 7.8 °C.

During the period, the area of open shrubland, most of which was converted to grassland, cropland, and non-vegetation land, declined sharply; the area of grassland increased most obviously, and the LST was lower. Although the area of cropland was small, the LST was higher; compared with land types with higher vegetation coverage such as grassland, the LST of non-vegetation land was higher, which may have been because a large amount of solar radiation was absorbed by vegetation, lowering its LST.

4. Discussion

4.1. Harmful Effects Caused by Changes in LST in Arid Regions

On the global scale, arid regions cover approximately one-third of the earth [49], and are one of the most sensitive regions to wind erosion phenomena [50]. In some recent research, a positive association was observed between the LST and wind erosion events [16,18]. The soil system has a synergistic relationship between various physical and chemical processes and environmental variables that affect its development [51]. Increasing the LST and decreasing the soil moisture, especially in the warm seasons over the arid regions, can be a factor in exacerbating soil erosion due to their effect on the salt accumulation on the soil surface [16,52]. Meanwhile, Different LUCC trends have different effects on the LST [53]. Heat sources (including bare land, urban and built-up land, and cropland) increase the LST, while cold sources (woodland, grassland, and water) reduce the LST [54]. Cropland and bare ground have a great impact on the regional LST, which is mainly reflected in the area and intensity of the impact [55]. This study found that between 1960 and 2017, the amount of open shrubs in the Ebinur Lake Watershed decreased sharply, and most of these areas were converted to grassland, cropland, and non-vegetation land. The vegetation coverage in the above-mentioned areas was low, and the LST was relatively high. This finding reasonably explains the variation in the LST of the Ebinur Lake Watershed from 1960 to 2017. However, the increased LST will lead to the reduction of regional soil moisture, vegetation degradation, and even increase the frequency of wind erosion and dust emission events. We believe this should attract the attention of local society.

4.2. Evaluation of VIC Model for LST Simulation

The LST is an important and complex hydrological state variable of land-atmosphere interactions. Land surface models (LSMs), such as the VIC model, are a key tool for enhancing the understanding of the process and providing predictions of the coupling between the terrestrial hydrosphere and its atmosphere. Distributed LSMs can be used to understand the hydrological status and flux of each grid cell, such as the LST. By comparing and analyzing the simulation accuracy of three different LSMs, Koch et al. [30] found that they all have certain spatial defects. The VIC has a distinct seasonality in its bias with a warm bias in winter (~3 °C) as opposed to a slight cool bias in summer (~−1 °C). The VIC model has a better simulation effect on the LST in warm months; in cold months, underestimation of the actual evapotranspiration will result in a high LST simulation. In this study, the same situation also appeared, but it may be because the study area is located in an arid area, that is, the actual evapotranspiration in autumn and winter, which are cold months, is inherently low, so except for high-altitude mountainous areas, the accuracy of the LST in cold months in plain areas is still good. However, the LST in high-altitude mountainous areas is obviously overestimated, only in November, December and January there is a big deviation in the mountainous area, and the LST in the areas close to the Tianshan Mountains is greater than zero in winter, which is unlikely to happen. This lack of spatial accuracy may be related to the occurrence of snow in mountainous areas, but further analysis is needed to study this in more detail.

5. Conclusions

This study used the VIC model to simulate the LST of the Ebinur Lake Watershed from 1960 to 2017, analyzed its temporal and spatial characteristics, conducted a multiscale cross-analysis of various meteorological elements and LST, analyzed the correlations between LUCC and LST, and obtained the following conclusions:

1. Although the model simulates the LST well in time and the space verification results are relatively good, the LST simulation in the high-altitude area of the cold month is seriously overestimated, which may be related to the occurrence of snowfall or to the altitude. Further research is needed.
2. The LST of the Ebinur Lake Watershed shows an overall increasing trend, and the annual average LST is higher in the central and eastern parts of the basin. On the temporal scale, the daily and monthly average LSTs showed unimodal trends. The interdecadal monthly changes are not obvious, and the monthly average LST from 2010 to 2017 fluctuates more than in other periods.
3. It is worth mentioning that there is a sudden change affected by the mean LST on a time scale of 1~2a (1980–1996); that is, there is a "strong-weak" transition in the LST.
4. From 1960 to 2017, the LUCC of the Ebinur Lake Watershed underwent major changes, and the reduction of open shrubs may have caused the LST increase in this area.

Author Contributions: N.A. conceived and designed the experiments, performed the experiments, analyzed the data, contributed reagents/materials/analysis tools, prepared figures and/or tables, authored or reviewed drafts of the paper, approved the final draft. J.D. conceived and designed the experiments, contributed reagents/materials/analysis tools, authored or reviewed drafts of the paper, approved the final draft. All authors have read and agreed to the published version of the manuscript.

Funding: This research was funded by National Natural Science Foundation of China (no. 42171269); Xinjiang Academician Workstation Cooperative Research Project (no. 2020.B-001); National Natural Science Foundation of China (no. 41771470); National Natural Science Foundation of China (no. 4196105).

Acknowledgments: Sincere thanks are extended to the teachers and students of the Key Laboratory of Oasis Ecology Department of Xinjiang University for their support and help. Furthermore, the authors would like to thank the reviewers and editors for providing helpful suggestions to improve this manuscript.

Conflicts of Interest: The authors declare no conflict of interest.

References

1. Stocker, T. *Climate Change 2013: The Physical Science Basis: Working Group I Contribution to the Fifth Assessment Report of the Intergovernmental Panel on Climate Change*; Cambridge University Press: Cambridge, UK, 2014.
2. Zhang, D.; Tang, R.; Zhao, W.; Tang, B.; Wu, H.; Shao, K.; Li, Z.-L. Surface Soil Water Content Estimation from Thermal Remote Sensing based on the Temporal Variation of Land Surface Temperature. *Remote Sens.* **2014**, *6*, 3170–3187. [CrossRef]
3. Harris, P.P.; Folwell, S.S.; Gallego-Elvira, B.; Rodríguez, J.; Milton, S.; Taylor, C.M. An Evaluation of Modeled Evaporation Regimes in Europe Using Observed Dry Spell Land Surface Temperature. *J. Hydrometeorol.* **2017**, *18*, 1453–1470. [CrossRef]
4. Vancutsem, C.; Ceccato, P.; Dinku, T.; Connor, S.J. Evaluation of MODIS land surface temperature data to estimate air temperature in different ecosystems over Africa. *Remote Sens. Environ.* **2010**, *114*, 449–465. [CrossRef]
5. Mustafa, E.K.; El-Hamid, H.T.A.; Tarawally, M. Spatial and temporal monitoring of drought based on land surface temperature, Freetown City, Sierra Leone, West Africa. *Arab. J. Geosci.* **2021**, *14*, 4. [CrossRef]
6. Panda, S.K.; Choudhury, S.; Saraf, A.K.; Das, J.D. MODIS land surface temperature data detects thermal anomaly preceding 8 October 2005 Kashmir earthquake. *Int. J. Remote Sens.* **2007**, *28*, 4587–4596. [CrossRef]
7. Zhao, N.; Han, S.; Xu, D.; Wang, J.; Yu, H. Cooling and Wetting Effects of Agricultural Development on Near-Surface Atmosphere over Northeast China. *Adv. Meteorol.* **2016**, *2016*, 6439276. [CrossRef]
8. Luo, D.; Jin, H.; Wu, Q.; Bense, V.; He, R.; Ma, Q.; Gao, S.; Jin, X.; Lü, L. Thermal regime of warm-dry permafrost in relation to ground surface temperature in the Source Areas of the Yangtze and Yellow rivers on the Qinghai-Tibet Plateau, SW China. *Sci. Total Environ.* **2018**, *618*, 1033–1045. [CrossRef]
9. He, C.Y.; Gao, B.; Huang, Q.X.; Ma, Q.; Dou, Y.Y. Environmental degradation in the urban areas of China: Evidence from multi-source remote sensing data. *Remote Sens. Environ.* **2017**, *193*, 65–75. [CrossRef]
10. Zhang, T.; Hoerling, M.P.; Perlwitz, J.; Sun, D.-Z.; Murray, D. Physics of U.S. Surface Temperature Response to ENSO. *J. Clim.* **2011**, *24*, 4874–4887. [CrossRef]
11. Winterdahl, M.; Erlandsson, M.; Futter, M.N.; Weyhenmeyer, G.A.; Bishop, K. Intra-annual variability of organic carbon concentrations in running waters: Drivers along a climatic gradient. *Glob. Biogeochem. Cycles* **2014**, *28*, 451–464. [CrossRef]
12. Li, R.; Wang, C.; Wu, D. Changes in precipitation recycling over arid regions in the Northern Hemisphere. *Theor. Appl. Clim.* **2018**, *131*, 489–502. [CrossRef]
13. Faramarzi, M.; Heidarizadi, Z.; Mohamadi, A.; Heydari, M. Detection of vegetation changes in relation to normalized dif-ference vegetation index (NDVI) in semi-arid rangeland in western Iran. *J. Agric. Sci. Technol.* **2018**, *20*, 51–60. Available online: http://hdl.handle.net/123456789/3627 (accessed on 29 November 2021).
14. Yang, L.; Sun, G.; Zhi, L.; Zhao, J. Negative soil moisture-precipitation feedback in dry and wet regions. *Sci. Rep.* **2018**, *8*, 1–9. [CrossRef]
15. Kok, J.F.; Ward, D.S.; Mahowald, N.M.; Evan, A.T. Global and regional importance of the direct dust-climate feedback. *Nat. Commun.* **2018**, *9*, 1–11. [CrossRef] [PubMed]
16. Ebrahimi-Khusfi, Z.; Sardoo, M.S. Recent changes in physical properties of the land surface and their effects on dust events in different climatic regions of Iran. *Arab. J. Geosci.* **2021**, *14*, 1–18. [CrossRef]
17. Khusfi, Z.E.; Roustaei, F.; Khusfi, M.E.; Naghavi, S. Investigation of the relationship between dust storm index, climatic parameters, and normalized difference vegetation index using the ridge regression method in arid regions of Central Iran. *Arid. Land Res. Manag.* **2020**, *34*, 239–263. [CrossRef]
18. Ryu, J.-H.; Hong, S.; Lyu Sang, J.; Chung, C.-Y.; Shi, I.; Cho, J. Effect of Hydro-meteorological and Surface Conditions on Variations in the Frequency of Asian Dust Events. *Korean J. Remote Sens.* **2018**, *34*, 25–43. [CrossRef]

19. Ji, F.; Wu, Z.; Huang, J.; Chassignet, E.P. Evolution of land surface air temperature trend. *Nat. Clim. Chang.* **2014**, *4*, 462–466. [CrossRef]
20. Huang, J.; Guan, X.; Ji, F. Enhanced cold-season warming in semi-arid regions. *Atmos. Chem. Phys. Discuss.* **2012**, *12*, 4627–4653. [CrossRef]
21. Liu, Y.; Peng, J.; Wang, Y. Efficiency of landscape metrics characterizing urban land surface temperature. *Landsc. Urban Plan.* **2018**, *180*, 36–53. [CrossRef]
22. Mahmood, R.; Pielke Sr, R.A.; Hubbard, K.G.; Niyogi, D.; Dirmeyer, P.A.; McAlpine, C.; Carleton, A.M.; Hale, R.; Gameda, S.; Beltrán-Przekurat, A. Land cover changes and their biogeophysical effects on climate. *Int. J. Climatol.* **2014**, *34*, 929–953. [CrossRef]
23. Pitman, A.; Arneth, A.; Ganzeveld, L. Regionalizing global climate models. *Int. J. Climatol.* **2012**, *32*, 321–337. [CrossRef]
24. Chen, L.; Dirmeyer, P. The relative importance among anthropogenic forcings of land use/land cover change in affecting temperature extremes. *Clim. Dyn.* **2019**, *52*, 2269–2285. [CrossRef]
25. Ahmed, B.; Kamruzzaman, M.; Zhu, X.; Rahman, M.S.; Choi, K. Simulating Land Cover Changes and Their Impacts on Land Surface Temperature in Dhaka, Bangladesh. *Remote Sens.* **2013**, *5*, 5969–5998. [CrossRef]
26. Turner II, B. Land system architecture for urban sustainability: New directions for land system science illustrated by application to the urban heat island problem. *J. Land Use Sci.* **2016**, *11*, 689–697. [CrossRef]
27. Pal, S.; Ziaul, S. Detection of land use and land cover change and land surface temperature in English Bazar urban centre. *Egypt J. Remote Sens. Space Sci.* **2017**, *20*, 125–145. [CrossRef]
28. Patz, J.A.; Campbell-Lendrum, D.; Holloway, T.; Foley, J.A. Impact of regional climate change on human health. *Nature* **2005**, *438*, 310–317. [CrossRef]
29. Change, I.C. *Mitigation of Climate Change. Contribution of Working Group III to the Fifth Assessment Report of the Intergovernmental Panel on Climate Change*; IPCC: Geneva, Switzerland, 2014; p. 1454.
30. Koch, J.; Siemann, A.; Stisen, S.; Sheffield, J. Spatial validation of large-scale land surface models against monthly land surface temperature patterns using innovative performance metrics. *J. Geophys. Res. Atmos.* **2016**, *121*, 5430–5452. [CrossRef]
31. Yuan, F.; Xie, Z.; Liu, Q.; Yang, H.; Su, F.; Liang, X.; Ren, L. An application of the VIC-3L land surface model and remote sensing data in simulating streamflow for the Hanjiang River basin. *Can. J. Remote Sens.* **2004**, *30*, 680–690. [CrossRef]
32. Wang, J.; Ding, J.; Yu, D.; Ma, X.; Zhang, Z.; Ge, X.; Teng, D.; Li, X.; Liang, J.; Lizaga, I.; et al. Capability of Sentinel-2 MSI data for monitoring and mapping of soil salinity in dry and wet seasons in the Ebinur Lake region, Xinjiang, China. *Geoderma* **2019**, *353*, 172–187. [CrossRef]
33. Wang, J.; Ding, J.; Yu, D.; Teng, D.; He, B.; Chen, X.; Ge, X.; Zhang, Z.; Wang, Y.; Yang, X.; et al. Machine learning-based detection of soil salinity in an arid desert region, Northwest China: A comparison between Landsat-8 OLI and Sentinel-2 MSI. *Sci. Total Environ.* **2020**, *707*, 136092. [CrossRef]
34. Wang, J.; Ding, J.; Li, G.; Liang, J.; Yu, D.; Aishan, T.; Zhang, F.; Yang, J.; Abulimiti, A.; Liu, J. Dynamic detection of water surface area of Ebinur Lake using multi-source satellite data (Landsat and Sentinel-1A) and its responses to changing environment. *Catena* **2019**, *177*, 189–201. [CrossRef]
35. Zhang, J.; Ding, J.; Wu, P.; Tan, J.; Huang, S.; Teng, D.; Cao, X.; Wang, J.; Chen, W. Assessing arid Inland Lake Watershed Area and Vegetation Response to Multiple Temporal Scales of Drought Across the Ebinur Lake Watershed. *Sci. Rep.* **2020**, *10*, 1354. [CrossRef] [PubMed]
36. Zhang, F.; Kung, H.; Johnson, V.C.; LaGrone, B.I.; Wang, J. Change Detection of Land Surface Temperature (LST) and some Related Parameters Using Landsat Image: A Case Study of the Ebinur Lake Watershed, Xinjiang, China. *Wetlands* **2018**, *38*, 65–80. [CrossRef]
37. Zhao, B.; Mao, K.; Cai, Y.; Shi, J.; Li, Z.; Qin, Z.; Meng, X.; Shen, X.; Guo, Z. A combined Terra and Aqua MODIS land surface temperature and meteorological station data product for China from 2003 to 2017. *Earth Syst. Sci. Data* **2020**, *12*, 2555–2577. [CrossRef]
38. Tang, C.; Piechota, T.C. Spatial and temporal soil moisture and drought variability in the Upper Colorado River Basin. *J. Hydrol.* **2009**, *379*, 122–135. [CrossRef]
39. Umair, M.; Kim, D.; Ray, R.L.; Choi, M. Estimating land surface variables and sensitivity analysis for CLM and VIC simulations using remote sensing products. *Sci. Total Environ.* **2018**, *633*, 470–483. [CrossRef] [PubMed]
40. Grinsted, A.; Moore, J.; Jevrejeva, S. Application of the cross wavelet transform and wavelet coherence to geophysical time series. *Nonlinear Process. Geophys.* **2004**, *11*, 561–566. [CrossRef]
41. Torrence, C.; Compo, G. A Practical Guide to Wavelet Analysis. *Bull. Am. Meteorol. Soc.* **1997**, *79*, 61–68. [CrossRef]
42. Aguiar-Conraria, L.; Soares, M. *The Continuous Wavelet Transform: A Primer*; NIPE Working Papers; Universidade do Minho: Braga, Portugal, 2010.
43. Hudgins, L.; Friehe, C.A.; Mayer, M.E. Wavelet transforms and atmospheric turbulence. *Phys. Rev. Lett.* **1993**, *71*, 3279–3282. [CrossRef]
44. Lau, K.M.; Weng, H. Climate Signal Detection Using Wavelet Transform: How to Make a Time Series Sing. *Bull. Am. Meteorol. Soc.* **1995**, *76*, 2391–2402. [CrossRef]
45. Huang, H.B.; Huang, X.R.; Yang, M.L.; Lim, T.C.; Ding, W.P. Identification of vehicle interior noise sources based on wavelet transform and partial coherence analysis. *Mech. Syst. Signal Process.* **2018**, *109*, 247–267. [CrossRef]

46. Troy, T.J.; Wood, E.; Sheffield, J. An efficient calibration method for continental-scale land surface modeling. *Water Resour. Res.* **2008**, *44*, W09411. [CrossRef]
47. Onderka, M.; Banzhaf, S.; Scheytt, T.; Krein, A. Seepage velocities derived from thermal records using wavelet analysis. *J. Hydrol.* **2013**, *479*, 64–74. [CrossRef]
48. Su, L.; Miao, C.; Borthwick, A.G.; Duan, Q. Wavelet-based variability of Yellow River discharge at 500-, 100-, and 50-year timescales. *Gondwana Res.* **2017**, *49*, 94–105. [CrossRef]
49. Wang, J.; Shi, T.; Yu, D.; Teng, D.; Ge, X.; Zhang, Z.; Yang, X.; Wang, H.; Wu, G. Ensemble machine-learning-based framework for estimating total nitrogen concentration in water using drone-borne hyperspectral imagery of emergent plants: A case study in an arid oasis, NW China. *Environ. Pollut.* **2020**, *266*, 115412. [CrossRef] [PubMed]
50. Reynolds, R.L.; Yount, J.C.; Reheis, M.; Goldstein, H.; Chavez, P., Jr.; Fulton, R.; Whitney, J.; Fuller, C.; Forester, R.M. Dust emission from wet and dry playas in the Mojave Desert, USA. *Earth Surf. Process. Landf.* **2007**, *32*, 1811–1827. [CrossRef]
51. Wang, J.; Hu, X.; Shi, T.; He, L.; Hu, W.; Wu, G. Assessing toxic metal chromium in the soil in coal mining areas via proximal sensing: Prerequisites for land rehabilitation and sustainable development. *Geoderma* **2022**, *405*, 115399. [CrossRef]
52. Balba, A.M. *Management of Problem Soils in Arid Ecosystems*; CRC Press: Boca Raton, FL, USA, 2018.
53. Fu, P.; Xie, Y.; Weng, Q.; Myint, S.; Meacham-Hensold, K.; Bernacchi, C. A physical model-based method for retrieving urban land surface temperatures under cloudy conditions. *Remote Sens. Environ.* **2019**, *230*, 111191. [CrossRef]
54. Shao, Z.; Zhang, L. Estimating Forest Aboveground Biomass by Combining Optical and SAR Data: A Case Study in Genhe, Inner Mongolia, China. *Sensors* **2016**, *16*, 834. [CrossRef]
55. Liu, T.; Zhang, S.; Yu, L.; Bu, K.; Yang, J.; Chang, L. Simulation of regional temperature change effect of land cover change in agroforestry ecotone of Nenjiang River Basin in China. *Theor. Appl. Clim.* **2017**, *128*, 971–981. [CrossRef]

remote sensing

MDPI

Article

Mapping Aquaculture Areas with Multi-Source Spectral and Texture Features: A Case Study in the Pearl River Basin (Guangdong), China

Yue Xu, Zhongwen Hu *, Yinghui Zhang, Jingzhe Wang, Yumeng Yin and Guofeng Wu

MNR Key Laboratory for Geo-Environmental Monitoring of Great Bay Area & Guangdong Key Laboratory of Urban Informatics & Shenzhen Key Laboratory of Spatial Smart Sensing and Services & School of Architecture and Urban Planning, Shenzhen University, 518060 Shenzhen, China; xuyue19@email.szu.edu.cn (Y.X.); zyhui@szu.edu.cn (Y.Z.); jingzhewang@szu.edu.cn (J.W.); yinyumeng2021@email.szu.edu.cn (Y.Y.); guofeng.wu@szu.edu.cn (G.W.)
* Correspondence: zwhoo@szu.edu.cn

Abstract: Aquaculture has grown rapidly in the field of food industry in recent years; however, it brought many environmental problems, such as water pollution and reclamations of lakes and coastal wetland areas. Thus, the evaluation and management of aquaculture industry are needed, in which accurate aquaculture mapping is an essential prerequisite. Due to the difference between inland and marine aquaculture areas and the difficulty in processing large amounts of remote sensing images, the accurate mapping of different aquaculture types is still challenging. In this study, a novel approach based on multi-source spectral and texture features was proposed to map simultaneously inland and marine aquaculture areas. Time series optical Sentinel-2 images were first employed to derive spectral indices for obtaining texture features. The backscattering and texture features derived from the synthetic aperture radar (SAR) images of Sentinel-1A were then used to distinguish aquaculture areas from other geographical entities. Finally, a supervised Random Forest classifier was applied for large scale aquaculture area mapping. To address the low efficiency in processing large amounts of remote sensing images, the proposed approach was implemented on the Google Earth Engine (GEE) platform. A case study in the Pearl River Basin (Guangdong Province) of China showed that the proposed approach obtained aquaculture map with an overall accuracy of 89.5%, and the implementation of proposed approach on GEE platform greatly improved the efficiency for large scale aquaculture area mapping. The derived aquaculture map may support decision-making services for the sustainable development of aquaculture areas and ecological protection in the study area, and the proposed approach holds great potential for mapping aquacultures on both national and global scales.

Keywords: multi-source remote sensing; aquaculture mapping; texture feature; Google Earth Engine; Pearl River Basin

Citation: Xu, Y.; Hu, Z.; Zhang, Y.; Wang, J.; Yin, Y.; Wu, G. Mapping Aquaculture Areas with Multi-Source Spectral and Texture Features: A Case Study in the Pearl River Basin (Guangdong), China. *Remote Sens.* **2021**, *13*, 4320. https://doi.org/10.3390/rs13214320

Academic Editor: Qiusheng Wu

Received: 16 August 2021
Accepted: 25 October 2021
Published: 27 October 2021

1. Introduction

Aquaculture has become one of the fastest-growing food industries [1], and the fishery products of China play an important role in the international seafood market [2], with over 60% of the fish farmed in the world [3]. However, according to the 2018 State of World Fisheries and Aquaculture report (SOFIA 2018) by the Food and Agriculture Organization of the United Nations (FAO), the proportion of marine fish resources within a biologically sustainable level showed a downward trend in recent years [4]. In China, many lakes and coastal wetlands were reclaimed in the past few years in order to support the fast development of fisheries [5], putting tremendous pressures on environments and hampering regional sustainable developments [6]. Accurately mapping aquaculture areas is an important support to policy development and implementation at regional, national, and global levels, and to measure progress towards sustainable developments [7].

Traditional field survey for aquaculture mapping suffers from low efficiency, and currently the satellite remote sensing technique is one of the most important methods due to its many advantages, such as low cost, wide monitoring range, high efficiency, and high repeated observations [8–10]. Optical and radar remote sensing images have been increasingly utilized to delineate aquaculture areas [11,12], and many methods have been developed for local [13], regional [14], and national scale [15] aquaculture mapping. Meanwhile, due to the periodic repeated observations of satellites, remote sensing images were used not only to map aquaculture areas at a single time [16], but also to map their time series distributions [17].

Previous studies can be roughly summarized into two categories according to basic mapping units: pixel-based and object-based approaches. The pixel-based methods are widely applied to the images with low and medium spatial resolutions. For example, artificially designed spectral and textural features were computed for each pixel, and a supervised machine learning classifier was used to map large scale aquaculture areas [18]; Sakamoto et al. [19] applied a wavelet-based filter for detecting inland-aquaculture areas from MODIS time series images, and deep-learning-based methods were used for aquaculture classification [20,21]. For the object-based classification method [22–24], images are segmented into many homogeneous segments, which are further classified through machine learning classifiers or classification rules derived from expert rules. For example, Wang et al. [25] segmented Landsat images into objects using multi-resolution segmentation method to extract raft-type aquaculture areas.

Considering data sources, most studies used only optical images for they are visually intuitive and easy to be understood. For mapping aquaculture facilities over small areas, high spatial resolution remote sensing images are frequently applied [26,27], whereas medium resolution images are generally used for mapping aquaculture facilities at a regional or national scale due to their wide coverage and better spectral resolution. For example, Ren et al. [28] combined Landsat series images and an object-based classification method to map the spatiotemporal distribution of aquaculture ponds in China's coastal zone. Synthetic aperture radar (SAR) images are also used for aquaculture mapping [14,29]. For examples, Hu et al. [29] detected floating raft aquaculture from SAR image using statistical region merging and contour feature; Ottinger et al. [14] employed time series Sentinel-1 images and object-based approach to map aquaculture ponds over river basins; and Zhang et al. [30] mapped marine raft aquaculture areas using a deep learning approach by enhancing the contour and orientation features of Sentinel-1 images.

Most studies applied satellite images at a certain time to map aquaculture areas, which may affect the accurate mapping of aquaculture areas for their dynamics. Although the influence of weather could be suppressed by carefully selecting images, other factors may also affect the accurate mapping of aquaculture areas. For example, some paddy fields are still water-dominated at the early stage of farming, and inland aquaculture ponds are drained during harvest time. Previous studies demonstrated that more accurate aquaculture maps could be obtained by using time series SAR images [31], and time series optical images was also proved to be effective in improving mapping accuracy [32,33]. Therefore, time series images are an ideal and reliable data source for aquaculture mapping.

The Google Earth Engine (GEE) platform provides a series of free remote sensing images, many kinds of image processing algorithms, and high-performance computing capabilities, and it can process huge amount of time series remote sensing images over a large-scale area [34–36]. Therefore, GEE has been widely used for mapping wetlands [37] and agricultural lands [38], and it also has the potentials to map aquaculture areas [39–42]. For examples, Xia et al. [39] proposed a framework for extracting aquaculture ponds by integrating existing multi-source remote sensing images on the GEE platform; Duan et al. mapped aquaculture ponds over coastal area of China using Landsat-8 images and GEE platform [41] and further analyzed their dynamic changes from 1990 to 2020 [42]. Existing studies were mainly focused on designing artificial image features or training deep leaning models to map a specific type of aquaculture area, such as aquaculture ponds. However, many different aquaculture types are found over a large area, and how to simultaneously extract multiple types of aquaculture areas over a large area has not been well studied.

This study proposed a novel approach for mapping aquaculture areas with multiple types over large areas. With the Pearl River Basin (Guangdong) of China as a case study, time series Sentinel images were used as a data source to overcome the accidental factors of single-time observation. The spectral indices (including normalized difference vegetation index (NDVI), normalized water index (NDWI), and normalized built-up index (NDBI)) derived from Sentinel-2A multispectral images, the VV and VH polarized data of Setinel-1 SAR images, and their derived texture features were used to map aquaculture areas using machine learning algorithms implemented in Google Earth Engine. The proposed method holds great potentials in simultaneously mapping different types of aquaculture areas over a large area.

2. Materials

2.1. Study Area

In this study, the Pearl River Basin (Guangdong) refers the part of the basin that located in Guangdong Province. The study area (Figure 1) covers the Pearl River Basin (Guangdong) and its 40 km buffer seaward from coastal line. The Pearl River with 2320 km long is the third-longest river in China, and it covers a region of subtropical maritime monsoon climate. The Pearl River Basin has been one of the most economically dynamic regions of China since ancient times. Local sea reclamation and marine aquaculture started very early, and especially the urbanization in the Pearl River Basin (Guangdong) is very fast since the launch of China's Reform Programme in 1979. Currently, this specific region is one of the most developed regions in China; however, limited river networks cannot meet the demands of aquaculture developments [43]. Therefore, the expansions of aquaculture land are mainly concentrated in the ocean and coastal region. With years of developments, this region has become one of the biggest aquaculture bases with massive abundant inland aquaculture ponds and mariculture areas. The aquaculture, including fish farming, shrimp farming, oyster farming, mariculture, algaculture (such as seaweed farming) and the cultivation of ornamental fish, is well-developed in this region.

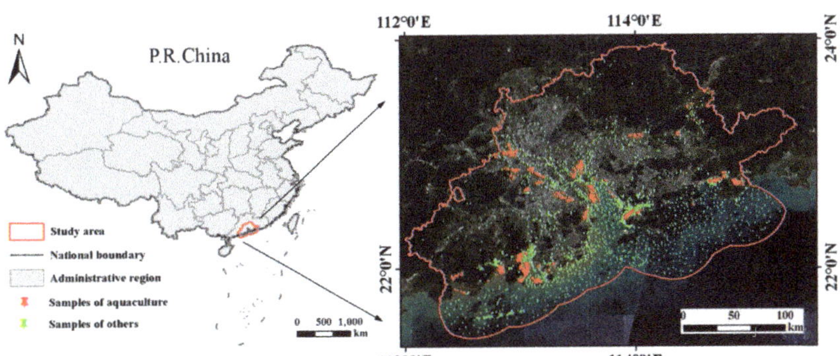

Figure 1. Geographical locations of the Pearl River Basin (Guangdong) and study area (annual median images of sentinel-2A (2020) after cloud removal).

2.2. Data

Sentinel series satellites are the important branches of the Copernicus Programme satellite constellation conducted by the European Space Agency (ESA) and an essential part of Global Monitoring for Environment and Security (GMES). Sentinel-1 satellite constellation is the first of this specific program, and it carries C-band synthetic-aperture radar (SAR) instrument with a temporal resolution of 6 days [44]. The Sentinel-2A/B satellites equipped with Multispectral Instrument (MSI) are capable of acquiring high spatial resolution (up to 10 m) optical images in 13 spectral bands every 5 days [45]. The ESA and European Commission's policies makes Sentinel's data easily accessible [46]. More detailed technical specifications of both Sentinel-1 SAR and Sentinel-2A MSI data are available in previous studies [47,48] and will not be repeated here.

The standard Level-1 (ground range detected, GRD) data product of Sentinel-1 in Interferometric Wide Swath (IW) mode and the calibrated Bottom of Atmosphere (BOA) products of Sentienel-2 used in this study were offered by GEE. Considering the data availability, this study was only focused on 2020. One arc-minute global relief model of the Earth's surface that integrates land topography and ocean bathymetry (ETOPO1) dataset was also collected to extract ocean bathymetry [49]. The Global Self-consistent, Hierarchical, High-resolution Geography Database (GSHHG) that comprises World Vector Shorelines (WVS) [50] was applied to extract shorelines of the study area. All these datasets were re-projected to the uniform coordinate reference system (WGS_1984_UTM_Zone_50N) and resampled to a same ground spatial resolution of 10 × 10 m.

3. Methods

The approach proposed in this study for large-scale aquaculture mapping with multi-source remotely sensed images and the GEE platform includes seven main steps (Figure 2): (1) data preprocessing, (2) feature extraction, (3) water surface extraction (including aquaculture land), (4) sample selection, (5) Random Forest classification in the GEE platform, (6) morphological post processing, and (7) accuracy assessment. These steps were repeated several times to find the optimal parameters.

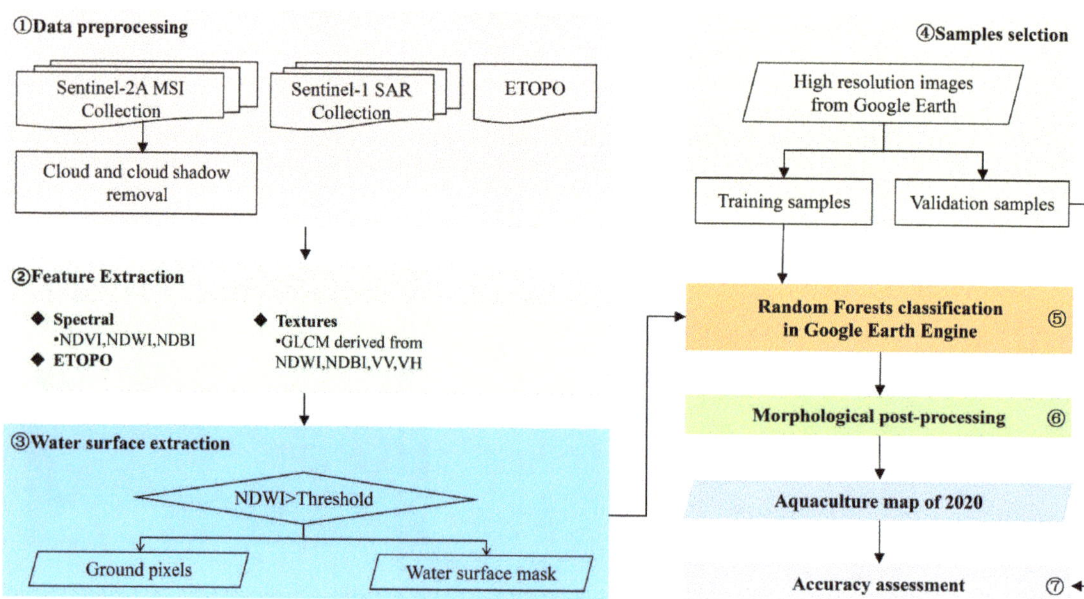

Figure 2. Approach proposed in this study.

3.1. Data Preprocessing

A bitmask band with cloud mask information (QA60) for the Sentinel-2A image is provided on the GEE platform. Cloud and cloud shadow affecting observation quality were identified by the code based on the Function of Mask (CFMask) [51], and they were marked in the QA60 band. The clear-sky pixels were selected according to QA60, and the Sentinel-2A images in 2020 with less than 20% cloud and cloud shadow cover were used in this study.

3.2. Feature Extraction for Each Image Pixel

3.2.1. Image Features of Aquaculture Areas

Only aquaculture ponds show regular shapes and textures on the Sentinel-2A multispectral images (Figure 3), because they are usually very large compared with medium-resolution image pixels. The spectral signatures of water ponds are also clearly presented in NDWI image, and texture features are clearly presented in all optical, spectral indices and SAR images.

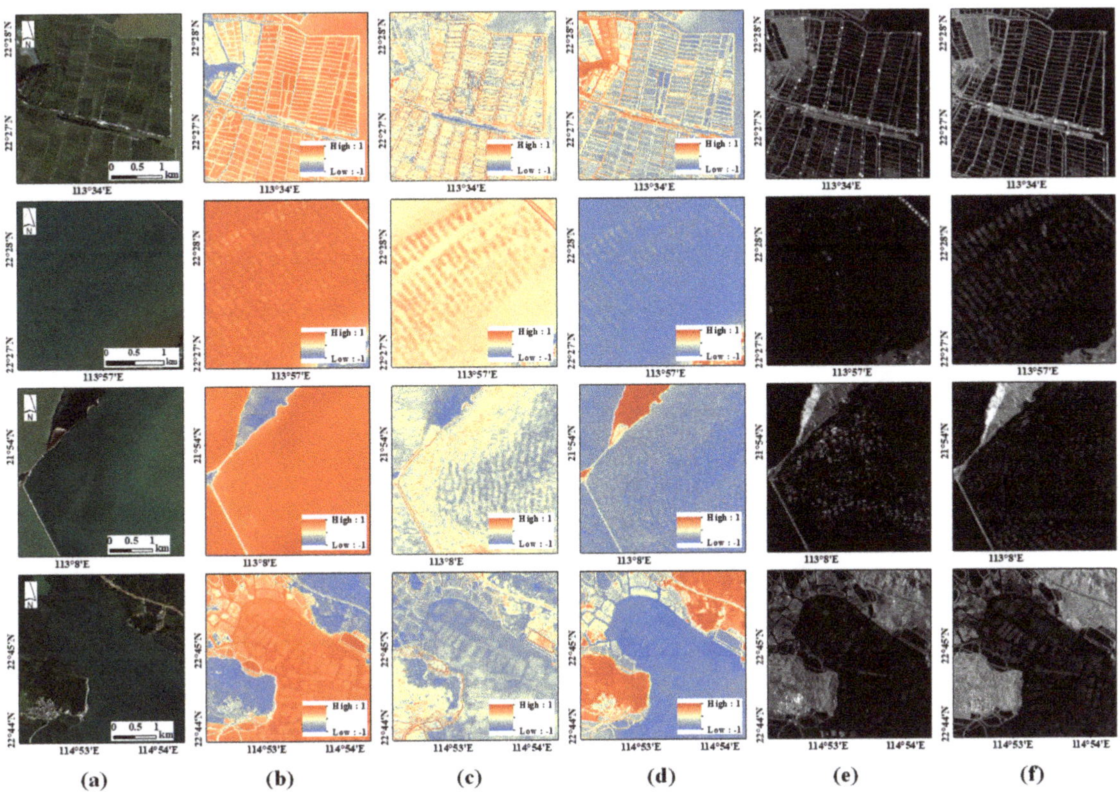

Figure 3. Images and features of typical aquaculture areas: (**a**) Sentinel-2A multispectral image, (**b**) NDWI, (**c**) NDBI, (**d**) NDVI, (**e**) VV Polarization image, and (**f**) VH Polarization image.

Mariculture facilities do not show obvious signatures in Sentinel-2A multispectral optical images, because they are relatively small compared with medium-resolution image pixels, or some are submerged in water. However, these aquaculture types show significantly different characteristics in NDWI, NDVI and NDBI images, because normalized difference image features amplify the differences. More specifically, the textures are very clear in Sentinel-1 SAR images, because the backscatter signature of the radar is sensitive to the roughness and structure of water surface.

Based on the above analyses, we found that NDVI, NDWI, and NDBI could better highlight the differences between aquaculture areas and other water bodies and the textures derived from index images; VV and VH images could further enhance the differences.

3.2.2. Normalized Difference Spectral Indices

In this study, NDWI, NDVI [52] and NDBI [53] were used as spectral features.

Normalized Difference Water Index (NDWI) (Equation (1)): NDWI was proposed by McFeeters [54], based on the fact of strong spectral absorption of water body in the near-infrared (NIR) band, and it is a robust method for detecting aquaculture water surface and has proved effective in segmenting muddy aquaculture ponds [39]. Correspondingly, NDWI can be used to distinguish land and water bodies.

$$\text{NDWI} = (\rho_{\text{green}} - \rho_{\text{nir}})/(\rho_{\text{green}} + \rho_{\text{nir}}) \tag{1}$$

where ρ_{green} is the reflectance of green band (B3 of MSI), and ρ_{nir} represents the reflectance of near-infrared band (B8 of MSI).

In this study, the images with annual cloud cover less than 20% in 2020 were used to calculate a series of NDWI values. The NDWI series were sorted from small to large, and the 25% maximum NDWI values (located in the 75–100% of the sequence) were selected to calculate the average NDWI image. This preprocessing takes full use of all the image features for the whole year and eliminates the seasonal differences of aquaculture areas.

Normalized difference vegetation index (NDVI) (Equation (2)): Generally, aquaculture ponds usually have higher nutrients than natural water bodies, and the algae concentration is also higher. The waterbodies with high algae concentration generally have high reflectance in NIR spectrum compared with natural waters (without aquacultures). Therefore, NDVI may separate aquaculture waters from natural waters at a certain degree.

$$\text{NDVI}=(\rho_{\text{nir}} - \rho_{\text{red}})/(\rho_{\text{nir}} + \rho_{\text{red}}) \tag{2}$$

where ρ_{red} represents the reflectance of red band (B4 of MSI), and ρ_{nir} represents the reflectance of near-infrared band (B8 of MSI). Referring to the processing of NDWI, we also sorted and filtered the NDVI values (75–100%) to compose the final average NDVI image.

Normalized difference built-up index (NDBI) (Equation (3)): NDBI is originally designed to detect built-up and barren areas. In practice, the extracted water mask often contains shadow noise with low reflectivity, which may be caused by tall buildings over built-up areas. The surrounding pixels of shadows usually have high NDBI values, and thus it is able to distinguish building shadows by analyzing their surrounding pixels.

$$\text{NDBI} = (\rho_{\text{swir1}} - \rho_{\text{nir}})/(\rho_{\text{swir1}} + \rho_{\text{nir}}) \tag{3}$$

where ρ_{swir1} represents the reflectance of the shortwave infrared band (B10 of MSI), and ρ_{nir} represents the reflectance of the near-infrared band (B8 of MSI). To further reduce the potential effects of shadow noise, we also sorted and filtered the NDBI values (75–100%) to compose the final average NDBI image. The morphological operation is an effective way to examine the neighbors of a pixel. Since a higher NDBI indicates a higher possibility of a built-up pixel, an inflation operator was conducted on the NDBI image to eliminate lathy and small building shadows, and a corrosion operator was further adopted to restore the image.

3.2.3. Backscatter Features

A single Sentinel-1 SAR image is usually contaminated by speckle noises, and previous studies demonstrated the effectiveness of SAR image enhancement using time series images [55]. Following Xia's work [39], we selected the mean values of the VV and VH time series images of 75–100% in ascending order at the pixel level to form the final VV and VH image over the whole study area.

3.2.4. Texture Features

Image texture provides relevant characteristics of spatial structures. In this study, the aquaculture lands hold relative regular shapes and arrangements, which are reflected in the regular change of gray levels of images pixels. Considering the results of previous studies, we combined different kinds of textures to address the problem of insufficient accuracy and to reduce the interference of the surroundings.

Gray Level Co-occurrence Matrix (GLCM) [56] is a method to analyze texture characteristics, and it calculates the correlation between two gray levels to reflect the comprehensive information about directions, intervals, and the changes in magnitude and speed. Some studies suggested that better recognition performance could be obtained by using extreme values to calculate texture features in different directions than in any direction or average direction. NDWI, NDBI, and backscattering coefficient of VV/VH were employed as input images for calculating GLCM, with gray level of 16, 150 × 150 m sliding windows and four directions (Figure 4). The NDVI image was mainly used to distinguish aquaculture ponds and natural waters, and it was not applied to compute texture features in this study.

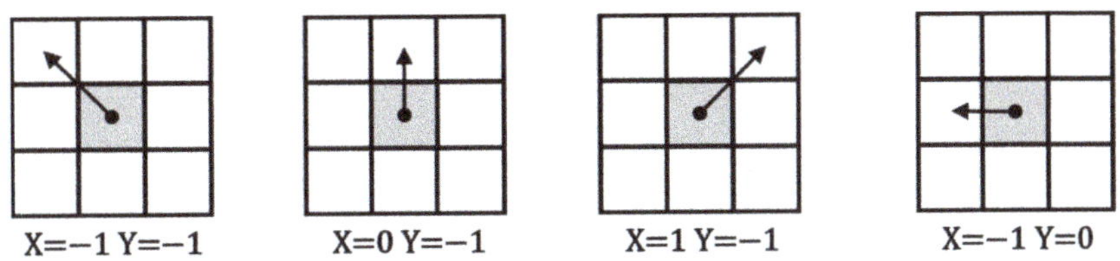

Figure 4. Direction measure of GLCM calculation in this study.

Numerous texture feature bands (n = 18) can be generated using GLCM method in GEE. Redundant textural measures are possibly complex and may reduce computing speed, which may influence the performance of classification. Following Yu's work [57], Angular Second Moment, Contrast, Correlation, Variance, Sum Average, Inverse Difference Moment, and Entropy were selected and computed from GLCM (Table 1), and their maximum and minimum values in four directions (Figure 4) were analyzed. Therefore, for each input image, a 14-dimension feature vector was obtained (7 measures × 2 values).

Table 1. Textural measures based on GLCM method in this study.

Parameter	Detailed Description
Angular Second Moment	Number of repeated pairs
Contrast	Local contrast of within an image window
Correlation	Correlation between pairs of pixels
Variance	Dispersion of gray-level distribution
Inverse Difference Moment	Homogeneity within window
Entropy	Randomness of a gray-level distribution
Sum Average	Sum Average of pixels within window

3.3. Water Surface Extraction

The pixels with aquaculture areas mainly show pure water or water-dominated spectral responses, and the probability of water surface can be simply reflected by NDWI. Inspired by [39], the water surface was first detected from NDWI image by that a pixel with a NDWI value higher than a threshold was classified as water. The aquaculture facilities in the sea, rivers, or lakes show different spectral signatures from pure water bodies. The sizes of aquaculture facilities are relatively smaller than medium spatial resolution image pixels and some aquaculture facilities are submerged in the water, and thus the pixels with aquaculture facilities are mixed pixels. The NDWI values of pixels with aquaculture facilities are usually lower than those of pure water pixels but higher than those of other objects, and thus the mixed objects can be classified as water surface using a threshold.

3.4. Sample Selection

The aquaculture areas in the Pearl River Basin (Guangdong) hold regular spatial morphological characteristics in remotely sensed images (Figure 5). In this study, the aquaculture areas were roughly divided into inland aquaculture ponds and marine aquaculture. Inland aquaculture ponds are mainly distributed in the junction zone between water body (river, lake and sea) and land. The inland aquaculture ponds are formed through the reclamation of coastal wetlands or inland lakes, and they are usually partitioned by embankments and have regular and compact shapes (Figure 5b,c). Inland aquaculture pond units are usually clustered together, showing regular texture structures.

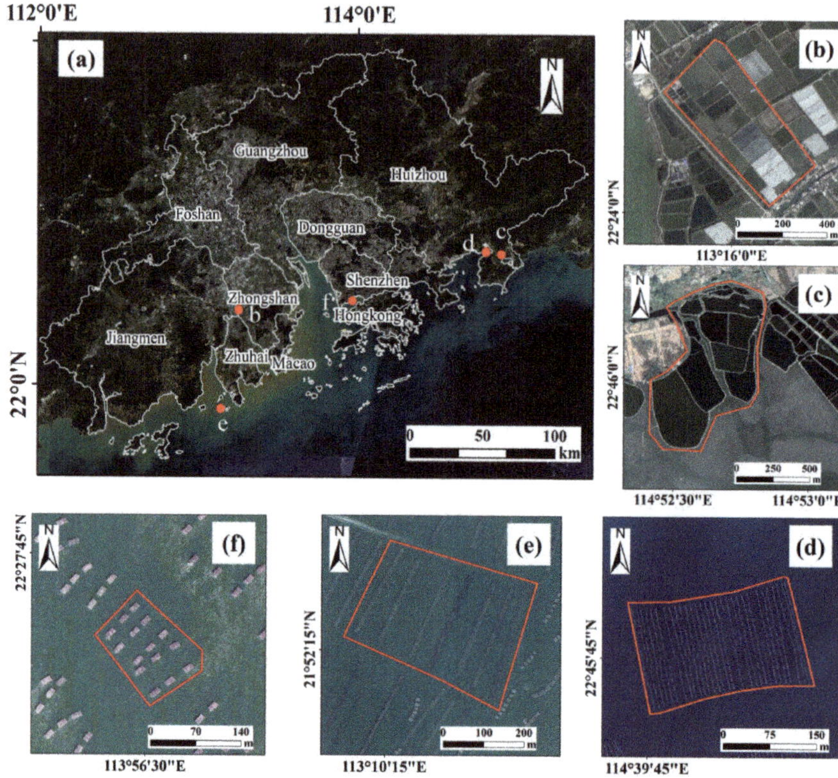

Figure 5. Typical aquaculture samples over the study area in high resolution Google Earth image (**a**): costal aquaculture pond (**b**,**c**), visible raft aquaculture (**d**,**e**), and concentrated-distributed cage culture area (**f**).

Marine aquaculture areas are often located in bays and inshore seawaters, and they include raft aquaculture (Figure 5d,e) and cage aquaculture (Figure 5f). The raft aquaculture facilities in neritic zone are composed of aquatic bamboo raft (for floating) and subaquatic thick rope (for fixing aquatic products). Correspondingly, the raft aquaculture areas are characterized by dull grey stripes (Figure 5d,e). The cage culture areas consist of aquatic plastic frames and suspended net cages (Figure 5f). Comparing with water bodies, cage aquaculture facilities appear as brighter colors in remotely sensed imagery, and their distributions are more concentrated, with regular rectangles.

In this study, a total number of 8317 samples were selected for classifier training ($n = 7317$) (Table 2) and accuracy assessment ($n = 1000$).

Table 2. Classification system used in this study.

Class	Description	Reference Samples
Aquaculture	Coastal ponds and marine aquaculture	4041
Water body	River network, lakes, and tidal wetlands	3067
Other	Building shadow and other small fragile patches with low reflectivity	209

3.5. Random Forest (RF) Classification on GEE Platform

GEE provides a serial of classification algorithms, including Minimum Distance (MD) [58], Random Forest (RF) [59], Support Vector Machine (SVM) [60], Regression Trees (CART) [61] and Naive Bayes (NB) [62]. According to previous studies, SVM and RF have better performance than other classification algorithms [63–65]. SVM algorithm performs good with limited training samples, however, RF performs better as training sets get larger [66]. RF is conceptually similar to tree-based learners, and they share same advantages [67]. RF can exhibit higher accuracy and efficiency even facing the unfavorable conditions of noise interference. Moreover, RF classifier shows better performance than other classifiers in land cover mapping using long time series images [68,69]. Therefore, RF classifier was employed in this study.

In this study, the composed NDVI, NDWI and NDBI products were also employed as spectral attributes of each pixel, the GLCM features of NDWI, NDBI, VV and VH were used as texture attributes, and the ETOPO1 value was applied as a topographic attribute. A feature vector of 60 dimensions (3 spectral + 4 images × 14 measures/image texture + 1 topographic) was obtained. A RF classifier on GEE platform was trained according to existing reference [70], in which the number of trees (n_{tree}) was set to 100, the maximum depth was unlimited, the minimum sample number of each tree node was 1, and the number of features for each tree was set as the square root of the variable number.

3.6. Morphological Post Processing

Based on the above process, the aquaculture areas in the Pearl River Basin (Guangdong) were obtained preliminarily. However, the texture difference between aquaculture land and streamway were not clearly detected, which may cause misclassification. Streamway is one of the equipped facilities of aquaculture ponds. The obvious morphological difference between open water and streamway can be observed, and thus the compactness of each extracted aquaculture pond was calculated for modifying misclassification with Equation (4), which reflects the geometric shape of the object. The threshold of compactness was set as 11.5 through a trial-and-error process.

$$\text{Compactness} = \frac{\text{W} \times \text{H}}{\text{Area}} \tag{4}$$

where W and H are the width and height of the minimum enclosing rectangle, respectively, and Area is the number of inner pixels of the object. The small holes ($\leq$1000 pixels) within aquaculture patches were filled, and the isolated small patches ($\leq$3 pixels) were removed. An illustration of post processes is presented in Figure 6.

3.7. Accuracy Assessment

A total of 1000 samples of aquacultures extracted from Google Earth images using visual interpretation were used as ground-truth references. The confusion matrix was first obtained, and then overall accuracy and Kappa coefficient were applied to assess the accuracy of the proposed approach.

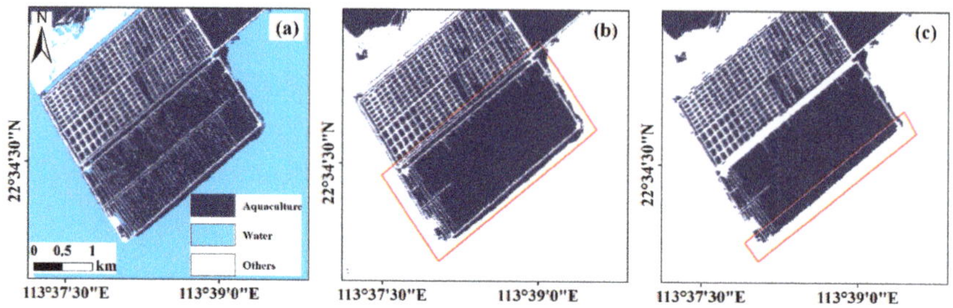

Figure 6. Overview of the morphological post processes of extracted aquaculture land in this study: (**a**) raw classification results based on RF, (**b**) exclusion of misclassified river, and (**c**) final classification map with postprocessing.

3.8. Optimal Parameter Selection

Several key factors affect the final classification performance: feature combination, window size to compute textures and threshold to extract water surface. The steps described in Sections 3.1–3.7 were repeated, and the values of these key factors maximizing overall accuracy were selected.

Feature Combination: Based on different types of features, including normalized difference spectral indices, backscatter features and image textures derived from spectral indices and SAR images, many kinds of feature combinations can be obtained. To investigate the optimal feature combination, six groups of features (Table 3) were used to obtain six aquaculture maps, respectively. The threshold $T = 0.06$ and texture window size with 15 were used. The determinations of T and window size are described in Sections 4.2 and 4.3, respectively. The 1000 random selected samples were used to evaluate their performances.

Table 3. Classification performances using different feature combinations.

Feature Group ID	ETOPO1	Indices	SAR	Textures of Indices	Textures of SAR	Accuracy	
						OA (%)	Kappa
A	√	√				81.5	0.609
B	√	√	√			83.1	0.639
C	√	√		√		88.2	0.748
D	√		√		√	86.9	0.719
E	√	√		√	√	89.5	0.776
F	√	√	√	√	√	89.0	0.765

Threshold to Extract Water Surface: Since aquaculture areas are first classified as a part of a water body, the threshold separating water and non-water surfaces plays an important role in the whole process. In this study, water surface was extracted by thresholding an NDWI image, and only one parameter T was used. The spectral indices, textures from indices and SAR images, as well as ETOPO1, were used. The window size was set at 15×15 pixels. The thresholds between -0.2 and 0.2 were tested, with a step of 0.02. The accuracies were evaluated using 1000 samples.

Window Size to Compute GLCM Features: The window size of gray-level co-occurrence matrix can directly affect the detection of aquaculture in a complex water environment. According to [71], a texture window should cover at least the minimum detectable target. Although inland aquaculture ponds have regular arrangements and clear textures, the ponds can be properly separated by thresholding NDWI image. However, the detection of mariculture facilities mainly depends on image textures. Based on our visual interpretation from Google Earth, the size of minimum detectable mariculture block is about 50 to 150 m. Therefore, a window size of 15×15 pixels was preferred. However, the experiments using a window of 5×5 pixels, 15×15 pixels, and 30×30 pixels were also carried out to determine the optimal size.

4. Results

4.1. Aquaculture Maps Obtained Using Multiple Features

Six maps were obtained using different feature combinations; their accuracies were assessed with 1000 validation samples (Table 3), and four typical sites of the maps were presented in Figure 7. The performances using only spectral indices were poor (Figure 7A and Table 3), and many water pixels were wrongly classified as aquaculture pixels, for some aquaculture facilities are submerged in waters and their spectral signatures are not obliviously different from water. The combination of spectral indices and SAR image features improved the performance (Figure 7B), especially for the second typical site, for SAR images reflect the backscatter characteristics of aquaculture facilities, which is totally different from a pure water body. However, many misclassifications still existed near shorelines, and the overall accuracy and Kappa coefficient are not satisfactory. The use of spectral indices and index-derived texture features greatly improved the classification performance (Figure 7C and Table 3), and most aquaculture ponds and mariculture areas were correctly detected, and an overall accuracy of 88.2% was obtained.

The results of combination D (Figure 7D) show the effectiveness of using SAR images and their corresponding textures. Comparing the results on the second site with those from combination C, it can be easily found that mariculture areas were better detected. However, there are many commission errors along coastlines, resulting in a lower OA (86.9%). The results in Figure 7 show that the combination of E and F obtained better performances than the combination of A, B, C, and D. However, the results of combination E had less holes, and the overall accuracy (89.5%) and Kappa coefficient (0.719) were both the highest compared with other combinations. Therefore, the feature in combination E was applied to obtain the final map.

4.2. Optimal Threshold for Water Body Segmentation

A total of 21 aquaculture maps were obtained using thresholds from -0.2 to 0.2; their overall accuracies and Kappa coefficients are presented in Figure 8, and five maps over a typical area are presented in Figure 9.

The overall accuracy and Kappa coefficient were low with a very small threshold ($T = -0.2$), due to many misclassified land pixels. With a small threshold, many mixed pixels containing waters were segmented as water surface, and further classified as aquaculture areas. As shown in Figure 9b, many embankment pixels were wrongly segmented as water surface and further misclassified as aquaculture areas. As the threshold increased, the misclassifications were gradually overcome (Figure 9b–d), the overall accuracy gradually increased, and finally the overall accuracy achieved the best result with $T = 0.06$. As the threshold further increased, the accuracy decreased for many mixed pixels and even some pure water pixels were segmented as land, and thus many aquaculture ponds were missed (Figure 9f). Generally, the optimal threshold is a balance of the two misclassification types, and the threshold 0.06 was preferred in our study.

4.3. Optimal Window Size for Texture Calculation

Figure 10 shows the maximum sum averaged texture derived from VH and the extracted aquaculture with a window size of 5×5 pixels, 15×15 pixels, and 30×30 pixels, respectively. Some small misclassifications were observed along shorelines with the window size of 5×5 pixels. With 15×15 windows, the textures better delineated the existing aquaculture areas, resulting in a satisfactory result (Figure 10f). With larger window size (30×30 pixels), the texture feature was over-smoothed. Moreover, the computational load dramatically increased with the increase in window size. Therefore, the window size of 15×15 pixels was used in this study.

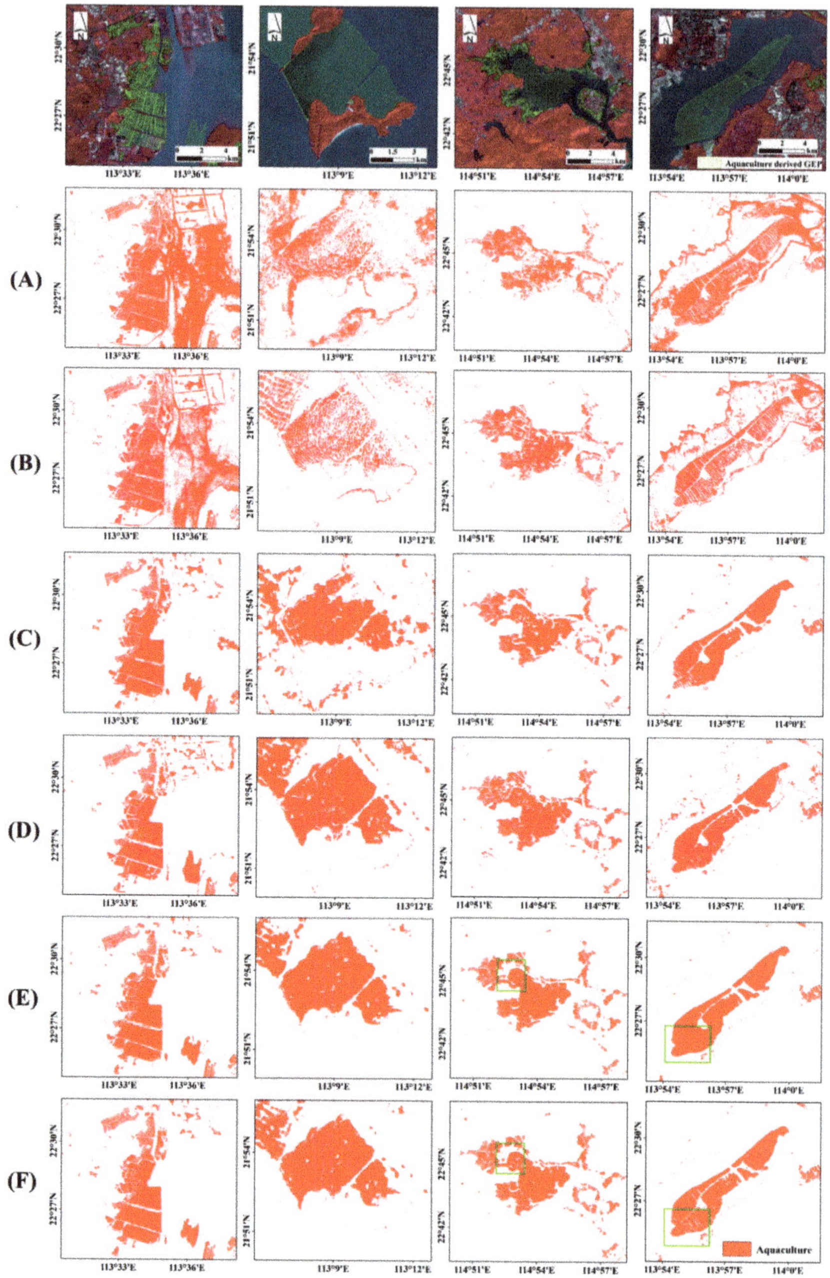

Figure 7. Aquaculture maps obtained using different feature combinations. The first-row present images and reference aquaculture areas at four typical sites, and the results of combinations (**A–F**) are presented in the second to seventh rows. The details of feature combinations are presented in Table 3.

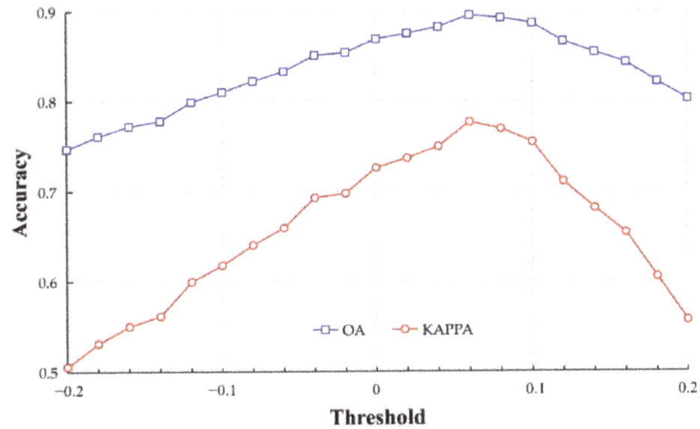

Figure 8. Overall accuracies and Kappa coefficients using different thresholds.

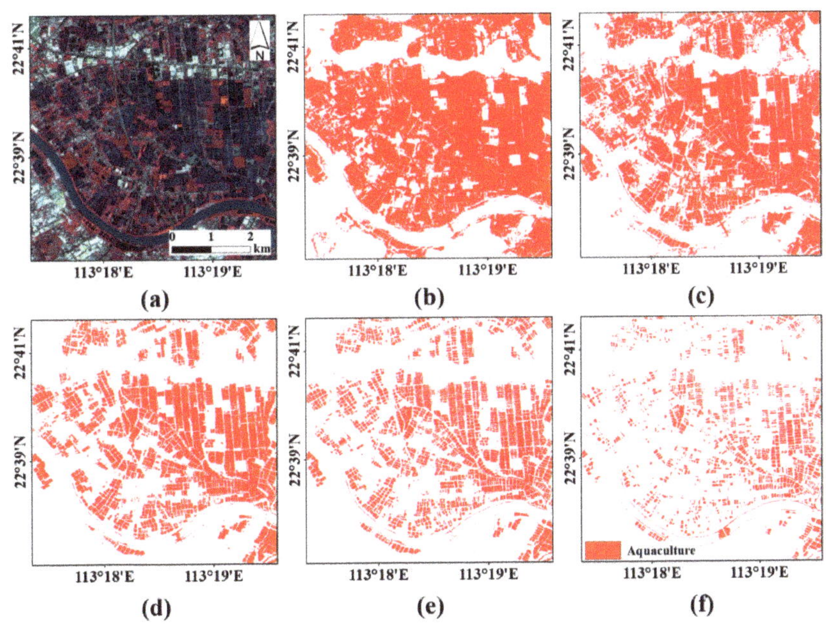

Figure 9. Typical classification maps using different thresholds: (**a**) Sentinel-2A MSI image, (**b**) $T = -0.2$, (**c**) $T = -0.1$, (**d**) $T = 0$, (**e**) $T = 0.1$, and (**f**) $T = 0.2$.

4.4. Aquaculture Map and Accuracy Assessment

With optimal feature combination, threshold, and window size, the final aquaculture maps were obtained (Figure 11). The aquacultures in the Pearl River Basin (Guangdong) had a total area of 1445.91 km^2 in 2020. The inland aquacultures were concentrated in the coast of bays, including Zhenhai Bay (Figure 11a), Yamen watercourse (Figure 11b), Pearl River Estuary (Figure 11g) and Niwanmen watercourse (Figure 11h). The area along the coast had the most concentrated aquacultures, which are mainly enclosed sea aquacultures. Moreover, a great deal of inland aquacultures were widely distributed in the inland river network.

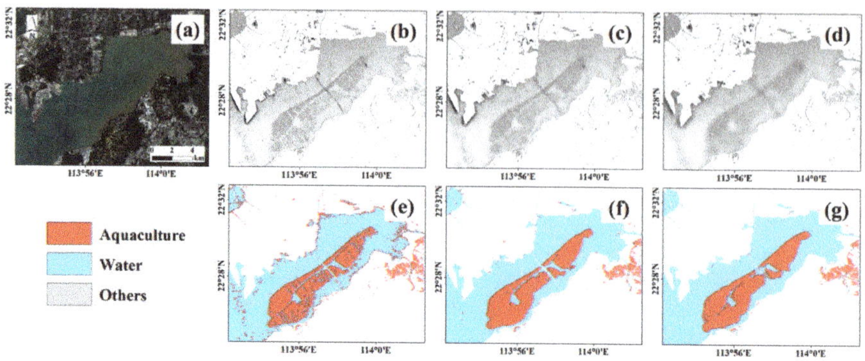

Figure 10. Maximum sum average texture of VH using different window sizes and the aquaculture maps: (**a**) Sentinel-2A false color images, (**b–d**) image features with a window size of 5 × 5 pixels, 15 × 15 pixels and 30 × 30 pixels, and (**e–g**) aquaculture maps.

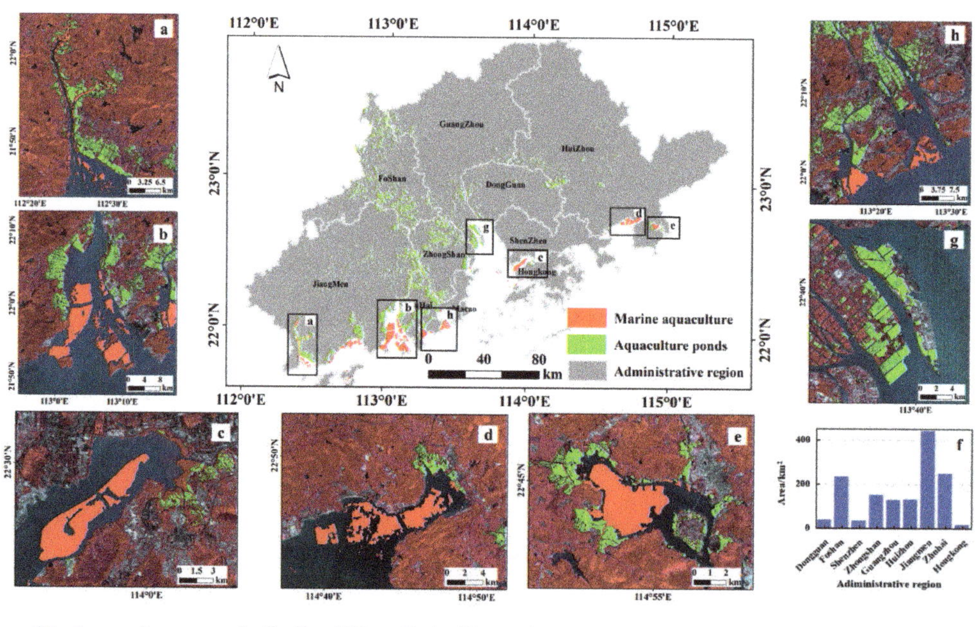

Figure 11. Aquaculture maps in the Pearl River Basin (Guangdong) in 2020: (**a**) Zhenhai Bay, (**b**)Yamen watercourse, (**c**) Shenzhen Bay, (**d**) Baisha Bay, (**e**) Baisha lake, (**f**) Histogram of aquaculture area in major administrative regions, (**g**) Pearl River Estuary, and (**h**) Niwanmen watercourse.

Marine aquacultures accounted for 22.8% of the total area of aquaculture in the Pearl River Basin (Guangdong), and they were concentrated in the bays along the coast. The raft cultures were mainly distributed in Yamen watercourse (Figure 11b) and Baisha Bay (Figure 11d), and cage culture in Shenzhen Bay (Figure 11d) and and Baisha lake (Figure 11e). The aquaculture areas of the major administrative regions in the Pearl River Basin (Guangdong) in 2020 are shown in Figure 11h. Jiangmen City had the largest area of aquaculture with about 443.56 km^2, which accounts for about 30.7% of the total aquaculture area in the study area. The second was Zhuhai City, with an aquaculture area of 251.52 km^2, accounting for 17.4% of the total aquaculture area.

The validation of aquaculture area map with the validation samples held an overall accuracy of 89.5% with a kappa coefficient of 0.776, a Producer's Accuracy of 82.82%, and a User's Accuracy of 89.47% (Table 4). Some commission and omissions errors for the proposed approach were still observed. Using a window of 150 × 150 m, some water bodies and tidal flats adjacent to aquaculture were misclassified as aquaculture because of their similar edge texture values. In addition, the aquacultures with high sediment content have similar spectral characteristics with non-aqueous bodies, and they were frequently misclassified.

Table 4. Confusion matrix of proposed approach using 1000 validation samples.

	Aquaculture	Others	Classified	User's Accuracy (%)
Aquaculture	0.323	0.038	0.358	89.47
Others	0.067	0.572	0.642	89.51
Reference	0.390	0.610	1	
Producer's accuracy (%)	82.82	93.77		
	Overall Accuracy = 89.5%, Kappa = 0.776			

The high user's and overall accuracies indicated the satisfactory performance of the proposed approach. However, the producer's accuracy was slightly lower than user's accuracy, for some mixed pixels containing parts of ponds were classified as water bodies at the stage of NDWI thresholding. This problem could be overcome by using sub-pixel mapping or high spatial resolution images.

5. Discussion

5.1. Comparison with Other Aquaculture Maps

The proposed approach in this study was compared with the methods proposed by Duan et al. [42] and Xia et al. [39]. The corresponding reference images were manually labeled. The maps are presented in Figure 12, and the confusion matrixes are presented in Tables 5–7. Duan et al. [42] adopted a spectrum, and spatial and morphological features of 30 m resolution Landsat images to build a Random Forest classifier to implement an automatic extraction of large-scale aquacultures. Xia's et al. [39] extracted the water surface from Sentinel-2A images using multiple thresholds, described the water patches using geometric and spectral features, and finally applied a Random Forest classifier to classify inland aquaculture ponds.

Table 5. Confusion matrix of Duan's method in typical sites [42].

Duan's Method	Aquaculture	Others	Classified	User's Accuracy (%)
Aquaculture	0.231	0.168	0.400	57.92
Others	0.032	0.568	0.600	94.65
Reference	0.264	0.736	1	
Producer's accuracy (%)	87.81	77.16		
	Overall Accuracy = 79.97%, Kappa = 0.557			

Table 6. Confusion matrix of Xia's method in typical sites [39].

Xia's Method	Aquaculture	Others	Classified	User's Accuracy (%)
Aquaculture.	0.250	0.071	0.320	77.97
Others	0.014	0.666	0.680	97.94
Reference	0.264	0.736	1	
Producer's accuracy (%)	94.70	90.42		
	Overall Accuracy = 91.55%, Kappa = 0.796			

Table 7. Confusion matrix of proposed approach in this study in typical sites.

Proposed Method	Aquaculture	Others	Classified	User's Accuracy (%)
Aquaculture	0.231	0.044	0.275	84.10
Others	0.032	0.693	0.725	95.54
Reference	0.264	0.736	1	
Producer's accuracy (%)	87.73	90.06		
	Overall Accuracy = 92.40%, Kappa = 0.867			

The map provided by Duan et al. [42] presented a general distribution of aquaculture ponds; however, almost all the details were missed (Figure 12b,j), for it was obtained using Landsat TM images with a spatial resolution of 30 m. Moreover, they conducted morphological closing and erosion operations post process, which further smoothed the details.

Xia's method and the proposed approach in this study were applied to two typical regions, and their derived aquaculture maps were compared. Xia's [39] and the proposed approach in this study produced similar aquaculture maps, and most of their details are clearer than Duan's map (Figure 12c,d,k,l) for Sentinel images of 10 m spatial resolution were applied and the post processing did not eliminate the details. The proposed method produced similar results to Xia's method (Figure 12c,d,k,l); however, less embankment pixels were misclassified. Therefore, the User's Accuracy of aquaculture areas (92.40%) was higher than that of Xia's method (77.97%). The main problem of the proposed method is that some isolated ponds were missed, resulting in a lower Producer's Accuracy (Table 7). However, the overall accuracy and Kappa coefficient indicated that the proposed method performed better over the tested areas.

The proposed approach in this study achieved better performance than Duan's and Xia's methods, which might be explained by that: (1) the texture features from normalized difference spectral index images increased the distinguishability of aquaculture area from other objects; and (2) radar images were integrated with optical ones, as they are sensitive to the texture structure of aquaculture areas, resulting in further improvement for aquaculture mapping. More important, the proposed approach was originally designed to simultaneously map aquaculture ponds and mariculture areas, whereas Xia's and Duan's methods can only map aquaculture ponds, and thus the proposed method has a better generalizability.

5.2. Impacts of Mixed-Pixels

The aquaculture maps in the Pear River Basin (Guangdong) were obtained with medium resolution Sentinel-1 SAR and Sentinel-2A multispectral images, and the results showed the advantages of using medium resolution images for large-scale thematic mapping. However, the accuracy and generalization of the proposed approach might be affected by mixed pixels of medium resolution images. The aquaculture pixels are often mixed with embankment among ponds, which often results in ambiguous boundaries and some inevitable errors. Thus, some parts of aquaculture ponds were classified as land, and some embankments were classified as ponds. It is difficult to overcome such errors using medium spatial resolution images, and very high spatial resolution images may easily overcome this problem.

The step of extracting water surface mask in our proposed approach is also affected by mixed pixels. Our method relies on the assumption that pixels covering mariculture facilities are still water-dominated, because the sizes of these facilities are usually relatively smaller than a pixel. Thus, they were first classified as waterbodies, and further been detected at later stages. However, this assumption may be not appropriate when using high spatial resolution images, because many pixels covering mariculture facilities are no longer water-dominated mixed pixels. In such case, these pixels will be segmented into non-aquaculture, affecting the accuracy of final map. Object-based methods might be an optimal solution for high spatial resolution images to overcome this limitation.

The proposed approach was designed for large scale aquaculture mapping using medium spatial resolution images. Although the boundaries are not detected very accurately, the aquaculture map over a large area can be efficiently obtained. In particularly, the medium spatial resolution images acquired by many satellite sensors (such as Landsat TM series and Sentinel-2A series) for several decades provide great convenience to monitor the development of the aquaculture industry.

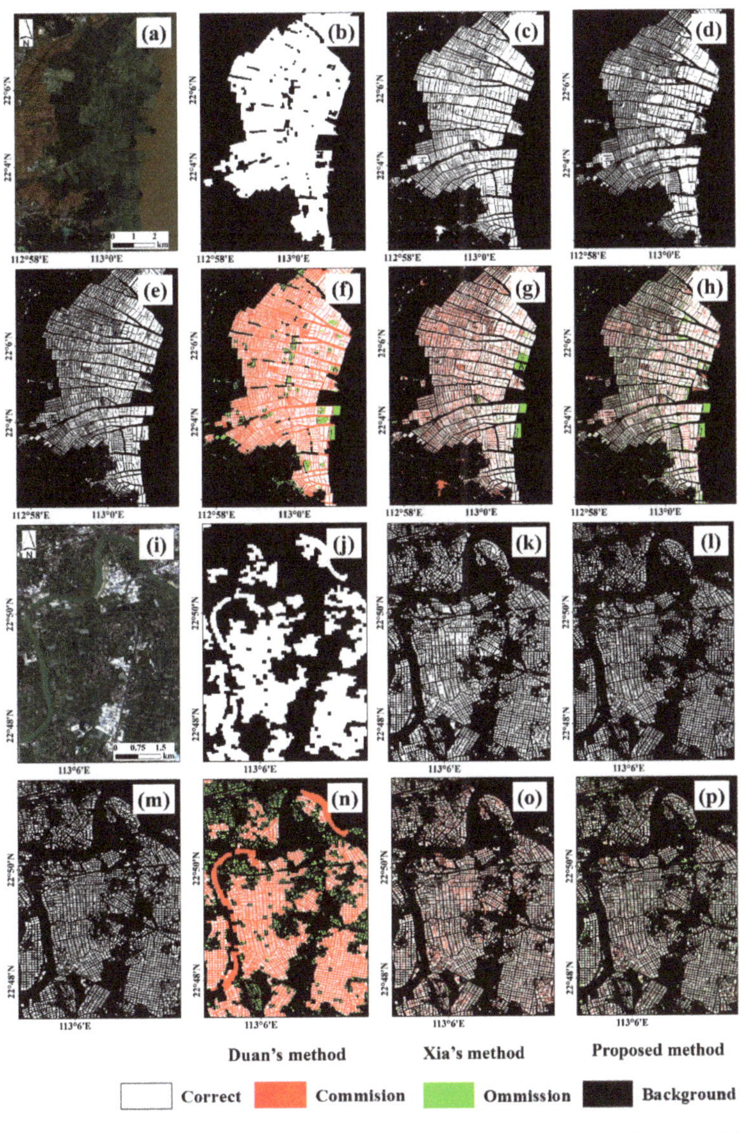

Figure 12. Aquaculture maps obtained using three methods. (**a**,**i**) present the Sentinel-2A true color images, and their corresponding reference maps are presented in (**e**,**m**). The results obtained using Duan's [42], Xia's [39], and the proposed approach are presented in (**b**–**d**) and (**j**–**l**), respectively. Their comparisons with the ground truth maps are presented in (**f**–**h**) and (**n**–**p**), respectively.

5.3. Selection of Time Series Images

In this study, time series images for a whole year were used for aquaculture mapping in order to eliminate the influence of accidental factors, such as water-dominated paddy fields and dry ponds during harvest period. Time series SAR images also are effective in suppressing speckle noise of radar image. Although there are many advantages, two problems should be noted.

First, the seasonal characteristics of study areas may affect the selection of time series images. For example, this study area is located in southern China, in which the temperature is usually high, and the water does not freeze; thus, the images with good quality acquired at any time can be used. However, in northern China, the pond water freezes in winter and its optical properties will change. Therefore, the time series images should be selected according to the specific seasonal characteristics of study areas.

Second, the assumption and basis of using time series images are that ground entities are not changed suddenly. If some aquaculture ponds are converted into agricultural land or built-up areas in winter, they may still be recognized as aquaculture areas. To improve the ability to respond to such sudden changes, it is necessary to shorten the time interval of time series images.

5.4. Limits and Future Works

An effective and efficient approach was proposed in this study for aquaculture area mapping over large areas; however, some issues should be further investigated. Firstly, some narrow rivers adjacent to aquaculture ponds were still misclassified. Thus, more accurate classifier and post-processing methods are still worth investigating, and the combination of post-processing and river vector boundaries may be a potential solution. Secondly, the aquaculture areas were roughly classified into two types in this study: aquaculture ponds and mariculture areas, and more specific aquaculture types are needed to be investigated. Finally, only one global threshold was used to extract waterbodies (containing aquaculture areas); it was not always optimal for different and complex image scenes, and thus a locally adaptive thresholding approach is a potential solution to improve the segmentation of the water surface. Only the aquaculture maps in 2020 were obtained in this study, and they are not sufficient for more potential applications. With historic earth observation images, it is necessary to analyze the long-time spatial–temporal changes of aquaculture areas and their impacts on economy and ecological systems and further to provide supports for the sustainable developments of the study area.

6. Conclusions

A novel approach was proposed in this study for simultaneously mapping multi-type aquaculture areas over large scale areas by combining spectral and texture features from optical (Sentinel-2A multispectral) and radar (Sentinel-1) images, and a case study in the Pear l River Basin (Guangdong) showed its efficiency. The main contribution of this work could be summarized as follows:

(1) We analyzed the spectral and textural features of aquaculture areas and demonstrated the effectiveness of fusing multiple image features for aquaculture mapping. We found that the use of textural features derived from the spectral indices can greatly improve the mapping accuracy and the use of textural features derived from SAR images can further improve mapping accuracy, as they are sensitive to marine aquaculture facilities.

(2) The proposed approach could generate a more accurate aquaculture map than previous studies. Moreover, the proposed approach was implemented on the GEE platform, and has great potential for national-scale and long-term aquaculture mapping.

Author Contributions: Conceptualization, Z.H.; methodology, Z.H., Y.X.; validation, Y.X., Y.Y.; formal analysis, Z.H., Y.X., J.W. and Y.Z.; investigation, Y.X. and Y.Y.; data curation, Y.X. and Y.Y.; writing—original draft preparation, Y.X. and Z.H.; writing—review and editing, Z.H., J.W., Y.Z. and G.W.; visualization, Y.X.; supervision, Z.H.; project administration, Z.H.; funding acquisition, Z.H. All authors have read and agreed to the published version of the manuscript.

Funding: This work was jointly supported by the National Natural Science Foundation of China (NSFC) (No. 41871227), the Natural Science Foundation of Guangdong Province (No. 2020A1515010678, 2020A1515111142) and the Basic Research Program of Shenzhen (No. JCYJ20190808122405692, 20200812112628001).

Institutional Review Board Statement: Not applicable.

Informed Consent Statement: Not applicable.

Data Availability Statement: The codes and data are available from this linkage: https://github.com/XuBenBen15045812917/Muilti-aquaculture_extraction.git (accessed on 25 October 2021).

Acknowledgments: We thank Xia at Eastern China Normal University for providing their code and Duan at Jiangsu Normal University for providing their aquaculture map for comparison.

Conflicts of Interest: The authors declare no conflict of interest.

References

1. Ottinger, M.; Clauss, K.; Kuenzer, C. Aquaculture: Relevance, distribution, impacts and spatial assessments—A review. *Ocean. Coast. Manag.* **2016**, *119*, 244–266. [CrossRef]
2. Cao, L.; Naylor, R.; Henriksson, P.; Leadbitter, D.; Metian, M.; Troell, M.; Zhang, W.B. China's aquaculture and the world's wild fisheries. *Science* **2015**, *347*, 133–135. [CrossRef] [PubMed]
3. FAO. *The State of World Fisheries and Aquaculture 2018-Meeting the Sustainable Development Goals*; FAO: Rome, Italy, 2018; pp. 1–227.
4. Pauly, D.; Zeller, D. Agreeing with FAO: Comments on SOFIA 2018. *Mar. Policy* **2019**, *100*, 332–333. [CrossRef]
5. Wang, Q.; Cheng, L.; Liu, J.; Li, Z.; Xie, S.; De Silva, S.S. Freshwater aquaculture in PR China: Trends and prospects. *Rev. Aquac.* **2015**, *7*, 283–302. [CrossRef]
6. Cao, L.; Wang, W.; Yang, Y.; Yang, C.; Yuan, Z.; Xiong, S.; Diana, J. Environmental impact of aquaculture and countermeasures to aquaculture pollution in China. *Environ. Sci. Pollut. Res.* **2007**, *14*, 452–462. [CrossRef]
7. Moffitt, C.M.; Cajas-Cano, L. Blue Growth: The 2014 FAO State of World Fisheries and Aquaculture. *Fisheries* **2014**, *39*, 552–553. [CrossRef]
8. Pham, T.D.; Yokoya, N.; Bui, D.T.; Yoshino, K.; Friess, D.A. Remote Sensing Approaches for Monitoring Mangrove Species, Structure, and Biomass: Opportunities and Challenges. *Remote Sens.* **2019**, *11*, 230. [CrossRef]
9. Alexandridis, T.K.; Topaloglou, C.A.; Lazaridou, E.; Zalidis, G.C. The performance of satellite images in mapping aquacultures. *Ocean. Coast. Manag.* **2008**, *51*, 638–644. [CrossRef]
10. Jayanthi, M.; Rekha, P.N.; Kavitha, N.; Ravichandran, P. Assessment of impact of aquaculture on Kolleru Lake (India) using remote sensing and Geographical Information System. *Aquac. Res.* **2006**, *37*, 1617–1626. [CrossRef]
11. Virdis, S.G.P. An object-based image analysis approach for aquaculture ponds precise mapping and monitoring: A case study of Tam Giang-Cau Hai Lagoon, Vietnam. *Environ. Monit. Assess.* **2014**, *186*, 117–133. [CrossRef]
12. Ottinger, M.; Clauss, K.; Kuenzer, C. Large-Scale Assessment of Coastal Aquaculture Ponds with Sentinel-1 Time Series Data. *Remote Sens.* **2017**, *9*, 440. [CrossRef]
13. Wang, M.; Cui, Q.; Wang, J.; Ming, D.P.; Lv, G.N. Raft cultivation area extraction from high resolution remote sensing imagery by fusing multi-scale region-line primitive association features. *ISPRS J. Photogramm. Remote Sens.* **2017**, *123*, 104–113. [CrossRef]
14. Ottinger, M.; Clauss, K.; Kuenzer, C. Opportunities and Challenges for the Estimation of Aquaculture Production Based on Earth Observation Data. *Remote Sens.* **2018**, *10*, 1076. [CrossRef]
15. Liu, Y.; Wang, Z.; Yang, X.; Zhang, Y.; Yang, F.; Liu, B.; Cai, P. Satellite-based monitoring and statistics for raft and cage aquaculture in China's offshore waters. *Int. J. Appl. Earth Obs. Geoinf.* **2020**, *91*, 102118. [CrossRef]
16. Aguilar-Manjarrez, J.; Travaglia, C. *Mapping Coastal Aquaculture and Fisheries Structures by Satellite Imaging Radar: Case Study of the Lingayen Gulf, the Philippines*; Food Agriculture Organization: Roma, Italy, 2004. [CrossRef]
17. Ren, C.; Wang, Z.; Zhang, B.; Li, L.; Chen, L.; Song, K.; Jia, M. Remote Monitoring of Expansion of Aquaculture Ponds Along Coastal Region of the Yellow River Delta from 1983 to 2015. *Chin. Geogr. Sci.* **2018**, *28*, 430–442. [CrossRef]
18. Xing, Q.; An, D.; Zheng, X.; Wei, Z.; Wang, X.; Li, L.; Tian, L.; Chen, J. Monitoring seaweed aquaculture in the Yellow Sea with multiple sensors for managing the disaster of macroalgal blooms. *Remote Sens. Environ.* **2019**, *231*, 111279. [CrossRef]
19. Sakamoto, T.; Van Phung, C.; Kotera, A.; Nguyen, K.D.; Yokozawa, M. Analysis of rapid expansion of inland aquaculture and triple rice-cropping areas in a coastal area of the Vietnamese Mekong Delta using MODIS time-series imagery. *Landsc. Urban Plan.* **2009**, *92*, 34–46. [CrossRef]
20. Liu, Y.; Yang, X.; Wang, Z.; Lu, C.; Li, Z.; Yang, F. Aquaculture area extraction and vulnerability assessment in Sanduao based on richer convolutional features network model. *J. Oceanol. Limnol.* **2019**, *37*, 1941–1954. [CrossRef]
21. Cui, B.; Fei, D.; Shao, G.; Lu, Y.; Chu, J. Extracting Raft Aquaculture Areas from Remote Sensing Images via an Improved U-Net with a PSE Structure. *Remote Sens.* **2019**, *11*, 2053. [CrossRef]
22. Zhang, T.; Yang, X.M.; Hu, S.S.; Su, F.Z. Extraction of coastline in aquaculture coast from multispectral remote sensing images: Object-based region growing integrating edge detection. *Remote Sens.* **2013**, *5*, 4470–4487. [CrossRef]

23. Zhang, T.; Li, Q.; Yang, X.; Zhou, C.; Su, F. Automatic Mapping Aquaculture in Coastal Zone from TM Imagery with OBIA Approach. In Proceedings of the 2010 18th International Conference on Geoinformatics, Beijing, China, 18–20 June 2010; pp. 1–4. [CrossRef]
24. Zeng, Z.; Wang, D.; Tan, W.; Huang, J. Extracting aquaculture ponds from natural water surfaces around inland lakes on medium resolution multispectral images. *Int. J. Appl. Earth Obs. Geoinf.* **2019**, *80*, 13–25. [CrossRef]
25. Wang, J.; Sui, L.; Yang, X.; Wang, Z.; Liu, Y.; Kang, J.; Lu, C.; Yang, F.; Liu, B. Extracting Coastal Raft Aquaculture Data from Landsat 8 OLI Imagery. *Sensors* **2019**, *19*, 1221. [CrossRef] [PubMed]
26. Komatsu, T.; Takahashi, M.; Ishida, K.; Suzuki, T.; Hiraishi, T.; Tameishi, H. Mapping of aquaculture facilities in Yamada Bay in Sanriku Coast, Japan, by IKONOS satellite imagery. *Fish. Sci.* **2002**, *68*, 584–587. [CrossRef]
27. Fan, J.; Zhao, J.; An, W.; Hu, Y. Marine Floating Raft Aquaculture Detection of GF-3 PolSAR Images Based on Col-lective Multikernel Fuzzy Clustering. *IEEE J. Sel. Top. Appl. Earth Obs. Remote Sens.* **2019**, *12*, 2741–2754. [CrossRef]
28. Ren, C.; Wang, Z.; Zhang, Y.; Zhang, B.; Chen, L.; Xi, Y.; Xiao, X.; Doughty, R.; Liu, M.; Jia, M.; et al. Rapid expansion of coastal aquaculture ponds in China from Landsat observations during 1984–2016. *Int. J. Appl. Earth Obs. Geoinf.* **2019**, *82*, 101902. [CrossRef]
29. Hu, Y.; Fan, J.; Wang, J. Target Recognition of Floating Raft Aquaculture in SAR Image Based on Statistical Region Merging. In Proceedings of the 2017 Seventh International Conference on Information Science and Technology (ICIST), Da Nang, Vietnam, 16–19 April 2017; pp. 429–432. [CrossRef]
30. Zhang, Y.; Wang, C.; Ji, Y.; Chen, J.; Deng, Y.; Chen, J.; Jie, Y. Combining Segmentation Network and Nonsubsampled Contourlet Transform for Automatic Marine Raft Aquaculture Area Extraction from Sentinel-1 Images. *Remote Sens.* **2020**, *12*, 4182. [CrossRef]
31. Prasad, K.A.; Ottinger, M.; Wei, C.; Leinenkugel, P. Assessment of Coastal Aquaculture for India from Sentinel-1 SAR Time Series. *Remote Sens.* **2019**, *11*, 357. [CrossRef]
32. Zhang, K.; Dong, X.; Liu, Z.; Gao, W.; Hu, Z.; Wu, G. Mapping Tidal Flats with Landsat 8 Images and Google Earth Engine: A Case Study of the China's Eastern Coastal Zone circa 2015. *Remote Sens.* **2019**, *11*, 924. [CrossRef]
33. Jia, M.; Wang, Z.; Mao, D.; Ren, C.; Wang, C.; Wang, Y. Rapid, robust, and automated mapping of tidal flats in China using time series Sentinel-2 images and Google Earth Engine. *Remote Sens. Environ.* **2021**, *255*, 112285. [CrossRef]
34. Halder, B.; Bandyopadhyay, J. Vegetation scenario of Indian part of Ganga Delta: A change analysis using Sentinel-1 time series data on Google earth engine platform. *Saf. Extreme Environ.* **2021**, 1–14. [CrossRef]
35. Th, A.; Emt, A.; Gang, C.B.; Gang, S.C.; Yz, D.; Yang, L.A.; Kz, A.; Yf, A. Mapping fine-scale human disturbances in a working landscape with Landsat time series on Google Earth Engine. *ISPRS J. Photogramm. Remote Sens.* **2021**, *176*, 250–261.
36. Gorelick, N.; Hancher, M.; Dixon, M.; Ilyushchenko, S.; Thau, D.; Moore, R. Google Earth Engine: Planetary-scale geospatial analysis for everyone. *Remote Sens. Environ.* **2017**, *202*, 18–27. [CrossRef]
37. Wang, C.; Liu, H.-Y.; Zhang, Y.; Li, Y.-F. Classification of land-cover types in muddy tidal flat wetlands using remote sensing data. *J. Appl. Remote Sens.* **2014**, *7*, 073457. [CrossRef]
38. Tieng, T.; Sharma, S.; Mackenzie, R.A.; Venkattappa, M.; Sasaki, N.K.; Collin, A. Mapping mangrove forest cover using Landsat-8 imagery, Sentinel-2, Very High Resolution Images and Google Earth Engine algorithm for entire Cambodia. *IOP Conf. Ser. Earth Environ. Sci.* **2019**, *266*, 012010. [CrossRef]
39. Xia, Z.; Guo, X.; Chen, R. Automatic extraction of aquaculture ponds based on Google Earth Engine. *Ocean. Coast. Manag.* **2020**, *198*, 105348. [CrossRef]
40. Sun, Z.; Luo, J.; Yang, J.; Yu, Q.; Zhang, L.; Xue, K.; Lu, L. Nation-Scale Mapping of Coastal Aquaculture Ponds with Sentinel-1 SAR Data Using Google Earth Engine. *Remote Sens.* **2020**, *12*, 3086. [CrossRef]
41. Duan, Y.; Li, X.; Zhang, L.; Chen, D.; Liu, S.; Ji, H. Mapping national-scale aquaculture ponds based on the Google Earth Engine in the Chinese coastal zone. *Aquaculture* **2020**, *520*, 734666. [CrossRef]
42. Duan, Y.; Tian, B.; Li, X.; Liu, D.; Sengupta, D.; Wang, Y.; Peng, Y. Tracking changes in aquaculture ponds on the China coast using 30 years of Landsat images. *Int. J. Appl. Earth Obs. Geoinf.* **2021**, *102*, 102383. [CrossRef]
43. Richards, R.; Ruddle, K.; Zhong, G. Integrated agriculture-aquaculture in South China: The dike-pond System of the Zhujiang Delta. *Geogr. Rev.* **1989**, *79*, 260–262. [CrossRef]
44. Snoeij, P.; Torres, R.; Geudtner, D.; Brown, M.; Deghaye, P.; Navastraver, I.; Ostergaard, A.; Rommen, B.; Floury, N.; Da-vidson, M. *Sentinel-1 Instrument Overview*; Esa Special Publication: Noordwijk, The Netherlands, 1 March 2013; pp. 2–5.
45. Martimort, P.; Fernandez, V.; Kirschner, V.; Isola, C.; Meygret, A. Sentinel-1 Multispectral Imager(MSI) and Calibration/Validation. In Proceedings of the IEEE International Geoscience and Remote Sensing Symposium (IGARSS), Munich, Germany, 22–27 July 2012; pp. 6999–7002.
46. Main-Knorn, M.; Pflug, B.; Louis, J.; Debaecker, V.; Müller-Wilm, U.; Gascon, F. Sen2Cor for Sentinel-2. In Proceedings of the Conference on Image and Signal Processing for Remote Sensing, Warsaw, Poland, 4 October 2017; p. 3. [CrossRef]
47. Li, Y.; Niu, Z.; Xu, Z.; Yan, X. Construction of High Spatial-Temporal Water Body Dataset in China Based on Sentinel-1 Archives and GEE. *Remote Sens.* **2020**, *12*, 2413. [CrossRef]
48. Sibanda, M.; Buthelezi, S.; Ndlovu, H.S.; Mothapo, M.C.; Mutanga, O. Mapping the Eucalyptus spp woodlots in communal areas of Southern Africa using Sentinel-2 Multi-Spectral Imager data for hydrological applications. *Phys. Chem. Earth* **2021**, *122*, 102999. [CrossRef]
49. Center, N. Hypsographic Curve of Earth's Surface from ETOPO1. Available online: Ngdc.noaa.gov (accessed on 26 October 2021).

50. Wessel, P.; Smith, W.H.F. A global, self-consistent, hierarchical, high-resolution shoreline database. *J. Geophys. Res. Solid Earth* **1996**, *101*, 8741–8743. [CrossRef]
51. Zhu, Z.; Wang, S.; Woodcock, C.E. Improvement and expansion of the Fmask algorithm: Cloud, cloud shadow, and snow detection for Landsats 4–7, 8, and Sentinel 2 images. *Remote Sens. Environ.* **2015**, *159*, 269–277. [CrossRef]
52. Pettorelli, N. Using the satellite-derived NDVI to assess ecological responses to environmental change. *Trends Ecol. Evol.* **2005**, *20*, 503–510. [CrossRef]
53. Osgouei, P.E.; Kaya, S.; Sertel, E.; Alganci, U. Separating built-up Areas from bare land in Mediterranean cities using Sentinel-2A imagery. *Remote Sens.* **2019**, *11*, 345. [CrossRef]
54. McFeeters, S.K. The use of the Normalized Difference Water Index (NDWI) in the delineation of open water features. *Int. J. Remote Sens.* **1996**, *17*, 1425–1432. [CrossRef]
55. Liew, S.C.; Kam, S.P.; Chen, P.; Muchlisin, Z.A. Mapping Tsunami-Affected Coastal Aquaculture Areas in Northern Sumatra Using High Resolution Satellite Imagery. In Proceedings of the Asian Association on Remote Sensing—26th Asian Conference on Remote Sensing and 2nd Asian Space Conference, Hanoi, Vietnam, 7–11 November 2005; pp. 116–120.
56. Haralick, R.M.; Shanmugam, K.; Dinstein, I. Textural Features for Image Classification. *IEEE Trans. Syst. Man Cybern.* **1973**, *SMC-3*, 610–621. [CrossRef]
57. Yu, G.; Zhou, X.; Hou, D.; Wei, D. Abnormal crowdsourced data detection using remote sensing image features. *Int. Arch. Photogramm. Remote Sens. Spat. Inf. Sci.* **2021**, 215–221. [CrossRef]
58. Hodgson, M.E. Reducing the computational requirements of the minimum-distance classifier. *Remote Sens. Environ.* **1988**, *25*, 117–128. [CrossRef]
59. Belgiu, M.; Drăguţ, L. Random forest in remote sensing: A review of applications and future directions. *ISPRS J. Photogramm. Remote Sens.* **2016**, *114*, 24–31. [CrossRef]
60. Halldorsson, G.H.; Benediktsson, J.A.; Sveinsson, J.R. Support Vector Machines in Multisource Classification. In Proceedings of the 23rd International Geoscience and Remote Sensing Symposium (IGARSS 2003), Toulouse, France, 21–25 July 2003; pp. 2054–2056.
61. Manno, A. CART: Classification and regression trees. *Int. J. Public Health* **2012**, *57*, 243–246.
62. Yang, B.; Yu, X. Remote Sensing Image Classification of Geoeye-1 High-Resolution Satellite. *ISPRS Int. Arch. Photogramm. Remote Sens. Spat. Inf. Sci.* **2014**, *XL-4*, 325–328. [CrossRef]
63. Clemente, J.P.; Fontanelli, G.; Ovando, G.G.; Roa, Y.L.B.; Lapini, A.; Santi, E. Google Earth Engine: Application of Algorithms for Remote Sensing of Crops in Tuscany (ITALY). In Proceedings of the IEEE Latin American GRSS and ISPRS Remote Sensing Conference (LAGIRS), Santiago, Chile, 21–26 March 2020; pp. 195–200.
64. Chung, L.C.H.; Xie, J.; Ren, C. Improved machine-learning mapping of local climate zones in metropolitan areas using composite Earth observation data in Google Earth Engine. *Build. Environ.* **2021**, *199*, 107879. [CrossRef]
65. Shaharum, N.S.N.; Shafri, H.Z.M.; Ghani, W.A.W.A.K.; Samsatli, S.; Al-Habshi, M.M.A.; Yusuf, B. Oil palm mapping over Peninsular Malaysia using Google Earth Engine and machine learning algorithms. *Remote Sens. Appl. Soc. Environ.* **2020**, *17*, 100287. [CrossRef]
66. Piragnolo, M.; Masiero, A.; Pirotti, F. Comparison of Random Forest and Support Vector Machine Classifiers Using UAV Remote Sensing Imagery. In Proceedings of the 19th Egu General Assembly Conference, Vienna, Austria, 23–28 April 2017; p. 15692.
67. Cutler, D.R.; Edwards, T.C.E., Jr.; Beard, K.H.; Cutler, A.; Hess, K.T.; Gibson, J.; Lawler, J.J. Random Forests for Classification in Ecology. *Ecology* **2007**, *88*, 2783–2792. [CrossRef] [PubMed]
68. Zhou, B.; Okin, G.S.; Zhang, J. Leveraging Google Earth Engine (GEE) and machine learning algorithms to incorporate in situ measurement from different times for rangelands monitoring. *Remote Sens. Environ.* **2020**, *236*, 111521. [CrossRef]
69. Pelletier, C.; Valero, S.; Inglada, J.; Champion, N.; Dedieu, G. Assessing the robustness of Random Forests to map land cover with high resolution satellite image time series over large areas. *Remote Sens. Environ.* **2016**, *187*, 156–168. [CrossRef]
70. Phan, T.N.; Kuch, V.; Lehnert, L. Land Cover Classification using Google Earth Engine and Random Forest Classifier—The Role of Image Composition. *Remote Sens.* **2020**, *12*, 2411. [CrossRef]
71. Pesaresi, M.; Gerhardinger, A.; Kayitakire, F. A robust built-up area presence index by anisotropic rotation-invariant textural measure. *J. Sel. Top. Appl. Earth Obs. Remote Sens.* **2008**, *1*, 180–192. [CrossRef]

MDPI
St. Alban-Anlage 66
4052 Basel
Switzerland
Tel. +41 61 683 77 34
Fax +41 61 302 89 18
www.mdpi.com

Remote Sensing Editorial Office
E-mail: remotesensing@mdpi.com
www.mdpi.com/journal/remotesensing

www.ingramcontent.com/pod-product-compliance
Lightning Source LLC
Chambersburg PA
CBHW061821210926
51850CB00006B/771

* 9 7 8 3 0 3 6 5 8 1 3 0 9 *